Gary G. Gundersen, Janne Heittokangas

Entire Functions

Also of Interest

Dynamics of Entire Functions
The Fractal Sets of Julia and Mandelbrot
Cándido Piñeiro, 2025
ISBN 978-3-11-168943-2, e-ISBN (PDF) 978-3-11-168968-5

Calculus of One Variable with Python and AI
Nikolaos Halidias, 2026
ISBN 978-3-11-914171-0, e-ISBN (PDF) 978-3-11-222828-9

Functional and Harmonic Analysis
Weighted Lebesgue Spaces, Muckenhoupt Weights, Convexity and BMO Spaces
René Erlin Castillo, Héctor Camilo Chaparro Gutiérrez, Julio César Ramos-Fernández, 2026
ISBN 978-3-11-914220-5, e-ISBN (PDF) 978-3-11-222324-6

Complex Analysis and Special Functions
Cauchy Formula, Elliptic Functions and Laplace's Method
Valery Serov, Markus Harju, 2025
ISBN 978-3-11-163211-7, e-ISBN (PDF) 978-3-11-163227-8

Variational Methods in Nonlinear Analysis
Optimization and Partial Differential Equations Applications
Dimitrios C. Kravvaritis, Athanasios N. Yannacopoulos, 2026
ISBN 978-3-11-133325-0, e-ISBN (PDF) 978-3-11-133329-8

Multigrid Methods
Axiomatic Convergence Theory for Linear and Weakly Nonlinear Problems
Abner J. Salgado, Steven M. Wise, Calvin Wong, 2026
ISBN 978-3-11-135419-4, e-ISBN (PDF) 978-3-11-135488-0

Gary G. Gundersen, Janne Heittokangas

Entire Functions

Classical and Modern Results, Special Functions and Complex Differential Equations

DE GRUYTER

Mathematics Subject Classification 2020
Primary: 30D15, 30D20, 30D35, 34M05, 34M10; Secondary: 30B10, 30D10, 33C10

Authors
Gary G. Gundersen
Department of Mathematics
University of New Orleans
New Orleans, LA 70148
USA
ggunders@uno.edu

Janne Heittokangas
Department of Physics and Mathematics
University of Eastern Finland
P.O. Box 111
80101 Joensuu
Finland
janne.heittokangas@uef.fi

ISBN 978-3-11-224912-3
ISBN 978-3-11-224913-0 (PDF)
ISBN 978-3-11-224914-7 (E-PUB)
DOI https://doi.org/10.1515/9783112249130

Library of Congress Control Number: 2026935977

Bibliographic information published by the Deutsche Nationalbibliothek
The Deutsche Nationalbibliothek lists this publication in the Deutsche Nationalbibliografie; detailed bibliographic data are available on the Internet at https://dnb.dnb.de.

De Gruyter and Walter de Gruyter GmbH are part of De Gruyter Brill.
www.degruyterbrill.com

Questions about General Product Safety Regulation: productsafety@degruyterbrill.com

Cover illustration: Shing Lok Che / iStock / Getty Images Plus

Dedicated to the wonderful memories of Minnie

Contents

Preface

The foundation of the theory of entire functions originated in the nineteenth century with the fundamental research of Cauchy, Weierstrass, Hadamard, Picard, Borel, and others. Older books on the classical theory of entire functions include Valiron [279], Saks and Zygmund [242], Boas [37], Cartwright [43], Markushevich [206], Holland [142], Levin [183], Jank-Volkmann [150], Gol'dberg, Levin and Ostrovskii [81], Zhang [305], and Rubel [237]. The only book that we have found in the twenty-first century that has a similar nature and spirit to these older books is the 2010 book of Gil' [79].

Interesting recent developments and important results from the 1960s, 1970s, and 1980s in the theory of entire functions do not appear in these books. We cover many of these results in this book, and for those results and proofs that are not easy to read in their original formats, we have added details and clarified the presentation. It should also be said that the classical theory is still useful and some older books are not easy to use by students.

This book originated from lecture notes for a course on entire functions taught at the University of Eastern Finland (UEF). These notes have been continuously updated with new material, some of which was recently published.

It is expected that the readers of this book have a basic knowledge of complex analysis, which, for example, consists of most of the material in Chapters 1–12 of the book [218] "Introduction to Complex Analysis" by Nevanlinna and Paatero.

In a natural way, this book can be thought of as having two parts. The first part consists of Chapters 1–5, which contains the classical theory of entire functions that can be found in older books. The second part consists of Chapters 6–12, which contains numerous results in the last 60 years up to the present day. The first part contains practice problems, whereas the second part does not have problems.

Chapters 1–5 can be used for a class on entire functions. The 137 problems in these sections can be used for this purpose, as the problems vary in difficulty from straightforward to more challenging. Some problems could be assigned as gradable turn-in problems or included on take-home exams. This setup has been successfully tested in lecture classes at the UEF.

We hope this modern presentation of entire functions will be of interest to a variety of readers, including (a) researchers who work with entire functions, (b) researchers in other areas who use entire functions in their work, (c) faculty who want to use the book to teach a class or conduct a seminar, and (d) students who want to learn about entire functions.

Chapter 1 contains preliminary results that are used throughout the book. If the reader is familiar with these results, then this chapter could be skipped and referred back to when needed. Chapters 2 and 3 deal with the classical theory of entire functions with its standard topics: order of growth, type, exponent of convergence, infinite products, canonical products, the factorizations of Weierstrass and Hadamard, and the theorems of Picard and Borel.

https://doi.org/10.1515/9783112249130-203

Chapter 4 contains classical topics, starting with growth properties of entire functions in the sense of Phragmén and Lindelöf and entire functions of order < 1. Then Wiman–Valiron theory, Nevanlinna's theory of meromorphic functions, and complex interpolation theory are presented. It is not necessary to have a background in Nevanlinna theory because all the needed definitions and results are covered.

The first part of the book concludes with Chapter 5 on entire functions with real zeros: the reciprocal of the gamma function, Mittag-Leffler functions, the Airy integral, solutions of $f'' + P(z)f = 0$, Lindelöf functions, and Laguerre–Pólya functions.

Chapter 6 is about exponential polynomials, where basic properties are discussed: order of growth, value distribution, and zero separation. A duality property of linear differential equations with a dominant exponential polynomial coefficient is given.

In Chapter 7, the Nevanlinna characteristic function $T(r,f)$ of an entire function f is compared to the logarithmic maximum modulus function $\log M(r,f)$. If f has Fejér or Fabry gaps, then $T(r,f) \sim \log M(r,f)$ as $r \to \infty$ outside a possible exceptional set. For any suitable growth function ϕ, Clunie proved that there exists an entire function f for which $T(r,f) \sim \log M(r,f) \sim \phi(r)$ as $r \to \infty$. Perfectly regular growth and completely regular growth are also covered in Chapter 7.

All discussions on differential equations prior to Chapter 8 are informal and elementary, whereas Chapters 8, 9, and 10 contain in-depth discussions on entire solutions of differential equations. Since differential equations is an enormous subject, we discuss some hand-picked topics and had to omit several topics.

Solutions of linear differential equations with polynomial coefficients are addressed in Chapter 8, which includes the topics of order of growth and oscillation of solutions, the asymptotic integration of second-order equations, generalizations of the Airy integral, contour integral solutions, and a new proof for a special case of a general Petrenko–Steinmetz theorem on completely regular growth.

Solutions of linear differential equations with entire coefficients are presented in Chapter 9, which begins with the topics of the multiplicities of zeros of solutions and the question of how many finite-order solutions are possible. We next discuss the important work of Bank and Laine on $f'' + A(z)f = 0$ and their useful Bank–Laine equation. Last, we give Bergweiler's recent example of an equation with completely regular growth coefficients that possesses a transcendental finite-order solution that does not have completely regular growth, which answered a 1994 question of Gol'dberg and Ostrovskii.

The focus of Chapter 10 is on entire solutions of nonlinear differential equations with entire function coefficients. Several classes of equations are discussed, including equations related to trigonometric identities and differential polynomials.

Chapter 11 presents different growth measures for entire functions. The concept of logarithmic order is useful for entire functions of zero order, whereas other growth measures are sometimes used for rapidly growing functions.

The final Chapter 12 contains various topics in the theory of entire functions: (a) asymptotic values, (b) Bank–Laine functions, (c) entire functions whose zeros and one-points all lie on a finite number of rays, (d) entire solutions of Fermat-type equations, and (e) the multiple points of entire functions.

Last, we would like to thank Walter Bergweiler, Alexandre Eremenko, Aimo Hinkkanen, and Ilpo Laine for many helpful discussions. The second author would like to thank the Magnus Ehrnrooth Foundation for partial financial support.

Gary G. Gundersen
Janne Heittokangas

1 Preliminaries

In this chapter, we discuss some auxiliary topics and key concepts from real and complex analysis, which will be needed throughout the book. Some of this material exists in research papers only. The reader could decide to skip this chapter on the first read and then refer back to Chapter 1 whenever needed, keeping in mind that the concepts of limit superior and limit inferior are needed from the beginning of Chapter 2 and onward.

1.1 Limit superior and limit inferior

Many topics in this book depend on the concepts of limit superior and limit inferior. We will first define them for sequences. Throughout the book, unless indicated otherwise, a sequence $\{z_n\}$ of points or functions $\{f_n(z)\}$ means that either (a) the sequence is infinite with $n \in \mathbb{N}$, i. e., $n = 1, 2, 3, \ldots$, or (b) the sequence is finite with $n = 1, \ldots, k$, where k is a positive integer.

Let $\{x_n\}$ be a sequence of real numbers. Define

$$a_n = \inf\{x_k : k \geq n\} \quad \text{and} \quad b_n = \sup\{x_k : k \geq n\}. \tag{1.1.1}$$

Since the supremum and infimum are unique, the points a_n and b_n are uniquely defined in $\mathbb{R} \cup \{\pm\infty\}$ for every n. If $x_n \leq M$ for all n, then $b_n \leq M$ for all n. If $\{x_n\}$ is not bounded from above, then $b_n = \infty$ for all n. Likewise, if $x_n \geq N$ for all n, then $a_n \geq N$ for all n, whereas if $\{x_n\}$ is not bounded from below, then $a_n = -\infty$ for all n.

Defining $A_n = \{x_k : k \geq n\}$, it is immediate that $A_{n+1} \subset A_n$ for every n. It follows that $\{a_n\}$ is a nondecreasing sequence, whereas $\{b_n\}$ is a nonincreasing sequence. Thus the limits

$$\lim_{n\to\infty} a_n \quad \text{and} \quad \lim_{n\to\infty} b_n$$

always exist as unique numbers in $\mathbb{R} \cup \{\pm\infty\}$. These limits have special importance.

Definition 1.1.1. Let $\{x_n\}$ be a sequence of real numbers. Then

$$\liminf_{n\to\infty} x_n = \lim_{n\to\infty} a_n \quad \text{and} \quad \limsup_{n\to\infty} x_n = \lim_{n\to\infty} b_n$$

are called the *limit inferior* and *limit superior* of $\{x_n\}$, respectively.

Example 1.1.2. Let $x_n = (-1)^n$ and $y_n = \sin(\frac{n\pi}{3})$. Then

$$\liminf_{n\to\infty} x_n = -1 \quad \text{and} \quad \limsup_{n\to\infty} x_n = 1,$$
$$\liminf_{n\to\infty} y_n = -\frac{\sqrt{3}}{2} \quad \text{and} \quad \limsup_{n\to\infty} y_n = \frac{\sqrt{3}}{2}.$$

https://doi.org/10.1515/9783112249130-001

From the definitions of the sequences $\{x_n\}$ and $\{y_n\}$, it is obvious that the ordinary limits $\lim_{n\to\infty} x_n$ and $\lim_{n\to\infty} y_n$ do not exist. Regarding the above calculations for lim inf and lim sup, see also Theorem 1.1.3.

In the literature, the notation

$$\liminf_{n\to\infty} x_n = \underline{\lim}_{n\to\infty} x_n \quad \text{and} \quad \limsup_{n\to\infty} x_n = \overline{\lim}_{n\to\infty} x_n$$

is also used. The limits superior and inferior are often called *upper and lower limits.*

Since $a_n \le b_n$ for every n, it follows that

$$\liminf_{n\to\infty} x_n \le \limsup_{n\to\infty} x_n. \tag{1.1.2}$$

Recall from real analysis that

$$\sup\{-x_k : k \ge n\} = -\inf\{x_k : k \ge n\}.$$

This gives rise to

$$\limsup_{n\to\infty}(-x_n) = -\liminf_{n\to\infty} x_n. \tag{1.1.3}$$

There are alternate ways to define the upper and lower limits of sequences; see, e. g., [245, 270]. For instance, in [270, pp. 647–648], the upper limit is defined as the supremum of the cluster points of the sequence, and the lower limit is defined as the infimum of the cluster points of the sequence.

The ordinary limit has a connection to the upper and lower limits.

Theorem 1.1.3. *Let $\{x_n\}$ be a sequence of real numbers, and set*

$$l = \liminf_{n\to\infty} x_n \quad \textit{and} \quad L = \limsup_{n\to\infty} x_n.$$

Then $\lim_{n\to\infty} x_n = M$ if and only if $l = L = M$. Here $M \in \mathbb{R} \cup \{\pm\infty\}$.

Proof. Suppose first that $x_n \to M$, where $M \in \mathbb{R}$. Then, for a given $\varepsilon > 0$, there exists $N_1 = N_1(\varepsilon) \in \mathbb{N}$ such that

$$M - \varepsilon < x_n < M + \varepsilon, \quad n \ge N_1.$$

But then from (1.1.2),

$$M - \varepsilon \le \inf\{x_k : k \ge N_1\} \le l \le L \le \sup\{x_k : k \ge N_1\} \le M + \varepsilon,$$

from which we deduce that $l = L = M$. If $M = \infty$, then for every large constant $T > 0$, there exists $N_2 = N_2(T) \in \mathbb{N}$ such that

$$x_n > T, \quad n \ge N_2.$$

But then $T \le \inf\{x_k : k \ge N_2\} \le l \le L$, from which $l = L = \infty$. Similarly, if $M = -\infty$, then it follows that $l = L = -\infty$.

Conversely, suppose that $l = L = M$ and first consider the case $M \in \mathbb{R}$. Then for every $\varepsilon > 0$, there exists $N_3 = N_3(\varepsilon) \in \mathbb{N}$ such that

$$M - \varepsilon < \inf\{x_k : k \ge n\} \le x_n \le \sup\{x_k : k \ge n\} < M + \varepsilon, \quad n \ge N_3.$$

This implies $x_n \to M$. If $M = \infty$, then for every large constant $T > 0$, there exists $N_4 = N_4(T) \in \mathbb{N}$ such that

$$x_n \ge \inf\{x_k : k \ge n\} > T, \quad n \ge N_4.$$

This implies $x_n \to \infty$. Similarly, if $M = -\infty$, then it follows that $x_n \to -\infty$. □

We proceed to state a list of properties satisfied by upper and lower limits.

Theorem 1.1.4 ([245, p. 409]). *Let $\{x_n\}$ and $\{y_n\}$ be sequences of real numbers.*

(a) *There are subsequences $\{x_{n_k}\}$ and $\{x_{n_j}\}$ of $\{x_n\}$ such that*

$$\lim_{k\to\infty} x_{n_k} = \liminf_{n\to\infty} x_n \quad \textit{and} \quad \lim_{j\to\infty} x_{n_j} = \limsup_{n\to\infty} x_n.$$

Moreover, any other limits L of subsequences of $\{x_n\}$ satisfy

$$\liminf_{n\to\infty} x_n \le L \le \limsup_{n\to\infty} x_n.$$

(b) *We have*

$$\liminf_{n\to\infty}(x_n + y_n) \ge \liminf_{n\to\infty} x_n + \liminf_{n\to\infty} y_n,$$
$$\limsup_{n\to\infty}(x_n + y_n) \le \limsup_{n\to\infty} x_n + \limsup_{n\to\infty} y_n$$

whenever the right-hand sides are not of the form $\infty - \infty$ or $-\infty + \infty$.

(c) *We have*

$$\liminf_{n\to\infty}(x_n + y_n) \le \liminf_{n\to\infty} x_n + \limsup_{n\to\infty} y_n \le \limsup_{n\to\infty}(x_n + y_n)$$

whenever the expression in the middle is not of the form $\infty - \infty$ or $-\infty + \infty$.

(d) *If the limit $\lim_{n\to\infty} x_n$ exists in $\mathbb{R}$, then*

$$\liminf_{n\to\infty}(x_n + y_n) = \lim_{n\to\infty} x_n + \liminf_{n\to\infty} y_n,$$
$$\limsup_{n\to\infty}(x_n + y_n) = \lim_{n\to\infty} x_n + \limsup_{n\to\infty} y_n.$$

(e) *If $\limsup_{n\to\infty} |x_n| = 0$, then $\lim_{n\to\infty} x_n = 0$.*

For more basic properties involving upper and lower limits, including those for products of sequences $\{x_n y_n\}$, see [245, p. 409]. Note that by (1.1.3) any statement in Theorem 1.1.4 involving lim sup can be transformed into a dual statement involving lim inf. Thus it suffices to prove the statements involving lim inf only. See [245] for details.

We now proceed to define upper and lower limits for functions. To this end, let $f : I_a \to \mathbb{R}$ be a function, where $I_a = (a, \infty)$ for $a \in \mathbb{R}$. Define two auxiliary functions $F, G : I_a \to \mathbb{R} \cup \{\pm\infty\}$ by

$$F(x) = \inf\{f(r) : r \geq x\} \quad \text{and} \quad G(x) = \sup\{f(r) : r \geq x\}.$$

It is immediate that $F(x)$ is a nondecreasing function of x, whereas $G(x)$ is a nonincreasing function of x. Therefore the limits

$$\lim_{x\to\infty} F(x) \quad \text{and} \quad \lim_{x\to\infty} G(x)$$

always exist as unique numbers in $\mathbb{R} \cup \{\pm\infty\}$.

Definition 1.1.5. Let $f : I_a \to \mathbb{R}$ be a function. Then

$$\liminf_{x\to\infty} f(x) = \lim_{x\to\infty} F(x) \quad \text{and} \quad \limsup_{x\to\infty} f(x) = \lim_{x\to\infty} G(x)$$

are called the *limit inferior* and *limit superior* of f, respectively.

As is the case with sequences, the notation

$$\liminf_{x\to\infty} f(x) = \underline{\lim}_{x\to\infty} f(x) \quad \text{and} \quad \limsup_{x\to\infty} f(x) = \overline{\lim}_{x\to\infty} f(x)$$

is also used in the literature. Properties analogous to those for upper and lower limits of sequences are also valid for upper and lower limits of functions. For example, a minor modification of the proof of Theorem 1.1.3 yields the following consequence.

Corollary 1.1.6. *Let $f : I_a \to \mathbb{R}$ be a function, and set*

$$l = \liminf_{x\to\infty} f(x) \quad \textit{and} \quad L = \limsup_{x\to\infty} f(x).$$

Then $\lim_{x\to\infty} f(x) = M$ if and only if $l = L = M$. Here $M \in \mathbb{R} \cup \{\pm\infty\}$.

Upper and lower limits can be defined locally as well. To this end, let $f : D_0(a, r) \to \mathbb{R}$ be a function, where $a \in \mathbb{R}$, $r > 0$, and

$$D_0(a, r) = \{x \in \mathbb{R} : 0 < |x - a| < r\}$$

is a punctured interval. For $\varepsilon \in (0, r)$, denote the auxiliary functions

$$\alpha(\varepsilon) = \inf\{f(x) : x \in D_0(a, \varepsilon)\} \quad \text{and} \quad \beta(\varepsilon) = \sup\{f(x) : x \in D_0(a, \varepsilon)\}.$$

Then $\alpha(\varepsilon)$ is nondecreasing, and $\beta(\varepsilon)$ is nonincreasing as $\varepsilon \to 0^+$.

Definition 1.1.7. Let $f : D_0(a,r) \to \mathbb{R}$ be a function. Then

$$\liminf_{x\to a} f(x) = \lim_{\varepsilon\to 0^+} \alpha(\varepsilon) \quad \text{and} \quad \limsup_{x\to a} f(x) = \lim_{\varepsilon\to 0^+} \beta(\varepsilon)$$

are called the *local limit inferior* and *local limit superior* of f at a, respectively.

Local upper and lower limits can also be considered as one-sided limits, where $D_0(a,\varepsilon)$ should be replaced by either $(a-\varepsilon, a)$ (left) or $(a, a+\varepsilon)$ (right).

Problems

Problem 1.1.8. Let $\{x_n\}$ be a sequence defined by

$$x_n = 2 + (-1)^n + \frac{(-1)^n}{n}.$$

Find $\limsup_{n\to\infty} x_n$, $\liminf_{n\to\infty} x_n$, $\sup\{x_n : n \in \mathbb{N}\}$, and $\inf\{x_n : n \in \mathbb{N}\}$. This example illustrates that these four numbers may all be different.

Problem 1.1.9. Let $\{x_n\}$ be a sequence defined by

$$x_n = \cos\left(\frac{n\pi}{4}\right) + \frac{1}{n}.$$

Find $\limsup_{n\to\infty} x_n$, $\liminf_{n\to\infty} x_n$, $\sup\{x_n : n \in \mathbb{N}\}$, and $\inf\{x_n : n \in \mathbb{N}\}$.

Problem 1.1.10. Let $\{x_n\}$ be the sequence defined as follows:

$$x_n = \cos\left(\frac{n\pi}{3}\right), \quad n \text{ odd}, \quad \text{and} \quad x_n = (-1)^{n/2}\sin\left(\frac{n\pi}{4}\right), \quad n \text{ even}.$$

Consider the alternate definitions of upper and lower limits and complete the following tasks:
(a) Find all the cluster points of $\{x_n\}$.
(b) Find $\liminf_{n\to\infty} x_n$ and $\limsup_{n\to\infty} x_n$.

Problem 1.1.11. Let $f, g : I_a \to \mathbb{R}$ be functions. Prove that

$$\liminf_{x\to\infty}\big(f(x)+g(x)\big) \geq \liminf_{x\to\infty} f(x) + \liminf_{x\to\infty} g(x),$$
$$\limsup_{x\to\infty}\big(f(x)+g(x)\big) \leq \limsup_{x\to\infty} f(x) + \limsup_{x\to\infty} g(x)$$

whenever the right-hand sides are not of the form $\infty - \infty$ or $-\infty + \infty$.

Problem 1.1.12. Let $f, g : I_a \to \mathbb{R}$ be functions such that the limit $\lim_{x\to\infty} f(x)$ exists in $\mathbb{R}$. Prove that

$$\liminf_{x\to\infty}\big(f(x)+g(x)\big) = \lim_{x\to\infty} f(x) + \liminf_{x\to\infty} g(x),$$
$$\limsup_{x\to\infty}\big(f(x)+g(x)\big) = \lim_{x\to\infty} f(x) + \limsup_{x\to\infty} g(x).$$

1.2 Measures and densities of sets

In the theory of entire functions (that is, functions that are analytic for all $z \in \mathbb{C}$), it often turns out to be necessary to express the size of subsets of $\mathbb{R}_+ = (0,\infty)$ in terms of measures. This is typically desired when one of the following situations occur: (a) If a property holds on a set, then it is of interest to know how large the set is, or (b) if a property holds on $\mathbb{R}_+$ except for a possible exceptional set, then it is of interest to know how large the exceptional set is.

The most commonly used measures are linear and logarithmic measures.

Definition 1.2.1. The *linear measure* of a set $E \subset (0,\infty)$ is defined by

$$\operatorname{meas}(E) = \int_E dr.$$

Since the integrand in the logarithmic measure of a set is $1/r$, we need to stay away from 0. The following definition is suitable for all the situations in the book.

Definition 1.2.2. Let E be a set satisfying $E \subset (c,\infty)$ for some constant $c > 0$. Then the *logarithmic measure* of E is defined by

$$\log\operatorname{meas}(E) = \int_E \frac{dr}{r}.$$

The only property of interest in these two definitions is whether $\operatorname{meas}(E) < \infty$ (finite linear measure) or whether $\log\operatorname{meas}(E) < \infty$ (finite logarithmic measure). Obviously, $\log\operatorname{meas}(E) < \infty$ whenever $\operatorname{meas}(E) < \infty$. However, the converse does not hold.

Example 1.2.3. Let

$$E = \bigcup_{n=1}^{\infty}\left(n, n + \frac{1}{n}\right).$$

Then

$$\operatorname{meas}(E) = \sum_{n=1}^{\infty} \frac{1}{n} = \infty.$$

On the other hand,

$$\log\operatorname{meas}(E) = \sum_{n=1}^{\infty}\left(\log\left(n + \frac{1}{n}\right) - \log n\right) = \sum_{n=1}^{\infty} \log\left(1 + \frac{1}{n^2}\right) < \sum_{n=1}^{\infty} \frac{1}{n^2} < \infty.$$

Thus E has finite logarithmic measure and infinite linear measure.

Linear and logarithmic measures are sufficient for many situations, but when a set E has infinite logarithmic measure, it is often not enough information, and this occurs in many situations as well. The following definition introduces a continuum of measures, where linear and logarithmic measures are special cases.

Definition 1.2.4. For $r_0 > 0$, let $\psi : (r_0, \infty) \to (0, \infty)$ be a differentiable, strictly increasing, and unbounded function. Then the *ψ-measure* of a set $E \subset (r_0, \infty)$ is the value of the integral

$$\int_E d\psi(t) = \int_E \psi'(t)\,dt. \tag{1.2.1}$$

If the integral in (1.2.1) is finite, then we say that E is of *finite ψ-measure*.

The important special cases where $\psi(t) = t$ and $\psi(t) = \log t$ correspond to linear and logarithmic measures, respectively. Even if the integral in (1.2.1) is not finite, the size of the set E can still be measured in terms of densities. The idea is to compare the size of the set $E \cap (r_0, r)$ to an unbounded function of r; see Definition 1.2.5. For the rest of this section, we assume that the function ψ and the set E have the properties in Definition 1.2.4, where $r_0 > 0$ and where the set E may or may not have finite ψ-measure.

Definition 1.2.5 ([15]). The *upper* and *lower ψ-densities* of E are defined, respectively, by

$$\overline{\psi\text{-dens}}(E) = \limsup_{r\to\infty} \frac{1}{\psi(r)} \int_{E\cap(r_0,r)} d\psi(t),$$

$$\underline{\psi\text{-dens}}(E) = \liminf_{r\to\infty} \frac{1}{\psi(r)} \int_{E\cap(r_0,r)} d\psi(t).$$

If these two limits are equal, then their common value is called the *ψ-density* of E and is denoted by

$$\psi\text{-dens}(E) = \lim_{r\to\infty} \frac{1}{\psi(r)} \int_{E\cap(r_0,r)} d\psi(t).$$

It is clear that $0 \le \underline{\psi\text{-dens}}(E) \le \overline{\psi\text{-dens}}(E) \le 1$. If $\psi(t) = t$ or $\psi(t) = \log t$, then we have the upper and lower linear/logarithmic densities of E, denoted by $\overline{\text{dens}}(E)$, $\underline{\text{dens}}(E)$, $\overline{\log\text{dens}}(E)$, and $\underline{\log\text{dens}}(E)$. It is known [278, p. 121] that

$$0 \le \underline{\text{dens}}(E) \le \underline{\log\text{dens}}(E) \le \overline{\log\text{dens}}(E) \le \overline{\text{dens}}(E) \le 1. \tag{1.2.2}$$

These inequalities are generalized to ψ-densities in Lemma 1.2.6, which is a special case of [15, Lemma 1]. Note that the e^{ψ}-density of E (replacing ψ with e^{ψ} in Definition 1.2.5) is equal to the linear density of E when $\psi(t) = \log t$.

Lemma 1.2.6 ([15]). *We have*

$$\overline{\psi\text{-dens}}(E^c) + \underline{\psi\text{-dens}}(E) = 1, \tag{1.2.3}$$

where E^c denotes the complement of E in (r_0, ∞). Moreover,

$$0 \le \underline{e^{\psi}\text{-dens}}(E) \le \underline{\psi\text{-dens}}(E) \le \overline{\psi\text{-dens}}(E) \le \overline{e^{\psi}\text{-dens}}(E) \le 1. \tag{1.2.4}$$

Proof. Using

$$\frac{1}{\psi(r)}\int_{(r_0,r)} d\psi(t) = \frac{1}{\psi(r)}\int_{E\cap(r_0,r)} d\psi(t) + \frac{1}{\psi(r)}\int_{E^c\cap(r_0,r)} d\psi(t)$$

and observing that the usual limit as $r \to \infty$ of the left-hand side exists and equals 1, we find that

$$\begin{aligned}\overline{\psi\text{-dens}}(E^c) &= \limsup_{r\to\infty}\frac{1}{\psi(r)}\int_{E^c\cap(r_0,r)} d\psi(t)\\ &= \lim_{r\to\infty}\frac{1}{\psi(r)}\int_{(r_0,r)} d\psi(t) + \limsup_{r\to\infty}\Big(-\frac{1}{\psi(r)}\int_{E\cap(r_0,r)} d\psi(t)\Big)\\ &= 1 - \liminf_{r\to\infty}\frac{1}{\psi(r)}\int_{E\cap(r_0,r)} d\psi(t) = 1 - \underline{\psi\text{-dens}}(E).\end{aligned}$$

This proves (1.2.3).

To prove the inequality $\overline{\psi\text{-dens}}(E) \le \overline{e^\psi\text{-dens}}(E)$ in (1.2.4), let $\varepsilon > 0$. By the definition of e^ψ-density, there exists $r_1 = r_1(\varepsilon) > r_0$ such that

$$\int_{E\cap(r_0,r)} de^{\psi(t)} \le (\overline{e^\psi\text{-dens}}(E) + \varepsilon)e^{\psi(r)}, \quad r > r_1. \tag{1.2.5}$$

Let $\chi_E(t)$ denote the characteristic function of E. We obtain for $r > r_1$,

$$\begin{aligned}\int_{E\cap(r_0,r)} d\psi(t) &= \int_{E\cap(r_0,r)} e^{-\psi(t)}\, de^{\psi(t)} = \int_{r_0}^{r} e^{-\psi(t)}\chi_E(t)\, de^{\psi(t)}\\ &= \int_{r_0}^{r} e^{-\psi(t)}\, d\Big(\int_{r_0}^{t}\chi_E(s)\, de^{\psi(s)}\Big).\end{aligned}$$

Integrating by parts and using (1.2.5) and the fact that $\psi(t)$ is strictly increasing, we get

$$\begin{aligned}\int_{E\cap[r_0,r]} d\psi(t) &= e^{-\psi(r)}\int_{r_0}^{r}\chi_E(s)\, de^{\psi(s)} - \int_{r_0}^{r}\Big(\int_{r_0}^{t}\chi_E(s)\, de^{\psi(s)}\Big)\, d(e^{-\psi(t)})\\ &= e^{-\psi(r)}\int_{E\cap(r_0,r)} de^{\psi(t)} + \int_{r_0}^{r}\Big(\int_{E\cap(r_0,t)} de^{\psi(s)}\Big)e^{-\psi(t)}\, d\psi(t)\\ &\le \overline{e^\psi\text{-dens}}(E) + \varepsilon + \int_{r_1}^{r}(\overline{e^\psi\text{-dens}}(E) + \varepsilon)\, d\psi(t) + \int_{r_0}^{r_1}\Big(\int_{E\cap(r_0,t)} de^{\psi(s)}\Big)e^{-\psi(t)}\, d\psi(t)\end{aligned}$$

$$= (\overline{e^{\psi}\text{-dens}}(E) + \varepsilon)\left(1 + \int_{r_1}^{r} d\psi(t)\right) + O(1).$$

This implies that $\overline{\psi\text{-dens}}(E) \leq \overline{e^{\psi}\text{-dens}}(E) + \varepsilon$. Letting $\varepsilon \to 0^+$ gives $\overline{\psi\text{-dens}}(E) \leq \overline{e^{\psi}\text{-dens}}(E)$. Similarly, we obtain

$$\overline{\psi\text{-dens}}(E^c) \leq \overline{e^{\psi}\text{-dens}}(E^c),$$

with E^c in place of E. Also, by replacing ψ with e^{ψ} in (1.2.3) we get

$$\overline{e^{\psi}\text{-dens}}(E^c) + \underline{e^{\psi}\text{-dens}}(E) = 1.$$

Therefore,

$$\begin{aligned}\underline{\psi\text{-dens}}(E) &= 1 - \overline{\psi\text{-dens}}(E^c) \geq 1 - \overline{e^{\psi}\text{-dens}}(E^c)\\ &= 1 - (1 - \underline{e^{\psi}\text{-dens}}(E)) = \underline{e^{\psi}\text{-dens}}(E),\end{aligned}$$

which proves the second inequality in (1.2.4). The three remaining inequalities in (1.2.4) are trivial. □

The following lemma from [121] allows us to avoid exceptional sets $E \subset (r_0, \infty)$ whose upper ψ-density satisfies $\overline{\psi\text{-dens}}(E) < 1$.

Lemma 1.2.7. *Let $g, h : (r_0, \infty) \to (0, \infty)$ be nondecreasing functions satisfying*

$$g(r) \leq h(r), \quad r \in (r_0, \infty) \setminus E,$$

where $\overline{\psi\text{-dens}}(E) < 1$. Let α be any fixed constant satisfying

$$\alpha > \frac{1}{1 - \overline{\psi\text{-dens}}(E)},$$

and then set

$$s(r) = \psi^{-1}(\alpha\psi(r)),$$

where ψ^{-1} is the inverse function of ψ. Then there exists $R \in (r_0, \infty)$ such that

$$g(r) \leq h(s(r)), \quad r > R.$$

Proof. Suppose that there exists a strictly increasing sequence $\{r_n\}$ in (r_0, ∞) such that $r_n \to \infty$ and $(r_n, s(r_n)) \subset E$ for every $n \in \mathbb{N}$. Denote

$$I = \bigcup_{n=1}^{\infty} (r_n, s(r_n)).$$

Then $I \subset E$, but

$$\overline{\psi\text{-dens}}(E) \geq \overline{\psi\text{-dens}}(I) \geq \limsup_{n\to\infty} \frac{1}{\psi(s(r_n))} \int\limits_{I\cap(r_0,s(r_n))} d\psi(t)$$

$$\geq \limsup_{n\to\infty} \frac{1}{\psi(s(r_n))} \int\limits_{r_n}^{s(r_n)} d\psi(t) = 1 - \frac{1}{\alpha} > \overline{\psi\text{-dens}}(E),$$

which is a contradiction. Thus there exists $R > r_0$ such that for all $r > R$, there exists $t \in (r, s(r)) \setminus E$. Since g and h are nondecreasing, it follows that

$$g(r) \leq g(t) \leq h(t) \leq h(s(r)), \quad r > R.$$

This completes the proof. □

Note that if $\psi(r) = r$ or if $\psi(r) = \log r$, then the function $s(r)$ in Lemma 1.2.7 reduces to $s(r) = \alpha r$ and $s(r) = r^\alpha$, respectively.

Next, the condition $\overline{\psi\text{-dens}}(E) < 1$ in Lemma 1.2.7 will be replaced with a more restrictive requirement that the set E is of finite ψ-measure. Such a set is clearly of zero upper ψ-density. The conclusion of the result becomes stronger in the sense that the function $s(r)$ grows slower than its analogue in Lemma 1.2.7.

Lemma 1.2.8. *Let $g, h : (r_0, \infty) \to (0, \infty)$ be nondecreasing functions satisfying*

$$g(r) \leq h(r), \quad r \in (r_0, \infty) \setminus E,$$

where $\int_E d\psi(t) < \infty$. Let α be any fixed constant satisfying $\alpha > \int_E d\psi(t)$, and then set $s(r) = \psi^{-1}(\alpha + \psi(r))$. Then

$$g(r) \leq h(s(r)), \quad r > r_0.$$

Proof. For $r > r_0$, we have

$$\int\limits_r^{s(r)} d\psi(t) = \psi(s(r)) - \psi(r) = \alpha > \int\limits_E d\psi(t).$$

In other words, the interval $(r, s(r))$ meets the complement of E for all $r > r_0$. Therefore, by taking $t \in (r, s(r)) \setminus E$ we get

$$g(r) \leq g(t) \leq h(t) \leq h(s(r)), \quad r > r_0,$$

where the monotonicity of g and h was used. □

Note that if $\psi(r) = r$ or if $\psi(r) = \log r$, then the function $s(r)$ in Lemma 1.2.8 reduces to $s(r) = \alpha + r$ and $s(r) = \beta r$, respectively, where $\beta = e^{\alpha} > 1$. While Lemma 1.2.8 gives a corollary regarding finite logarithmic measure, the next lemma goes slightly further.

Lemma 1.2.9 ([87]). *Let g and h be nondecreasing functions on $[0, \infty)$ such that $g(r) \leq h(r)$ for all $r \notin E \cup [0, 1]$, where $E \subset (0, \infty)$ is a set of finite logarithmic measure. Let $\alpha > 1$ be a constant. Then there exists $R = R(\alpha) > 0$ such that $g(r) \leq h(\alpha r)$ for all $r > R$.*

Proof. We reason as on p. 519 of [9]. From the hypothesis on E, it follows that there exists a constant $R = R(\alpha) > 0$ such that for all $r \geq R$, the interval $(r, \alpha r)$ must contain a point $t \notin E \cup [0, 1)$. Indeed, if this were not true, then it would follow that there exists a strictly increasing sequence $\{r_n\}$ of positive real numbers tending to infinity such that $(r_n, \alpha r_n) \subset E$, where the intervals $(r_n, \alpha r_n)$ are disjoint. Then

$$\int_E \frac{dr}{r} \geq \sum_{n=1}^{\infty} \int_{r_n}^{\alpha r_n} \frac{dt}{t} = \sum_{n=1}^{\infty} \log \frac{\alpha r_n}{r_n} = \sum_{n=1}^{\infty} \log \alpha = \infty,$$

which would give the contradiction that E has infinite logarithmic measure. Hence if $R > 0$ is large enough, then for all $r > R$, the interval $(r, \alpha r)$ meets the complement of E. Therefore, by taking $t \in (r, \alpha r) \setminus E$ we get

$$g(r) \leq g(t) \leq h(t) \leq h(\alpha r), \quad r > R,$$

where the monotonicity of g and h was used. □

Lemma 1.2.9 will be used later in the book. Clearly, Lemma 1.2.9 holds if "logarithmic measure" is replaced by "linear measure," and this is enough in some situations.

Theorem 1.2.10 shows how limit superior and limit inferior can be illustrated in terms of ψ-densities. The proof uses Lemma 1.2.7, but the details are omitted because a decent proof with a sufficient amount of details can be found in the original reference [121].

Theorem 1.2.10 ([121]). *Let $\varphi : (r_0, \infty) \to [0, \infty)$ be a function with*

$$\limsup_{r\to\infty} \varphi(r) = K \quad \textit{and} \quad \liminf_{r\to\infty} \varphi(r) = k,$$

where $0 \leq k < K < \infty$. Suppose that there exists a differentiable, strictly increasing, and unbounded function $\psi : (r_0, \infty) \to \mathbb{R}_+$ such that $\varphi(r)\psi(r)$ is nondecreasing on (r_0, ∞). Then for all $0 < \varepsilon \leq K$, the sets

$$F_\varepsilon = \{r \in (r_0, R) : |\varphi(r) - K| < \varepsilon\} \quad \textit{and} \quad G_\varepsilon = \{r \in (r_0, R) : |\varphi(r) - k| < \varepsilon\}$$

satisfy

$$\overline{\psi\text{-dens}}(F_\varepsilon) \geq \frac{\varepsilon}{K}, \quad \overline{\psi\text{-dens}}(G_\varepsilon) \geq \frac{\varepsilon}{k+\varepsilon}, \tag{1.2.6}$$

$$\underline{\psi\text{-dens}}(F_\varepsilon) \leq \frac{k}{k+\varepsilon}, \quad \underline{\psi\text{-dens}}(G_\varepsilon) \leq \frac{K-\varepsilon}{K}. \tag{1.2.7}$$

In addition,

$$\overline{e^\psi\text{-dens}}(F_\varepsilon) = \overline{e^\psi\text{-dens}}(G_\varepsilon) = 1, \quad \underline{e^\psi\text{-dens}}(F_\varepsilon) = \underline{e^\psi\text{-dens}}(G_\varepsilon) = 0.$$

If $K = +\infty$, then for every large $M > 0$, the set $H_M = \{r \in (r_0, R) : \varphi(r) > M\}$ satisfies

$$\overline{\psi\text{-dens}}(H_M) = 1 \quad \textit{and} \quad \underline{\psi\text{-dens}}(H_M) \leq \frac{k}{M}. \tag{1.2.8}$$

One possible application for Theorem 1.2.10 is in the case where $\psi(r) = \log r$ and $\varphi(r) = \frac{\log\log M(r,f)}{\log r}$, where $M(r,f)$ is the maximum modulus of an entire function f. Then K and k correspond to the order and lower order of f, respectively. These topics will be discussed in Chapter 2.

Problems

Problem 1.2.11. Suppose that a set $E \subset [1,\infty)$ has finite logarithmic measure. Prove that $\text{dens}(E) = 0$.

Problem 1.2.12. Calculate the linear and logarithmic densities of the set

$$E = \bigcup_{n=1}^{\infty} [n, n+1/2].$$

1.3 Convex functions

We begin by recalling the definitions of convex and concave functions.

Definition 1.3.1. A function $f : I \to \mathbb{R}$ is said to be *convex* if

$$f(\alpha x + (1-\alpha)y) \leq \alpha f(x) + (1-\alpha)f(y) \tag{1.3.1}$$

for all $\alpha \in [0,1]$ and $x, y \in I$, where I is a real interval. The function f is *midpoint convex* if

$$f\left(\frac{x+y}{2}\right) \leq \frac{f(x)+f(y)}{2}, \quad x, y \in I.$$

Geometrically, a convex function f satisfies the following property [17, p. 192]: If $x < s < y$, where $x, y \in I$, then the point $(s, f(s))$ lies below or on the line segment joining $(x, f(x))$ and $(y, f(y))$. A convex function is sometimes called concave upward or convex downward.

Definition 1.3.2. A function $f : I \to \mathbb{R}$ is said to be *concave* if inequality (1.3.1) is reversed, that is, if

$$f(ax + (1-a)y) \geq af(x) + (1-a)f(y) \tag{1.3.2}$$

for all $a \in [0,1]$ and $x, y \in I$, where I is a real interval.

Obviously, f is concave if and only if $-f$ is convex. A concave function is sometimes called concave downward or convex upward.

If f is continuous on a real interval I, then f is midpoint convex if and only if f is convex [220, Theorem 1.1.4]. We mention that a convex function on a open interval I is continuous, whereas if I is not an open interval, then f is not necessarily continuous [240, pp. 61–62]. A convex function need not be differentiable at every point; see Problem 1.3.12.

Proposition 1.3.3 ([220, Theorem 1.3.1]). *Let $f : I \to \mathbb{R}$ be differentiable, where I is an open interval. Then f is convex if and only if the derivative f' is a nondecreasing function.*

Proposition 1.3.4 ([17, Theorem 6.4.6]). *Let $f : I \to \mathbb{R}$ be twice differentiable, where I is an open interval. Then f is convex if and only if $f''(x) \geq 0$ for all $x \in I$.*

The function $f(x) = e^x$ is convex on $I = (-\infty, \infty)$. If f and g are convex functions and f is nondecreasing, then $f \circ g$ is clearly convex. The assumption that f is nondecreasing cannot be omitted. For example, $f(x) = -x$ and $g(x) = x^2$ are both convex, but $(f \circ g)(x) = -x^2$ is not convex.

There are many ways of generalizing the notion of convexity. We will discuss some of them in this section, starting with a definition.

Definition 1.3.5. A function $f : (0, \infty) \to (0, \infty)$ is said to be
(a) *logarithmically convex*, or *log-convex* for short, if $\log \circ f$ is convex,
(b) *convex in* $\log x$ if $f \circ \exp$ is convex.

Remark. (a) A log-convex function f is convex because it is a composition of the strictly increasing convex function exp and the function $\log \circ f$, which is convex by definition. Conversely, the function $g(x) = x^2$ is convex, but not log-convex.
(b) It is easy to see that if f is convex and nondecreasing, then f is convex in $\log x$. Conversely, the function $f(x) = \log^2 x$ is convex in $\log x$ but not convex.
(c) If f is log-convex and nondecreasing, then f is convex in $\log x$ by Remarks (a) and (b). The function $f_1(x) = x^2$ is convex and convex in $\log x$ but not log-convex. The function $f_2(x) = 1/x$ is convex, log-convex, and convex in $\log x$.

Definition 1.3.5 can be easily generalized.

Definition 1.3.6. Let $\psi : (0, \infty) \to \mathbb{R}$ be a function. Then a function $f : (0, \infty) \to (0, \infty)$ is said to be
(a) *ψ-convex* if $\psi \circ f$ is convex,
(b) *convex in* $\psi(x)$ if ψ is strictly increasing and if $f \circ \psi^{-1}$ is convex.

These new concepts can be expressed in terms of derivatives as follows.

Lemma 1.3.7. *Let $f : (0,\infty) \to (0,\infty)$ and $\psi : (0,\infty) \to \mathbb{R}$ be differentiable functions. Then the following two statements hold:*
(a) *f is ψ-convex if and only if $\psi'(f(x))f'(x)$ is a nondecreasing function of x.*
(b) *Suppose that $\psi : (0,\infty) \to (0,\infty)$ is a strictly increasing bijection. Then f is convex in $\psi(x)$ if and only if $\frac{df(x)}{d\psi(x)}$ is a nondecreasing function of x.*

Proof. (a) Set $g = \psi \circ f$. Clearly, g is differentiable, so that g is convex if and only if $g'(x) = \psi'(f(x))f'(x)$ is a nondecreasing function of x.
(b) Set $h = f \circ \psi^{-1}$, which is clearly a well-defined differentiable function. By making a change of variable $r = \psi^{-1}(x)$ we obtain

$$\begin{aligned} h'(x) &= \frac{d}{dx} f(\psi^{-1}(x)) = f'(\psi^{-1}(x)) \frac{d}{dx}\psi^{-1}(x) \\ &= \frac{f'(\psi^{-1}(x))}{\psi'(\psi^{-1}(x))} = \frac{f'(r)}{\psi'(r)} = \frac{df(r)}{d\psi(r)}. \end{aligned}$$

Noting that ψ^{-1} is also a strictly increasing bijection, it follows that $h'(x)$ is a nondecreasing function of x if and only if $\frac{df(r)}{d\psi(r)}$ is a nondecreasing function of r. This proves the statement. □

The following lemma is an elementary integral representation for a differentiable function f that is convex in $\psi(r)$. The result will be needed in Section 7.3.

Lemma 1.3.8. *Let $f : (0,\infty) \to (0,\infty)$ be a differentiable function, and let $\psi : (0,\infty) \to (0,\infty)$ be a continuously differentiable function. Suppose that ψ is a strictly increasing bijection, and let $x_0 > 0$. If f is convex in $\psi(x)$, then*

$$f(x) = f(x_0) + \int_{x_0}^{x} \varphi(t)\, d\psi(t),$$

where $\varphi : (0,\infty) \to (0,\infty)$ is a nondecreasing function.

Proof. Set $\varphi(t) = \frac{df(t)}{d\psi(t)}$, which is a nondecreasing function of t by Lemma 1.3.7. It follows that φ is integrable [239, Theorem 6.9]. The statement then follows by the fundamental theorem of calculus. □

Convex functions in general have a more complicated integral representation. We will state such a result without a proof.

Lemma 1.3.9 ([230, p. 7]). *Let $f : (0,\infty) \to \mathbb{R}$ be a function. Then f is convex if and only if there exists a nondecreasing left-continuous function $\varphi : (0,\infty) \to \mathbb{R}$ such that for all closed intervals $[a,b] \subset (0,\infty)$, we have*

$$f(x) = f(a) + \int_a^x \varphi(t)\, dt, \quad a \le x \le b.$$

Moreover, f has one-sided derivatives at each point of $(0,\infty)$ *and they are equal to* φ *except perhaps for at most a countable number of points.*

Suppose that $f(x)$ is convex in $\psi(x)$, where $\psi(x)$ is continuously differentiable. Then $f \circ \psi^{-1}$ is convex and (by Lemma 1.3.9) has the integral representation

$$(f \circ \psi^{-1})(y) = (f \circ \psi^{-1})(a) + \int_a^y \varphi(t)\, dt, \quad y \ge a > 0,$$

where the function $\varphi(t)$ possesses the properties in Lemma 1.3.9. Writing $x = \psi^{-1}(y)$ and $c = \psi^{-1}(a)$, we obtain the following integral representation for f:

$$f(x) = f(c) + \int_a^{\psi(x)} \varphi(t)\, dt = f(c) + \int_c^x \varphi(\psi(s))\, d\psi(s), \quad x \ge c > 0, \tag{1.3.3}$$

where $c > 0$ is large enough so that $\psi(c) \ge 0$.

For further generalizations of the notion of convexity, see [3].

Problems

Problem 1.3.10. If f is a convex function on an interval I, then show that for any three points a, b, c on I satisfying $a \le b \le c$, we have

$$f(b) \le \max\{f(a), f(c)\}.$$

Problem 1.3.11. Prove the following statements:
(a) $f(x) = (x+1)^c$ is a convex function on $(0,\infty)$ when $c \ge 1$.
(b) $f(x) = \sqrt{x}$ is a concave function on $(0,\infty)$.

Problem 1.3.12. Give an example of a convex function on $[-1,1]$ that is not differentiable at $x = 0$. Justify your answer.

Problem 1.3.13. Answer the following questions:
(a) Determine if $f(x) = \sqrt{x}$ is log-convex.
(b) Determine if $g(x) = 2 + \sin\sqrt{x}$ is convex in $\log x$.

Problem 1.3.14. Suppose that $f, g : (0,\infty) \to (0,\infty)$ are twice differentiable and unbounded functions such that $g'(x) > 0$ for all $x > 0$. Prove that $f(x)$ is convex in $g(x)$ if and only if

$$\frac{f''(x)}{f'(x)} \ge \frac{g''(x)}{g'(x)}, \quad x > 0.$$

Problem 1.3.15. Let f and g be as in Problem 1.3.14. Use the result in Problem 1.3.14 to prove that if $f(x)$ is convex in $g(x)$, then $f'(x)/g'(x)$ is a nondecreasing function.

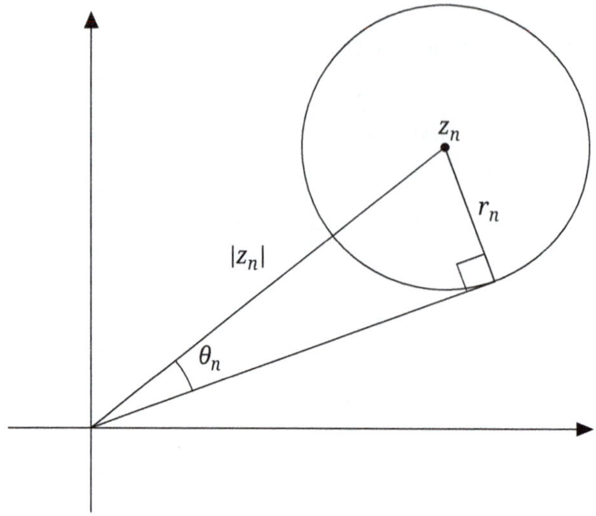

Figure 1.1: A view of the disc $D(z_n, r_n)$ from the origin.

1.4 $\mathcal{E}$-sets and R-sets

In 1960, Hayman [105] introduced an *$\mathcal{E}$-set* as a countable collection of Euclidean discs $D(z_n, r_n) = \{z : |z - z_n| < r_n\}$ that do not include the origin, for which $z_n \in \mathbb{C}$, $|z_n| \to \infty$, $r_n > 0$, and whose subtending angles at the origin have a finite sum. Any such disc can be seen from the origin at the subtending angle $2\theta_n$, where $\sin\theta_n = r_n/|z_n|$; see Figure 1.1. Thus, by the definition,

$$\sum_{n=1}^{\infty} \theta_n < \infty. \tag{1.4.1}$$

From (1.4.1) we get $\theta_n \to 0$ as $n \to \infty$. Thus by the standard limit $\lim_{\theta\to 0} \frac{\sin\theta}{\theta} = 1$ we conclude the following: There exists $N \in \mathbb{N} \setminus \{1\}$ such that

$$\theta_n/2 \le \sin\theta_n \le 2\theta_n, \quad n \ge N. \tag{1.4.2}$$

Therefore, by (1.4.1) and (1.4.2),

$$\sum_{n=1}^{\infty} \frac{r_n}{|z_n|} = \sum_{n=1}^{N-1} \frac{r_n}{|z_n|} + \sum_{n=N}^{\infty} \frac{r_n}{|z_n|} \le O(1) + \sum_{n=N}^{\infty} 2\theta_n < \infty. \tag{1.4.3}$$

Due to the convergence in (1.4.3), we have $r_n/|z_n| \to 0$ as $n \to \infty$.

Let E denote the circular projection of the discs $D(z_n, r_n)$ on $(0, \infty)$. Since the discs $D(z_n, r_n)$ do not contain the origin and since there are at most finitely many discs near the origin, there exists a fixed constant $c > 0$ such that $|z_n| - r_n > c > 0$ for all n. We find that E has finite logarithmic measure, since

$$\int_E \frac{dx}{x} \leq \sum_{n=1}^{\infty} \int_{|z_n|-r_n}^{|z_n|+r_n} \frac{dx}{x} \leq \sum_{n=1}^{\infty} \frac{2r_n}{|z_n| - r_n} = \sum_{n=1}^{\infty} \frac{2r_n/|z_n|}{1 - r_n/|z_n|} < \infty.$$

The essence of $\mathcal{E}$-sets lies in the following lemma.

Lemma 1.4.1 ([105]). *The set $\Theta \subset [0, 2\pi)$ of values θ for which the ray $\arg z = \theta$ meets infinitely many discs of a given $\mathcal{E}$-set has linear measure zero.*

Proof. Suppose that the $\mathcal{E}$-set in question consists of Euclidean discs $D(z_n, r_n)$, and let $2\theta_n$ be the subtending angle in which each such disc can be seen from the origin. Recall that $\sin\theta_n = r_n/|z_n|$. Given $\varepsilon > 0$, we conclude from (1.4.3) that there exists $M(\varepsilon) \in \mathbb{N}$ for which

$$\sum_{n=M(\varepsilon)}^{\infty} \frac{r_n}{|z_n|} < \frac{\varepsilon}{4}.$$

Without loss of generality, we may assume that $M(\varepsilon) \geq N$, where N is the integer in (1.4.2). Thus the set $\Theta_\varepsilon \subset [0, 2\pi)$ of values θ for which the ray $\arg z = \theta$ meets at least one of the discs $D(z_n, r_n)$ for $n \geq M(\varepsilon)$ has the linear measure

$$\int_{\Theta_\varepsilon} dt \leq \sum_{n=M(\varepsilon)}^{\infty} 2\theta_n \leq \sum_{n=M(\varepsilon)}^{\infty} 4\sin\theta_n = \sum_{n=M(\varepsilon)}^{\infty} \frac{4r_n}{|z_n|} < \varepsilon.$$

Clearly, $\Theta \subset \Theta_\varepsilon$ for every $\varepsilon > 0$, and hence $\int_\Theta dt = 0$. □

The set Θ in Lemma 1.4.1 can be countably infinite; see [59, Example 1.1]. An important special case of an $\mathcal{E}$-set, introduced by Bank, Laine, and Langley [14] in 1986, is an *R-set* consisting of Euclidean discs $D(\zeta_n, \rho_n)$ for which $\zeta_n \in \mathbb{C}$, $|\zeta_n| \to \infty$, $\rho_n > 0$, and whose diameters have a finite sum, that is,

$$\sum_n \rho_n < \infty. \tag{1.4.4}$$

Thus the circular projection E of an R-set onto the positive real axis has finite linear measure, that is, $\int_E dt < \infty$.

The conclusion in Lemma 1.4.1 is obviously valid for any given R-set. We proceed to state and prove a result that is more specific to R-sets.

Lemma 1.4.2 ([59]). *Let $b \in \mathbb{R}$. Then the set $C \subset \mathbb{R}$ of values c for which the line $y = bx + c$ or the line $x = c$ meets infinitely many discs of a given R-set has linear measure zero.*

Proof. We prove the case $b = 0$. The general case then follows from this by a rotation and a translation. Suppose that the R-set in question consists of Euclidean discs $D(\zeta_n, \rho_n)$. Given $\varepsilon > 0$, we conclude from (1.4.4) that there exists $N(\varepsilon) \in \mathbb{N}$ for which

$$\sum_{n=N(\varepsilon)}^{\infty} \rho_n < \frac{\varepsilon}{2}.$$

Let $C_\varepsilon \subset \mathbb{R}$ denote the set of values $c \in \mathbb{R}$ for which the line $y = c$ meets at least one of the discs $D(\zeta_n, \rho_n)$ for $n \geq N(\varepsilon)$. We have

$$\int_{C_\varepsilon} dt \leq \sum_{n=N(\varepsilon)}^{\infty} 2\rho_n < \varepsilon.$$

Since $C \subset C_\varepsilon$ for every $\varepsilon > 0$, it follows that $\int_C dt = 0$. □

The set C in Lemma 1.4.2 can be uncountable; see [59, Example 2.4].

Remark. We have seen that the circular projection of an $\mathcal{E}$-set/R-set onto $(0, \infty)$ has finite logarithmic/linear measure. Moreover, almost every ray/line, respectively, avoids infinitely many discs constituting these sets.

As in [59], for $x_0 \geq 0$, let $K : (x_0, \infty) \to (0, \infty)$ be a concave, continuous, nondecreasing, and subadditive function. (For the definition of a subadditive function, see Section 11.2.) Following [59], a countable collection of Euclidean discs $D(z_n, r_n)$ that do not include the origin for which $z_n \in \mathbb{C}$, $|z_n| \to \infty$, $r_n > 0$, and

$$\sum_n \frac{r_n}{K(|z_n|)} < \infty$$

is called a *K-set*. It is proved in [59] that the set $C \subset \mathbb{R}$ of values c for which the curves $y = cK(x)$ meet infinitely many discs of a K-set has linear measure zero. Moreover, the circular projection E of a K-set on $(0, \infty)$ satisfies $\int_E \frac{dt}{K(t)} < \infty$. Choosing $K(r) = r$ and $K(r) \equiv 1$, we get an $\mathcal{E}$-set and an R-set as special cases.

Problems

Problem 1.4.3. Let $D(z_n, r_n)$ with $z_n \in \mathbb{C}$, $|z_n| \to \infty$, and $r_n > 0$ be Euclidean discs not containing the origin and satisfying

$$\sum_{n=1}^{\infty} \frac{r_n}{\sqrt{|z_n|}} < \infty.$$

Let $E \subset (0, \infty)$ denote the circular projection of the collection of discs $D(z_n, r_n)$. Give a direct proof of the fact that $\int_E \frac{dt}{\sqrt{t}} < \infty$.

1.5 Stieltjes integral

The Stieltjes integral is a generalization of the Riemann integral, named after the Dutch mathematician T. J. Stieltjes (1856–1894). Proofs are often similar to those for Riemann integrals. Since Stieltjes integrals are used in this book, we give this short section for the convenience of the reader, using the presentation in [270, Section 18.9]; see also [4, 239]. Examples and the integration-by-parts formula are given. Proofs are often omitted.

The Stieltjes integral

$$\int_a^b f(x)\,dg(x) \tag{1.5.1}$$

is defined for two real-valued functions f and g defined on a finite real interval $[a,b]$. In the case $g(x) = x$, the Stieltjes integral becomes the Riemann integral

$$\int_a^b f(x)\,dx.$$

Also, when f is Riemann integrable and g has a continuous derivative, we have

$$\int_a^b f(x)\,dg(x) = \int_a^b f(x)g'(x)\,dx,$$

which is also a Riemann integral. Since it is assumed that the reader is familiar with Riemann integrals, we focus on Stieltjes integrals when $g(x)$ is a step function, which is how Stieltjes integrals are used in this book.

Definition 1.5.1. Let f and g be two real-valued functions on the interval $[a,b]$, let $P = \{x_0, x_1, \dots, x_n\}$ be a partition of $[a,b]$, and let $x_1', x_2', \dots, x_n'$ be points satisfying $x_{k-1} \le x_k' \le x_k$ for $1 \le k \le n$. Then the Stieltjes integral (1.5.1) is defined as

$$\int_a^b f(x)\,dg(x) = \lim_{\|P\|\to 0} \sum_{k=1}^{n} f(x_k')(g(x_k) - g(x_{k-1})),$$

provided that the sums converge to a unique limit that is independent of the partition P, where $\|P\| = \max_{1\le k\le n} |x_k - x_{k-1}|$ is the *norm* of the partition P.

Remark. Although a and b in the integral (1.5.1) are finite numbers, we can have $a = -\infty$ or $b = +\infty$ (or both) in (1.5.1) by using the standard limit process that is used for improper Riemann integrals.

Proposition 1.5.2. *If $\int_a^b dg(x)$ exists, then $\int_a^b dg(x) = g(b) - g(a)$.*

Proof. For every $\varepsilon > 0$, there exists a partition $P_\varepsilon = \{x_0, x_1, \dots, x_n\}$ of $[a, b]$ such that

$$\left| \int_a^b dg(x) - \sum_{k=1}^{n} (g(x_k) - g(x_{k-1})) \right| < \varepsilon.$$

By a telescoping sum we have

$$\left| \int_a^b dg(x) - (g(b) - g(a)) \right| < \varepsilon.$$

Since $\varepsilon > 0$ is arbitrary, the statement follows. □

We give two examples presented in [270].

Example 1.5.3. Let f be a continuous function on $[a, b]$. Suppose that $g(x) = c$ for $a < x < b$, where c is a real number, and let $g(a)$ and $g(b)$ have any real values. It is easy to see that [270, pp. 561–562]

$$\int_a^b f(x)\, dg(x) = f(a)\big(c - g(a)\big) + f(b)\big(g(b) - c\big). \tag{1.5.2}$$

Example 1.5.3 can be generalized as follows.

Example 1.5.4. Let f be a continuous function on $[a, b]$. Suppose that $a_0 = a < a_1 < a_2 < \cdots < a_N = b$, where $g(x) = c_k$ for $a_{k-1} < x < a_k$ $(1 \le k \le N)$, such that $g(a_k)$ can be any values for $0 \le k \le N$. It can be shown that [270, p. 562]

$$\int_a^b f(x)\, dg(x) = f(a)\big(c_1 - g(a)\big) + \sum_{k=1}^{N-1} f(a_k)(c_{k+1} - c_k) + f(b)\big(g(b) - c_N\big). \tag{1.5.3}$$

In the case $N = 1$, where the summation does not exist, (1.5.3) reduces to (1.5.2).

For an entire function h in the complex plane $\mathbb{C}$, let $n(r, 0, h)$ denote the number of zeros of h that lie in the disc $|z| \le r$, counted according to multiplicities.

Example 1.5.5. By using Example 1.5.4 we will show that

$$\int_0^6 x^2\, dg(x) = 5\pi^2,$$

where $g(x) = n(x, 0, \cos z)$. Since the zeros of $\cos z$ are all simple zeros, which occur at the points $z = (2k+1)\pi/2$, $k \in \mathbb{Z}$, we have $g(x) = 0$ for $0 \le x < \pi/2$, $g(x) = 2$ for $\pi/2 \le x < 3\pi/2$, and $g(x) = 4$ for $3\pi/2 \le x \le 6$. Then $a_0 = 0$, $a_1 = \pi/2$, $a_2 = 3\pi/2$, and $a_3 = 6$, and thus from (1.5.3) we obtain

$$\int_0^6 x\, dg(x) = 0(0-0) + (\pi/2)^2(2-0) + (3\pi/2)^2(4-2) + 6^2(4-4) = 5\pi^2.$$

Last, we mention the *integration-by-parts* formula for Stieltjes integrals,

$$\int_a^b f(x)\,dg(x) = f(b)g(b) - f(a)g(a) - \int_a^b g(x)\,df(x). \tag{1.5.4}$$

Part of proving formula (1.5.4) is to show that if the integral on the left exists, then the integral on the right also exists; see [4, Theorem 7.6].

Problems

Problem 1.5.6. Suppose that $\int_a^b f(x)\,dg(x) = 0$ for every monotonic function $f(x)$ on $[a,b]$. Prove that $g(x)$ must be a constant function on $[a,b]$.

Problem 1.5.7. Find the numerical value of the Stieltjes integral

$$\int_{1/2}^{3} \frac{dg(x)}{x},$$

where $g(x) = n(x, 0, (z-2)^2 \sin \pi z)$.

1.6 Uniform convergence

Many deductions in this book are based on uniform convergence on compact subsets of $\mathbb{C}$. Here a *compact set* is a closed and bounded set, whereas a *domain* is an open and connected set.

First, for the convenience of the reader, we state and prove several well-known convergence theorems concerning uniform convergence on compact subsets of a domain $G \subset \mathbb{C}$. Second, we observe that a sequence of functions converges uniformly on compact subsets of a domain G if and only if the sequence converges in the Cartan metric on G.

Definition 1.6.1. Let $\{f_n(z)\}$ be a sequence of functions on a domain $G \subset \mathbb{C}$.

(a) We say that $\{f_n(z)\}$ *converges pointwise* in G to a function $f(z)$ if for all $z \in G$ and $\varepsilon > 0$, there exists $N = N(\varepsilon, z) \in \mathbb{N}$ such that

$$|f_n(z) - f(z)| < \varepsilon, \quad n \geq N.$$

(b) We say that $\{f_n(z)\}$ *converges uniformly on compact subsets* of G to a function $f(z)$ if for every compact set $K \subset G$ and all $\varepsilon > 0$, there exists $N = N(\varepsilon, K) \in \mathbb{N}$ such that

$$|f_n(z) - f(z)| < \varepsilon, \quad z \in K,\ n \geq N.$$

The next well-known result from topology, called the *Heine–Borel theorem*, is used in some parts of the book, including the proof of Theorem 1.6.11.

Theorem 1.6.2 ([204, Theorem 3.14]). *Suppose that a compact set $K \subset \mathbb{C}$ is contained in the union of a collection of open sets. Then K is already contained in the union of a finite collection of these open sets.*

Due to the Heine–Borel theorem, the uniform convergence on compact subsets of a domain $G \subset \mathbb{C}$ is sometimes called the *local uniform convergence* on G.

The *Cauchy criterion* for ordinary sequences has a counterpart for uniform convergence of a sequence of functions.

Theorem 1.6.3. *Let $\{f_n(z)\}$ be a sequence of functions on a domain $G \subset \mathbb{C}$. Then $\{f_n(z)\}$ is uniformly convergent to a function $f(z)$ on compact subsets of G if and only if for every compact set $K \subset G$ and all $\varepsilon > 0$, there exists $N = N(\varepsilon, K) \in \mathbb{N}$ such that*

$$|f_n(z) - f_m(z)| < \varepsilon, \quad z \in K,\ n, m \geq N. \tag{1.6.1}$$

Proof. Suppose that $\{f_n(z)\}$ is uniformly convergent to a function $f(z)$ on compact subsets of G. Then applying

$$|f_n(z) - f_m(z)| \leq |f_n(z) - f(z)| + |f(z) - f_m(z)|$$

together with Definition 1.6.1(b), we obtain statement (1.6.1).

Conversely, suppose that for every compact set $K \subset G$ and all $\varepsilon > 0$, there exists $N = N(\varepsilon, K) \in \mathbb{N}$ such that (1.6.1) holds. Then the sequence $\{f_n(z)\}$ converges pointwise to a function $f(z)$ by the sequence version of the Cauchy criterion. Fixing n and letting $m \to \infty$ in (1.6.1), we obtain

$$|f_n(z) - f(z)| \leq \varepsilon, \quad z \in K,\ n \geq N,$$

which establishes the uniform convergence on compact sets. □

Definition 1.6.4. Let $\{f_n(z)\}$ be a sequence of functions on a domain $G \subset \mathbb{C}$, and set

$$S_k(z) = f_1(z) + \cdots + f_k(z), \quad z \in G.$$

We say that the infinite series $\sum_{n=1}^{\infty} f_n(z)$ *converges pointwise* in G or *converges uniformly on compact subsets* of G if the sequence $\{S_k(z)\}$ of partial sums has the corresponding property.

The absolute (pointwise and uniform) convergence of a series $\sum_{n=1}^{\infty} f_n(z)$ is defined in the usual way by replacing $\{S_k(z)\}$ in the definition above with $\{T_k(z)\}$, where

$$T_k(z) := |f_1(z)| + \cdots + |f_k(z)|, \quad z \in G.$$

We now recall the *Weierstrass M-test* for a series of functions.

Theorem 1.6.5. *Let $\{f_n(z)\}$ be a sequence of functions on a domain $G \subset \mathbb{C}$, and let $\{M_n\}$ be a sequence of positive numbers such that $\sum_{n=1}^{\infty} M_n$ converges. If for every compact set $K \subset G$, there exists $N = N(K) \in \mathbb{N}$ such that*

$$|f_n(z)| \leq M_n, \quad z \in K, \ n \geq N,$$

then the series $\sum_{n=1}^{\infty} f_n(z)$ converges absolutely and uniformly on compact subsets of G.

Proof. Let $K \subset G$ be a compact set, let $N = N(K) \in \mathbb{N}$ be the number provided by the assumption, and let $\varepsilon > 0$. Denote $R_k = \sum_{n=1}^{k} M_n$. Since $\sum_{n=1}^{\infty} M_n$ converges, the sequence version of the Cauchy criterion implies that there exists $N_0 \in \mathbb{N}$ such that

$$M_{j+1} + \cdots + M_k = |R_k - R_j| < \varepsilon, \quad k > j \geq N_0.$$

Now

$$|T_k(z) - T_j(z)| = |f_{j+1}(z)| + \cdots + |f_k(z)| \leq M_{j+1} + \cdots + M_k < \varepsilon$$

whenever $z \in K$ and $k > j \geq \max\{N, N_0\}$. Since $K \subset G$ was an arbitrary compact set, it follows from Theorem 1.6.3 that $\{T_k(z)\}$ converges uniformly on compact subsets of G. This implies the statement. □

In this book, we work with sequences $\{f_n(z)\}$ of analytic functions. In most of the applications, we will work in the complex plane $\mathbb{C}$ instead of a general domain G. It is natural to ask if the limit function of a convergent sequence of analytic functions is always analytic. This is indeed the case if the convergence is uniform on compact subsets of the given domain, but if the convergence is pointwise, the limit function may be nonanalytic [20]. Regarding this question, we now state and prove the *Weierstrass theorem*, which carries over to the derivatives as well.

Theorem 1.6.6. *Let $\{f_n(z)\}$ be a sequence of analytic functions in a domain G such that the limit*

$$\lim_{n\to\infty} f_n(z) = f(z) \tag{1.6.2}$$

exists uniformly on compact subsets of G, where $f(z)$ is a function in G. Then $f(z)$ is analytic in G. Moreover, for each $k \in \mathbb{N}$, $f^{(k)}(z) = \lim_{n\to\infty} f_n^{(k)}(z)$ uniformly on compact subsets of G.

Proof. Let $z_0 \in G$ be arbitrary, and choose $r > 0$ such that $\overline{D}(z_0, r) \subset G$, where

$$D(z_0, r) = \{z \in \mathbb{C} : |z - z_0| < r\}$$

is an open disc, and $\overline{D}(z_0, r)$ is its closure. By the Cauchy integral formula,

$$f_n(z) = \frac{1}{2\pi i} \int_{\partial D(z_0,r)} \frac{f_n(\zeta)}{\zeta - z}\, d\zeta, \quad z \in D(z_0, r), \ n \in \mathbb{N}. \tag{1.6.3}$$

Let $\varepsilon > 0$. Since the convergence is uniform on compact subsets of G, it is uniform on $\partial D(z_0, r)$, and there exists $n_\varepsilon \in \mathbb{N}$ such that

$$|f_n(\zeta) - f(\zeta)| < \varepsilon$$

for all $\zeta \in \partial D(z_0, r)$ and $n \geq n_\varepsilon$. For each fixed $z \in D(z_0, r)$, there exists $\beta \in (0, 1]$, depending on z, such that $|\zeta - z| \geq \beta r$ for all $\zeta \in \partial D(z_0, r)$. Therefore,

$$\begin{aligned} &\left|\frac{1}{2\pi i} \int\limits_{\partial D(z_0,r)} \frac{f_n(\zeta)}{\zeta - z}\, d\zeta - \frac{1}{2\pi i} \int\limits_{\partial D(z_0,r)} \frac{f(\zeta)}{\zeta - z}\, d\zeta\right| \\ &\quad \leq \frac{1}{2\pi} \int\limits_{\partial D(z_0,r)} \frac{|f_n(\zeta) - f(\zeta)|}{|\zeta - z|}\, |d\zeta| \leq \frac{2\pi r\varepsilon}{2\pi\beta r} = \frac{\varepsilon}{\beta}, \quad n \geq n_\varepsilon, \end{aligned}$$

and so

$$\lim_{n\to\infty} \frac{1}{2\pi i} \int\limits_{\partial D(z_0,r)} \frac{f_n(\zeta)}{\zeta - z}\, d\zeta = \frac{1}{2\pi i} \int\limits_{\partial D(z_0,r)} \frac{f(\zeta)}{\zeta - z}\, d\zeta, \quad z \in D(z_0, r).$$

By (1.6.2), (1.6.3), and the uniqueness of the limit,

$$f(z) = \frac{1}{2\pi i} \int\limits_{\partial D(z_0,r)} \frac{f(\zeta)}{\zeta - z}\, d\zeta, \quad z \in D(z_0, r).$$

If $z, z + h \in D(z_0, r)$ where $h \neq 0$, then

$$\begin{aligned} \frac{f(z+h) - f(z)}{h} &= \frac{1}{2\pi h i} \int\limits_{\partial D(z_0,r)} \left(\frac{f(\zeta)}{\zeta - (z+h)} - \frac{f(\zeta)}{\zeta - z}\right) d\zeta \\ &= \frac{1}{2\pi i} \int\limits_{\partial D(z_0,r)} \frac{f(\zeta)}{(\zeta - z)(\zeta - (z+h))}\, d\zeta \\ &\to \frac{1}{2\pi i} \int\limits_{\partial D(z_0,r)} \frac{f(\zeta)}{(\zeta - z)^2}\, d\zeta, \quad h \to 0. \end{aligned}$$

This proves that

$$f'(z) = \frac{1}{2\pi i} \int\limits_{\partial D(z_0,r)} \frac{f(\zeta)}{(\zeta - z)^2}\, d\zeta$$

exists in $D(z_0, r)$ and, consequently, in G. Hence $f(z)$ is analytic in G.

Similarly as above, we obtain that for $k \in \mathbb{N}$,

$$\left|\frac{1}{2\pi i} \int\limits_{\partial D(z_0,r)} \frac{f_n(\zeta)}{(\zeta - z)^{k+1}}\, d\zeta - \frac{1}{2\pi i} \int\limits_{\partial D(z_0,r)} \frac{f(\zeta)}{(\zeta - z)^{k+1}}\, d\zeta\right| < \frac{\varepsilon}{\beta^{k+1} r^k}, \quad n \geq n_\varepsilon,$$

and so

$$\lim_{n\to\infty} \frac{1}{2\pi i} \int_{\partial D(z_0,r)} \frac{f_n(\zeta)}{(\zeta - z)^{k+1}}\, d\zeta = \frac{1}{2\pi i} \int_{\partial D(z_0,r)} \frac{f(\zeta)}{(\zeta - z)^{k+1}}\, d\zeta, \quad z \in D(z_0, r).$$

This proves the statement that $f^{(k)}(z) = \lim_{n\to\infty} f_n^{(k)}(z)$ in $D(z_0, r)$. Since $z_0 \in G$ was arbitrarily chosen, it follows by the Heine–Borel theorem that the identity $f^{(k)}(z) = \lim_{n\to\infty} f_n^{(k)}(z)$ holds uniformly on compact subsets of $\mathbb{C}$. □

The following well-known result, called *Rouché's theorem*, will be used in the proof of Theorem 1.6.8.

Theorem 1.6.7 ([218, p. 178], [1, p. 263]). *Let f and g be analytic functions on and inside a simple closed contour C, and suppose that $|f(z) - g(z)| < |f(z)|$ for z on C. Then f and g have the same number of zeros inside C, where the zeros are counted by multiplicities.*

We proceed to state and prove *Hurwitz' theorem*, which associates the zeros of a convergent sequence of analytic functions with those of the limit function.

Theorem 1.6.8. *Let $\{f_n(z)\}$ be a sequence of analytic functions in a domain G, converging uniformly on compact subsets of G to a nonconstant analytic function $f(z)$. If $f(z_0) = 0$ for some $z_0 \in G$, then for each sufficiently small $r > 0$, there exists $N = N(r) \in \mathbb{N}$ such that for each $n \geq N$, the function $f_n(z)$ has the same number of zeros as $f(z)$ in $D(z_0, r)$.*

Proof. Choose $r > 0$ so small that f has no zeros in $\overline{D}(z_0, r)$ except at z_0. Such a value $r > 0$ exists by the uniqueness theorem for analytic functions; see Theorem 3.1.1. Then by continuity there exists $\delta > 0$ such that $|f(z)| > \delta$ on the circle $|z - z_0| = r$. Since $f_n \to f$ uniformly on $|z - z_0| = r$, we have

$$\left|f_n(z) - f(z)\right| < \delta < \left|f(z)\right|, \quad |z - z_0| = r,$$

for all n sufficiently large. The conclusion now follows from Rouché's theorem. □

We next focus on locally bounded families and sequences of analytic functions. The definition itself does not require the functions to be analytic.

Definition 1.6.9. A family $\mathcal{F}$ of functions is *locally bounded* on a domain G if for each $z_0 \in G$, there exist a constant $M = M(z_0) > 0$ and a disc $D(z_0, r) \subset G$ such that

$$\left|f(z)\right| \leq M, \quad z \in D(z_0, r),\ f \in \mathcal{F}. \tag{1.6.4}$$

Theorem 1.6.10. *Suppose that $\mathcal{F}$ is a locally bounded family of analytic functions in a domain G. Then the family $\mathcal{F}' = \{f' : f \in \mathcal{F}\}$ of derivatives form a locally bounded family on G.*

Proof. For each $z_0 \in G$, there exist a constant $M = M(z_0) > 0$ and a disc $D(z_0, r) \subset G$ such that (1.6.4) holds. Set $C_r = \{z : |z - z_0| = r/2\}$. Then, for $z \in D(z_0, r/4)$, Cauchy's formula gives

$$|f'(z)| \leq \frac{1}{2\pi} \int_{C_r} \frac{|f(\zeta)||d\zeta|}{|\zeta - z|^2} \leq \frac{8M}{\pi r^2} \int_{C_r} |d\zeta| = \frac{8M}{r}$$

for all $f' \in \mathcal{F}'$, so that $\mathcal{F}'$ is locally bounded. □

The converse of Theorem 1.6.10 is false, which is illustrated by the family $\mathcal{F} = \{n : n \in \mathbb{N}\}$ of constant functions.

The following *Vitali–Porter theorem* shows that local boundedness and pointwise convergence imply uniform convergence on compact subsets.

Theorem 1.6.11. *Suppose that $\{f_n(z)\}$ is a locally bounded sequence of analytic functions on a domain G such that*

$$\lim_{n\to\infty} f_n(z) = f(z) \in \mathbb{C}, \quad z \in G.$$

Then $f_n \to f$ uniformly on compact subsets of G, and f is analytic in G.

Proof. We deduce from Theorem 1.6.10 and the Heine–Borel theorem that the sequence $\{f_n'(z)\}$ of derivatives is uniformly bounded on compact subsets of G. Hence, given a compact set $K \subset G$, there exists $M > 0$ such that

$$|f_n'(z)| \leq M, \quad z \in K,\ n \in \mathbb{N}.$$

Let $\varepsilon > 0$. For each $z \in K$, define

$$U_z = \left\{\zeta \in G : |\zeta - z| < \frac{\varepsilon}{3M}\right\}.$$

Then $\bigcup_{z\in K} U_z$ forms an open cover of K. Since K is compact, there exist finitely many points $z_1, \ldots, z_k \in K$ such that $K \subset \bigcup_{j=1}^k U_{z_j}$. As each $z \in K$ belongs to some set U_{z_i}, we have $|z - z_i| < \frac{\varepsilon}{3M}$ for each such z. Since $\{f_n\}$ converges pointwise at each z_i, there exists $N \in \mathbb{N}$ such that $n, m \geq N$ implies

$$|f_n(z_i) - f_m(z_i)| < \frac{\varepsilon}{3}, \quad 1 \leq i \leq k, \tag{1.6.5}$$

by Cauchy's criterion for convergence. Then for $n, m \geq N$ and $z \in K$,

$$|f_n(z) - f_m(z)| \leq |f_n(z) - f_n(z_i)| + |f_n(z_i) - f_m(z_i)| + |f_m(z_i) - f_m(z)|,$$

where $1 \leq i \leq k$. The second term on the right-hand side is $< \frac{\varepsilon}{3}$ by (1.6.5). To deal with the first and third terms on the right-hand side, note that for $z \in U_{z_i}$,

$$|f_j(z) - f_j(z_i)| = \left| \int_{z_i}^{z} f_j'(\zeta)\, d\zeta \right| \le M|z - z_i| < \frac{\varepsilon}{3}, \quad j \in \mathbb{N}.$$

We conclude that $|f_n(z) - f_m(z)| < \varepsilon$ for all $n, m \ge N$. This gives $f_n \to f$ uniformly on K. Since $K \subset G$ is an arbitrary compact set, the limit function f is analytic on G by Theorem 1.6.6. □

The next theorem, which is stated in various forms in the literature, is a standard result concerning analytic functions defined by an integral.

Theorem 1.6.12 ([169, p. 308], [221, pp. 31–32]). *Let $t \in I$, where I is a real interval, which can be finite or infinite, and let $z \in G$, where G is a domain in $\mathbb{C}$. Assume that $f(z,t)$ is a function that satisfies the following conditions:*
(a) *$f(z,t)$ is a continuous function of both variables.*
(b) *For each fixed t, $f(z,t)$ is an analytic function of z.*
(c) *For each compact subset K of G, the integral*

$$F(z) = \int_I f(z,t)\, dt$$

converges uniformly for $z \in K$.

Then $F(z)$ is analytic in G, and its derivatives of all orders can be found by differentiating under the integral sign.

A lemma is required for the proof.

Lemma 1.6.13 ([57, p. 83]). *Let γ be a closed path in a domain G, and let h be a continuous function on γ. For each $n \in \mathbb{N}$, set*

$$H_n(z) = \int_\gamma \frac{h(w)}{(w - z)^n}\, dw, \quad z \notin \gamma.$$

Then each H_n is analytic in $G \setminus \gamma$, and $H_n'(z) = nH_{n+1}(z)$.

Proof. Let $z, a \in \mathbb{C} \setminus \gamma$, $z \ne a$, and let $w \in \gamma$. We begin with the factorization

$$\frac{1}{(w-z)^n} - \frac{1}{(w-a)^n} = (z-a) \sum_{k=1}^{n} \frac{1}{(w-z)^{n-k+1}(w-a)^k}, \tag{1.6.6}$$

which is certainly true for $n = 1$. Assuming that (1.6.6) is true for some $n = m \ge 1$, we obtain

$$\frac{1}{(w-z)^{m+1}} - \frac{1}{(w-a)^{m+1}} = \frac{1}{(w-z)(w-z)^m} - \frac{1}{(w-a)(w-a)^m}$$

$$= \frac{1}{w-z}\left(\frac{1}{(w-z)^m} - \frac{1}{(w-a)^m}\right) + \left(\frac{1}{w-z} - \frac{1}{w-a}\right)\frac{1}{(w-a)^m}$$
$$= \frac{z-a}{w-z}\sum_{k=1}^{m}\frac{1}{(w-z)^{m-k+1}(w-a)^k} + \frac{z-a}{(w-z)(w-a)}\cdot\frac{1}{(w-a)^m}$$
$$= (z-a)\sum_{k=1}^{m}\frac{1}{(w-z)^{(m+1)-k+1}(w-a)^k} + \frac{z-a}{(w-z)(w-a)^{m+1}}$$
$$= (z-a)\sum_{k=1}^{m+1}\frac{1}{(w-z)^{(m+1)-k+1}(w-a)^k}.$$

Hence, by induction, (1.6.6) holds for all $n \in \mathbb{N}$. The points on the curve γ form a compact set, and since h is continuous on γ, there exists $M > 0$ such that $|h(w)| \le M$ for all $w \in \gamma$. Moreover, there is $s > 0$ such that $|w-a| \ge s$ for all $w \in \gamma$. Let L denote the length of γ, and suppose that $|z-a| < s/2$. Then the factorization (1.6.6) implies

$$\begin{aligned}|H_n(z) - H_n(a)| &\le M\int_\gamma\left|\frac{1}{(w-z)^n} - \frac{1}{(w-a)^n}\right||dw| \\ &\le M|z-a|\int_\gamma\sum_{k=1}^{n}\frac{1}{|w-z|^{n-k+1}|w-a|^k}|dw| \\ &\le ML|z-a|\sum_{k=1}^{n}\frac{2^{n-k+1}}{s^{n+1}} \le C|z-a|,\end{aligned}$$

where $C = (\frac{2}{s})^{n+1}ML$. Letting $z \to a$, we find that H_n is continuous at a. Consequently, each H_n is (Lipschitz) continuous in $G \setminus \gamma$.

Next, let $z, a \in G \setminus \gamma$, $z \ne a$. From (1.6.6) we have

$$\frac{H_n(z) - H_n(a)}{z-a} = \sum_{k=1}^{n}\int_\gamma\frac{h(w)}{(w-z)^{n-k+1}(w-a)^k}\,dw. \tag{1.6.7}$$

Since $a \notin \gamma$, the function $h(w)(w-a)^{-k}$ is continuous on γ for each k. By the first part of the proof each integral in (1.6.7) is a continuous function of z in $G \setminus \gamma$. Hence the limit in (1.6.7) as $z \to a$ exists, and we find that

$$H_n'(a) = n\int_\gamma\frac{h(w)}{(w-a)^{n+1}}\,dw = nH_{n+1}(a),$$

which completes the proof. □

Proof of Theorem 1.6.12. Let $\{I_n\}$ be a sequence of nonempty closed intervals such that $I_n \subset I_{n+1}$ for all n and $I_n \to I$ as $n \to \infty$. Let $D(z_0, R)$ be an open disc with center z_0 and radius $R > 0$, whose closure $\overline{D}(z_0, R)$ lies in the domain G. Let γ be the curve $\partial D(z_0, R)$

oriented in the counterclockwise direction. Then for each $z \in D(z_0, R)$, Cauchy's integral formula gives

$$f(z,t) = \frac{1}{2\pi i} \int_\gamma \frac{f(w,t)}{w-z}\, dw$$

and

$$\frac{\partial^k}{\partial z^k} f(z,t) = \frac{k!}{2\pi i} \int_\gamma \frac{f(w,t)}{(w-z)^{k+1}}\, dw.$$

Thus

$$F(z) = \frac{1}{2\pi i} \int_I \int_\gamma \frac{f(w,t)}{w-z}\, dw\, dt.$$

We now restrict ourselves to points z satisfying $|z - z_0| \le R/2$, in which case

$$\frac{1}{|w-z|} \le \frac{2}{R}, \quad w \in \gamma. \tag{1.6.8}$$

For each $n \in \mathbb{N}$, set

$$F_n(z) = \frac{1}{2\pi i} \int_{I_n} \int_\gamma \frac{f(w,t)}{w-z}\, dw\, dt.$$

Then

$$F(z) - F_n(z) = \frac{1}{2\pi i} \int_{I \setminus I_n} \int_\gamma \frac{f(w,t)}{w \quad z}\, dw\, dt, \tag{1.6.9}$$

where the term $\frac{1}{w-z}$ in the integrand has a bounded modulus by (1.6.8). Hence, by the assumptions, the double integral on the right-hand side of (1.6.9) converges uniformly on $|z - z_0| \le R/2$. Since $I_n \to I$, we find that $F_n(z) \to F(z)$ as $n \to \infty$ uniformly on $|z - z_0| \le R/2$.

We proceed to consider the analyticity of the functions $F_n(z)$ and their limit function $F(z)$. By (1.6.8) and the assumptions,

$$\int_{I_n} \int_\gamma \left| \frac{f(w,t)}{w-z} \right| |dw|\, |dt| \le \frac{2}{R} \int_{I_n} \int_\gamma |f(w,t)|\, |dw|\, |dt| < \infty.$$

Indeed, the function $f(w,t)$ is continuous in both variables, and the sets I_n and γ are compact. Hence by Fubini's theorem [214, p. 207], we may interchange the order of integration in the definition of $F_n(z)$ and obtain

$$F_n(z) = \frac{1}{2\pi i} \int_\gamma \frac{1}{w-z} \left(\int_{I_n} f(w,t)\,dt \right) dw.$$

Since $\int_{I_n} f(w,t)\,dt$ is a continuous function of w on γ, Lemma 1.6.13 implies that each $F_n(z)$ is analytic in $G \setminus \gamma$. Moreover, by a repeated use of Lemma 1.6.13,

$$\begin{aligned} F_n^{(k)}(z) &= \frac{k!}{2\pi i} \int_\gamma \frac{1}{(w-z)^{k+1}} \left(\int_{I_n} f(w,t)\,dt \right) dw \\ &= \frac{k!}{2\pi i} \int_{I_n} \int_\gamma \frac{f(w,t)}{(w-z)^{k+1}}\,dw\,dt, \end{aligned} \tag{1.6.10}$$

where the order of integration has been interchanged again by Fubini's theorem. Now Theorem 1.6.6 shows that $F(z)$ is analytic and $F_n^{(k)}(z) \to F^{(k)}(z)$ as $n \to \infty$ uniformly on $|z - z_0| \le R/2$. Moreover, if $\varepsilon > 0$ is arbitrarily small, then (1.6.10) and the uniform convergence imply that there exists $N \in \mathbb{N}$ such that

$$\begin{aligned} \left| F^{(k)}(z) - \int_I \frac{\partial^k}{\partial z^k} f(z,t)\,dt \right| &\le \left| F^{(k)}(z) - F_n^{(k)}(z) \right| + \left| F_n^{(k)}(z) - \int_I \frac{\partial^k}{\partial z^k} f(z,t)\,dt \right| \\ &\le \frac{\varepsilon}{2} + \frac{k!}{2\pi} \left| \int_{I \setminus I_n} \int_\gamma \frac{f(w,t)}{(w-z)^{k+1}}\,dw\,dt \right| < \varepsilon \end{aligned}$$

whenever $n \ge N$ and $|z - z_0| \le R/2$. Hence the derivatives of all orders of $F(z)$ can be found by differentiating under the integral sign. Since $z_0 \in G$ was arbitrarily chosen, this property holds in all of G. □

Finally, we discuss uniform convergence from the *Cartan metric* point of view. Although we do not use Cartan's metric in this book, we give this brief discussion of it because it deserves to be better known. Some preparations are needed.

An *exhaustion by compact sets* of a nonempty set $S \subset \mathbb{C}$ is a nested sequence $\{K_n\}$ of compact subsets of S with the following two properties:

(1) K_n is contained in the interior of K_{n+1} for all n.

(2) $S = \bigcup_{n=1}^{\infty} K_n$.

It follows immediately that if S is exhaustible by the compact sets K_n and if $K \subset S$ is a compact set, then there exists an index $n \in \mathbb{N}$ such that $K \subset K_n$.

Example 1.6.14. The complex plane $\mathbb{C}$ is exhaustible by the compact sets $K_n = \{z \in \mathbb{C} : |z| \le n\}$, and the unit disc $\mathbb{D} := \{z \in \mathbb{C} : |z| < 1\}$ is exhaustible by the compact sets $K_n = \{z \in \mathbb{C} : |z| \le 1 - 1/n\}$.

Our focus is on the case where the set S is a domain. It turns out that every domain is exhaustible by compact sets. We showed that $\mathbb{C}$ is exhaustible by compact sets.

Example 1.6.15. Every domain $G \subsetneq \mathbb{C}$ is exhaustible by compact sets. To see this, set
$$K_n = \{z \in \mathbb{C} : \operatorname{dist}(z, \mathbb{C} \setminus G) \geq 1/n \text{ and } |z| \leq n\}, \quad n \in \mathbb{N},$$
where $\operatorname{dist}(z, \mathbb{C} \setminus G) = \inf\{|z - w| : w \in \mathbb{C} \setminus G\}$ and $\mathbb{C} \setminus G \neq \emptyset$. If $z \notin G$, then $\operatorname{dist}(z, \mathbb{C} \setminus G) = 0 < 1/n$, and hence $z \notin K_n$. Using contraposition, this proves that $K_n \subset G$. Moreover, each K_n is a closed set as an intersection of two closed sets, and each K_n is bounded by the disc $|z| \leq n$. Thus each K_n is a compact subset of G. Finally, it is easy to see that properties (1) and (2) above hold.

Definition 1.6.16 ([42, p. 147]). Let $A(G)$ be the vector space of all analytic functions in a domain $G \subset \mathbb{C}$, and let $\{K_n\}$ be an exhaustion of G. Define the distance function $d : A(G) \times A(G) \to [0, \infty)$ by
$$d(f, g) = \sum_{n=1}^{\infty} \frac{1}{2^n} \min\Big\{1, \sup_{z \in K_n} |f(z) - g(z)|\Big\}, \quad f, g \in A(G).$$

It is easy to see that the distance function d in the previous definition is a normalized metric on $A(G)$ such that $0 \leq d(f, g) \leq 1$ for all $f, g \in A(G)$. This metric was introduced by Cartan [42].

Uniform convergence on compact subsets of a domain is mainly a topological concept. The following theorem shows that this topological concept can be seen as a metric concept, which gives more flexibility.

Theorem 1.6.17 ([244, Theorem 1.3.2]). *Uniform convergence on compact subsets of a domain $G \subset \mathbb{C}$ is equivalent to convergence in the Cartan metric d.*

Proof. Let $\{K_n\}$ be an exhaustion of G.

Suppose first that $d(f_k, f) \to 0$ as $k \to \infty$ on G. Let $K \subset G$ be compact, and let K_ν be a fixed member of the exhaustion such that $K \subset K_\nu$. For an arbitrary but fixed $\varepsilon \in (0, 1)$, choose $\delta > 0$ satisfying $2^\nu \delta < \varepsilon$. Then there exists $N_1 \in \mathbb{N}$ such that
$$d(f_k, f) = \sum_{n=1}^{\infty} \frac{1}{2^n} \min\Big\{1, \sup_{z \in K_n} |f_k(z) - f(z)|\Big\} < \delta, \quad k \geq N_1.$$

Therefore,
$$\min\Big\{1, \sup_{z \in K_\nu} |f_k(z) - f(z)|\Big\} < 2^\nu \delta < \varepsilon, \quad k \geq N_1,$$
and since $0 < \varepsilon < 1$ and $K \subset K_\nu$, we conclude that
$$\sup_{z \in K} |f_k(z) - f(z)| < \varepsilon, \quad k \geq N_1,$$
as desired.

Conversely, suppose that $f_k \to f$ uniformly on compact subsets of G. For each $k \in \mathbb{N}$, the terms of the series
$$\sum_{n=1}^{\infty} \frac{1}{2^n} \min\Big\{1, \sup_{z \in K_n} |f_k(z) - f(z)|\Big\}$$

are dominated by $1/2^n$. Hence, the Weierstrass M-test implies that there exists $N_2 \in \mathbb{N}$, independent of k, such that

$$\sum_{n=N_2}^{\infty} \frac{1}{2^n} \min\Big\{1, \sup_{z \in K_n} |f_k(z) - f(z)|\Big\} < \frac{\varepsilon}{2}.$$

Since by assumption, $f_k \to f$ uniformly on compact subsets of G, there exists $N_3 \in \mathbb{N}$ such that

$$\sum_{n=1}^{N_2-1} \frac{1}{2^n} \min\Big\{1, \sup_{z \in K_n} |f_k(z) - f(z)|\Big\} < \frac{\varepsilon}{2}, \quad k \geq N_3.$$

In other words, $d(f_k, f) < \varepsilon/2 + \varepsilon/2 = \varepsilon$ for every $k \geq N_3$. □

Problems

Problem 1.6.18. Show that the series

$$\sum_{n=0}^{\infty} \frac{1}{z^n}$$

represents an analytic function on the domain $G = \{z : |z| > 1\}$.

Problem 1.6.19. Show that the series

$$\sum_{n=1}^{\infty} \frac{z^{2n}}{1 + z^n}$$

represents an analytic function on the domain $G = \{z : |z| < 1\}$.

Problem 1.6.20. Let $G = \{z : |z| > 1\}$, and let

$$f_n(z) = \frac{z^n}{1 + z^n}, \quad n = 1, 2, \ldots.$$

Show that $f_n(z) \to 1$ uniformly on compact subsets of G.

Problem 1.6.21. If a series $\sum_{n=1}^{\infty} f_n(z)$ of functions $\{f_n(z)\}$ on a set A converges uniformly to a function $f(z)$ on A, then prove that the sequence of functions $\{f_n(z)\}$ converges uniformly to 0 on A.

For the next two problems, we use the following standard definition. A function $f : S \to \mathbb{C}$ on a set $S \subset \mathbb{C}$ is said to be *bounded on* S if there exists a constant $M > 0$ such that $|f(z)| \leq M$ for all $z \in \mathbb{C}$.

Problem 1.6.22. Suppose that $\{f_n(z)\}$ and $\{g_n(z)\}$ are sequences of functions on a set $S \subset \mathbb{C}$ such that $f_n(z)$ converges uniformly to $f(z)$ on S and $g_n(z)$ converges uniformly to $g(z)$ on S. If $f(z)$ and $g(z)$ are both bounded on S, then show that $f_n(z)g_n(z)$ converges uniformly to $f(z)g(z)$ on S.

Hint: Use $f_n(g_n - g) + g(f_n - f)$.

Problem 1.6.23. Let $\{f_n(z)\}$ be a sequence of functions that converges uniformly to a function $f(z)$ on a set $S \subset \mathbb{C}$. Set $g_n(z) = h(z)f_n(z)$ and $g(z) = h(z)f(z)$, where $h(z)$ is bounded on S. Show that $g_n(z) \to g(z)$ uniformly on S.

1.7 Fundamentals of linear differential equations

In this section, we discuss the analytic nature of solutions of linear differential equations. We begin by proving the following theorem.

Theorem 1.7.1 ([144, Section 12.22]). *Suppose that functions $A_0(z), \ldots, A_{n-1}(z)$ and $B(z)$ are analytic in a disc $D(z_0, r)$ of radius r centered at $z_0 \in \mathbb{C}$, and let $w_0, \ldots, w_{n-1} \in \mathbb{C}$, where $n \in \mathbb{N}$. Then the initial value problem*

$$\begin{cases} f^{(n)} + A_{n-1}(z)f^{(n-1)} + \cdots + A_0(z)f = B(z), \\ f(z_0) = w_0, f'(z_0) = w_1, \ldots, f^{(n-1)}(z_0) = w_{n-1} \end{cases} \tag{1.7.1}$$

admits a unique solution f that is analytic in $D(z_0, r)$.

A consequence of this theorem is the following result, which is of particular interest in this book.

Corollary 1.7.2. *If $A_0(z), \ldots, A_{n-1}(z)$, and $B(z)$ are all entire functions, then every solution f of the differential equation*

$$f^{(n)} + A_{n-1}(z)f^{(n-1)} + \cdots + A_0(z)f = B(z)$$

is an entire function.

Proof of Theorem 1.7.1. Let $A_j^*(z) = A_j(z + z_0)$ and $B^*(z) = B(z + z_0)$, and suppose that g is a unique analytic solution of the initial value problem

$$\begin{cases} g^{(n)} + A_{n-1}^*(z)g^{(n-1)} + \cdots + A_0^*(z)g = B^*(z), \\ g(0) = w_0, g'(0) = w_1, \ldots, g^{(n-1)}(0) = w_{n-1} \end{cases}$$

in $D(0, r)$. Then $f(z) := g(z - z_0)$ is a unique analytic solution of the initial value problem (1.7.1) in $D(z_0, r)$. Therefore, we may assume that $z_0 = 0$ in (1.7.1).

Suppose that the Taylor series

$$f(z) = \sum_{k=0}^{\infty} \frac{f^{(k)}(0)}{k!} z^k = \sum_{k=0}^{\infty} c_k z^k \tag{1.7.2}$$

is a formal solution of the initial value problem (1.7.1). The word "formal" is associated with uniform convergence of the series f on compact subsets of $D(0, r)$, which we will prove later on. The uniform convergence enables, for example, term-by-term differentiation. Now

$$f^{(k)}(0) = k!c_k = w_k, \quad k = 0, \dots, n-1, \tag{1.7.3}$$

and further values of $f^{(k)}(0)$ for $k \geq n$ can be obtained recursively by differentiating the differential equation in (1.7.1) with respect to z, that is,

$$f^{(k)}(0) = \left(-A_{n-1}(z)f^{(n-1)}(z) - \dots - A_0(z)f(z) + B(z)\right)^{(k-n)}\Big|_{z=0}, \quad k \geq n.$$

This method defines the Taylor coefficients of f uniquely, each value of $f^{(k)}(0)$ for $k \geq n$ being dependent on $f(0), f'(0), \dots, f^{(k-1)}(0)$. Thus we may write

$$f^{(k)}(0) = \sum_{i=1}^{k} d_{ki} f^{(k-i)}(0), \quad k \geq n, \tag{1.7.4}$$

where each d_{ki} is a constant. If $w_0 = w_1 = \dots = w_{n-1} = 0$, then from (1.7.3) and (1.7.4) we deduce that $f^{(k)}(0) = 0$ for every $k \geq 0$. Consequently, $f \equiv 0$, in which case $B(z) \equiv 0$ would also hold. Thus we may suppose that at least one of the initial values w_j is nonzero.

Let $\varrho \in (0, r)$ be arbitrary, and let $M > 0$ be a constant such that

$$\max\{|A_0(z)|, \dots, |A_{n-1}(z)|, |B(z)|\} \leq M \quad \text{for } z \in \overline{D}(0, \varrho). \tag{1.7.5}$$

In $D(0, r)$, the functions $A_0(z), \dots, A_{n-1}(z), B(z)$ have the power series representations

$$A_j(z) = \sum_{k=0}^{\infty} a_{jk} z^k \quad \text{and} \quad B(z) = \sum_{k=0}^{\infty} b_k z^k. \tag{1.7.6}$$

By Cauchy's inequalities (see (2.1.3)),

$$\max\{|a_{jk}|, |b_k|\} \leq \frac{M}{\varrho^k}, \quad k \geq 0. \tag{1.7.7}$$

Denote

$$C(z) = \frac{M}{1 - \frac{z}{\varrho}} = \sum_{k=0}^{\infty} \frac{M}{\varrho^k} z^k.$$

Since the coefficients of $C(z)$ are positive real numbers, the maximum of $|C(z)|$ on $|z| = t \in (0, \varrho)$ is attained at $z = t$. Thus from (1.7.7) it follows that

$$\max\{M(t, A_j), M(t, B)\} \leq M(t, C), \quad 0 < t < \varrho,$$

where $M(t, A_j)$, $M(t, B)$, $M(t, C)$ are the standard maximum modulus functions; see Section 2.1.

Consider the differential equation

$$\frac{d^n F}{dz^n} = C(z)\left(\frac{d^{n-1}F}{dz^{n-1}} + \cdots + \frac{dF}{dz} + F + 1\right), \quad z \in D(0,\varrho). \tag{1.7.8}$$

Suppose that (1.7.8) has a formal solution

$$F(z) = \sum_{k=0}^{\infty} \frac{F^{(k)}(0)}{k!} z^k = \sum_{k=0}^{\infty} C_k z^k. \tag{1.7.9}$$

From the above discussion we are allowed to choose the first n Taylor coefficients any way we want, and the rest of the coefficients are determined by these initial values after differentiating (1.7.8) with respect to z. Hence we may choose

$$C_0 = |c_0|, \quad C_1 = |c_1|, \ldots, C_{n-1} = |c_{n-1}|,$$

where the constants c_k are as in (1.7.2). Denote the recurrence relation determining the values $F^{(k)}(0)$ for $k \geq n$ by

$$F^{(k)}(0) = \sum_{i=1}^{k} D_{ki} F^{(k-i)}(0), \quad k \geq n. \tag{1.7.10}$$

Since the coefficients of the series expansion $C(z)$ are positive real numbers, and since each D_{ki} is derived from those coefficients and from $C_0, C_1, \ldots, C_{n-1} \geq 0$ (of which at least one is positive) by addition and multiplication, it follows that $D_{ki} > 0$ for all $k \geq n$ and $i = 1, \ldots, k$. The process of finding the constants D_{ki} in (1.7.10) is the same as that of finding the constants d_{ki} in (1.7.4). Thus

$$|d_{ki}| \leq D_{ki}, \quad k \geq n, \ i = 1, \ldots, k.$$

Consequently,

$$\left|f^{(k)}(0)\right| \leq F^{(k)}(0), \quad k \geq 0,$$

and hence $|c_k| \leq C_k$ for all $k \geq 0$.

We proceed to prove that the radius of convergence of the majorant series F is ϱ, where $\varrho \in (0, r)$ was chosen arbitrarily. Thus the function f in (1.7.2) is a unique analytic solution of the initial value problem (1.7.1) in $D(z_0, r)$, where $z_0 = 0$ has been assumed. To simplify the situation, we make the substitution $z = \varrho\zeta$. Then

$$C(z) = \frac{M}{1-\zeta} \quad \text{and} \quad \frac{d^j}{dz^j} = \frac{1}{\varrho^j}\frac{d^j}{d\zeta^j}.$$

Equation (1.7.8) now becomes

$$(1-\zeta)\frac{d^n F}{d\zeta^n} = M\left(\varrho\frac{d^{n-1}F}{d\zeta^{n-1}} + \cdots + \varrho^{n-1}\frac{dF}{d\zeta} + \varrho^n(F+1)\right). \tag{1.7.11}$$

Suppose that a power series $w(\zeta) = \sum_{i=0}^{\infty} a_i \zeta^i$ is a formal solution of this equation. Differentiating term by term, for $j = 1, \dots, n$, we obtain

$$\begin{aligned} w^{(j)}(\zeta) &= \sum_{i=j}^{\infty} i(i-1)\cdots(i-j+1)a_i\zeta^{i-j} \\ &= \sum_{k=0}^{\infty}(j+k)(j+k-1)\cdots(k+1)a_{j+k}\zeta^k \\ &= \sum_{k=0}^{\infty}\frac{(j+k)!}{k!}a_{j+k}\zeta^k. \end{aligned}$$

Then (1.7.11) can be written in the form

$$\begin{aligned} &\sum_{k=0}^{\infty}\frac{(n+k)!}{k!}a_{n+k}\zeta^k - \sum_{k=1}^{\infty}\frac{(n+k-1)!}{(k-1)!}a_{n+k-1}\zeta^k \\ &\quad = M\sum_{i=1}^{n}\varrho^i\sum_{k=0}^{\infty}\frac{(n+k-i)!}{k!}a_{n+k-i}\zeta^k + M\varrho^n \\ &\quad = \sum_{k=0}^{\infty}\sum_{i=1}^{n}M\varrho^i\frac{(n+k-i)!}{k!}a_{n+k-i}\zeta^k + M\varrho^n. \end{aligned}$$

This leads to the recurrence relations

$$\begin{cases} n!a_n = \sum_{i=1}^{n} M\varrho^i(n-i)!a_{n-i} + M\varrho^n, \\ (n+k)!a_{n+k} - k(n+k-1)!a_{n+k-1} = \sum_{i=1}^{n} M\varrho^i(n+k-i)!a_{n+k-i}, \quad k \geq 1. \end{cases}$$

In order for $\sum_{k=0}^{\infty} a_k\zeta^k$ to be formally equal to $\sum_{k=0}^{\infty} C_k z^k$, we must insist that

$$a_k = \varrho^k C_k, \quad k = 0, \dots, n-1.$$

The coefficients a_k of $\sum_{k=0}^{\infty} a_k\zeta^k$ are now uniquely determined for all $k \geq 0$. It follows by induction that $a_k \geq 0$ for every $k \geq 0$. Moreover, since we have assumed that at least one of the initial values w_j is nonzero, it follows by the construction that $a_k > 0$ for all $k \geq n$.

The latter recurrence relation yields

$$a_{n+k} = \frac{k + M\varrho}{n+k}a_{n+k-1} + \varepsilon_{n+k-2}, \quad k \geq 1, \tag{1.7.12}$$

where $\varepsilon_{n+k-2} \geq 2$ is an “error term” defined by

$$\varepsilon_{n+k-2} = \sum_{i=2}^{n}\frac{M\varrho^i(n+k-i)!}{(n+k)!}a_{n+k-i} \geq 0.$$

So far the only restrictions for the constant $M > 0$ are stated in (1.7.5). From now on we require, in addition to (1.7.5), that $M\varrho > n$. Then by (1.7.12) we have

$$\alpha_{n+k} > \alpha_{n+k-1}, \quad k \geq 1,$$
$$\alpha_{n+k-1} \geq \frac{k-1+M\varrho}{n+k-1}\alpha_{n+k-2} > \alpha_{n+k-2}, \quad k \geq 2,$$

and, inductively,

$$\alpha_{n+k-1} > \alpha_{n+k-i}, \quad 2 \leq i \leq n,\ k \geq n.$$

Therefore, since M, ϱ, and n are fixed,

$$0 \leq \frac{\varepsilon_{n+k-2}}{\alpha_{n+k-1}} \leq \sum_{i=2}^{n} \frac{M\varrho^i(n+k-i)!}{(n+k)!} \to 0, \quad k \to \infty,$$

and by (1.7.12),

$$\frac{\alpha_{n+k}}{\alpha_{n+k-1}} = \frac{k+M\varrho}{n+k} + \frac{\varepsilon_{n+k-2}}{\alpha_{n+k-1}} \to 1, \quad k \to \infty.$$

By the ratio test the series $w(\zeta) = \sum_{i=0}^{\infty} \alpha_i \zeta^i$ converges for $|\zeta| < 1$, and since $z = \rho\zeta$, the dominant series F in (1.7.9) has the radius of convergence equal to ϱ. This completes the proof. □

Remark. Note that if $w_0 = \cdots = w_{n-1} = 0$ and $B(z) \equiv 0$ in Theorem 1.7.1, then the unique solution f is $f \equiv 0$ in $D(z_0, r)$. Also, if the coefficients in (1.7.1) are entire, then the unique solution f is also entire.

For the rest of this section, we discuss a few known properties of solutions of equations of the form

$$f^{(n)} + A_{n-1}(z)f^{(n-1)} + \cdots + A_0(z)f = 0, \tag{1.7.13}$$
$$f^{(n)} + A_{n-1}(z)f^{(n-1)} + \cdots + A_0(z)f = B(z), \tag{1.7.14}$$

where the coefficients $A_0(z), \ldots, A_{n-1}(z)$, and $B(z)$ are entire. This discussion is parallel to that of differential equations in a real variable [40, Section 5.2].

(1) If $f_1, \ldots, f_n$ are n entire solutions of (1.7.13), then it is easy to verify by a direct substitution that the linear combination $f = c_1 f_1 + \cdots + c_n f_n$ is also an entire solution for any choice of the constants $c_1, \ldots, c_n \in \mathbb{C}$.
(2) Let $f_1, \ldots, f_n$ be n entire solutions of (1.7.13). Then $f_1, \ldots, f_n$ are linearly independent if and only if the *Wronskian determinant* $W = W(f_1, \ldots, f_n)$ is nonzero at each point of $\mathbb{C}$. (The Wronskian is defined in Section 6.3.) This is a consequence of basic linear algebra, because the equation

$$c_1 f_1(z) + \cdots + c_n f_n(z) = 0$$

can be differentiated $n-1$ times, which then gives a linear system of n equations in the n unknowns $c_1, \dots, c_n$, whose coefficient matrix has W as its determinant.

(3) Let $f_1, \dots, f_n$ be linearly independent entire solutions of (1.7.13), and let f be an entire solution of (1.7.13). Then there exist constants $c_1, \dots, c_n \in \mathbb{C}$ such that

$$f(z) = c_1 f_1(z) + \cdots + c_n f_n(z). \tag{1.7.15}$$

Indeed, let $z_0 \in \mathbb{C}$ be arbitrary, and denote

$$w_0 = f(z_0), \quad w_1 = f'(z_0), \quad \dots, \quad w_{n-1} = f^{(n-1)}(z_0).$$

The determinant of the coefficients of the linear system

$$\begin{aligned} c_1 f_1(z_0) + \cdots + c_n f_n(z_0) &= w_0, \\ c_1 f_1'(z_0) + \cdots + c_n f_n'(z_0) &= w_1, \\ &\vdots \\ c_1 f_1^{(n-1)}(z_0) + \cdots + c_n f_n^{(n-1)}(z_0) &= w_{n-1}, \end{aligned}$$

in the unknowns $c_1, \dots, c_n$ is $W(f_1, \dots, f_n)(z_0)$. Since $W(f_1, \dots, f_n)(z_0) \neq 0$, we can use Cramer's rule to solve for each c_j. With these values of c_j, we denote

$$g(z) = c_1 f_1(z) + \cdots + c_n f_n(z),$$

which we know to be an entire solution of (1.7.13) by (1). Moreover,

$$g(z_0) = w_0, \quad g'(z_0) = w_1, \dots, g^{(n-1)}(z_0) = w_{n-1}.$$

The corresponding initial value problem has a unique solution by Theorem 1.7.1, and hence $g(z) \equiv f(z)$. This proves (1.7.15).

(4) Suppose that f_p is any particular solution of the nonhomogeneous equation (1.7.14), and let f be another solution of (1.7.14). An easy substitution shows that the *complementary function* $f_c = f - f_p$ solves the homogeneous equation (1.7.13). It follows from (3) that there exist constants $c_1, \dots, c_n \in \mathbb{C}$ such that

$$f_c = c_1 f_1 + \cdots + c_n f_n.$$

Thus we have proved that an arbitrary solution of (1.7.14) is a sum of the complementary function f_c and a single particular solution f_p of (1.7.14).

Problems

Problem 1.7.3. Knowing that the functions

$$f_1(z) = e^{-3z}, \quad f_2(z) = \cos 2z, \quad f_3(z) = \sin 2z$$

are linearly independent entire solutions of the third-order equation

$$f''' + 3f'' + 4f' + 12f = 0,$$

find an entire solution f satisfying $f(0) = 0, f'(0) = 5$, and $f''(0) = -39$.

2 Growth of entire functions

An analytic function $f : \mathbb{C} \to \mathbb{C}$ is called *entire*. These functions have many interesting properties. In this chapter, we focus on their growth properties. Typically, the growth of an entire function f is expressed in terms of its maximum modulus $M(r,f)$, which is defined in Section 2.1. It turns out that the growth of f is also determined by its Taylor coefficients, or even by its real part. Further, we prove that f and its derivative f' belong to the same growth class. An entire function that is not a polynomial is called *transcendental*.

2.1 Order and type

We begin this section by stating the *maximum modulus principle*, which is a complex analysis result that is useful in the theory and applications of entire functions.

Theorem 2.1.1 ([1, p. 97]). *Let f be an analytic function on a domain G. Then the following statements hold:*
(a) *If f is nonconstant, then the function $|f(z)|$ cannot have a maximum in G.*
(b) *If G is a bounded domain and if the function $|f(z)|$ is continuous on the compact set $\overline{G}$, then $|f(z)|$ assumes its maximum on the boundary ∂G of G.*

The *maximum modulus* of an entire function f in the disc $|z| \le r$ is the quantity

$$M(r,f) = \max_{|z|\le r} |f(z)|, \quad r > 0.$$

The reason the maximum can be used here instead of the supremum is because the function $|f(z)|$ is continuous on the compact set $\{z : |z| \le r\}$. Furthermore, from the *maximum modulus principle* it follows that this maximum value must occur on the boundary $|z| = r$ of $|z| \le r$. Hence we may write

$$M(r,f) = \max_{|z|=r} |f(z)|, \quad r > 0.$$

Proposition 2.1.2. *If f is a nonconstant entire function, then $M(r,f)$ is a strictly increasing and continuous function for $r > 0$.*

Proof. Suppose that $M(r,f)$ is not strictly increasing. Then there exist values r_1 and r_2 satisfying $0 < r_1 < r_2$ and $M(r_1,f) \ge M(r_2,f)$. Since $|z| = r_1$ is a compact set, there exists a point z_1 satisfying $|z_1| = r_1$ and

$$|f(z_1)| = M(r_1,f) \ge M(r_2,f). \tag{2.1.1}$$

Since $|z| \le r_2$ is a compact set, $|f(z)|$ must have a maximum in $|z| \le r_2$, but from (2.1.1) we see that this maximum of $|f(z)|$ in $|z| \le r_2$ must occur in the domain $|z| < r_2$, which

https://doi.org/10.1515/9783112249130-002

contradicts the maximum modulus principle. This contradiction proves that $M(r,f)$ is strictly increasing.

To prove that $M(r,f)$ is continuous, take $0 < r_1 < r_2$, and choose $\theta_0 \in \mathbb{R}$ such that $|f(r_2 e^{i\theta_0})| = M(r_2,f)$. Then by the first part of the proof

$$0 < M(r_2,f) - M(r_1,f) \leq |f(r_2 e^{i\theta_0})| - |f(r_1 e^{i\theta_0})|. \tag{2.1.2}$$

Let $\varepsilon > 0$. By the continuity of $|f(z)|$ there exists $\delta > 0$ such that the right-hand side of (2.1.2) is bounded above by ε, provided that $0 < r_2 - r_1 < \delta$. □

The following well-known result, called *Liouville's theorem*, is very useful in the theory of entire functions.

Theorem 2.1.3. *A bounded entire function is a constant.*

Proof. Let f be a bounded entire function. From Cauchy's formula for derivatives we obtain

$$f^{(n)}(0) = \frac{n!}{2\pi r^n} \int_0^{2\pi} \frac{f(re^{i\theta})}{e^{in\theta}}\, d\theta, \quad r > 0,$$

which gives

$$|f^{(n)}(0)| \leq \frac{n!}{2\pi r^n} \int_0^{2\pi} |f(re^{i\theta})|\, d\theta \leq \frac{n!}{r^n} M(r,f), \quad r > 0. \tag{2.1.3}$$

Since f is bounded, there exists a real constant C such that $M(r,f) \leq C$ for all r. Then from (2.1.3),

$$|f^{(n)}(0)| \leq \frac{Cn!}{r^n}, \quad r > 0.$$

By considering $n \geq 1$ and letting $r \to \infty$ we see that $f^{(n)}(0) = 0$ for all $n \geq 1$. By expanding f into its Taylor series about $z_0 = 0$,

$$f(z) = \sum_{n=0}^{\infty} \frac{f^{(n)}(0)}{n!} z^n, \quad z \in \mathbb{C},$$

we obtain that $f(z) \equiv f(0)$. Hence f is a constant. □

Remark. From Liouville's theorem and the maximum modulus principle it follows that the maximum modulus function of a nonconstant entire function f satisfies $M(r,f) \to \infty$ as $r \to \infty$.

Lemma 2.1.4 ([150, Satz 1.2]). *Let $P(z) = a_n z^n + a_{n-1} z^{n-1} + \cdots + a_0$, $a_n \neq 0$, be a polynomial. Then for all $\varepsilon > 0$, there exists $r(\varepsilon) > 0$ such that*

$$(1-\varepsilon)|a_n||z|^n \le |P(z)| \le (1+\varepsilon)|a_n||z|^n, \quad |z| \ge r(\varepsilon).$$

Proof. We may write $|P(z)| = |a_n||z|^n|1 + r_n(z)|$, where

$$r_n(z) = \frac{a_{n-1}}{a_n}\frac{1}{z} + \cdots + \frac{a_0}{a_n}\frac{1}{z^n}.$$

Let $\varepsilon > 0$. Clearly, there exists $r(\varepsilon) > 0$ such that

$$|r_n(z)| \le \left|\frac{a_{n-1}}{a_n}\frac{1}{z}\right| + \cdots + \left|\frac{a_0}{a_n}\frac{1}{z^n}\right| < \varepsilon, \quad |z| \ge r(\varepsilon).$$

Finally,

$$\begin{aligned}(1-\varepsilon)|a_n||z|^n &\le (1 - |r_n(z)|)|a_n||z|^n \le |P(z)| \\ &\le (1 + |r_n(z)|)|a_n||z|^n \le (1+\varepsilon)|a_n||z|^n\end{aligned}$$

for all $|z| \ge r(\varepsilon)$. □

If f is a nonconstant polynomial, then by Lemma 2.1.4 we have the inequalities

$$C_1 \log r \le \log M(r,f) \le C_2 \log r \tag{2.1.4}$$

for some constants $0 < C_1 < C_2 < \infty$ and all r large enough. Alternatively, this can be expressed as

$$\log r \lesssim \log M(r,f) \lesssim \log r.$$

The following result for transcendental entire functions is not surprising.

Lemma 2.1.5. *An entire function f is transcendental if and only if*

$$\lim_{r\to\infty} \frac{\log M(r,f)}{\log r} = \infty. \tag{2.1.5}$$

Proof. Suppose that (2.1.5) holds. If f is a polynomial, then (2.1.4) holds, which contradicts (2.1.5). Hence f must be transcendental.

Conversely, suppose that f is transcendental. Then we may write f as a Taylor series

$$f(z) = \sum_{n=0}^{\infty} a_n z^n,$$

where infinitely many coefficients a_n are nonzero. Cauchy's integral formula for derivatives yields (2.1.3), that is,

$$\left|f^{(n)}(0)\right| \le \frac{n!}{r^n} M(r,f), \quad r > 0.$$

Since $a_n = \frac{f^{(n)}(0)}{n!}$, we obtain Cauchy's inequalities

$$|a_n| r^n \le M(r,f), \quad r > 0,\ n \in \mathbb{N} \cup \{0\}. \tag{2.1.6}$$

Let $k \in \mathbb{N}$ be arbitrarily large, and choose any fixed $N \in \mathbb{N}$ such that $a_N \neq 0$ and $N \ge k+1$. Then from (2.1.6),

$$|a_N| r \le |a_N| r^{N-k} \le \frac{M(r,f)}{r^k}, \quad r \ge 1.$$

The squeeze theorem for unbounded limits now gives us

$$\lim_{r\to\infty} \frac{M(r,f)}{r^k} = \infty.$$

Thus, for all $C > 1$, there exists $r(C,k) > 1$ such that

$$\frac{M(r,f)}{r^k} \ge C, \quad r \ge r(C,k).$$

In other words,

$$\frac{\log M(r,f)}{\log r} \ge \frac{\log C}{\log r} + k \ge k, \quad r \ge r(C,k),$$

and statement (2.1.5) follows. □

Remark. Sometimes it is convenient to express the growth between two expressions, such as in (2.1.4), by means of the *big-O-notation*, also known as the Bachmann–Landau notation, named after Paul Bachmann (1837–1920) and Edmund Landau (1877–1938). For $0 \le r < \infty$, let $F(r)$ be a function, either real or complex-valued, and let $G(r)$ be a real-valued positive function. We write $F(r) = O(G(r))$ if there exist constants $C > 0$ and $R > 0$ such that $|F(r)| \le CG(r)$ for all $r \ge R$. Equivalently, this can be expressed as $|F(r)| \lesssim G(r)$ or as

$$\limsup_{r\to\infty} \frac{|F(r)|}{G(r)} < \infty.$$

The analogous *little-o-notation* $F(r) = o(G(r))$ means that

$$\lim_{r\to\infty} \frac{|F(r)|}{G(r)} = 0.$$

The expressions $G(r) = O(1)$ and $G(r) = o(1)$ mean, respectively, that $G(r)$ is bounded or tends to zero as $r \to \infty$. Sometimes the notation $|F(r)| \sim G(r)$ is used for brevity instead of the equivalent notation $|F(r)| = (1 + o(1))G(r)$.

Lemma 2.1.5 suggests that we need a more specific way to express the growth of transcendental entire functions.

Definition 2.1.6. The *order of growth* of an entire function f is the number

$$\rho = \rho(f) = \limsup_{r\to\infty} \frac{\log\log M(r,f)}{\log r} \in [0,\infty]$$

when f is nonconstant, and $\rho(f) = 0$ when f is a constant.

Polynomials are clearly of order zero, and it is easy to see that $\rho(e^z) = 1$, since $M(r,e^z) = e^r$. If $\rho(f) \in (0,\infty)$, then the growth of f can be expressed more precisely in terms of its type.

Definition 2.1.7. Let f be an entire function of order $\rho \in (0,\infty)$. Then the *type* of f is the number

$$\tau = \tau(f) = \limsup_{r\to\infty} \frac{\log M(r,f)}{r^\rho} \in [0,\infty].$$

If $\tau = \infty$, $0 < \tau < \infty$, or $\tau = 0$, then the function f is said to be of *maximum* (or infinite), *mean* (or normal), or *minimum* (or zero) type of order ρ, respectively.

Obviously, $M(r,e^{2z}) = e^{2r}$, so that $\tau(e^{2z}) = 2$.

Lemma 2.1.8. *Let f be an entire function, and define*

$$\alpha = \inf\left\{\beta > 0 : M(r,f) \le e^{r^\beta} \text{ for all } r \ge r(\beta)\right\}.$$

Then $\rho(f) = \alpha$. If such a $\beta > 0$ does not exist, then $\rho(f) = \infty$.

Proof. Suppose first that $\rho = \rho(f) < \infty$. Let $\varepsilon > 0$. By the definition of order, there exists $r(\varepsilon) > 1$ such that

$$\frac{\log\log M(r,f)}{\log r} \le \rho + \varepsilon, \quad r \ge r(\varepsilon).$$

It follows that

$$M(r,f) \le e^{r^{\rho+\varepsilon}}, \quad r \ge r(\varepsilon).$$

This gives us $\alpha \le \rho + \varepsilon$. Since $\varepsilon > 0$ was chosen arbitrarily, we obtain $\alpha \le \rho$.

On the other hand, if $\beta > 0$ is such that the condition in the definition of the constant α is satisfied, then

$$\frac{\log\log M(r,f)}{\log r} \le \frac{\log\log e^{r^\beta}}{\log r} = \beta, \quad r \ge r(\beta). \tag{2.1.7}$$

The existence of such β is clear because $\rho < \infty$. The upper bound in (2.1.7) is independent of r, so we may take the limsup on the lower bound, which gives us $\rho \le \beta$. Since α is the infimum of all such constants β, we deduce that $\rho \le \alpha$. This completes the proof that $\rho = \alpha$ when $\rho < \infty$.

Suppose now that $\rho = \rho(f) = \infty$. If $\alpha < \infty$, then for each $\varepsilon > 0$, there exists $r(\varepsilon) > 0$ such that

$$M(r,f) \le e^{r^{\alpha+\varepsilon}}, \quad r \ge r(\varepsilon),$$

which gives $\rho \le \alpha + \varepsilon < \infty$, a contradiction. Thus $\alpha = \infty$, and hence $\rho = \alpha = \infty$. □

We proceed with an analogue of Lemma 2.1.8 for the type of an entire function. The proof is similar to that of Lemma 2.1.8 and is left as an exercise.

Lemma 2.1.9. *Let f be an entire function of order $\rho \in (0,\infty)$, and define*

$$\gamma = \inf\left\{\beta > 0 : M(r,f) \le e^{\beta r^{\rho}} \text{ for all } r \ge r(\beta)\right\}.$$

Then $\tau(f) = \gamma$. If such a $\beta > 0$ does not exist, then $\tau(f) = \infty$.

We now turn to the growth of sums and products of entire functions.

Theorem 2.1.10. *Let f and g be entire functions. Then the following statements hold:*
(a) $\rho(f+g) \le \max\{\rho(f), \rho(g)\}$.
(b) $\rho(fg) \le \max\{\rho(f), \rho(g)\}$.
(c) *If $\rho(f) > \rho(g)$, then $\rho(f+g) = \rho(f)$.*

Proof. Let $\varepsilon > 0$. Part (a) obviously holds if either $\rho(f) = \infty$ or $\rho(g) = \infty$. Now assume that $\rho(f) < \infty$ and $\rho(g) < \infty$. Then there exists $r(\varepsilon) > 1$ such that

$$M(r,f) \le e^{r^{\rho(f)+\varepsilon/2}} \quad \text{and} \quad M(r,g) \le e^{r^{\rho(g)+\varepsilon/2}}$$

for all $r \ge r(\varepsilon)$. By elementary estimates,

$$\begin{aligned} M(r,f+g) &= \max_{|z|\le r}|f(z)+g(z)| \le \max_{|z|\le r}|f(z)| + \max_{|z|\le r}|g(z)| \\ &= M(r,f) + M(r,g) \le e^{r^{\rho(f)+\varepsilon/2}} + e^{r^{\rho(g)+\varepsilon/2}} \\ &\le 2e^{r^{\max\{\rho(f),\rho(g)\}+\varepsilon/2}} = e^{\log 2 + r^{\max\{\rho(f),\rho(g)\}+\varepsilon/2}}, \quad r \ge r(\varepsilon). \end{aligned}$$

We may suppose that $r(\varepsilon) \ge \exp(\frac{2\log 2}{\varepsilon})$. Then $r(\varepsilon)^{\varepsilon/2} \ge 2 \ge \log 2$ and

$$\begin{aligned} \log 2 + r^{\max\{\rho(f),\rho(g)\}+\varepsilon/2} &\le r^{\varepsilon/2} + r^{\max\{\rho(f),\rho(g)\}+\varepsilon/2} \\ &\le 2r^{\max\{\rho(f),\rho(g)\}+\varepsilon/2} \le r^{\max\{\rho(f),\rho(g)\}+\varepsilon} \end{aligned}$$

for all $r \ge r(\varepsilon)$. In particular,

$$M(r,f+g) \le e^{r^{\max\{\rho(f),\rho(g)\}+\varepsilon}}, \quad r \ge r(\varepsilon).$$

Hence $\rho(f+g) \leq \max\{\rho(f), \rho(g)\}+\varepsilon$, where $\varepsilon > 0$ is arbitrarily small. This proves part (a). Note that the inequality in (a) can be strict, e. g., when $g = -f$.

Part (b) is proved similarly by first noting that (b) is clearly true when either $\rho(f) = \infty$ or $\rho(g) = \infty$ and then relying on the following estimates in the case where $\rho(f) < \infty$ and $\rho(g) < \infty$:

$$\begin{aligned} M(r,fg) &= \max_{|z|\leq r} |f(z)g(z)| \leq \max_{|z|\leq r} |f(z)| \cdot \max_{|z|\leq r} |g(z)| \\ &= M(r,f) \cdot M(r,g) \leq e^{r^{\rho(f)+\varepsilon/2}} \cdot e^{r^{\rho(g)+\varepsilon/2}} \\ &\leq e^{2r^{\max\{\rho(f),\rho(g)\}+\varepsilon/2}} \leq e^{r^{\max\{\rho(f),\rho(g)\}+\varepsilon}} \end{aligned}$$

for all $r \geq r(\varepsilon)$. The inequality in (b) can also be strict, e. g., when $g = 1/f$ and f has no zeros.

To prove part (c), first suppose that $\rho(g) < \rho(f) < \infty$. Since

$$\begin{aligned} M(r,f+g) &\leq M(r,f) + M(r,g) \leq e^{r^{\rho(f)+\varepsilon/2}} + e^{r^{\rho(g)+\varepsilon/2}} \\ &\leq 2e^{r^{\rho(f)+\varepsilon/2}} \leq e^{r^{\rho(f)+\varepsilon}}, \quad r \geq r(\varepsilon), \end{aligned}$$

it follows that $\rho(f+g) \leq \rho(f)$.

We proceed to prove that $\rho(f+g) \geq \rho(f)$. By the definition of order, there exists an increasing sequence $\{r_n\}$ of positive real numbers tending to infinity such that

$$\rho(f) = \lim_{n\to\infty} \frac{\log\log M(r_n,f)}{\log r_n}.$$

Hence there exists $N(\varepsilon) \in \mathbb{N}$ such that

$$\rho(f) - \varepsilon/2 \leq \frac{\log\log M(r_n,f)}{\log r_n}, \quad n \geq N(\varepsilon).$$

In other words,

$$M(r_n,f) \geq e^{r_n^{\rho(f)-\varepsilon/2}}, \quad n \geq N(\varepsilon). \tag{2.1.8}$$

Consider any r_n with $n \geq N(\varepsilon)$. Since f is continuous on $|z| = r_n$ and since $|z| = r_n$ is compact, we can find a point z_n such that $|z_n| = r_n$ and

$$|f(z_n)| = M(r_n,f) \geq e^{r_n^{\rho(f)-\varepsilon/2}}, \quad n \geq N(\varepsilon),$$

where (2.1.8) was used. We may take a slightly larger $N(\varepsilon)$, if needed, so that $r_n \geq r(\varepsilon)$ for all $n \geq N(\varepsilon)$. Then

$$|(f+g)(z_n)| \geq |f(z_n)| - |g(z_n)| \geq e^{r_n^{\rho(f)-\varepsilon/2}} - e^{r_n^{\rho(g)+\varepsilon/2}}, \quad n \geq N(\varepsilon).$$

Also, we can assume that $\varepsilon > 0$ is small enough so that

$$\rho(f) - \varepsilon/2 > \rho(g) + \varepsilon/2 > 0.$$

Now

$$r_n^{\rho(g)+\varepsilon/2} - r_n^{\rho(f)-\varepsilon/2} = r_n^{\rho(f)-\varepsilon/2}(r_n^{\rho(g)-\rho(f)+\varepsilon} - 1) \to -\infty$$

as $n \to \infty$. Thus we may suppose that

$$r_n^{\rho(g)+\varepsilon/2} - r_n^{\rho(f)-\varepsilon/2} \leq -\log 2, \quad n \geq N(\varepsilon).$$

Therefore,

$$\begin{aligned} M(r_n, f+g) &\geq |(f+g)(z_n)| \geq e^{r_n^{\rho(f)-\varepsilon/2}} - e^{r_n^{\rho(g)+\varepsilon/2}} \\ &= e^{r_n^{\rho(f)-\varepsilon/2}}(1 - e^{r_n^{\rho(g)+\varepsilon/2} - r_n^{\rho(f)-\varepsilon/2}}) \geq \frac{1}{2} e^{r_n^{\rho(f)-\varepsilon/2}} \geq e^{r_n^{\rho(f)-\varepsilon}} \end{aligned}$$

for all $n \geq N(\varepsilon)$. This implies $\rho(f+g) \geq \rho(f)$, which shows that $\rho(f+g) = \rho(f)$ whenever $\rho(g) < \rho(f) < \infty$.

Now assume that $\rho(g) < \rho(f) = \infty$. Let μ be an arbitrarily chosen fixed constant satisfying $\rho(g) + \varepsilon < \mu < \infty$. By the definition of order, there exists a strictly increasing sequence $\{r_n\}$ of positive real numbers tending to infinity such that

$$M(r_n, f) \geq e^{r_n^{\mu}}, \quad n \geq N(\mu).$$

We also have

$$M(r, g) \leq e^{r^{\rho(g)+\varepsilon}}, \quad r \geq R(\varepsilon).$$

Since

$$M(r, f) \leq M(r, f+g) + M(r, g),$$

we obtain

$$e^{r_n^{\mu}} \leq M(r_n, f+g) + M(r_n, g) \leq M(r_n, f+g) + o(1)e^{r_n^{\mu}}, \quad n \geq N(\mu).$$

Hence

$$(1 + o(1))e^{r_n^{\mu}} \leq M(r_n, f+g), \quad n \geq N(\mu).$$

It follows that $\rho(f+g) \geq \mu$. By letting $\mu \to \infty$ we get $\rho(f+g) = \infty$, which completes the proof of part (c). □

Remark. If $\rho(f) > \rho(g)$, then $\rho(fg) = \rho(f)$ is also true. Similarly as above,

$$\begin{aligned} M(r,fg) &\le M(r,f) \cdot M(r,g) \le e^{r^{\rho(f)+\varepsilon/2}} \cdot e^{r^{\rho(g)+\varepsilon/2}} \\ &\le e^{2r^{\rho(f)+\varepsilon/2}} \le e^{r^{\rho(f)+\varepsilon}}, \quad r \ge r(\varepsilon), \end{aligned}$$

and so $\rho(fg) \le \rho(f)$. The reverse inequality $\rho(fg) \ge \rho(f)$ will be proved in Theorem 3.8.3, where we have all the necessary tools at our disposal.

Problems

Problem 2.1.11. Find the order and type of $f(z) = 2e^{-iz} + 3e^{4iz^2}$.

Problem 2.1.12. Let f be an entire function that is bounded on the set

$$A = \{z \in \mathbb{C} : |z| = n,\ n \in \mathbb{N}\}.$$

Show that f is a constant function.

Problem 2.1.13. Suppose that f is an entire function that satisfies a polynomial equation of the form

$$f(z)^n + c_{n-1} f(z)^{n-1} + \cdots + c_0 = 0, \quad z \in \mathbb{C},$$

where $n \in \mathbb{N}$ and $c_0, c_1, \ldots, c_{n-1}$ are constants. Prove that f is a constant.

Problem 2.1.14. If f is an entire function that satisfies $f(2z) = f(z)$ for every $z \in \mathbb{C}$, then prove that f is a constant.

Problem 2.1.15. If f is an entire function that satisfies

$$\frac{f(z)}{z} \to 0$$

as $z \to \infty$, then prove that f is a constant. (Observe that this is a stronger result than Liouville's theorem.)

Problem 2.1.16. If f is an entire function such that $|f(z)| \le |z|^2$ for all $z \in \mathbb{C}$, then show that f is a polynomial of degree at most two.

Problem 2.1.17. Suppose that f is a nonconstant entire function that has no zeros. Prove that there exists a sequence $\{z_n\}$ that satisfies $z_n \to \infty$ such that the sequence $\{|f(z_n)|\}$ is bounded.

Problem 2.1.18. If f is an entire function satisfying

$$|f(z)| \le |z| e^{-|\operatorname{Im}(z)|}, \quad z \in \mathbb{C},$$

then show that $f(z) \equiv 0$.

Problem 2.1.19. Prove Lemma 2.1.9 (by modifying the proof of Lemma 2.1.8).

2.2 Growth in terms of the Taylor coefficients

In the last section, the order $\rho(f)$ and type $\tau(f)$ of an entire function f were defined in terms of the maximum modulus function $M(r,f)$. In this section, we show that $\rho(f)$ and $\tau(f)$ can also be expressed in terms of the Taylor coefficients of the Taylor series of f.

We begin with the following result, known as the *Cauchy–Hadamard theorem*, which involves the *radius of convergence* R of a power series.

Theorem 2.2.1. *Given a power series*

$$\sum_{n=0}^{\infty} a_n z^n, \quad a_n \in \mathbb{C}, \tag{2.2.1}$$

let $R = 1/L \in [0,\infty]$, *where*

$$L = \limsup_{n\to\infty} \sqrt[n]{|a_n|}.$$

Then there are three possibilities:
(a) *If* $R = 0$, *then the series* (2.2.1) *is divergent for all* $z \neq 0$.
(b) *If* $R \in (0,\infty)$, *then for* $|z| < R$, *the series* (2.2.1) *is absolutely convergent and analytic, and for* $|z| > R$, *the series* (2.2.1) *is divergent.*
(c) *If* $R = \infty$, *then the series* (2.2.1) *is an entire function and is absolutely convergent for all* $z \in \mathbb{C}$.

Proof. (a) Suppose that $R = 0$, and fix $z \neq 0$. Since now $L = \infty$, there exists a strictly increasing sequence $\{n_k\}$ of natural numbers satisfying

$$\sqrt[n_k]{|a_{n_k}|} > \frac{1}{|z|}, \quad k \in \mathbb{N}. \tag{2.2.2}$$

This gives $|a_{n_k} z^{n_k}| > 1$ for all $k \in \mathbb{N}$, so that the condition

$$\lim_{n\to\infty} a_n z^n = 0, \tag{2.2.3}$$

which is necessary for the convergence of the series (2.2.1), cannot hold. Hence (2.2.1) diverges for every $z \neq 0$.

(b) Suppose that $R \in (0,\infty)$. Since (2.2.1) clearly converges for $z = 0$, we first assume that $z \neq 0$ and $|z| < R$. In this case, there exists $\alpha \in (0,1)$ such that

$$\alpha^2/L = |z| < R = 1/L.$$

Then

$$L < L/\alpha = \alpha/|z|,$$

and hence $\sqrt[n]{|a_n|} < \alpha/|z|$ for all n large enough, say $n \geq N$. In other words,

$$|a_n z^n| < \alpha^n, \quad n \geq N.$$

Since the majorant $\sum_{n=0}^{\infty} \alpha^n$ is a convergent geometric series, the power series (2.2.1) converges absolutely for $|z| < R$.

Next suppose that K is a compact subset of $|z| < R$. It follows that there exists r satisfying $0 < r < R$ such that $K \subset \{z : |z| \leq r\}$. Since

$$\sum_{n=0}^{\infty} |a_n| r^n < \infty \tag{2.2.4}$$

from what was just proven, and since $|a_n z^n| \leq |a_n r^n|$ for all $z \in K$, it follows from (2.2.4) and the Weierstrass M-test that the series (2.2.1) converges uniformly on K. Since K was an arbitrary compact subset of $|z| < R$, by the Weierstrass theorem it follows that the series (2.2.1) is analytic on $|z| < R$.

Finally, assume that $|z| > R$. Now $L > 1/|z|$, and similarly as in case (a), there is an infinite sequence $\{n_k\}$ of natural numbers such that (2.2.2) holds. Then the necessary condition (2.2.3) for the convergence of the power series (2.2.1) does not hold, and thus the series diverges.

(c) Suppose that $R = \infty$, and let $z \neq 0$. In this case,

$$L = 0 < \alpha/|z| \quad \text{for all } \alpha \in (0,1).$$

Therefore the power series (2.2.1) converges absolutely by exactly the same argument as in case (b). Then also using the argument in case (b), we obtain that (2.2.1) is an entire function. □

An entire function f can be written as a power series

$$f(z) = \sum_{n=0}^{\infty} a_n z^n, \quad z \in \mathbb{C}, \tag{2.2.5}$$

where it is well known that the coefficients are the Taylor coefficients, that is,

$$a_n = \frac{f^{(n)}(0)}{n!}, \quad n = 0, 1, 2, \ldots.$$

Thus power series and Taylor series have the same meaning and can be used interchangeably. Since the Weierstrass theorem was applied in the proof of Theorem 2.2.1(c), we obtain that the power series (2.2.5) can be differentiated term-by-term.

If $|z| \leq r$, then from (2.2.5),

$$|f(z)| \leq \sum_{n=0}^{\infty} |a_n||z|^n \leq \sum_{n=0}^{\infty} |a_n| r^n. \tag{2.2.6}$$

The majorant series in (2.2.6) converges for any fixed $r \in (0, \infty)$ by the Cauchy–Hadamard theorem. Moreover, by the divergence test we have $|a_n|r^n \to 0$ as $n \to \infty$. Thus $|a_n| \to 0$ as $n \to \infty$. In particular, we may suppose that $|a_n| < 1$ whenever n is large enough.

Theorem 2.2.2 ([37, Theorem 2.2.2]). *Let f be an entire function with Taylor series (2.2.5). Then*

$$\rho(f) = \limsup_{n\to\infty} \frac{n \log n}{-\log |a_n|},$$

where the expression $\frac{n\log n}{-\log|a_n|}$ is considered to be zero whenever $a_n = 0$.

Proof. For convenience, set

$$\sigma = \limsup_{n\to\infty} \frac{n \log n}{-\log |a_n|}.$$

(1) First, we prove that $\rho(f) \geq \sigma$. If $\sigma = 0$, then this inequality is trivial, so we may suppose that $\sigma > 0$. Choose $\mu \in (0, \sigma)$. By the definition of the constants μ and σ, there exist infinitely many natural numbers n such that $0 < |a_n| < 1$ and

$$\frac{n \log n}{-\log |a_n|} \geq \mu \quad \Longleftrightarrow \quad \log |a_n| \geq -\frac{n \log n}{\mu}.$$

From Cauchy's inequalities (2.1.6),

$$\log M(r,f) \geq \log(r^n |a_n|) = n \log r + \log |a_n| \geq n \log r - \frac{n \log n}{\mu}.$$

The above infinitely many n will be used to define an infinite sequence of strictly increasing r values, $r_n = (en)^{1/\mu}$. Then $n = r_n^\mu / e$, and

$$\log M(r_n, f) \geq \frac{n}{\mu} \log(en) - \frac{n}{\mu} \log n = \frac{n}{\mu} = \frac{r_n^\mu}{e\mu}.$$

Consequently,

$$\log\log M(r_n, f) \geq \mu \log r_n + \log \frac{1}{e\mu}.$$

This inequality is true for infinitely many n, that is, for a subsequence of $\{r_n\}$. For simplicity, we denote this subsequence again by $\{r_n\}$, so that the above inequality now holds for all $n \in \mathbb{N}$. This gives us

$$\rho(f) = \limsup_{r\to\infty} \frac{\log\log M(r,f)}{\log r} \geq \limsup_{n\to\infty} \frac{\log\log M(r_n,f)}{\log r_n} \geq \limsup_{n\to\infty} \frac{\mu \log r_n + \log \frac{1}{e\mu}}{\log r_n} = \mu.$$

Since $\mu \in (0, \sigma)$ was chosen arbitrarily, it follows that $\rho(f) \geq \sigma$.

(2) Next, we prove that $\rho(f) \le \sigma$. We may assume that $\sigma < \infty$. Let now $\mu \in (\sigma, \infty)$. Then there exists $N(\mu) \in \mathbb{N}$ such that for all $n \ge N(\mu)$ with $a_n \ne 0$, we have

$$0 \le \frac{n \log n}{-\log |a_n|} \le \mu.$$

Therefore,

$$\log |a_n| \le -\frac{n}{\mu} \log n = \log(n^{-\frac{n}{\mu}}) \quad \text{whenever } a_n \ne 0, n \ge N(\mu).$$

We obtain $|a_n| \le n^{-\frac{n}{\mu}}$ for all $n \ge N(\mu)$, even if $a_n = 0$. Set $A = \max_{0 \le n \le N(\mu)-1} |a_n|$. Now

$$\begin{aligned} M(r,f) = \max_{|z| \le r} \left| \sum_{n=0}^{\infty} a_n z^n \right| &\le \sum_{n=0}^{N(\mu)-1} |a_n| r^n + \sum_{n=N(\mu)}^{\infty} |a_n| r^n \\ &\le AN(\mu) r^{N(\mu)} + \sum_{n=1}^{\infty} n^{-\frac{n}{\mu}} r^n \\ &= AN(\mu) r^{N(\mu)} + \sum_{1 \le n < (2r)^\mu} n^{-\frac{n}{\mu}} r^n + \sum_{n \ge (2r)^\mu} n^{-\frac{n}{\mu}} r^n \\ &= AN(\mu) r^{N(\mu)} + S_1 + S_2, \quad r \ge 1. \end{aligned}$$

Since $n^{1/\mu} \ge 2r$ in S_2, we have $rn^{-1/\mu} \le \frac{1}{2}$, and so

$$S_2 = \sum_{n \ge (2r)^\mu} (rn^{-\frac{1}{\mu}})^n \le \sum_{n \ge (2r)^\mu} \left(\frac{1}{2}\right)^n \le \sum_{n=1}^{\infty} \left(\frac{1}{2}\right)^n = 1.$$

To estimate S_1, we first find that $n^{-\frac{n}{\mu}} \le \frac{1}{n^2}$ for all $n \ge 2\mu$. Consequently,

$$\sum_{n=1}^{\infty} n^{-\frac{n}{\mu}} = \sum_{1 \le n < 2\mu} n^{-\frac{n}{\mu}} + \sum_{n \ge 2\mu} n^{-\frac{n}{\mu}} = K < \infty.$$

Then

$$S_1 = \sum_{1 \le n < (2r)^\mu} n^{-\frac{n}{\mu}} r^n \le r^{(2r)^\mu} \sum_{1 \le n < (2r)^\mu} n^{-\frac{n}{\mu}} \le Kr^{(2r)^\mu}, \quad r \ge 1.$$

Finally,

$$\log M(r,f) = O(r^\mu \log r),$$

so that $\rho(f) \le \mu$. Since $\mu \in (\sigma, \infty)$ was chosen arbitrarily, we obtain $\rho(f) \le \sigma$. □

The following result, called *Stirling's formula*, is often useful when making calculations with Taylor coefficients.

Theorem 2.2.3 ([218, p. 280]). *If $n \geq 0$ is an integer, then*

$$n! = (1 + o(1))\sqrt{2\pi n}\left(\frac{n}{e}\right)^n$$

as $n \to \infty$.

Example 2.2.4. Consider the function $g(z) = \cos(z^2)$ with the Taylor series

$$g(z) = \sum_{n=0}^{\infty}(-1)^n \frac{(z^2)^{2n}}{(2n)!} = \sum_{n=0}^{\infty}(-1)^n \frac{z^{4n}}{(2n)!}.$$

By Theorem 2.2.2 we have

$$p(g) = \limsup_{n\to\infty} \frac{n \log n}{-\log|a_n|} = \limsup_{n\to\infty} \frac{4n\log(4n)}{\log((2n)!)}. \tag{2.2.7}$$

By Stirling's formula,

$$\log\big((2n)!\big) \sim \log\left(\sqrt{4\pi n}\left(\frac{2n}{e}\right)^{2n}\right) = \frac{1}{2}\log(4\pi n) + 2n\log(2n) - 2n. \tag{2.2.8}$$

Substituting (2.2.8) into (2.2.7), we obtain $p(g) = 2$.

A more general statement of Stirling's formula involving the gamma function is given in Theorem 5.1.4. A rigorous proof of Stirling's formula is rather lengthy. Instead of discussing the asymptotic comparability, we prove upper and lower estimates for the factorial, which are sufficient to use in later sections.

Lemma 2.2.5 ([242, pp. 314–315]). *For all $n \in \mathbb{N}$, we have*

$$\left(\frac{n}{e}\right)^n < n! < (2n+1)\left(\frac{n}{e}\right)^n. \tag{2.2.9}$$

Proof. From the power series for e^x we have

$$e^n = \sum_{k=0}^{\infty} \frac{n^k}{k!}. \tag{2.2.10}$$

Hence it is immediate that $e^n > n^n/n!$. This proves the left inequality.

To prove the right inequality, first observe from (2.2.10) that

$$e^n = 1 + \frac{n}{1!} + \cdots + \frac{n^{n-1}}{(n-1)!} + \frac{n^n}{n!}\left(1 + \frac{n}{n+1} + \frac{n^2}{(n+1)(n+2)} + \cdots\right).$$

Therefore,

$$e^n < n\frac{n^n}{n!} + \frac{n^n}{n!}\sum_{k=0}^{\infty}\left(\frac{n}{n+1}\right)^k = (2n+1)\frac{n^n}{n!}.$$

This gives the right inequality. □

Since $\sqrt[n]{n} \to 1$ as $n \to \infty$, we see from either Theorem 2.2.3 or Lemma 2.2.5 that the following result holds.

Corollary 2.2.6. *We have*

$$\sqrt[n]{n!} = (1 + o(1))\frac{n}{e}, \quad n \to \infty.$$

Lemma 2.2.7 ([142, Lemma 4.13]). *Let f be analytic in an open disc about $z = 0$, where it has the Taylor expansion*

$$f(z) = \sum_{n=0}^{\infty} a_n z^n. \tag{2.2.11}$$

Suppose that there exist constants $\lambda > 0$, $\mu > 0$, and $N = N(\lambda, \mu) \in \mathbb{N}$ such that

$$|a_n| \le \left(\frac{e\mu\lambda}{n}\right)^{n/\mu}, \quad n \ge N. \tag{2.2.12}$$

Then the Taylor expansion in (2.2.11) *converges absolutely in the complex plane, and f is entire. Moreover, for every $\varepsilon > 0$, there exists $R_0 = R_0(\varepsilon) > 0$ such that*

$$M(r,f) \le e^{(\lambda+\varepsilon)r^{\mu}}, \quad r \ge R_0. \tag{2.2.13}$$

Proof. The radius of convergence R for the power series in (2.2.11) satisfies

$$\frac{1}{R} = \limsup_{n\to\infty} \sqrt[n]{|a_n|} \le \limsup_{n\to\infty} \left(\frac{e\mu\lambda}{n}\right)^{1/\mu} = 0,$$

so that $R = \infty$. Then by Theorem 2.2.1(c) the power series (2.2.11) converges absolutely in $\mathbb{C}$, and f is an entire function.

Finally, we will prove that (2.2.13) holds. For fixed $r > 0$, we define the following auxiliary function:

$$h(x) = \left(\frac{e\mu\lambda}{x}\right)^{x/\mu} r^x, \quad x > 0.$$

Then

$$h'(x) = h(x)\left(\frac{1}{\mu}\left(\log\left(\frac{e\mu\lambda}{x}\right) - 1\right) + \log r\right) = 0$$

if and only if $x = \mu\lambda r^{\mu}$. We see that h reaches its maximum at this point. Therefore,

$$\left(\frac{e\mu\lambda}{x}\right)^{x/\mu} r^x = h(x) \le h(\mu\lambda r^{\mu}) = e^{\lambda r^{\mu}}, \quad x > 0. \tag{2.2.14}$$

Moreover, if $n \geq N(r) := \max\{N, 2^\mu e\mu\lambda r^\mu\}$ and $r \geq 1$ is large enough, then by (2.2.12) we have

$$\sqrt[n]{|a_n| r^n} \leq \left(\frac{e\mu\lambda}{n}\right)^{1/\mu} r = \exp\left(\frac{1}{\mu}\log\frac{e\mu\lambda}{n} + \log r\right)$$
$$\leq \exp\left(\log\frac{1}{2r} + \log r\right) = \frac{1}{2},$$

and hence

$$|a_n| r^n \leq \left(\frac{1}{2}\right)^n, \quad n \geq N(r).$$

We can assume that r is large enough so that $N(r) \geq N + 1$. We can now estimate the maximum modulus by writing

$$M(r,f) = \max_{|z|\leq r}\left|\sum_{n=0}^{\infty} a_n z^n\right| \leq \sum_{n=0}^{\infty} |a_n| r^n$$
$$= \sum_{n=0}^{N} |a_n| r^n + \sum_{n=N+1}^{N(r)} |a_n| r^n + \sum_{n=N(r)+1}^{\infty} |a_n| r^n = \Sigma_1 + \Sigma_2 + \Sigma_3.$$

Denoting $b = \sum_{n=0}^{N} |a_n|$, we have that for large r,

$$\Sigma_1 \leq br^N \quad \text{and} \quad \Sigma_3 \leq \sum_{n=1}^{\infty}\left(\frac{1}{2}\right)^n = 1. \tag{2.2.15}$$

Finally, by (2.2.12),

$$\Sigma_2 \leq (N(r) - N) \max_{N+1\leq n\leq N(r)} |a_n| r^n \leq (N(r) - N) \max_{n\geq N}\left(\left(\frac{e\mu\lambda}{n}\right)^{n/\mu} r^n\right),$$

which, combined with (2.2.14), gives

$$\Sigma_2 \leq (N(r) - N)e^{\lambda r^\mu}. \tag{2.2.16}$$

Putting together estimates (2.2.15) and (2.2.16), we obtain (2.2.13). □

Theorem 2.2.8 ([37, Theorem 2.2.10], [205, Theorem 9.5]). *Let $f(z) = \sum_{n=0}^{\infty} a_n z^n$ be an entire function of order $\rho \in (0,\infty)$. Then the type τ of f satisfies*

$$\tau = \frac{1}{e\rho}\limsup_{n\to\infty} n|a_n|^{\rho/n}.$$

Proof. We denote

$$L = \limsup_{n\to\infty} n|a_n|^{\rho/n}, \tag{2.2.17}$$

and proceed to prove that $L = e\tau\rho$.

We will first prove that $e\tau\rho \leq L$. Since this is trivial if $L = \infty$, we may suppose that $L < \infty$. Choose any $K > L/e\rho$. By the definition of the limit superior L, there exists $N \in \mathbb{N}$ such that

$$n|a_n|^{\rho/n} < e\rho K, \quad n \geq N.$$

In other words,

$$|a_n| \leq \left(\frac{e\rho K}{n}\right)^{n/\rho}, \quad n \geq N.$$

By Lemma 2.2.7, for every $\varepsilon > 0$, there exists $R_0 = R_0(\varepsilon) > 0$ such that

$$M(r,f) \leq e^{(K+\varepsilon)r^\rho}, \quad r \geq R_0. \tag{2.2.18}$$

By Definition 2.1.7 we have $\tau \leq K + \varepsilon$. Since $\varepsilon > 0$ is arbitrary, we get $\tau \leq K$, and since $K > L/e\rho$ is arbitrary, we may replace K with $L/e\rho$ in the inequality $\tau \leq K$ to conclude that $e\tau\rho \leq L$.

Second, we will prove that $e\tau\rho \geq L$. Since this is trivial if $L = 0$, we may suppose that $L > 0$. Choose any $M \in (0, L)$. By the definition of L again, there are infinitely many indices n such that

$$n|a_n|^{\rho/n} \geq M \quad \Longleftrightarrow \quad |a_n| \geq \left(\frac{M}{n}\right)^{n/\rho}.$$

For these indices n, we define the sequence $\{r_n\}$ of positive real numbers by

$$r_n^\rho = \frac{ne}{M}.$$

Thus $\{r_n\}$ tends to infinity. By Cauchy's inequalities $|a_k|r^k \leq M(r,f)$ (see (2.1.6)), we obtain an estimate

$$M(r_n,f) \geq |a_n|r_n^n \geq \left(\frac{M}{n}\right)^{n/\rho}\left(\frac{ne}{M}\right)^{n/\rho} = e^{n/\rho} = e^{\frac{M}{\rho e}r_n^\rho}. \tag{2.2.19}$$

Therefore,

$$\tau = \limsup_{r\to\infty} \frac{\log M(r,f)}{r^\rho} \geq \limsup_{n\to\infty} \frac{\log M(r_n,f)}{r_n^\rho} = \frac{M}{\rho e}.$$

Since $M \in (0, L)$ is arbitrary, we deduce that $e\tau\rho \geq L$. □

The following theorem is a converse to Theorem 2.2.8, the starting point being the constant L defined in (2.2.17). The result is contained in [37, Theorem 2.2.10], but it is stated and proved slightly differently than in [37].

Theorem 2.2.9. *Let $f(z) = \sum_{n=0}^{\infty} a_n z^n$ be a transcendental entire function, let $\rho \in (0, \infty)$ be any constant, and let L be the constant defined in* (2.2.17).
(a) *If $L = 0$, then $\rho(f) \le \rho$, and if $\rho(f) = \rho$, then $\tau(f) = 0$.*
(b) *If $L = \infty$, then $\rho(f) \ge \rho$, and if $\rho(f) = \rho$, then $\tau(f) = \infty$.*
(c) *If $0 < L < \infty$, then $\rho(f) = \rho$ and $\tau(f) = \frac{L}{e\rho}$.*

Proof. (a) If $L = 0$, then we may replace the constant K in the proof of Theorem 2.2.8 with an arbitrary $\varepsilon > 0$. Then estimate (2.2.18) becomes

$$M(r,f) \le e^{2\varepsilon r^\rho}, \quad r \ge R_0,$$

from which (a) follows.

(b) If $L = \infty$, then the constant $M > 0$ in the proof of Theorem 2.2.8 can be arbitrarily large, and (b) follows from estimate (2.2.19).

(c) Finally, we suppose that $0 < L < \infty$. Let $\varepsilon \in (0, L)$. First, we consider the order of f. Clearly, there exists $N \in \mathbb{N}$ such that

$$|a_n| \le \left(\frac{L+\varepsilon}{n}\right)^{n/\rho}, \quad n \ge N. \tag{2.2.20}$$

Hence

$$\frac{n \log n}{\log(1/|a_n|)} \le \frac{\rho \log n}{\log(\frac{n}{L+\varepsilon})} \le \rho + \varepsilon$$

whenever $a_n \ne 0$ and $n > \max\{L+\varepsilon, (L+\varepsilon)^{1+\rho/\varepsilon}\}$. By Theorem 2.2.2, f is of order $\le \rho + \varepsilon$. Letting $\varepsilon \to 0^+$ gives $\rho(f) \le \rho$. On the other hand, by the definition of limit superior we have

$$\left(\frac{L-\varepsilon}{n}\right)^{n/\rho} \le |a_n| \tag{2.2.21}$$

for infinitely many indices n. Similarly as above, we may prove that $\rho(f) \ge \rho$. Thus $\rho(f) = \rho$.

Second, we consider the type $\tau(f)$ of f, now knowing that $\rho(f) = \rho$. For infinitely many indices n for which (2.2.21) holds, we define a sequence $\{r_n\}$ of positive real numbers by

$$r_n^\rho = \frac{ne}{L-\varepsilon}.$$

Thus $\{r_n\}$ tends to infinity. By Cauchy's inequalities,

$$M(r_n,f) \ge |a_n| r_n^n \ge \left(\frac{L-\varepsilon}{n}\right)^{n/\rho} \left(\frac{ne}{L-\varepsilon}\right)^{n/\rho} = e^{n/\rho} = \exp((L-\varepsilon) r_n^\rho / e\rho),$$

from which $\tau(f) \ge \frac{L-\varepsilon}{e\rho}$. By letting $\varepsilon \to 0^+$ we get $\tau(f) \ge \frac{L}{e\rho}$.

Since adding a polynomial to f has no affect on the order or type of f, we may suppose, instead of (2.2.20), that

$$a_0 = 0 \quad \text{and} \quad |a_n| \le \left(\frac{L+\varepsilon}{n}\right)^{n/\rho}, \qquad n \in \mathbb{N}.$$

Then for $r = |z| > 0$,

$$|f(z)| \le \sum_{n=0}^{\infty} |a_n| r^n \le \sum_{n=1}^{\infty} \left(\frac{(L+\varepsilon)r^\rho}{n}\right)^{n/\rho}. \tag{2.2.22}$$

Similarly as in the proof of Lemma 2.2.7, we define the auxiliary function

$$h(x) = \left(\frac{(L+\varepsilon)r^\rho}{x}\right)^{x/\rho} = \exp\left(\frac{x}{\rho}\log\left(\frac{(L+\varepsilon)r^\rho}{x}\right)\right), \qquad x > 0.$$

It is easy to see that

$$h(x) \le h\left(\frac{(L+\varepsilon)r^\rho}{e}\right) = \exp\left(\frac{L+\varepsilon}{e\rho} r^\rho\right).$$

Thus none of the terms in the upper bound for $|f(z)|$ in (2.2.22) exceeds $\exp(\frac{L+\varepsilon}{e\rho} r^\rho)$, attained for $n = (L+\varepsilon)r^\rho/e$ (which may not be an integer). We are then able to estimate

$$\begin{aligned} |f(z)| &\le \sum_{1 \le n \le (L+2\varepsilon)r^\rho} \left(\frac{(L+\varepsilon)r^\rho}{n}\right)^{n/\rho} + \sum_{n > (L+2\varepsilon)r^\rho} \left(\frac{(L+\varepsilon)r^\rho}{n}\right)^{n/\rho} \\ &\le (L+2\varepsilon)r^\rho \exp\left(\frac{L+\varepsilon}{e\rho} r^\rho\right) + \sum_{n=0}^{\infty} \left(\frac{L+\varepsilon}{L+2\varepsilon}\right)^{n/\rho} \\ &= O(\exp((L+2\varepsilon)r^\rho/e\rho)), \end{aligned}$$

from which $\tau(f) \le \frac{L+2\varepsilon}{e\rho}$. Letting $\varepsilon \to 0^+$ gives $\tau(f) \le \frac{L}{e\rho}$. □

Problems

Problem 2.2.10. Let $0 < \alpha < \infty$ and $0 < \sigma < \infty$. Prove that the functions

$$f_1(z) = \sum_{n=2}^{\infty} \left(\frac{e\sigma\alpha}{n \log n}\right)^{n/\alpha} z^n, \quad f_2(z) = \sum_{n=2}^{\infty} \left(\frac{e\alpha \log n}{n}\right)^{n/\alpha} z^n,$$

$$f_3(z) = \sum_{n=2}^{\infty} \left(\frac{1}{\log n}\right)^n z^n, \quad f_4(z) = \sum_{n=0}^{\infty} e^{-n^2} z^n$$

are entire and find their order. When possible, find their type.

Problem 2.2.11. In the literature, the symbols $\cos\sqrt{z}$ and $\frac{\sin\sqrt{z}}{\sqrt{z}}$ are often used to denote the power series

$$\cos\sqrt{z} = \sum_{n=0}^{\infty} \frac{(-1)^n z^n}{(2n)!} \quad \text{and} \quad \frac{\sin\sqrt{z}}{\sqrt{z}} = \sum_{n=0}^{\infty} \frac{(-1)^n z^n}{(2n+1)!},$$

obtained from the power series of $\cos z$ and $\sin z$ via formal computation. Prove that $\cos\sqrt{z}$ and $\frac{\sin\sqrt{z}}{\sqrt{z}}$ are entire and find their order and type.

Problem 2.2.12. Let

$$g(z) = \begin{cases} \frac{1-\cos z}{z^2}, & z \neq 0, \\ \frac{1}{2}, & z = 0. \end{cases}$$

Justify that g is an entire function.

Problem 2.2.13. Show that the function

$$f(z) = \sum_{n=1}^{\infty} \frac{\sinh n}{n!} z^n$$

is entire and find its order $\rho(f)$.

Problem 2.2.14. If the power series

$$f(z) = \sum_{n=0}^{\infty} a_n z^n$$

is an entire function of order one and minimum type, then show that

$$\lim_{n\to\infty} \sqrt[n]{n!|a_n|} = 0.$$

Problem 2.2.15. If $f(z) = \sum_{n=0}^{\infty} a_n z^n$ is an entire function of order 3, then what is the order of the entire function $g(z) = \sum_{n=0}^{\infty} |a_n|^5 z^n$?

Problem 2.2.16. Show that

$$g(z) = \sum_{n=0}^{\infty} \frac{z^{3n}}{3^{3n}(n!)^2}$$

is an entire function and find the order $\rho(g)$.

2.3 Order and type of a derivative

We will give a direct proof of the following result, which can also be proved by using the Taylor coefficient formulas; see Problems 2.3.2 and 2.3.3.

Theorem 2.3.1 ([37, Theorem 2.4.1]). *Any entire function f and its derivative f' are of the same order and type (whenever the type is defined for f).*

Proof. Denote $z = re^{i\theta}$. Integrating along a line segment, we have

$$f(z) = \int_0^z f'(w)\,dw + f(0) = e^{i\theta}\int_0^r f'(se^{i\theta})\,ds + f(0).$$

Hence

$$|f(z)| \le \int_0^r |f'(se^{i\theta})|\,ds + |f(0)| \le rM(r,f') + |f(0)|.$$

Since the upper bound does not depend on θ, the lower bound can be replaced with $M(r,f)$. This shows that $\rho(f) \le \rho(f')$.

On the other hand, by Cauchy's formula,

$$f'(z) = \frac{1}{2\pi i}\int_{|w|=2r} \frac{f(w)}{(w-z)^2}\,dw = \frac{r}{\pi}\int_0^{2\pi} \frac{f(2re^{i\varphi})}{(2re^{i\varphi}-z)^2}\,e^{i\varphi}d\varphi. \tag{2.3.1}$$

Thus

$$|f'(z)| \le \frac{r}{\pi}\int_0^{2\pi} \frac{|f(2re^{i\varphi})|}{|2re^{i\varphi}-z|^2}\,d\varphi \le \frac{2}{r}M(2r,f). \tag{2.3.2}$$

Once again, the upper bound does not depend on θ, so the lower bound can be replaced with $M(r,f')$. This shows that $\rho(f') \le \rho(f)$.

Finally, if $0 < \rho(f) < \infty$, then $\tau(f) \in [0,\infty]$ is well-defined, and the above reasoning gives $\tau(f) \le \tau(f')$ from Definition 2.1.7. To prove that $\tau(f') \le \tau(f)$, we integrate in (2.3.1) over $|w| = (1+\varepsilon)r$ for a small fixed $\varepsilon > 0$ instead of $|w| = 2r$. This leads to

$$|f'(z)| \le \frac{2}{r\varepsilon^2}M((1+\varepsilon)r,f)$$

instead of (2.3.2). It follows that $\tau(f') \le (1+\varepsilon)^{\rho}\tau(f)$. Since $\varepsilon > 0$ can be arbitrarily small, we obtain $\tau(f') \le \tau(f)$. □

Problems

Problem 2.3.2. Let f be an entire function of order $\rho \in (0,\infty)$. Use the Taylor coefficients formula to prove that $\rho(f) = \rho(f')$.

Problem 2.3.3. Let f be an entire function of order $\rho \in (0,\infty)$. Use the Taylor coefficients formula to prove that $\tau(f) = \tau(f')$.

2.4 Growth of the real part

An entire function f can be written in terms of its real and imaginary parts as $f = u + iv$. We will show that the real part u dominates the growth of f in a certain sense. This result is known as the Borel–Carathéodory theorem. Since $if = -v + iu$, an analogous result holds for the imaginary part v. For the proof, we need an auxiliary result known as the *Schwarz lemma.*

Lemma 2.4.1. *Suppose that f is analytic in $|z| < R$ and continuous on $|z| \le R$, $f(0) = 0$, and $|f(z)| \le M$ on $|z| = R$. Then with $z = re^{i\theta}$, we have*

$$|f(re^{i\theta})| \le \frac{Mr}{R}, \qquad 0 \le r \le R.$$

Proof. Define

$$\phi(z) = \begin{cases} f(z)/z, & z \ne 0, \\ f'(0), & z = 0. \end{cases}$$

Then ϕ is analytic in $|z| < R$ and continuous on $|z| \le R$, same as f. Moreover,

$$|\phi(z)| \le M/R \quad \text{on } |z| = R.$$

Then by the maximum modulus principle, $|\phi(z)| \le M/R$ for all $|z| \le R$. This yields the result. □

If f is analytic in $|z| < R$, then we denote the maximum of its real part by

$$A(r,f) = \max_{|z|\le r} \operatorname{Re}(f(z)), \qquad 0 \le r < R.$$

Clearly,

$$e^{A(r,f)} = \max_{|z|\le r} |e^{f(z)}| = M(r, e^f), \tag{2.4.1}$$

where the right-hand side is a strictly increasing function of r whenever f is nonconstant. Thus it follows from (2.4.1) that $A(r,f)$ is also a strictly increasing function of r whenever f is nonconstant.

We are now ready to state and prove the *Borel–Carathéodory theorem.*

Theorem 2.4.2 ([273, p. 174]). *Suppose that f is analytic in $|z| < R$ and continuous on $|z| \le R$. Then*

$$M(r,f) \le \frac{2r}{R-r} A(R,f) + \frac{R+r}{R-r} |f(0)|, \qquad 0 \le r < R.$$

Proof. The result is obvious if f is a constant function, so we may suppose that f is nonconstant.

Consider first the case where $f(0) = 0$. Since f is nonconstant, $A(r,f)$ is strictly increasing in r, which means that $A(R,f) > A(0,f) = 0$. Define

$$\phi(z) = \frac{f(z)}{2A(R,f) - f(z)}, \quad |z| \le R.$$

Since the real part of the denominator does not vanish, we see that ϕ is analytic in $|z| < R$ and continuous on $|z| \le R$, same as f. Moreover, $\phi(0) = 0$, and if $f = u + iv$, then

$$|\phi(z)|^2 = \frac{u^2 + v^2}{(2A(R,f) - u)^2 + v^2} \le 1, \quad |z| \le R.$$

Hence the Schwarz lemma gives

$$|\phi(z)| \le \frac{r}{R}, \quad |z| = r \le R.$$

Consequently,

$$|f(z)| = \left|\frac{2A(R,f)\phi(z)}{1 + \phi(z)}\right| \le \frac{2rA(R,f)}{(1 - |\phi(z)|)R} \le \frac{2r}{R - r}A(R,f), \quad 0 \le r < R,$$

and the statement follows.

Finally, suppose that $f(0) \neq 0$. We apply the previously obtained result for $g(z) = f(z) - f(0)$, which satisfies $g(0) = 0$. Thus

$$|f(z)| - |f(0)| \le |g(z)| \le \frac{2r}{R - r}A(R,g) \le \frac{2r}{R - r}(A(R,f) + |f(0)|),$$

from which the statement follows. □

We state an immediate consequence of Theorem 2.4.2.

Corollary 2.4.3. *Suppose that f is analytic in $|z| < R$ and continuous on $|z| \le R$. If $A(R,f) \ge 0$, then*

$$M(r,f) \le \frac{R + r}{R - r}(A(R,f) + |f(0)|), \quad 0 \le r < R.$$

Example 2.4.4. Let $f(z) = -\frac{1+z}{1-z}$. It is clear that f is analytic in the unit disc $\mathbb{D}$. The function $z \to \frac{1+z}{1-z}$ maps $\mathbb{D}$ onto the right half-plane, which is then rotated to the left half-plane. Hence $A(r,f) < 0$ for all $0 < r < 1$. This shows that the assumption $A(R,f) \ge 0$ is necessary in Corollary 2.4.3 and that the term $|f(0)|$ cannot be dropped in Theorem 2.4.2.

Remark. (a) If f is entire, then we can make a standard substitution $R = 2r$, and Theorem 2.4.2 gives us $M(r,f) \le 2A(2r,f) + 3|f(0)|$. This shows that $A(r,f) \to \infty$ as $r \to \infty$ for nonconstant entire functions f.

(b) The upper bound in Theorem 2.4.2 needs to contain some term tending to infinity as $r \to R^-$. Indeed, let

$$f(z) = i \log \frac{1 + z}{1 - z} = -\arg\left(\frac{1 + z}{1 - z}\right) + i \log\left|\frac{1 + z}{1 - z}\right|, \tag{2.4.2}$$

where $|z| = r < R < 1$, and where the principal branch for the logarithm is chosen. Then $f(0) = 0$ and $|A(R,f)| < \pi/2$ regardless of how near R is to 1. However, $M(r,f) \to \infty$ as $r \to 1^-$.

It was proved by K. Ramachandra in 1989 that the unbounded factors in Theorem 2.4.2 can be replaced with logarithmic terms. The function f in (2.4.2) shows that this factor is essentially the best possible.

Theorem 2.4.5 ([229]). *Suppose that f is analytic in $|z| < R$ and continuous on $|z| \le R$. Further, let*

$$C_1 = \frac{e}{\log(2+\sqrt{3})} \approx 2.07 \quad and \quad C_2 = 4 + 2\log 2 \approx 5.39.$$

Then

$$M(r,f) \le C_1\left(\log\frac{R}{R-r} + C_2\right)\left(|A(R,f)| + |f(0)|\right), \quad 0 \le r < R.$$

Problems

Problem 2.4.6. Prove the following extension of Corollary 2.4.3: Suppose that f is analytic in $|z| < R$ and continuous on $|z| \le R$ such that $A(R,f) \ge 0$. Let $n \in \mathbb{N} \cup \{0\}$. Then

$$M\left(r,f^{(n)}\right) \le \frac{2^{n+2}n!R}{(R-r)^{n+1}}\left(A(R,f) + |f(0)|\right), \quad 0 \le r < R.$$

Hint: Use Cauchy's formula

$$f^{(n)}(z) = \frac{n!}{2\pi i}\int_C \frac{f(w)}{(w-z)^{n+1}}\,dw,$$

where C is the circle with center z and radius $\delta = (R-r)/2$, and $|z| = r$.

2.5 Lower order and lower type

We define a lower order and a lower type of an entire function by replacing lim sup with lim inf in the usual definitions of order and type.

Definition 2.5.1. The *lower order of growth* of an entire function f is the number

$$\underline{\rho} = \underline{\rho}(f) = \liminf_{r\to\infty}\frac{\log\log M(r,f)}{\log r} \in [0,\infty].$$

If $\underline{\rho} \in (0,\infty)$, then the *lower type* of f is the number

$$\underline{\tau} = \underline{\tau}(f) = \liminf_{r\to\infty}\frac{\log M(r,f)}{r^{\underline{\rho}}} \in [0,\infty].$$

It is clear that $\underline{\rho}(f) \le \rho(f)$ and $\underline{\tau}(f) \le \tau(f)$ and that the strict inequalities are possible. Moreover, for any fixed $\underline{\rho}$ and ρ satisfying $0 \le \underline{\rho} \le \rho \le \infty$, there exists an entire function of order ρ and lower order $\underline{\rho}$ [82, p. 238].

Let f be an entire function such that $0 \le \underline{\rho}(f) < \rho(f) < \infty$, and let $\varepsilon > 0$. By the definitions of order and lower order there exist strictly increasing sequences $\{r_n\}$ and $\{s_n\}$ of positive real numbers tending to infinity such that

$$r_n^{\rho(f)-\varepsilon} \le \log M(r_n,f) \quad \text{and} \quad \log M(s_n,f) \le s_n^{\underline{\rho}(f)+\varepsilon}.$$

In the following theorem, we show that there exists a much larger set of r-values than just a sequence where $\log M(r,f)$ is either near the maximal growth or near the minimal growth. For the most general form of the result, see [121, Corollary 4.1], the proof of which is based on Theorem 1.2.10. The version below, however, is technically less challenging, yet enough for the purposes of this book.

Theorem 2.5.2 ([121, Corollary 4.1]). *Let f be a transcendental entire function such that $0 \le \underline{\rho}(f) < \rho(f) \le \infty$, and let a and b be constants satisfying $\underline{\rho}(f) < a < b < \rho(f)$. Define*

$$\begin{aligned} E &= \{r > r_0 : \log M(r,f) < r^a\}, \\ F &= \{r > r_0 : \log M(r,f) > r^b\}, \end{aligned} \tag{2.5.1}$$

where $r_0 > 1$ is chosen so that $M(r_0,f) > 1$. Then

$$\overline{\log \operatorname{dens}}(E) \ge 1 - \frac{\underline{\rho}(f)}{a}, \quad \underline{\log \operatorname{dens}}(E) \le \frac{a}{\rho(f)}, \tag{2.5.2}$$

$$\overline{\log \operatorname{dens}}(F) \ge 1 - \frac{b}{\rho(f)}, \quad \underline{\log \operatorname{dens}}(F) \le \frac{\underline{\rho}(f)}{b}. \tag{2.5.3}$$

Proof. To prove the first inequality in (2.5.2), we suppose on the contrary that $\overline{\log \operatorname{dens}}(E) < \frac{a-\underline{\rho}(f)}{a}$. Then we can find $\varepsilon > 0$ small enough such that

$$\overline{\log \operatorname{dens}}(E) < \frac{a - \underline{\rho}(f) - 2\varepsilon}{a - \varepsilon} \quad \text{and} \quad \frac{a - \varepsilon}{\underline{\rho}(f) + \varepsilon} > 1.$$

Set $\alpha = \frac{a-\varepsilon}{\underline{\rho}(f)+\varepsilon}$. Since $\alpha > 1/(1 - \overline{\log \operatorname{dens}}(E))$ and

$$r^a \le \log M(r,f), \quad r \in (r_0,\infty) \setminus E,$$

we conclude from Lemma 1.2.7 with $\psi(t) = \log t$ that for some $R > r_0$,

$$r^a \le \log M(r^\alpha,f), \quad r > R.$$

It follows that $\underline{\rho}(f) \ge \frac{a}{\alpha} > \underline{\rho}(f)$, a contradiction. Thus our original assumption is false, which proves the first inequality in (2.5.2).

To prove the second inequality in (2.5.2), we observe that

$$\log M(r,f) < r^{\alpha}, \quad r \in (r_0,\infty) \setminus E^c, \tag{2.5.4}$$

where $E^c = (r_0,\infty) \setminus E$. Suppose first that $\rho(f) < \infty$ and $\overline{\log\operatorname{dens}}(E^c) < \frac{\rho(f)-a}{\rho(f)}$. Then we can find $\varepsilon > 0$ small enough such that

$$\overline{\log\operatorname{dens}}(E^c) < \frac{\rho(f) - a - 2\varepsilon}{\rho(f) - \varepsilon} \quad \text{and} \quad \frac{\rho(f) - \varepsilon}{a + \varepsilon} > 1.$$

Set $\alpha = \frac{\rho(f)-\varepsilon}{a+\varepsilon}$. Then by applying Lemma 1.2.7 with $\psi(t) = \log t$ to (2.5.4) we obtain that $\rho(f) \le a\alpha < \rho(f)$, a contradiction. Thus $\overline{\log\operatorname{dens}}(E^c) \ge \frac{\rho(f)-a}{\rho(f)}$. Then we conclude from (1.2.3) that $\underline{\log\operatorname{dens}}(E) \le \frac{a}{\rho(f)}$.

Suppose then that $\rho(f) = \infty$ and $\overline{\log\operatorname{dens}}(E^c) < 1$. By the above reasoning, choosing $\alpha > 1/(1 - \overline{\log\operatorname{dens}}(E^c))$, it is easy to see that $\rho(f) \le a\alpha < \infty$, a contradiction. Hence, if $\rho(f) = \infty$, then we have $\overline{\log\operatorname{dens}}(E^c) = 1$, and thus $\underline{\log\operatorname{dens}}(E) = 0$. This completes the proof of (2.5.2).

The proof of (2.5.3) is similar to that of (2.5.2). □

Next, we prove an analogue of Theorem 2.5.2 for the type and lower type.

Theorem 2.5.3 ([121, Corollary 4.6]). *Let f be a transcendental entire function of order $\rho \in (0,\infty)$ such that $0 \le \underline{\tau}(f) < \tau(f) \le \infty$, and let a and b be constants satisfying $\underline{\tau}(f) < a < b < \tau(f)$. Define*

$$\begin{aligned} G &= \{r > r_0 : \log M(r,f) < ar^{\rho}\},\\ H &= \{r > r_0 : \log M(r,f) > br^{\rho}\}, \end{aligned} \tag{2.5.5}$$

where $r_0 > 1$ is chosen so that $M(r_0,f) > 1$. Then

$$\overline{\operatorname{dens}}(G) \ge 1 - \left(\frac{\underline{\tau}(f)}{a}\right)^{\frac{1}{\rho}}, \quad \underline{\operatorname{dens}}(G) \le \left(\frac{a}{\tau(f)}\right)^{\frac{1}{\rho}}, \tag{2.5.6}$$

$$\overline{\operatorname{dens}}(H) \ge 1 - \left(\frac{b}{\tau(f)}\right)^{\frac{1}{\rho}}, \quad \underline{\operatorname{dens}}(H) \le \left(\frac{\underline{\tau}(f)}{b}\right)^{\frac{1}{\rho}}. \tag{2.5.7}$$

Proof. To prove the first inequality in (2.5.6), we suppose on the contrary that $\overline{\operatorname{dens}}(G) < 1 - \left(\frac{\underline{\tau}(f)}{a}\right)^{\frac{1}{\rho}}$. Choose $\varepsilon > 0$ small enough such that

$$\overline{\operatorname{dens}}(G) < 1 - \left(\frac{\underline{\tau}(f)}{a - 2\varepsilon}\right)^{\frac{1}{\rho}} \quad \text{and} \quad \frac{a - \varepsilon}{\underline{\tau}(f)} > 1.$$

Set $\alpha = \left(\frac{a-\varepsilon}{\underline{\tau}(f)}\right)^{\frac{1}{\rho}}$. Then $\alpha > 1/(1 - \overline{\operatorname{dens}}(G))$, and

$$ar^\rho \le \log M(r,f), \quad r \in (r_0,\infty) \setminus G.$$

From Lemma 1.2.7 with $\psi(t) = t$ it follows that $ar^\rho \le \log M(\alpha r,f)$ for all $r > R$. Hence $\underline{\tau}(f) \ge \frac{a}{\alpha^\rho} > \underline{\tau}(f)$, a contradiction.

To prove the second inequality in (2.5.6), it suffices to prove that

$$\overline{\operatorname{dens}}(G^c) \ge 1 - \left(\frac{a}{\tau(f)}\right)^{\frac{1}{\rho}}$$

and then apply (1.2.3). The reasoning should be divided into two cases, $\tau(f) < \infty$ and $\tau(f) = \infty$. The proof of (2.5.7) is similar to that of (2.5.6). □

Recall from Theorem 2.2.2 that if $f(z) = \sum_{n=0}^{\infty} a_n z^n$ is entire, then

$$\rho(f) = \limsup_{n\to\infty} \frac{n\log n}{-\log|a_n|},$$

where the expression $\frac{n\log n}{-\log|a_n|}$ is considered to be zero if $a_n = 0$. If, furthermore, $\rho(f) \in (0,\infty)$, then by Theorem 2.2.8,

$$\tau(f) = \frac{1}{e\rho(f)} \limsup_{n\to\infty} n|a_n|^{\rho(f)/n}.$$

Hence it is natural to think that the corresponding results might hold for the lower order and the lower type, where each lim sup is replaced with a lim inf. However, this is not always true.

Example 2.5.4 ([250]). Let

$$f(z) = e^{z^2} + e^z = 2 + \sum_{k=1}^{\infty} \frac{1}{k!} z^{2k} + \sum_{k=1}^{\infty} \frac{1}{k!} z^k.$$

Then $f(z) = 2 + \sum_{n=1}^{\infty} a_n z^n$, where

$$a_n = \begin{cases} \frac{1}{k!} + \frac{1}{(2k)!}, & n = 2k, \\ \frac{1}{(2k-1)!}, & n = 2k-1. \end{cases}$$

Since $\log M(r,f) = (1+o(1))r^2$, it is clear that $\underline{\rho}(f) = \rho(f) = 2$ and $\underline{\tau}(f) = \tau(f) = 1$. However, by Stirling's formula, $n! = (1+o(1))(\frac{n}{e})^n$ as $n \to \infty$, from which we obtain

$$\liminf_{n\to\infty} \frac{n\log n}{-\log|a_n|} = 1 \quad \text{and} \quad \frac{1}{2e} \liminf_{n\to\infty} n|a_n|^{2/n} = 0.$$

The next two theorems contain lower bounds for the lower order and lower type in terms of Taylor coefficients, as well as upper bounds for the usual order and type. These estimates will be useful when discussing entire functions of regular growth and of perfectly regular growth in Section 7.4.

Theorem 2.5.5 ([37, 153, 250, 251]). *Let $f(z) = \sum_{n=0}^{\infty} a_n z^n$ be an entire function of order ρ and lower order $\underline{\rho}$. Then*

$$\underline{\rho} \geq \liminf_{n\to\infty} \frac{n\log n}{-\log|a_n|}. \tag{2.5.8}$$

If, furthermore, $a_n \neq 0$ for all large n, then

$$\underline{\rho} \geq \liminf_{n\to\infty} \frac{n\log n}{-\log|a_n|} \geq \liminf_{n\to\infty} \frac{\log n}{\log|\frac{a_n}{a_{n+1}}|} \tag{2.5.9}$$

and

$$\rho \leq \limsup_{n\to\infty} \frac{\log n}{\log|\frac{a_n}{a_{n+1}}|}. \tag{2.5.10}$$

As in Theorem 2.2.2, the expression $\frac{n\log n}{-\log|a_n|}$ in (2.5.8) is considered to be equal to zero whenever $a_n = 0$.

Theorem 2.5.6 ([37, 153, 250, 251]). *Let $f(z) = \sum_{n=0}^{\infty} a_n z^n$ be an entire function of order $\rho \in (0,\infty)$, type τ, and lower type $\underline{\tau}$. Then*

$$\underline{\tau} \geq \frac{1}{e\rho} \liminf_{n\to\infty} n|a_n|^{\rho/n}. \tag{2.5.11}$$

If, furthermore, $a_n \neq 0$ for all large n, then

$$\underline{\tau} \geq \frac{1}{e\rho} \liminf_{n\to\infty} n|a_n|^{\rho/n} \geq \liminf_{n\to\infty} \frac{n}{\rho}\left|\frac{a_{n+1}}{a_n}\right|^{\rho} \tag{2.5.12}$$

and

$$\tau \leq \limsup_{n\to\infty} \frac{n}{\rho}\left|\frac{a_{n+1}}{a_n}\right|^{\rho}. \tag{2.5.13}$$

Example 2.5.4 shows that it is possible to have strict inequalities in (2.5.8) and (2.5.11). On the other hand, equalities hold in (2.5.8)–(2.5.13) for $f(z) = e^z$.

To prove Theorem 2.5.5, we need to prove (2.5.8), (2.5.9), and (2.5.10). These three proofs will be given separately.

Proof of (2.5.8). Following the reasoning in [250], set

$$C := \liminf_{n\to\infty} \frac{n\log n}{-\log|a_n|}.$$

Our task is to prove that $\underline{\rho} \geq C$. This holds trivially when $C = 0$. We separate the proof for $0 < C \leq \infty$ into two cases.

Case 1. Suppose that $0 < C < \infty$. Let $\varepsilon > 0$ be an arbitrarily small fixed constant satisfying $C - \varepsilon > 0$. There exists $N = N(\varepsilon) \in \mathbb{N}$ such that

$$\frac{n \log n}{-\log |a_n|} \geq C - \varepsilon > 0, \quad n \geq N. \tag{2.5.14}$$

Since $\frac{n \log n}{-\log |a_n|} = 0$ when $a_n = 0$, we observe from (2.5.14) that $0 < |a_n| < 1$ and $-\log |a_n| > 0$ for $n \geq N$. By (2.5.14) we have

$$|a_n| \geq n^{-\frac{n}{C-\varepsilon}}, \quad n \geq N.$$

Set $r_n = 2n^{\frac{1}{C-\varepsilon}}$. If $r_n \leq r \leq r_{n+1}$ and $n \geq N$, then by Cauchy's inequalities,

$$\begin{aligned} M(r,f) &\geq |a_n| r^n \geq |a_n| r_n^n \geq n^{-\frac{n}{C-\varepsilon}} \exp(n \log r_n) \\ &= n^{-\frac{n}{C-\varepsilon}} \exp(n \log 2 + n \log n^{\frac{1}{C-\varepsilon}}) = \exp(n \log 2). \end{aligned}$$

Using

$$n = \left(\frac{r_{n+1}}{2}\right)^{C-\varepsilon} - 1 \geq \left(\frac{r}{2}\right)^{C-\varepsilon} - 1,$$

we obtain

$$\log M(r,f) \geq n \log 2 \geq \left(\left(\frac{r}{2}\right)^{C-\varepsilon} - 1\right) \log 2, \quad n \geq N,$$

from which $\underline{\rho} \geq C - \varepsilon$. By letting $\varepsilon \to 0^+$ we obtain (2.5.8) for $0 < C < \infty$.

Case 2. Finally, suppose that $C = \infty$. Let $\varepsilon > 0$. There exists $K = K(\varepsilon) \in \mathbb{N}$ such that

$$\frac{n \log n}{-\log |a_n|} \geq \frac{1}{\varepsilon}, \quad n \geq K,$$

where we can again assume that $0 < |a_n| < 1$ for $n \geq K$. It follows that

$$|a_n| \geq n^{-\varepsilon n}, \quad n \geq K.$$

Set $r_n = 2n^{\varepsilon}$. If $r_n \leq r \leq r_{n+1}$ and $n \geq K$, then, similarly as in Case 1,

$$M(r,f) \geq |a_n| r^n \geq |a_n| r_n^n \geq n^{-\varepsilon n} \exp(n \log r_n) = \exp(n \log 2).$$

Consequently,

$$\log M(r,f) \geq n \log 2 \geq \left(\left(\frac{r}{2}\right)^{1/\varepsilon} - 1\right) \log 2, \quad n \geq K,$$

from which $\underline{\rho} \geq \frac{1}{\varepsilon}$. By letting $\varepsilon \to 0^+$ we see that (2.5.8) holds when $C = \infty$. This completes the proof of (2.5.8). □

Proof of (2.5.9). We assume that $a_n \neq 0$ for all large n, say $n \geq N_0$. Modifying the reasoning in [153], set

$$p := \liminf_{n\to\infty} \frac{\log n}{\log |a_n/a_{n+1}|}.$$

Since we showed that (2.5.8) holds, our task is to prove that

$$\liminf_{n\to\infty} \frac{n \log n}{-\log |a_n|} \geq p. \tag{2.5.15}$$

Obviously, (2.5.15) holds for $p = 0$. We separate the proof for $0 < p \leq \infty$ into two cases.

Case 1. Suppose that $0 < p < \infty$. Let $\varepsilon > 0$ be an arbitrarily small fixed number satisfying $p - \varepsilon > 0$. Then there exists $N = N(\varepsilon) \geq N_0$ such that

$$p - \varepsilon/2 \leq \frac{\log m}{\log |a_m/a_{m+1}|}, \qquad m \geq N.$$

Since $p - \varepsilon/2 > 0$, we have $|a_m/a_{m+1}| > 1$ for $m \geq N$, and

$$\left|\frac{a_m}{a_{m+1}}\right|^{p-\varepsilon/2} \leq m, \qquad m \geq N.$$

Using this inequality for $m = N, N+1, \ldots, n-1$, when n is large, and multiplying these $n - N$ inequalities, we obtain

$$\left|\frac{a_N}{a_n}\right|^{p-\varepsilon/2} \leq N(N+1)\cdots(n-1) = \frac{(n-1)!}{(N-1)!}.$$

Taking logarithms and noting that $(p - \varepsilon/2)\log |a_N|$ and $-\log(N-1)!$ are bounded terms as $n \to \infty$, we get

$$(p - \varepsilon/2)\log \frac{1}{|a_n|} \leq \log(n-1)! + O(1), \quad n \to \infty.$$

From Stirling's formula $n! \sim (\frac{n}{e})^n$,

$$\begin{aligned}\log(n-1)! &= \log n! - \log n \sim (n-1)\log n - n \\ &= (1 + o(1))n \log n, \quad n \to \infty.\end{aligned} \tag{2.5.16}$$

Hence

$$(p - \varepsilon/2)\log \frac{1}{|a_n|} \leq (1 + o(1))n \log n, \quad n \to \infty. \tag{2.5.17}$$

Since we know that $a_n \to 0$, it follows that $0 < |a_n| < 1$ and $-\log |a_n| > 0$ for n sufficiently large. Dividing (2.5.17) by $-\log |a_n|$ and replacing $\varepsilon/2$ with ε, we deduce that

$$\frac{n \log n}{-\log |a_n|} \geq p - \varepsilon$$

for all n large enough. Thus

$$\liminf_{n\to\infty} \frac{n \log n}{-\log |a_n|} \geq p - \varepsilon.$$

By letting $\varepsilon \to 0^+$ we see that (2.5.15) holds for $0 < p < \infty$.

Case 2. Finally, suppose that $p = \infty$. For any small fixed $\varepsilon > 0$, there exists $K = K(\varepsilon) \in \mathbb{N}$ such that

$$\frac{\log m}{\log |a_m/a_{m+1}|} \geq \frac{1}{\varepsilon}, \quad m \geq K,$$

which gives

$$\left|\frac{a_m}{a_{m+1}}\right| \leq m^{\varepsilon}, \quad m \geq K.$$

Using this inequality with $m = K, K+1, \ldots, n-1$, when n is large, and multiplying these $n - K$ inequalities, we obtain

$$\left|\frac{a_K}{a_n}\right| \leq \left(\frac{(n-1)!}{(K-1)!}\right)^{\varepsilon}.$$

Using (2.5.16), we get

$$-\log |a_n| \leq \varepsilon(1 + o(1)) n \log n, \quad n \to \infty,$$

which gives

$$\frac{n \log n}{-\log |a_n|} \geq \frac{1}{\varepsilon}(1 + o(1)), \quad n \to \infty.$$

By letting $\varepsilon \to 0^+$ we see that

$$\liminf_{n\to\infty} \frac{n \log n}{-\log |a_n|} = \infty.$$

This proves (2.5.15) for $p = \infty$, completing the proof of (2.5.9). □

Proof of (2.5.10). We assume that $a_n \neq 0$ for all large n, say $n \geq N_0$. Modifying the reasoning in [153], set

$$P := \limsup_{n\to\infty} \frac{\log n}{\log |a_n/a_{n+1}|}.$$

By Theorem 2.2.2, our task is to prove that

$$\limsup_{n\to\infty} \frac{n\log n}{-\log|a_n|} \le P. \tag{2.5.18}$$

The case $P = \infty$ being trivial, suppose that $0 \le P < \infty$. Let $\varepsilon > 0$. Then there exists $M = M(\varepsilon) \ge N_0$ such that

$$\frac{\log m}{\log|a_m/a_{m+1}|} \le P + \varepsilon/2, \quad m \ge M. \tag{2.5.19}$$

Since $a_m \to 0$, there exists $N \ge M$ such that

$$0 < |a_m| < 1 \quad \text{and} \quad |a_m/a_{m+1}| > 1, \quad m \ge N. \tag{2.5.20}$$

The latter statement holds because if it were not true, then there would exist a subsequence $\{b_k\}$ of $\{a_m\}$ satisfying $0 < |b_1| \le |b_2| \le |b_3| \le \cdots$, which would contradict $a_m \to 0$. From (2.5.19) and (2.5.20) we obtain

$$m \le \left|\frac{a_m}{a_{m+1}}\right|^{P+\varepsilon/2}, \quad m \ge N.$$

Using this inequality for $m = N, N+1, \ldots, n-1$, when n is large, and multiplying these $n - N$ inequalities, we obtain

$$N(N+1)\cdots(n-1) = \frac{(n-1)!}{(N-1)!} \le \left|\frac{a_N}{a_n}\right|^{P+\varepsilon/2}.$$

By taking logarithms we get

$$\log(n-1)! \le (P+\varepsilon/2)\log\frac{1}{|a_n|} + O(1), \quad n \to \infty.$$

From (2.5.16) we have

$$(1+o(1))n\log n \le (P+\varepsilon/2)\log\frac{1}{|a_n|}, \quad n \to \infty. \tag{2.5.21}$$

Dividing (2.5.21) by $-\log|a_n| > 0$ and replacing $\varepsilon/2$ with ε, we deduce that

$$\frac{n\log n}{-\log|a_n|} \le P + \varepsilon$$

for all n large enough. Thus

$$\limsup_{n\to\infty} \frac{n\log n}{-\log|a_n|} \le P + \varepsilon.$$

By letting $\varepsilon \to 0^+$ we see that (2.5.18) holds for $0 \le P < \infty$. This completes the proof of (2.5.10). □

The proof of Theorem 2.5.5 is now complete. To prove Theorem 2.5.6, we need to prove (2.5.11), (2.5.12), and (2.5.13).

Proof of (2.5.11). Following the reasoning in [251], set

$$D := \frac{1}{e\rho} \liminf_{n\to\infty} n|a_n|^{\rho/n}.$$

Our task is to prove that $\underline{\tau} \geq D$. Obviously, this holds when $D = 0$. We separate the proof for $0 < D \leq \infty$ into two cases.

Case 1. Suppose that $0 < D < \infty$. Let $\varepsilon > 0$ be a small fixed constant satisfying $D - \varepsilon > 0$. There exists $N = N(\varepsilon) \in \mathbb{N}$ such that

$$0 < |a_n| < 1 \quad \text{and} \quad n|a_n|^{\rho/n} \geq (D-\varepsilon)e\rho, \quad n \geq N.$$

Hence

$$|a_n| \geq \left(\frac{(D-\varepsilon)e\rho}{n}\right)^{n/\rho}, \quad n \geq N.$$

Using this together with Cauchy's inequalities, we get

$$\begin{aligned}\log M(r,f) &\geq \log(|a_n|r^n) = \log|a_n| + n\log r\\ &\geq n\log r + \frac{n}{\rho}\log((D-\varepsilon)e\rho) - \frac{n}{\rho}\log n, \quad n \geq N.\end{aligned}$$

Set $A = \frac{1}{\rho D}$ and $s_n = (nA)^{1/\rho}$. If $s_n \leq r \leq s_{n+1}$ and $n \geq N$, then

$$\log r \geq \log s_n = \frac{1}{\rho}\log(nA)$$

and

$$\begin{aligned}\log M(r,f) &\geq \frac{n}{\rho}(\log(nA) - \log n) + \frac{n}{\rho}\log((D-\varepsilon)e\rho)\\ &= \frac{n}{\rho}\log(A(D-\varepsilon)e\rho) = \frac{n}{\rho}\log\frac{(D-\varepsilon)e}{D}.\end{aligned}$$

Then

$$\frac{\log M(r,f)}{r^\rho} \geq \frac{\log M(r,f)}{s_{n+1}^\rho} \geq \frac{nD}{n+1}\log\frac{(D-\varepsilon)e}{D} \sim D\log\frac{(D-\varepsilon)e}{D},$$

from which $\underline{\tau} \geq D\log\frac{(D-\varepsilon)e}{D}$. Letting $\varepsilon \to 0^+$ gives $\underline{\tau} \geq D$, which proves (2.5.11) for $0 < D < \infty$.

Case 2. Suppose that $D = \infty$. Let $\varepsilon > 0$. There exists $K = K(\varepsilon) \in \mathbb{N}$ such that

$$0 < |a_n| < 1 \quad \text{and} \quad n|a_n|^{\rho/n} \geq \frac{e\rho}{\varepsilon}, \qquad n \geq K.$$

This can be rewritten as

$$|a_n| \geq \left(\frac{e\rho}{\varepsilon n}\right)^{n/\rho}, \qquad n \geq K.$$

Set $s_n = (\frac{\varepsilon n}{\rho})^{1/\rho}$. If $s_n \leq r \leq s_{n+1}$ and $n \geq K$, then, similarly as in Case 1,

$$\log M(r,f) \geq \log |a_n| + n \log r \geq \frac{n}{\rho} \log\left(\frac{e\rho}{\varepsilon n}\right) + n \log s_n = \frac{n}{\rho},$$

so that

$$\frac{\log M(r,f)}{r^\rho} \geq \frac{\log M(r,f)}{s_{n+1}^\rho} \geq \frac{n}{\varepsilon(n+1)} \sim \frac{1}{\varepsilon},$$

from which $\underline{\tau} \geq \frac{1}{\varepsilon}$. This shows that $\underline{\tau} = \infty$, which proves (2.5.11) for $D = \infty$. This completes the proof of (2.5.11). □

Proof of (2.5.12). We assume that $a_n \neq 0$ for all large n, say $n \geq N_0$. Following the reasoning in [153], set

$$q := \liminf_{n\to\infty} \frac{n}{\rho}\left|\frac{a_{n+1}}{a_n}\right|^\rho.$$

Since we showed (2.5.11), our task is to prove that

$$\frac{1}{e\rho} \liminf_{n\to\infty} n|a_n|^{\rho/n} \geq q. \tag{2.5.22}$$

This clearly holds for $q = 0$. We separate the proof for $0 < q \leq \infty$ into two cases.

Case 1. Suppose that $0 < q < \infty$. Let $\varepsilon > 0$ be a small fixed constant satisfying $q - \varepsilon > 0$. There exists $N = N(\varepsilon) \geq N_0$ such that

$$q - \varepsilon/2 \leq \frac{m}{\rho}\left|\frac{a_{m+1}}{a_m}\right|^\rho, \qquad m \geq N. \tag{2.5.23}$$

Considering $m = N, N+1, \ldots, n-1$ in (2.5.23) for large n and multiplying these $n - N$ inequalities, we obtain

$$(q - \varepsilon/2)^{n-N} \leq \frac{(n-1)!}{(N-1)!} \cdot \frac{1}{\rho^{n-N}} \cdot \left|\frac{a_n}{a_N}\right|^\rho,$$

that is,

$$\frac{B_N \rho^n}{(n-1)!}(q-\varepsilon/2)^{n-N} \le |a_n|^\rho, \quad n \ge N,$$

where $B_N = (N-1)!\rho^{-N}|a_N|^\rho$. Then using Stirling's formula and the familiar limit $\lim_{n\to\infty} n^{1/n} = 1$, and by replacing $\varepsilon/2$ with ε, we deduce that

$$e\rho(q-\varepsilon) \le n|a_n|^{\rho/n}$$

for all n large enough. This implies (2.5.22) for $0 < q < \infty$.

Case 2. Finally, suppose that $q = \infty$. For small fixed $\varepsilon > 0$, there exists $K = K(\varepsilon) \in \mathbb{N}$ such that

$$\frac{m}{\rho}\left|\frac{a_{m+1}}{a_m}\right|^\rho \ge \frac{1}{\varepsilon}, \quad m \ge K.$$

Using this inequality for $m = K, K+1, \dots, n-1$, when n is large, and then multiplying these $n-K$ inequalities, we get

$$\frac{(n-1)!}{(K-1)!} \cdot \frac{1}{\rho^{n-K}} \cdot \left|\frac{a_n}{a_K}\right|^\rho \ge \frac{1}{\varepsilon^{n-K}},$$

that is,

$$|a_n|^\rho \ge |a_K|^\rho \cdot \frac{\rho^{n-K}}{\varepsilon^{n-K}} \cdot \frac{(K-1)!}{(n-1)!}.$$

By Stirling's formula we see that

$$n|a_n|^{\rho/n} \ge e\rho(1+o(1))\frac{1}{\varepsilon}, \quad n \to \infty.$$

This implies (2.5.22) for $q = \infty$ and completes the proof of (2.5.12). □

Proof of (2.5.13). We assume that $a_n \ne 0$ for all large n, say $n \ge N_0$. Following the reasoning in [153], set

$$Q := \limsup_{n\to\infty} \frac{n}{\rho}\left|\frac{a_{n+1}}{a_n}\right|^\rho.$$

By Theorem 2.2.8 our task is to prove that

$$\frac{1}{e\rho} \limsup_{n\to\infty} n|a_n|^{\rho/n} \le Q. \tag{2.5.24}$$

The case $Q = \infty$ being trivial, suppose that $0 \le Q < \infty$. Let $\varepsilon > 0$. Then there exists $N = N(\varepsilon) \ge N_0$ such that

$$\frac{m}{\rho}\left|\frac{a_{m+1}}{a_m}\right|^{\rho} \le Q + \varepsilon/2, \quad m \ge N. \tag{2.5.25}$$

Considering $m = N, N+1, \ldots, n-1$ in (2.5.25) and multiplying these $n - N$ inequalities, we obtain

$$\frac{(n-1)!}{(N-1)!} \cdot \frac{1}{\rho^{n-N}} \cdot \left|\frac{a_n}{a_N}\right|^{\rho} \le (Q + \varepsilon/2)^{n-N},$$

that is,

$$|a_n|^{\rho} \le \frac{B_N \rho^n}{(n-1)!}(Q + \varepsilon/2)^{n-N}, \quad n \ge N,$$

where $B_N = (N-1)!\rho^{-N}|a_N|^{\rho}$. Then using Stirling's formula and the familiar limit $\lim_{n\to\infty} n^{1/n} = 1$, and by replacing $\varepsilon/2$ with ε, we deduce that

$$n|a_n|^{\rho/n} \le e\rho(Q + \varepsilon)$$

for all n large enough. This implies (2.5.24) for $0 \le Q < \infty$ and completes the proof of (2.5.13). This completes the proof of Theorem 2.5.6. □

Problems

Problem 2.5.7. Let f be a transcendental entire function such that $0 \le \underline{\rho}(f) < \rho(f) \le \infty$. Prove that the sets E and F in (2.5.1) are of lower linear density zero and of upper linear density one.

Problem 2.5.8. Prove the two inequalities in (2.5.3).

Problem 2.5.9. Prove the two inequalities in (2.5.7).

---

3 Zeros of entire functions

We begin this chapter by studying the exponent of convergence, which measures the frequency of zeros (or more generally, a-points) of entire functions, just as the order of growth measures the growth. As we might expect, the two quantities are related to one another.

If f is a polynomial of degree $n \geq 1$ and if $r > 0$ is large enough, then the equation $f(z) = 0$ has n zeros in the disc $|z| < r$ by the fundamental theorem of algebra (Theorem 3.9.1). Moreover, $M(r,f)$ is comparable to the function r^n; see Lemma 2.1.4. The connection between the frequency of zeros of f and the growth of $M(r,f)$ carries over to transcendental entire functions as well. This is a consequence of Jensen's theorem, which will be discussed in Section 3.2.

We then proceed to discuss the convergence of infinite products and to factorize a given entire function f into a product of a power of z, an exponential factor, and an infinite product P. It turns out that the zeros of f determine the factors in the product P. In the remaining part of this chapter, we take a closer look at the geometry of the zeros of complex polynomials in terms of convex hulls.

3.1 Exponent of convergence

We begin by stating and proving the *uniqueness theorem* for analytic functions on simply connected domains.

Theorem 3.1.1. *Let f and g be analytic in a simply connected domain G, let $z_0 \in G$, and let $\{z_n\}$ be a sequence of points in G such that $z_n \neq z_0$ and $z_n \to z_0$ as $n \to \infty$. If $f(z_n) = g(z_n)$ for every $n \geq 1$, then $f(z) \equiv g(z)$ in G.*

Proof. Define $h(z) = f(z) - g(z)$. We divide the proof into two parts.

(1) Since h is analytic at z_0, it has a series expansion

$$h(z) = a_0 + a_1(z - z_0) + a_2(z - z_0)^2 + \cdots,$$

which converges in some (possibly small) open disc $D_0 = D_0(z_0, r_0)$ centered at z_0. As an analytic function, h is continuous at z_0, and hence

$$h(z_0) = \lim_{n\to\infty} h(z_n) = 0.$$

Therefore $a_0 = 0$, and

$$h(z) = (z - z_0)(a_1 + a_2(z - z_0) + \cdots) = (z - z_0)w(z).$$

https://doi.org/10.1515/9783112249130-003

Since $h(z_n) = 0$ and $z_n \neq z_0$, it follows that $w(z_n) = 0$. As an analytic function, w is continuous at z_0. Hence $w(z_n) \to w(z_0)$, which means that $w(z_0) = a_1 = 0$. Continuing in this fashion, we conclude that $a_n = 0$ for all $n \in \mathbb{N} \cup \{0\}$. Hence $h(z) \equiv 0$ in D_0.

(2) Now let $\zeta_0 \in G$. We connect the two points z_0 and ζ_0 by a "chain of open discs" as follows. Since G is simply connected, there exists a curve Γ that connects z_0 to ζ_0, where Γ lies entirely in G.

In the case where $G = \mathbb{C}$, we can choose Γ to be the line segment between z_0 and ζ_0. In the case where $G \neq \mathbb{C}$, we proceed as follows. Let $d > 0$ denote the distance from the curve Γ to the boundary ∂G of G, that is,

$$d = \min_{z \in \Gamma,\, w \in \partial(G)} |z - w|.$$

Now for each $z \in \Gamma$, consider the collection of open discs with center at z and radius d. Each such disc lies entirely in G from the definition of d. Since the curve Γ is a compact set, by the Heine–Borel theorem it follows that there exists a finite number of these open discs, say $D_1, D_2, \ldots, D_m$, such that

$$\Gamma \subset \bigcup_{k=1}^{m} D_k.$$

Furthermore, we can assume that these discs have the following properties: (a) $D_k = D(c_k, d)$, where $c_k \in \Gamma$ with $c_1 = z_0$ and $c_m = \zeta_0$, (b) $D_k \cap D_{k+1} \neq \emptyset$, and (c) the center c_{k+1} of D_{k+1} also lies in the previous disc D_k.

Since we showed in part (1) that $h \equiv 0$ on D_0 and $D_1 \subset D_0$ since $d \leq r_0$, we have $h \equiv 0$ on D_1. Since $D_1 \cap D_2 \neq \emptyset$ and the center c_2 of D_2 also lies in D_1, we can use a similar argument as in part (1) to obtain that $h \equiv 0$ on D_2. By continuing in this manner we obtain that $h \equiv 0$ on D_k for all $k = 1, 2, \ldots, m$. Hence $h(\zeta_0) = 0$, or $f(\zeta_0) = g(\zeta_0)$. Since ζ_0 is an arbitrary point in G, we have $f \equiv g$ on G. □

A consequence of the uniqueness theorem is that the zeros of an entire function $f \not\equiv 0$ cannot accumulate to any finite point in $\mathbb{C}$ because $\mathbb{C}$ is simply connected. This property leads us to define the concept of exponent of convergence.

Definition 3.1.2. Let $\{r_n\}$ be a nondecreasing sequence of positive real numbers tending to infinity. The *exponent of convergence* of $\{r_n\}$ is defined as

$$\lambda = \inf\left\{ a > 0 : \sum_{n=1}^{\infty} r_n^{-a} < \infty \right\}.$$

If $\sum_{n=1}^{\infty} r_n^{-a} = \infty$ for every $a > 0$, then we set $\lambda = \infty$.

Definition 3.1.3. Let f be an entire function with an infinite number of zeros, where $\{z_n\}$ is the sequence of nonzero zeros of f, repeated according to their multiplicity and ordered by increasing moduli, i. e., $0 < |z_1| \leq |z_2| \leq \cdots$. The *exponent of convergence* of the sequence of zeros of f is defined as

$$\lambda(f) = \inf\left\{\alpha > 0 : \sum_{n=1}^{\infty} |z_n|^{-\alpha} < \infty\right\}.$$

If $\sum_{n=1}^{\infty} |z_n|^{-\alpha} = \infty$ for every $\alpha > 0$, then we set $\lambda(f) = \infty$. If f has only a finite number of zeros (or no zeros), then we set $\lambda(f) = 0$.

Let $n(r) = n(r, 1/f)$ denote the number of zeros z_n of f in $|z| \le r$, each zero counted according to its multiplicity. For the rest of this section, we assume that $f(0) \ne 0$. This is not an essential restriction since we may always replace $n(r)$ with $n(r) - n(0)$ if $f(0) = 0$. Moreover, finitely many zeros at the origin (or at any point for that matter) have no significant impact on the growth of the function $n(r)$ when f has an infinite number of zeros.

Lemma 3.1.4. *Let f be an entire function with an infinite sequence $\{z_n\}$ of nonzero zeros, and let $\alpha > 0$ be a constant. Then the series $\sum_{n=1}^{\infty} |z_n|^{-\alpha}$ converges if and only if the integral $\int_0^{\infty} \frac{n(t)}{t^{\alpha+1}}\, dt$ converges.*

Proof. Since we assume that $f(0) \ne 0$ and since the origin cannot be an accumulation point of the zeros of f by the uniqueness theorem, there exists a number $\delta > 0$ such that f has no zeros in the disc $|z| \le \delta$. Thus $n(\delta) = 0$, and so the integrand $\frac{n(t)}{t^{\alpha+1}}$ does not have a singularity at the origin.

By Riemann–Stieltjes integration,

$$\sum_{|z_n|<r} |z_n|^{-\alpha} = \int_0^r \frac{dn(t)}{t^\alpha}.$$

This identity can also be seen as follows: We find by the uniqueness theorem that $n(t)$ is a step function because the zeros of f are located on countably many circles centered at $z = 0$. Between these radii, $n(t)$ is a constant, and so $dn(t) = 0$ on these intervals. When passing over these radii, $dn(t)$ jumps by a positive integer, which is equal to the number of zeros of f on the particular circle.

Integrating by parts, we have

$$\begin{aligned}\sum_{|z_n|<r} |z_n|^{-\alpha} &= \left[\frac{n(t)}{t^\alpha}\right]_0^r + \alpha \int_0^r \frac{n(t)}{t^{\alpha+1}}\, dt \\ &= \frac{n(r)}{r^\alpha} + \alpha \int_0^r \frac{n(t)}{t^{\alpha+1}}\, dt. \end{aligned} \tag{3.1.1}$$

Suppose that the sum on the left of (3.1.1) converges as $r \to \infty$. Since the term $\frac{n(r)}{r^\alpha}$ on the right is nonnegative, it follows that the nonnegative integral on the right converges as $r \to \infty$.

Conversely, suppose that the integral $\int_0^r \frac{n(t)}{t^{\alpha+1}}\,dt$ on the right of (3.1.1) converges as $r \to \infty$. Since $n(r)$ is nondecreasing, we have

$$\frac{1}{\alpha r^\alpha}\left(1-\frac{1}{2^\alpha}\right)n(r) = n(r)\int_r^{2r}\frac{dt}{t^{\alpha+1}} \le \int_r^{2r}\frac{n(t)}{t^{\alpha+1}}\,dt \le \int_0^{\infty}\frac{n(t)}{t^{\alpha+1}}\,dt < \infty.$$

This gives $n(r) = O(r^\alpha)$, and hence the sum on the left of (3.1.1) converges as $r \to \infty$. □

Corollary 3.1.5. *Let f be an entire function with $f(0) \ne 0$. Then*

$$\lambda(f) = \inf\left\{\alpha > 0 : \int_0^\infty \frac{n(t)}{t^{\alpha+1}}\,dt < \infty\right\}.$$

Theorem 3.1.6. *Let f be an entire function that has at least one zero. Then*

$$\lambda(f) = \limsup_{r\to\infty}\frac{\log n(r)}{\log r}.$$

Proof. Since finitely many possible zeros at the origin have no effect on the exponent of convergence, we may suppose that $f(0) \ne 0$. For convenience, set

$$\sigma = \limsup_{r\to\infty}\frac{\log n(r)}{\log r}. \tag{3.1.2}$$

We first prove that $\lambda(f) \le \sigma$. We may suppose that $\sigma < \infty$. For any small $\varepsilon > 0$, there exists $r_\varepsilon > 0$ such that

$$n(r) \le r^{\sigma+\varepsilon}, \quad r \ge r_\varepsilon.$$

Then for $M > r_\varepsilon$,

$$\begin{aligned}\int_0^M \frac{n(t)}{t^{\alpha+1}}\,dt &= \int_0^{r_\varepsilon}\frac{n(t)}{t^{\alpha+1}}\,dt + \int_{r_\varepsilon}^M \frac{n(t)}{t^{\alpha+1}}\,dt \\ &\le \int_0^{r_\varepsilon}\frac{n(t)}{t^{\alpha+1}}\,dt + \int_{r_\varepsilon}^\infty \frac{dt}{t^{\alpha+1-\sigma-\varepsilon}} < \infty,\end{aligned}$$

provided that $\alpha + 1 - \sigma - \varepsilon > 1$, that is, if $\alpha > \sigma + \varepsilon$. We may then let $M \to \infty$ on the left-hand side of the inequality. In particular, $\int_0^\infty \frac{n(t)}{t^{\alpha+1}}\,dt$ converges for all $\alpha > 0$ such that $\alpha > \sigma + \varepsilon$. Therefore it follows that

$$\inf\left\{\alpha > 0 : \int_0^\infty \frac{n(t)}{t^{\alpha+1}}\,dt < \infty\right\} \le \sigma + \varepsilon.$$

By Corollary 3.1.5 we have $\lambda(f) \le \sigma + \varepsilon$. Letting $\varepsilon \to 0^+$ gives $\lambda(f) \le \sigma$.

To prove $\lambda(f) \ge \sigma$, we may suppose that $\sigma > 0$. There are two cases to consider. First suppose that $\sigma < \infty$. Let $\varepsilon \in (0, \sigma)$. By the definition of σ in (3.1.2) there exists a strictly increasing sequence $\{r_n\}$ of positive real numbers tending to ∞ such that

$$\frac{\log n(r_n)}{\log r_n} \ge \sigma - \varepsilon, \quad n \in \mathbb{N}.$$

In other words, $n(r_n) \ge r_n^{\sigma-\varepsilon}$ for $n \in \mathbb{N}$. Choose any $\alpha \in (0, \sigma - \varepsilon)$. For each $n \in \mathbb{N}$, choose

$$s_n \ge 2^{1/\alpha} r_n.$$

Since $n(r)$ is nondecreasing, we get

$$\int_{r_n}^{s_n} \frac{n(t)}{t^{\alpha+1}}\, dt \ge n(r_n) \int_{r_n}^{s_n} \frac{dt}{t^{\alpha+1}} \ge \frac{r_n^{\sigma-\varepsilon}}{\alpha}\left(\frac{1}{r_n^\alpha} - \frac{1}{s_n^\alpha}\right) \ge \frac{r_n^{\sigma-\alpha-\varepsilon}}{2\alpha}.$$

Since $\sigma - \alpha - \varepsilon > 0$, we obtain that

$$\lim_{n\to\infty} \int_{r_n}^{s_n} \frac{n(t)}{t^{\alpha+1}}\, dt = \infty.$$

Therefore $\int_0^\infty \frac{n(t)}{t^{\alpha+1}}\, dt$ diverges for every $\alpha \in (0, \sigma - \varepsilon)$. Consequently,

$$\inf\left\{\alpha > 0 : \int_0^\infty \frac{n(t)}{t^{\alpha+1}}\, dt < \infty\right\} \ge \sigma - \varepsilon.$$

By Corollary 3.1.5 we have $\lambda(f) \ge \sigma - \varepsilon$. Letting $\varepsilon \to 0^+$ gives $\lambda(f) \ge \sigma$. Hence $\lambda(f) = \sigma$.

Thus it remains to consider the case $\sigma = \infty$. Let $M > 0$ be arbitrarily large. Then there exists a strictly increasing sequence $\{u_n\}$ of positive real numbers tending to ∞ such that

$$\frac{\log n(u_n)}{\log u_n} \ge M, \quad n \in \mathbb{N},$$

that is, $n(u_n) \ge u_n^M$ for $n \in \mathbb{N}$. Analogously as above, it follows that

$$\lambda(f) = \inf\left\{\alpha > 0 : \int_0^\infty \frac{n(t)}{t^{\alpha+1}}\, dt < \infty\right\} \ge M.$$

Since $M > 0$ is arbitrarily large, we obtain $\lambda(f) = \infty = \sigma$. □

Problems

Problem 3.1.7. If $f(z) = (e^z - 1)e^{z^3}$, then show that $\rho(f) = 3$ and $\lambda(f) = 1$.

Problem 3.1.8. Give an example of an entire function $f(z)$ that satisfies $\rho(f) = \infty$ and $\lambda(f) = 2$. Justify your answer.

Problem 3.1.9. Let h be an entire function with triple zeros at $4, 8, 12, \ldots$ such that h has no other zeros. Show that there exists a constant $c > 0$ such that $n(r, 1/h) = (1 + o(1))cr$ as $r \to \infty$. Find c.

Problem 3.1.10. Let g be the function defined by

$$g(z) = \sum_{n=1}^{\infty} \left(\frac{z}{z-2} \right)^n .$$

(a) Show that g is analytic in the half-plane $\operatorname{Re} z < 1$.
(b) Explain why g can be analytically continued from $\operatorname{Re} z < 1$ to an entire function f.
(c) Prove that the entire function f in (b) is unique.

Problem 3.1.11. Suppose that f is an entire function with simple zeros at the points $z = k + mi$, $k, m \in \mathbb{N}$, such that f has no other zeros. Find $\lambda(f)$.

Problem 3.1.12. If f is an entire function such that $f(z) \not\equiv 0$, then show that f has at most a countable number of zeros.

Problem 3.1.13. Let f be an entire function that satisfies $f(x+1) = f(x)$ for all $x \in \mathbb{R}$. Prove that $f(z+1) = f(z)$ for all $z \in \mathbb{C}$. Find an entire function f that satisfies this equation.

For the next two problems, we recall the following result, which is a corollary of the *reflection principle*: If f is an entire function that is real on the real axis, then

$$\overline{f(z)} = f(\bar{z}), \quad z \in \mathbb{C}.$$

Further, we recall that an entire function f is called an *even function* if $f(-z) = f(z)$ for all $z \in \mathbb{C}$ and an *odd function* if $f(z) = -f(-z)$ for all $z \in \mathbb{C}$.

Problem 3.1.14. If f is an entire function that is real on the real axis and real on the imaginary axis, then show that f is an even function. State a transcendental entire function that satisfies this statement.

Problem 3.1.15. If f is an entire function that is real on the real axis and purely imaginary on the imaginary axis, then show that f is an odd function. State a transcendental entire function that satisfies this statement.

Problem 3.1.16. If f is an entire function that maps the real axis into the imaginary axis, then show that $f(\bar{z}) = -\overline{f(z)}$ for all $z \in \mathbb{C}$.

3.2 Jensen's theorem

The number $n(r) = n(r, 1/f)$ of zeros of f that lie in $|z| \le r$ is called the *nonintegrated counting function*, where the zeros are counted by multiplicities. If $f(0) \neq 0$, then the corresponding *integrated counting function* is defined as

$$N(r) = N(r, 1/f) = \int_0^r \frac{n(t)}{t}\,dt.$$

By the uniqueness theorem, $n(\delta) = 0$ for some $\delta > 0$, and hence, in fact,

$$N(r) = \int_\delta^r \frac{n(t)}{t}\,dt.$$

In particular, the origin is not a singular point of this integral. On the other hand, if $f(0) = 0$, then the origin would be a singular point of the integral, and in this case the integrated counting function is defined by

$$N(r) = N(r, 1/f) = \int_0^r \frac{n(t) - n(0)}{t}\,dt + n(0)\log r. \tag{3.2.1}$$

If $f(0) \neq 0$, then this definition agrees with the previous definition, because $n(0) = 0$. Similarly, we define $n(r, a, f)$ as the number of a-points of f that lie in $|z| \leq r$ and define the corresponding integrated counting function $N(r, a, f)$ accordingly. The corresponding notation $n(r, 1/(f - a))$ and $N(r, 1/(f - a))$ can also be used. Further, we denote the exponent of convergence of a-points of f by $\lambda(a, f)$. For the zeros of f, it is customary to abbreviate $\lambda(0, f) = \lambda(f)$.

Lemma 3.2.1. *Let f be an entire function. Then the following statements hold:*
(a) *$N(r)$ is continuous for $r \geq 0$ when $f(0) \neq 0$ and for $r > 0$ when $f(0) = 0$.*
(b) *$N(r)$ is differentiable for $r > 0$ except for the sequence $\{r_k\}$ of positive real numbers that are the jump discontinuities of $n(r)$. At each r_k, the one-sided derivatives of $N(r)$ exist, but they are not equal.*

Proof. If f has no zeros, then $N(r) \equiv 0$, and the statements clearly hold. Suppose then that f has at least one zero.

First, we assume that $f(0) \neq 0$ and that f has an infinite number of zeros. By the uniqueness theorem, the zeros of f lie on circles $|z| = r_k$, where $\{r_k\}$ is a strictly increasing sequence of positive real numbers tending to infinity such that $n(r) = m_k$ for $r_k \leq r < r_{k+1}$, where each m_k is a positive integer such that

$$0 < m_1 < m_2 < m_3 < \cdots < m_k < m_{k+1} < \cdots.$$

Since $f(0) \neq 0$, we have $n(r) = 0$ for $0 \leq r < r_1$. Hence for $r_k \leq r < r_{k+1}$, we have

$$N(r) = \int_{r_1}^{r_2} \frac{n(t)}{t}\,dt + \int_{r_2}^{r_3} \frac{n(t)}{t}\,dt + \cdots + \int_{r_k}^{r} \frac{n(t)}{t}\,dt.$$

This gives

$$N(r) = m_k \log r - m_1 \log r_1 + \sum_{j=2}^{k} (m_{j-1} - m_j) \log r_j, \quad r_k \le r < r_{k+1}, \tag{3.2.2}$$

where the summation does not exist when $k = 1$. Then for $k \ge 2$, we have

$$N(r) = m_{k-1} \log r - m_1 \log r_1 + \sum_{j=2}^{k-1} (m_{j-1} - m_j) \log r_j, \quad r_{k-1} \le r < r_k, \tag{3.2.3}$$

where the summation does not exist when $k = 2$.

It follows from $N(r) \equiv 0$ for $0 \le r < r_1$ and from (3.2.3) that $N(r)$ is differentiable for all $r > 0$ satisfying $r \ne r_k$ ($k \in \mathbb{N}$), and thus $N(r)$ is continuous at these points and also at $r = 0$. It remains to consider the points r_k.

First consider r_1. Since $N(r) \equiv 0$ for $0 \le r < r_1$ and since (3.2.2) holds for $k = 1$, we see that $N(r) \to N(r_1) = 0$ as $r \to r_1$. Hence $N(r)$ is continuous at $r = r_1$. On the other hand, from (3.2.2) and from $N(r) \equiv 0$ for $0 \le r < r_1$ we obtain

$$\lim_{r \to r_1^+} \frac{N(r) - N(r_1)}{r - r_1} = \frac{m_1}{r_1} \ne 0 = \lim_{r \to r_1^-} \frac{N(r) - N(r_1)}{r - r_1}.$$

This proves parts (a) and (b) for the point r_1.

Next consider r_k when $k \ge 2$. From (3.2.2) and (3.2.3) we obtain

$$\begin{aligned} N(r_k) &= m_k \log r_k - m_1 \log r_1 + \sum_{j=2}^{k} (m_{j-1} - m_j) \log r_j \\ &= m_{k-1} \log r_k - m_1 \log r_1 + \sum_{j=2}^{k-1} (m_{j-1} - m_j) \log r_j \end{aligned}$$

and

$$\lim_{r \to r_k^-} N(r) = N(r_k) = \lim_{r \to r_k^+} N(r).$$

This proves that $N(r)$ is continuous at any r_k, which shows that $N(r)$ is continuous for all $r \ge 0$. Regarding differentiability, we see that

$$\lim_{r \to r_k^-} \frac{N(r) - N(r_k)}{r - r_k} = \frac{m_{k-1}}{r_k} \ne \frac{m_k}{r_k} = \lim_{r \to r_k^+} \frac{N(r) - N(r_k)}{r - r_k}.$$

This proves part (b) when $f(0) \ne 0$ and f has infinitely many zeros.

Finally, when either f has finitely many zeros or $f(0) = 0$, then (a) and (b) can be proved with simple adjustments in the above reasoning. In the case where $f(0) = 0$, we use the definition of $N(r)$ in (3.2.1) and $N(r) = n(0) \log r$ for $0 < r < r_1$. □

The following important result relates the frequency of the zeros and the growth of an entire function and is known as *Jensen's theorem.*

Theorem 3.2.2. *Let f be an entire function such that $f(0) \neq 0$. Then*

$$N(r) = \frac{1}{2\pi} \int_0^{2\pi} \log|f(re^{i\theta})|\, d\theta - \log|f(0)|, \quad r \geq 0. \tag{3.2.4}$$

Proof. We prove (3.2.4) in five steps.

(1) Since the statement is clearly true when $r = 0$, we assume that $r > 0$ in the rest of the proof.

(2) Suppose that f has no zeros in $|z| \leq r$. By the uniqueness theorem, there exists a constant $R > r$ such that f has no zeros in $|z| < R$. Thus $\log f(z)$ is analytic in $|z| < R$. The branch of the logarithm is irrelevant. Hence by Cauchy's formula,

$$\log f(0) = \frac{1}{2\pi i} \int_{|z|=r} \frac{\log f(z)}{z}\, dz = \frac{1}{2\pi} \int_0^{2\pi} \log f(re^{i\theta})\, d\theta. \tag{3.2.5}$$

Since $N(r) = 0$, (3.2.4) follows by taking the real parts of this formula.

(3) Suppose now that f has at least one zero in $|z| < r$ but not on $|z| = r$. Let $a_1, \dots, a_n$ be the zeros of f in $|z| < r$, listed with respect to multiplicities and organized according to increasing modulus. Define

$$g(z) = f(z) \prod_{k=1}^{n} \frac{r^2 - \bar{a}_k z}{r(z - a_k)}.$$

If $|z| = r$, then

$$\left|\frac{r^2 - \bar{a}_k z}{r(z - a_k)}\right| = \left|\frac{r^2 - \bar{a}_k re^{i\varphi}}{r(re^{i\varphi} - a_k)}\right| = \left|\frac{r - \bar{a}_k e^{i\varphi}}{r - a_k e^{-i\varphi}}\right| = \left|\frac{\overline{r - a_k e^{-i\varphi}}}{r - a_k e^{-i\varphi}}\right| = 1,$$

and so $|g(z)| = |f(z)| \neq 0$. Clearly, g has no zeros in $|z| < r$ either, and by the uniqueness theorem, g has no zeros in $|z| < R$ for some $R > r$. It follows that $\log|g(z)|$ is harmonic in $|z| < R$ (exercise). By the mean value property of harmonic functions,

$$\log|g(0)| = \frac{1}{2\pi} \int_0^{2\pi} \log|g(re^{i\varphi})|\, d\varphi.$$

Since

$$|g(0)| = |f(0)| \prod_{k=1}^{n} \frac{r}{|a_k|},$$

we get

$$\frac{1}{2\pi}\int_0^{2\pi}\log|f(re^{i\varphi})|\,d\varphi = \frac{1}{2\pi}\int_0^{2\pi}\log|g(re^{i\varphi})|\,d\varphi = \log|g(0)|$$
$$= \log|f(0)| + \sum_{k=1}^{n}\log\frac{r}{|a_k|}.$$

Comparing this to (3.2.4), we find that

$$\int_0^r \frac{n(t)}{t}\,dt = \sum_{k=1}^{n}\log\frac{r}{|a_k|}$$

remains to be proved. But this follows from

$$\sum_{k=1}^{n}\log\frac{r}{|a_k|} = \sum_{k=1}^{n}\log\frac{r}{r_k} = \log\left(\prod_{k=1}^{n}\frac{r}{r_k}\right) = \log\frac{r^n}{r_1\cdots r_n}$$
$$= n\log r - \sum_{k=1}^{n}\log r_k$$
$$= \sum_{k=1}^{n-1}k(\log r_{k+1} - \log r_k) + n(\log r - \log r_n)$$
$$= \sum_{k=1}^{n-1}k\int_{r_k}^{r_{k+1}}\frac{dt}{t} + n\int_{r_n}^{r}\frac{dt}{t} = \int_0^r\frac{n(t)}{t}\,dt,$$

where we have used the notation $|a_k| = r_k$.

(4) Suppose that f has no zeros in $|z| < r$ but has at least one zero on $|z| = r$. We will integrate $(\log f(z))/z$ along $|z| = r$ with the exception of making small indentations around the zeros of f, and then letting the indentations get smaller and smaller, we will show that the value of the integral along the indentations goes to zero. This will mean that the path of integration can be replaced by the whole circle $|z| = r$, and we can then use (3.2.5). The details are as follows.

By the uniqueness theorem, there exists a constant $R > r$ such that the only zeros of f in $|z| < R$ lie on $|z| = r$. Let $b_1, b_2, \ldots, b_m$, $m \geq 1$, denote the zeros of f that lie on $|z| = r$. Let $\delta > 0$ be a small constant satisfying

$$0 < r - \delta < r < r + \delta < R$$

and

$$\{z : |z - b_k| = \delta\} \cap \{z : |z - b_j| = \delta\} = \emptyset, \quad k \neq j.$$

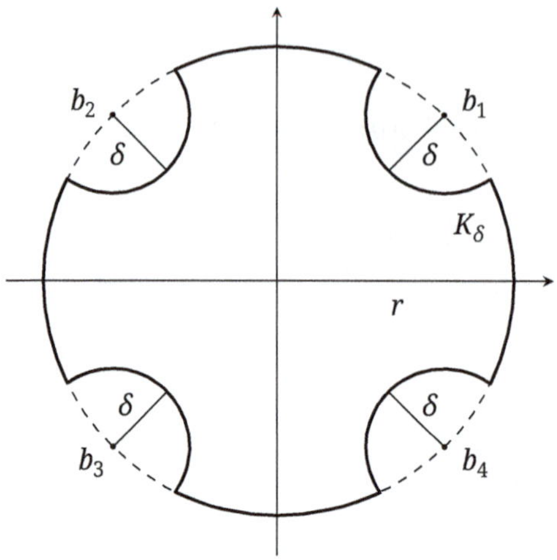

Figure 3.1: Contour K_δ that avoids the zeros b_1, b_2, b_3, b_4 of f on $|z| = r$.

Define the simple closed contour K_δ consisting of the arcs on the circle $|z| = r$ between the circles $|z - b_k| = \delta$, together with the "lower arcs" on the circles $|z - b_k| = \delta$ whose points z satisfy $|z| \leq r$. Then f has no zeros on K_δ and no zeros in the interior of K_δ; see Figure 3.1. Since f is analytic in $|z| < R$ and K_δ lies in $|z| < R$, it follows that $\log f(z)$ is analytic on and inside K_δ. Then by the Cauchy integral formula,

$$\log f(0) = \frac{1}{2\pi i} \int_{K_\delta} \frac{\log f(z)}{z}\, dz. \tag{3.2.6}$$

We now show that the curve K_δ in (3.2.6) can be replaced by the circle $|z| = r$ by letting $\delta \to 0^+$. To this end, let D_δ represent the lower arc of one of the indentations, that is, for some k, let D_δ consist of the points z satisfying $|z - b_k| = \delta$ and $|z| \leq r$. Since f has a zero of order $j \geq 1$ at b_k, we have $f(z) = (C + o(1))(z - b_k)^j$ as $z \to b_k$, where $C \neq 0$ is a constant. Then

$$\left| \int_{D_\delta} \frac{\log f(z)}{z}\, dz \right| \leq \int_{D_\delta} \frac{|\log(|C + o(1)||z - b_k|^j)| + M}{|z|}\, |dz|,$$

where $M > 0$ is a constant that depends on which branch of the logarithm was chosen. In the last integral, we have $|z - b_k| = \delta$, $|z| \geq r - \delta > 0$, and the length of the arc D_δ is less than $2\pi\delta$. Hence we obtain

$$\left| \int_{D_\delta} \frac{\log f(z)}{z}\, dz \right| \leq \frac{-B\delta \log \delta}{r - \delta},$$

where $B > 0$ is a constant. Since $\delta \log \delta = \log \delta / \delta^{-1} \to 0$ as $\delta \to 0^+$ by L'Hospital's rule, this shows that

$$\int_{D_\delta} \frac{\log f(z)}{z}\,dz \to 0, \quad \delta \to 0^+.$$

Therefore the contributions to the integral in (3.2.6) from the indentations D_δ tend to zero as δ tends to zero. Thus K_δ can be replaced by $|z| = r$ in (3.2.6), i. e.,

$$\log f(0) = \frac{1}{2\pi i} \int_{|z|=r} \frac{\log f(z)}{z}\,dz.$$

Taking the real parts gives

$$0 = \frac{1}{2\pi} \int_0^{2\pi} \log|f(re^{i\theta})|\,d\theta - \log|f(0)|, \quad r \geq 0. \tag{3.2.7}$$

Since $n(t) = 0$ for $0 \leq t < r$, we have $N(t) = 0$ for $0 \leq t < r$, and by continuity, $N(r) = 0$. Then (3.2.7) gives (3.2.4).

(5) The last case to consider is when f has at least one zero in $|z| < r$ and at least one zero on $|z| = r$. Let $a_1, \ldots, a_n$ be the zeros of f in $|z| < r$, listed with respect to multiplicities and organized according to increasing modulus. Define the function g by

$$g(z) = f(z) \prod_{k=1}^{n} \frac{r^2 - \bar{a}_k z}{r(z - a_k)}.$$

As in case (3), g has no zeros in $|z| < r$, $|g(z)| = |f(z)|$ on $|z| = r$, and g has the same zeros as f on $|z| = r$. Since this is case (4) for g and we showed that (3.2.4) holds in case (4), we apply (3.2.4) to g to obtain

$$\begin{aligned}\frac{1}{2\pi} \int_0^{2\pi} \log|f(re^{i\varphi})|\,d\varphi &= \frac{1}{2\pi} \int_0^{2\pi} \log|g(re^{i\varphi})|\,d\varphi = \log|g(0)| \\ &= \log|f(0)| + \sum_{k=1}^{n} \log \frac{r}{|a_k|}.\end{aligned}$$

Since we showed in case (3) that

$$\sum_{k=1}^{n} \log \frac{r}{|a_k|} = \int_0^r \frac{n(t)}{t}\,dt,$$

this proves (3.2.4). This completes the proof of Jensen's theorem. □

Jensen's theorem (also known as Jensen's formula) can be generalized to meromorphic functions and to the case where $\log|f(0)|$ is replaced with $\log|f(z)|$. Such a generalization is known as the Poisson–Jensen formula [108], and it is the foundation of

Nevanlinna theory; see Section 4.6. In addition to being a mathematician, Johan Jensen (1859–1925) was a telephone engineer [140, p. 114]. Although he was the President of the Danish Mathematical Society, Jensen never held a typical academic position and did mathematics in his spare time.

Regarding the proof of Jensen's formula, the details of cases (4) and (5) are often omitted in standard textbooks. Since the result is very important for the theory of entire functions, we will now give an alternative proof for cases (4) and (5), based on a continuity argument established in [273, pp. 126–127].

Second proof of Jensen's theorem

We assume that cases (1), (2), and (3) have been proved. In the first proof of Jensen's theorem, cases (4) and (5) were proved separately. Here we will take a different point of view and prove cases (4) and (5) together as one case.

Let s_1 and s_2 be constants satisfying $0 < s_1 < s_2$ such that f has zeros on both circles $|z| = s_1$ and $|z| = s_2$ but no zeros in the annulus $s_1 < |z| < s_2$. From cases (1)–(3) we know that (3.2.4) is true for all $r \in (s_1, s_2)$. Statement (3.2.4) is also true for $r = 0$ and for all r such that f has no zeros in $|z| \le r$. Further, in the special case where f has all its zeros on one circle $|z| = s$, the statement is true for $r < s$ and $r > s$. Hence it suffices to prove (3.2.4) for $r = s$, where $s > 0$ is such that f has at least one zero on $|z| = s$.

Suppose that f has at least one zero on $|z| = s$ and possibly one or more zeros in $|z| < s$. The function $N(r)$ is continuous for $r > 0$ by Lemma 3.2.1. Thus we proceed to prove the continuity of the integral

$$\int_0^{2\pi} \log|f(re^{i\theta})|\, d\theta \tag{3.2.8}$$

in the statement when r lies in a small interval I that contains s. Indeed, by the uniqueness theorem there exist constants ϱ, R satisfying $0 < \varrho < s < R$ such that f has no zeros in the annulus $\varrho < |z| < R$ other than those on the circle $|z| = s$. Therefore we may choose $I = (c_1, c_2)$, where c_1, c_2 are constants such that $\varrho < c_1 < s < c_2 < R$. Without loss of generality, we may further suppose that $I = (c_1, c_2)$ is contained in $(s/2, 2s)$.

Let z_0 be a zero of f of multiplicity k on $|z| = s$. Without loss of generality, we may assume that $z_0 = s$. For the time being, we suppose that f has no other zeros on $|z| = s$. Then the function $g(z) = f(z)/(1 - z/s)^k$ is entire and has no zeros on $|z| = s$ but can have zeros in $|z| \le \varrho < s$. We may now write

$$\int_0^{2\pi} \log|f(re^{i\theta})|\, d\theta = k \int_0^{2\pi} \log\left|1 - \frac{r}{s} e^{i\theta}\right| d\theta + \int_0^{2\pi} \log|g(re^{i\theta})|\, d\theta,$$

where the integral involving g is continuous for $r \in I$. Thus we turn our attention to the continuity of the integral in the middle.

Clearly, $r \in I$ implies $r/s \le 2$ and

$$\sin^2\theta \le \sin^2\theta + \left(\cos\theta - \frac{r}{s}\right)^2 = 1 - 2\frac{r}{s}\cos\theta + \frac{r^2}{s^2} = \left|1 - \frac{r}{s}e^{i\theta}\right|^2 \le 9.$$

Hence, for a constant $\delta \in (0, \pi/4)$, we have

$$\begin{aligned}\left|\int_{-\delta}^{\delta} \log\left|1 - \frac{r}{s}e^{i\theta}\right| d\theta\right| &\le \int_{-\delta}^{\delta} (\log 3 + |\log|\sin\theta||)\, d\theta \\ &= 2\int_0^{\delta}\left(\log 3 + \log\frac{1}{\sin\theta}\right) d\theta \le 2\int_0^{\delta}\left(\log 3 + \log\frac{2}{\theta}\right) d\theta \\ &= 2[\theta\log 6 + \theta - \theta\log\theta]_0^{\delta} \le 4\delta\log\frac{6}{\delta} \le 36\delta\log\frac{1}{\delta}.\end{aligned}$$

Here the lower bound of substitution should be considered as a limit as $\theta \to 0^+$, in which case $\theta\log\theta \to 0$ by L'Hospital's rule. The last inequality on the right follows from the elementary estimates $8\log\frac{1}{\delta} \ge 8\log\frac{4}{\pi} \ge \log 6$. Again by L'Hospital's rule, we have $36\delta\log\frac{1}{\delta} \to 0$ as $\delta \to 0^+$ for the upper bound. Let $\varepsilon > 0$. Writing

$$\int_0^{2\pi} \log\left|1 - \frac{r}{s}e^{i\theta}\right| d\theta = \left(\int_{|\theta|<\delta} + \int_{\delta\le|\theta|\le\pi}\right) \log\left|1 - \frac{r}{s}e^{i\theta}\right| d\theta =: J_1 + J_2$$

and choosing $\delta = \delta(\varepsilon) > 0$ small enough, we see that $|J_1| < \varepsilon$ for $r \in I$ and that J_2 is continuous for $r \in I$. Thus the whole integral is continuous for $r \in I$, which in turn implies the continuity of the integral in (3.2.8) for all $r \in I$.

We have been under a restriction that f has only one zero on $|z| = s$. Now suppose that f has zeros $s_1, \dots, s_m$ of corresponding multiplicities $k_1, \dots, k_m$ on $|z| = s$ and possibly one or more zeros in $|z| < s$. Then the entire function

$$g(z) = \frac{f(z)}{(1 - z/s_1)^{k_1}(1 - z/s_2)^{k_2}\cdots(1 - z/s_m)^{k_m}}$$

has no zeros on $|z| = s$ but can have zeros in $|z| < s$. The process discussed above can be repeated for each of the zeros $s_1, \dots, s_m$, proving the continuity of the integral in (3.2.8) for all values of r near s. This proves the statement. □

Jensen's theorem ties the frequency of the zeros of an entire function f to the growth of its maximum modulus $M(r, f)$. We now state and prove two results in this category.

Theorem 3.2.3. *Let f be entire and of finite order ρ. Then for each $\varepsilon > 0$, we have $n(r) = O(r^{\rho+\varepsilon})$ as $r \to \infty$.*

Proof. Assuming that $f(0) \ne 0$, we may further assume that $|f(0)| \ge 1$ by multiplying f by a constant, if needed. By Jensen's theorem,

$$N(r) \leq \frac{1}{2\pi}\int_0^{2\pi} \log|f(re^{i\theta})|\,d\theta \leq \log M(r,f). \tag{3.2.9}$$

By the definition of order, $\log M(r,f) \leq r^{\rho+\varepsilon}$ for all sufficiently large r. Since $n(r)$ is nondecreasing,

$$\begin{aligned} n(r)\log 2 = n(r)\int_r^{2r} \frac{dt}{t} &\leq \int_r^{2r} \frac{n(t)}{t}\,dt \leq \int_0^{2r} \frac{n(t)}{t}\,dt \\ &= N(2r) \leq \log M(2r,f) \leq 2^{\rho+\varepsilon} r^{\rho+\varepsilon} \end{aligned}$$

for all sufficiently large r. The statement follows. □

Theorem 3.2.4. *For every entire function f, we have $\lambda(a,f) \leq \rho(f)$ for all $a \in \mathbb{C}$.*

Proof. The statement is clear if $\rho(f) = \infty$, so we suppose that $\rho = \rho(f) < \infty$. First consider the case $a = 0$. By Theorem 3.2.3, given $\varepsilon > 0$, there exist constants $C > 0$ and $R > 0$ such that

$$n(r) \leq Cr^{\rho+\varepsilon}, \quad r \geq R.$$

By Theorem 3.1.6, $\lambda(f) \leq \rho(f) + \varepsilon$, where we may let $\varepsilon \to 0^+$. If $a \neq 0$, then the reasoning above can be applied to the function $g(z) = f(z) - a$. □

Example 3.2.5. For $f_1(z) = \sin z$ and $f_2(z) = e^{z^2}\sin z$, we have $\lambda(f_1) = 1 = \rho(f_1)$ and $\lambda(f_2) = 1 < 2 = \rho(f_2)$. Indeed, $\sin z = 0$ if and only if $z = n\pi$, $n \in \mathbb{Z}$, where all these zeros are simple. Hence

$$\lambda(f_1) = \lambda(f_2) = \inf\left\{\alpha > 0 : 2\sum_{n=1}^{\infty} |n\pi|^{-\alpha} < \infty\right\} = 1.$$

Moreover, the function $f_3(z) = \exp(e^z)$ is of infinite order and has no zeros. This shows that both the equality $\lambda(f) = \rho(f)$ and the strict inequality $\lambda(f) < \rho(f)$ may hold in Theorem 3.2.4.

An entire function f is called *periodic* if there exists a constant $c \in \mathbb{C} \setminus \{0\}$ such that

$$f(z) = f(z + c), \quad z \in \mathbb{C}.$$

If c is a period of f, then all integer multiples of c are periods of f. For a nonconstant periodic function f, a constant c for which $|c|$ is the minimum amongst all the periods is called a *fundamental period* of f. For example, 2π and -2π are fundamental periods for $\sin z$, and $2\pi i$ and $-2\pi i$ are fundamental periods for e^z.

Theorem 3.2.6. *A nonconstant periodic entire function f satisfies $\rho(f) \geq 1$.*

Proof. Denote $f(0) = a$, and let c denote a period of f. Then, by periodicity,

$$f(nc) = a, \quad n \in \mathbb{Z}.$$

Let z be a point such that $|z| = r \geq |c|$. Then there exists $k \in \mathbb{N}$ such that

$$k|c| \leq |z| \leq (k+1)|c|.$$

This gives us a lower bound for the number of a-points of f:

$$n(r, a, f) \geq 2k \geq 2\left(\frac{r}{|c|} - 1\right).$$

Then $\lambda(a, f) \geq 1$ by Theorem 3.1.6, and finally $\rho(f) \geq 1$ by Theorem 3.2.4. □

Theorem 3.2.3 shows how the growth of $n(r)$ is dominated by the order of f. Next, we will give a refinement of this result for functions of finite type.

Theorem 3.2.7 ([37, Theorem 2.5.13]). *If f is an entire function of finite positive order ρ and finite type τ, then*

$$\limsup_{r\to\infty} \frac{n(r)}{r^\rho} \leq e\rho\tau. \tag{3.2.10}$$

Proof. Given $\varepsilon > 0$, the definition of type gives

$$\log M(r, f) \leq (\tau + \varepsilon/2) r^\rho, \quad r \geq R_0.$$

If $f(0) \neq 0$, then Jensen's theorem implies

$$N(r) \leq \log M(r, f) + O(1).$$

If $f(0) = 0$, then there exists a positive integer m such that $g(z) = z^{-m} f(z)$ is an entire function that satisfies $g(0) \neq 0$. Applying Jensen's theorem to g and then going back to f gives

$$N(r) \leq \log M(r, f) + O(\log r).$$

Hence, in both cases,

$$N(r) \leq (\tau + \varepsilon) r^\rho, \quad r \geq R_1 \geq R_0.$$

This implies, for $\beta > 1$,

$$n(r) \log \beta \leq \int_r^{\beta r} \frac{n(t)}{t}\, dt \leq N(\beta r) \leq \beta^\rho (\tau + \varepsilon) r^\rho, \quad r \geq R_1.$$

In other words,

$$\frac{n(r)}{r^\rho} \leq \frac{(\tau + \varepsilon)\beta^\rho}{\log \beta}, \qquad r \geq R_1,$$

where the right-hand side attains its minimal value when $\beta = e^{1/\rho} > 1$. Taking the limit superiors and then letting $\varepsilon \to 0^+$ proves the statement. □

The inequality (3.2.10) in Theorem 3.2.7 is sharp. Indeed, Boas [38, Theorem 6] has shown that for every $\rho > 0$, there exists an entire function f of order ρ and type $\tau = \frac{1}{e\rho}$ such that

$$\limsup_{r \to \infty} \frac{n(r)}{r^\rho} = 1,$$

which gives an equality in (3.2.10). Another example where inequality (3.2.10) becomes an equality was given by Buck [39]. On the other hand, it is easy to see that (3.2.10) can also be a strict inequality; for instance, consider $f(z) = e^z - 1$ or $f(z) = \sin z$ (see Problem 3.2.8).

Problems

Problem 3.2.8. If $f(z) = \sin z$, then verify that

$$n(r, 1/f) = (1 + o(1))\frac{2r}{\pi}.$$

Problem 3.2.9. Let $f \not\equiv 0$ be an entire function such that $f(\sqrt{n}) = 0$ for $n \in \mathbb{N}$.

(a) Prove that $\rho(f) \geq 2$.

(b) Prove that the result in (a) is the best possible by giving an example of an entire function f of order $\rho(f) = 2$ that has zeros at the points $\sqrt{n}$.

Problem 3.2.10. Use Jensen's theorem to find the exact value of the real integral

$$\int_0^{2\pi} \log\left|2 - e^{i\theta}\right| d\theta.$$

Problem 3.2.11. Let f be an entire function such that $f(0) = 1$. Show that $n(r, 1/f) \leq \log M(er, f)$ for all $r \geq 0$.

Problem 3.2.12. Suppose that f is an entire function that satisfies $|f(z)| \leq e^{|z|}$ for all $z \in \mathbb{C}$. If f has zeros at all the integers, then prove that $f(z) \equiv 0$ on $\mathbb{C}$.

Problem 3.2.13. Suppose that g is analytic in the disc $|z| < R$ and has no zeros there. Prove that $\log|g(z)|$ is harmonic in $|z| < R$. This result is needed in the proof of Jensen's theorem.

3.3 Infinite products

In Section 3.4, we will separate the zeros of an entire function f as a product component of f. This product component is known as a *Weierstrass product*, and for an entire function whose zero sequence has finite exponent of convergence, it has a useful unique form called the *canonical product*. If f has infinitely many zeros, then the product component is an infinite product, and for background, we discuss infinite products in this section. The presentation is parallel to that in [142, Section 1.16].

Recall that an infinite sum $\sum_{k=1}^{\infty} a_k$ of complex numbers a_k converges if the limit of partial sums

$$\lim_{n\to\infty} \sum_{k=1}^{n} a_k$$

exists as a finite number. The convergence of an infinite product is defined in a similar manner. The sequence $\{P_n\}$ of partial products of an infinite product $\prod_{k=1}^{\infty} b_k$ of complex numbers b_k is defined by $P_n = \prod_{k=1}^{n} b_k$. The general definition for the convergence of an infinite product is as follows.

Definition 3.3.1. We say that a given infinite product $\prod_{k=1}^{\infty} b_k$ of complex numbers b_k *converges* if the following conditions hold:
(1) At most finitely many factors b_k are zero.
(2) The partial products $\{P_n\}$ obtained after removing the zero factors (if any) approach a finite limit as $n \to \infty$.
(3) The limit in (2) is not zero.

If the product is convergent, its value is defined to be the limit in (2), provided that no factor is zero; otherwise, its value is zero. If one of conditions (1)–(3) does not hold, then the product in question *diverges*.

Remark. Writing $P_n = \prod_{k=1}^{n} b_k$, we see that $\prod_{k=1}^{\infty} b_k$ converges to a nonzero value if and only if the sequence $\{P_n\}$ converges and $\lim_{n\to\infty} P_n \neq 0$. Then $b_n = P_n/P_{n-1}$, and

$$\lim_{n\to\infty} b_n = \frac{\lim_{n\to\infty} P_n}{\lim_{n\to\infty} P_{n-1}} = 1.$$

This is why infinite products are often written in the form $\prod_{k=1}^{\infty}(1 + a_k)$, where $\lim_{k\to\infty} a_k = 0$ whenever the product converges to a nonzero value.

Example 3.3.2. We have

$$\prod_{k=2}^{\infty}\left(1 - \frac{1}{k^2}\right) = \frac{1}{2},$$

since the partial product

$$P_n = \prod_{k=2}^{n}\left(1 - \frac{1}{k^2}\right) = \prod_{k=2}^{n} \frac{(k-1)}{k} \cdot \frac{(k+1)}{k}$$

satisfies

$$P_n = \frac{1}{2} \cdot \frac{3}{2} \cdot \frac{2}{3} \cdot \frac{4}{3} \cdot \frac{3}{4} \cdot \frac{5}{4} \cdots \frac{(n-1)}{n} \cdot \frac{(n+1)}{n} = \frac{1}{2} \cdot \frac{(n+1)}{n} \to \frac{1}{2}$$

as $n \to \infty$.

Definition 3.3.3. The infinite product $\prod_{k=1}^{\infty}(1 + a_k)$ with $a_k \in \mathbb{C}$ is said to be *absolutely convergent* if $\prod_{k=1}^{\infty}(1 + |a_k|)$ converges.

Theorem 3.3.4. *If $a_k \in \mathbb{C} \setminus \{-1\}$ for $k \in \mathbb{N}$, then the product $\prod_{k=1}^{\infty}(1 + a_k)$ is absolutely convergent if and only if the sum $\sum_{k=1}^{\infty} |a_k|$ converges.*

Example 3.3.5. The product

$$\prod_{n=2}^{\infty}\left(1 - \frac{(-1)^n}{n}\right) = \frac{1}{2}$$

converges but does not converge absolutely by Theorem 3.3.4, because the harmonic series diverges.

For the proof of Theorem 3.3.4, we need *Cauchy's criterion* for products.

Lemma 3.3.6. *If $b_k \in \mathbb{C}$, then the product $\prod_{k=1}^{\infty} b_k$ is convergent if and only if for every $\varepsilon > 0$, there exists $N = N(\varepsilon) \in \mathbb{N}$ such that*

$$\left|\prod_{k=n+1}^{m} b_k - 1\right| < \varepsilon, \quad m > n \geq N. \tag{3.3.1}$$

Proof. Suppose that the product $\prod_{k=1}^{\infty} b_k$ converges. Since a finite number of zero factors can be ignored, we may assume that $b_k \neq 0$ for every k. Then

$$P_n = \prod_{k=1}^{n} b_k \to P \neq 0, \quad n \to \infty.$$

In other words, the sequence $\{P_n\}$ of partial products converges to a nonzero number P. By the definition of the limit and by Cauchy's criterion for ordinary sequences, given $\varepsilon > 0$, there exists $N = N(\varepsilon) \in \mathbb{N}$ such that

$$|P_n| > \frac{|P|}{2} \quad \text{and} \quad |P_m - P_n| < \frac{|P|}{2}\varepsilon$$

whenever $m > n \geq N$. Thus

$$\left|\prod_{k=n+1}^{m} b_k - 1\right| = \left|\frac{P_m}{P_n} - 1\right| = \frac{|P_m - P_n|}{|P_n|} < \frac{\frac{|P|}{2}\varepsilon}{\frac{|P|}{2}} = \varepsilon$$

whenever $m > n \geq N$.

Conversely, suppose that for every $\varepsilon > 0$, there exists $N = N(\varepsilon) \in \mathbb{N}$ such that (3.3.1) holds. Then condition (1) in Definition 3.3.1 is clearly satisfied. Set $\varepsilon = \frac{1}{2}$. Then by the assumption there exists $N_1 \in \mathbb{N}$ such that

$$\left|\prod_{k=N}^{m} b_k - 1\right| < \frac{1}{2}, \quad m \geq N_1,$$

that is,

$$\frac{1}{2} < \left|\prod_{k=N}^{m} b_k\right| < \frac{3}{2}, \quad m \geq N_2.$$

We conclude that if the sequence $\{\prod_{k=N}^{m} b_k\}$ has a limit as $m \to \infty$, then this limit is finite and nonzero. Hence we need to prove the convergence of this sequence. To this end, let $\varepsilon > 0$ be arbitrary. For this ε, there exists $N_2 \in \mathbb{N}$ such that

$$\left|\prod_{k=n+1}^{m} b_k - 1\right| < \frac{2\varepsilon}{3}$$

whenever $m > n \geq N_2$. This implies

$$\left|\prod_{k=N}^{m} b_k - \prod_{k=N}^{n} b_k\right| = \left|\prod_{k=N}^{n} b_k\right| \cdot \left|\prod_{k=n+1}^{m} b_k - 1\right| < \varepsilon$$

whenever $m > n \geq \max\{N_1, N_2\}$. Thus by Cauchy's criterion for ordinary sequences, the sequence $\{\prod_{k=N}^{m} b_k\}$ converges. Since this limit is nonzero, the original product is also convergent. □

Proof of Theorem 3.3.4. Suppose that $\sum_{k=1}^{\infty} |a_k|$ converges, and set

$$Q_n = \prod_{k=1}^{n} (1 + |a_k|).$$

Then $Q_n \geq 1$ for all $n \in \mathbb{N}$, and the sequence $\{Q_n\}$ is nondecreasing. By the monotone convergence theorem it suffices to prove that $\{Q_n\}$ is bounded from above. However,

$$Q_n \leq \prod_{k=1}^{n} \exp(|a_k|) = \exp\left(\sum_{k=1}^{n} |a_k|\right) \leq \exp\left(\sum_{k=1}^{\infty} |a_k|\right) < \infty,$$

where we have used the inequality $1 + x \leq e^x$ for $x \geq 0$.

Conversely, suppose that $\prod_{k=1}^{\infty}(1 + |a_k|)$ converges. For $m > n$, we can prove by induction that

$$\sum_{k=n+1}^{m} |a_k| \leq \prod_{k=n+1}^{m} (1 + |a_k|) - 1. \tag{3.3.2}$$

Indeed, (3.3.2) is clearly true for $m = n+1$, and if (3.3.2) holds for some particular $m > n$, then

$$\begin{aligned}\sum_{k=n+1}^{m+1} |a_k| &\le \prod_{k=n+1}^{m} (1+|a_k|) - 1 + |a_{m+1}| \\ &\le \prod_{k=n+1}^{m} (1+|a_k|) - 1 + |a_{m+1}| \prod_{k=n+1}^{m} (1+|a_k|) \\ &= (1+|a_{m+1}|) \prod_{k=n+1}^{m} (1+|a_k|) - 1 = \prod_{k=n+1}^{m+1} (1+|a_k|) - 1,\end{aligned}$$

which means that (3.3.2) also holds for $m+1$. Let $\varepsilon > 0$. By the assumption, estimate (3.3.2), and Lemma 3.3.6 there exists $N \in \mathbb{N}$ such that

$$\sum_{k=n+1}^{m} |a_k| < \varepsilon, \quad m > n \ge N.$$

Then by the Cauchy criterion for series we find that $\sum_{k=1}^{\infty} |a_k|$ converges. □

It is well known from real analysis that an absolutely convergent infinite series with real terms is convergent. This is also true for series with complex terms.

Lemma 3.3.7 ([74, p. 131]). *An absolutely convergent infinite series of complex terms is convergent.*

The following theorem shows that the property in Lemma 3.3.7 holds for infinite products as well.

Theorem 3.3.8. *An absolutely convergent infinite product $\prod_{k=1}^{\infty}(1+a_k)$ with $a_k \ne -1$ for $k \in \mathbb{N}$ is convergent.*

Proof. For $m > n$, we can prove by induction that

$$\left|\prod_{k=n+1}^{m} (1+a_k) - 1\right| \le \prod_{k=n+1}^{m} (1+|a_k|) - 1. \tag{3.3.3}$$

Indeed, (3.3.3) is clearly true for $m = n+1$, and if (3.3.3) holds for some particular $m > n$, then

$$\begin{aligned}\left|\prod_{k=n+1}^{m+1} (1+a_k) - 1\right| &= \left|(1+a_{m+1}) \prod_{k=n+1}^{m} (1+a_k) - 1\right| \\ &\le \left|\prod_{k=n+1}^{m} (1+a_k) - 1\right| + \left|a_{m+1} \prod_{k=n+1}^{m} (1+a_k)\right| \\ &\le \prod_{k=n+1}^{m} (1+|a_k|) - 1 + |a_{m+1}| \cdot \prod_{k=n+1}^{m} (1+|a_k|)\end{aligned}$$

$$= \prod_{k=n+1}^{m+1} (1 + |a_k|) - 1,$$

which means that (3.3.3) also holds for $m + 1$. The statement then follows by applying Lemma 3.3.6 two times to (3.3.3). □

Definition 3.3.9. Let $\{f_k(z)\}$ be a sequence of analytic functions in a domain $G \subset \mathbb{C}$. We say that the infinite product

$$\prod_{k=1}^{\infty}(1 + f_k(z))$$

converges uniformly on compact subsets of G if for every compact subset $K \subset G$, there are at most finitely many factors in the product that have zeros in K, and if these factors (if any) are removed, then the partial products of the remaining infinite product converge uniformly on K to a function that does not vanish on K.

Theorem 3.3.10. *Let* $\{f_n(z)\}$ *be a sequence of analytic functions in a domain* G *such that* $\sum_{k=1}^{\infty} |f_k(z)|$ *converges uniformly on compact subsets of* G. *Then the infinite product* $\prod_{k=1}^{\infty}(1 + f_k(z))$ *converges uniformly on compact subsets of* G *and is analytic in* G.

Proof. Suppose that $\sum_{k=1}^{\infty} |f_k(z)|$ is uniformly convergent on a compact set $K \subset G$. Then $\sum_{k=1}^{\infty} |f_k(z)|$ is continuous on K by [204, Theorem 15.3], and hence there exists a constant $M > 0$ such that

$$\sum_{k=1}^{\infty} |f_k(z)| \le M, \quad z \in K.$$

For the partial product $P_n(z)$, we have

$$\begin{aligned} |P_n(z)| &= \left|\prod_{k=1}^{n}(1 + f_k(z))\right| \le \prod_{k=1}^{n}(1 + |f_k(z)|) \\ &\le \prod_{k=1}^{n} \exp(|f_k(z)|) = \exp\left(\sum_{k=1}^{n} |f_k(z)|\right) \le e^M \end{aligned}$$

for all $n \in \mathbb{N}$ and $z \in K$. Therefore

$$|P_{n+1}(z) - P_n(z)| = |P_n(z) f_{n+1}(z)| \le e^M |f_{n+1}(z)|, \quad n \in \mathbb{N},\ z \in K.$$

Let $\varepsilon > 0$. By the uniform convergence there exists $N = N(\varepsilon, K) \in \mathbb{N}$ such that

$$\sum_{k=N}^{\infty} |f_k(z)| < e^{-M}\varepsilon, \quad z \in K.$$

Hence, for $m > n \ge N$ and $z \in K$, we have

$$\begin{aligned}|P_n(z) - P_m(z)| &\leq |P_n(z) - P_{n+1}(z)| + |P_{n+1}(z) - P_{n+2}(z)| + \cdots + |P_{m-1}(z) - P_m(z)|\\ &\leq e^M \sum_{k=n}^{m} |f_k(z)| < e^M e^{-M} \varepsilon = \varepsilon.\end{aligned}$$

Since the compact set $K \subset G$ was arbitrarily chosen, it follows by Theorem 1.6.3 that the sequence $\{P_n(z)\}$ of partial products converges uniformly on compact subsets of G to the function $P(z) = \prod_{k=1}^{\infty}(1 + f_k(z))$. The fact that $P(z)$ is analytic in G is a consequence of Theorem 1.6.6. □

Problems

Problem 3.3.11. Show that
(a) $\prod_{k=1}^{\infty}\left(1 - \frac{2}{(k+1)(k+2)}\right) = \frac{1}{3}$.
(b) $\prod_{k=3}^{\infty}\frac{k^2-4}{k^2-1} = \frac{1}{4}$.

Problem 3.3.12. Determine the exact value of the products:
(a) $\prod_{k=2}^{\infty}\frac{k^3-1}{k^3+1}$.
(b) $\prod_{k=1}^{\infty}\left(1 + \frac{1}{k(k+2)}\right)$.

Problem 3.3.13. Explain why the following products converge or diverge:
(a) $\prod_{k=1}^{\infty}\left(1 + \frac{1}{k^{1/k}}\right)$.
(b) $\prod_{k=1}^{\infty}\left(1 - \frac{1}{k^{1/k}}\right)$.

Problem 3.3.14. Let $\mathbb{D} = \{z \in \mathbb{C} : |z| < 1\}$ be the unit disc. Prove the following:
(a) $\prod_{k=0}^{\infty}(1 + z^{2^k})$ is analytic in $\mathbb{D}$ and has no zeros in $\mathbb{D}$.
(b) $(1 - z)\prod_{k=0}^{N}(1 + z^{2^k}) = 1 - z^{2^{N+1}}$ for all $N \in \mathbb{N}$ and $z \in \mathbb{D}$.
(c) $\prod_{k=0}^{\infty}(1 + z^{2^k}) = \frac{1}{1-z}$ for all $z \in \mathbb{D}$.

Note. The function $\frac{1}{1-z}$ is rational and is defined everywhere in the complex plane except at the point $z = 1$, but the associated product in part (c) is defined only for $z \in \mathbb{D}$.

Problem 3.3.15. If $\{b_n\}$ is a sequence of constants such that the series $\sum_{n=1}^{\infty}|b_n|$ converges, then show that

$$f(z) = \prod_{n=1}^{\infty}(1 + b_n z)$$

is an entire function.

3.4 Weierstrass factorization theorem

If g is an entire function, then $f = e^g$ is entire and has no zeros. Conversely, we have the following result.

Theorem 3.4.1. *If f is an entire function that has no zeros, then there exists an entire function g such that $f = e^g$. In addition, if f is of finite order $\rho = \rho(f)$, then g is a polynomial satisfying* $\deg g = \rho$.

Proof. Since f is entire with no zeros, the function f'/f is entire. Let w be an entire function that satisfies $f'/f = w'$. By integration we see that

$$f(z) = e^{w(z)+C},$$

where C is some constant. Set $g = w + C$.

Suppose then that $\rho < \infty$, and let $\varepsilon > 0$. Keeping in mind that $f = e^g$, there exists $R = R(\varepsilon) > 0$ such that

$$A(r,g) = \log M(r,f) \leq r^{\rho+\varepsilon}, \quad r > R,$$

where, as in Section 2.4, $A(r,g)$ is the maximum of the real part of $g(z)$ on $|z| = r$. Hence by Theorem 2.4.2,

$$M(r,g) \leq 2A(2r,g) + 3|g(0)| = O(r^{\rho+\varepsilon}).$$

This shows that g is a polynomial of degree $\leq \rho$. Since $\log M(r,f) \leq M(r,g)$, the strict inequality $\deg g < \rho$ leads to an immediate contradiction. Therefore $\deg g = \rho$. □

The function f'/f that appears in the previous proof is called a *logarithmic derivative* because $(\log f)' = f'/f$. It has a simple pole at every zero of f.

Corollary 3.4.2. *If f is an entire function of order $\rho(f) < 1$ and has no zeros, then f is a constant function.*

Consider next a polynomial $P(z)$ that has finitely many zeros $z_1, \ldots, z_n \in \mathbb{C} \setminus \{0\}$. By the fundamental theorem of algebra (Theorem 3.9.1), there exists a constant $C \in \mathbb{C}$ such that

$$\begin{aligned} P(z) &= C(z_1 - z) \cdots (z_n - z) \\ &= Cz_1 \cdots z_n \left(1 - \frac{z}{z_1}\right) \cdots \left(1 - \frac{z}{z_n}\right) \\ &= P(0)\left(1 - \frac{z}{z_1}\right) \cdots \left(1 - \frac{z}{z_n}\right). \end{aligned}$$

In other words, the zeros of a polynomial can be factored in different ways. Now suppose that f is an entire function with exactly the zeros $z_1, \ldots, z_n$ and no other zeros. Then f/P is an entire function with no zeros. Hence by Theorem 3.4.1 we obtain $f = Pe^g$ for some entire function g.

So far we have discussed entire functions with no zeros and with finitely many zeros. Suppose next that $f \not\equiv 0$ is entire with an infinite number of zeros. Let $\{z_n\}$ be the sequence of nonzero zeros of f listed according to multiplicities and arranged by increasing moduli, i. e.,

$$0 < |z_1| \leq |z_2| \leq \cdots \leq |z_n| \leq |z_{n+1}| \leq \cdots .$$

By the uniqueness theorem of analytic functions we have $\lim_{n\to\infty} |z_n| = \infty$. Observe that factoring these zeros of f like we did for the polynomial P above is not an obvious thing to do, because the product

$$\prod_{n=1}^{\infty}\left(1 - \frac{z}{z_n}\right)$$

may diverge. In fact, we would like this product to converge uniformly on closed discs $|z| \leq R$ (or equivalently on compact subsets of $\mathbb{C}$), but, according to Theorem 3.3.10, this happens when

$$\sum_{n=1}^{\infty} \frac{1}{|z_n|} < \infty.$$

Some sequences $\{z_n\}$ satisfy this (for example, $z_n = n^2$) but not all (for example, $z_n = n$). Thus we must modify the situation somehow. For this purpose, we define the *Weierstrass primary factors* (a. k. a. convergence factors) by

$$\begin{aligned} E_0(z) &= 1 - z, \\ E_n(z) &= (1 - z)\exp\left(z + \frac{z^2}{2} + \cdots + \frac{z^n}{n}\right), \qquad n \in \mathbb{N}. \end{aligned}$$

Lemma 3.4.3. *We have* $|1 - E_n(z)| \leq |z|^{n+1}$ *for* $|z| \leq 1$ *and* $n \in \mathbb{N} \cup \{0\}$.

Proof. The statement is trivial for $n = 0$, so we suppose that $n \geq 1$. Then

$$\begin{aligned} E_n'(z) &= -\exp\left(z + \frac{z^2}{2} + \cdots + \frac{z^n}{n}\right) + (1 - z)(1 + z + \cdots + z^{n-1})\exp\left(z + \frac{z^2}{2} + \cdots + \frac{z^n}{n}\right) \\ &= -z^n \exp\left(z + \frac{z^2}{2} + \cdots + \frac{z^n}{n}\right). \end{aligned}$$

This shows that the Taylor coefficients (at the origin) of the entire function

$$\begin{aligned} 1 - E_n(z) = E_n(0) - E_n(z) &= -\int_0^z E_n'(t)\,dt \\ &= \int_0^z t^n \exp\left(t + \frac{t^2}{2} + \cdots + \frac{t^n}{n}\right) dt \end{aligned}$$

are all nonnegative real numbers and that $1 - E_n(z)$ has a zero of multiplicity $n+1$ at the origin. Therefore,

$$f(z) = \frac{1 - E_n(z)}{z^{n+1}}$$

is also an entire function whose Taylor coefficients (at the origin) are also nonnegative real numbers. By the maximum modulus principle, the maximal value of $|f(z)|$ in $|z| \leq 1$ is attained on the unit circle $|z| = 1$. Since the Taylor coefficients of f are nonnegative, the maximum modulus is attained at the point $z = 1$. In other words, if $|z| \leq 1$, then

$$\left|\frac{1 - E_n(z)}{z^{n+1}}\right| = |f(z)| \leq f(1) = \frac{1 - E_n(1)}{1^{n+1}} = 1,$$

which establishes the lemma. □

The Weierstrass primary factors can be used to construct a convergent product based on an arbitrary sequence of points.

Theorem 3.4.4. *Let $\{z_n\}$ be an arbitrary infinite sequence of nonzero complex numbers (repetition allowed) having no accumulation points in the finite complex plane $\mathbb{C}$. Then there exists a sequence $\{n_k\}$ of nonnegative integers such that*

$$F(z) = \prod_{k=1}^{\infty} E_{n_k}\left(\frac{z}{z_k}\right) \tag{3.4.1}$$

is an entire function having zeros precisely at the points z_n with multiplicity equal to the number of times z_n appears in the sequence.

Proof. Without loss of generality, we may suppose that the points $\{z_n\}$ are organized by increasing moduli. We want to show that the product F converges uniformly on compact subsets of $\mathbb{C}$ for an appropriately chosen sequence $\{n_k\}$. By Lemma 3.4.3,

$$\left|1 - E_{n_k}\left(\frac{z}{z_k}\right)\right| \leq \left|\frac{z}{z_k}\right|^{n_k+1}, \quad |z| \leq |z_k|.$$

Therefore, for each k, we choose a nonnegative integer n_k such that the series

$$\sum_{k=1}^{\infty} \left|\frac{z}{z_k}\right|^{n_k+1} \tag{3.4.2}$$

converges uniformly on compact subsets of $\mathbb{C}$. For example, $n_k+1 = k$ will always suffice. Indeed, for any $R > 0$, if $|z| \leq R$, then

$$\sum_{k=1}^{\infty} \left|\frac{z}{z_k}\right|^k = \sum_{|z_k| \leq 2R} \left|\frac{z}{z_k}\right|^k + \sum_{|z_k| > 2R} \left|\frac{z}{z_k}\right|^k \leq \sum_{|z_k| \leq 2R} \left|\frac{z}{z_k}\right|^k + \sum_{|z_k| > 2R} \left(\frac{1}{2}\right)^k.$$

Since $z_k \to \infty$, the first sum on the right has only finitely many terms, whereas the second sum on the right converges as a geometric series. Having chosen a sequence $\{n_k\}$ with the required property, we have, for $|z_k| > 2R$,

$$\left|1 - E_{n_k}\left(\frac{z}{z_k}\right)\right| \le \left|\frac{z}{z_k}\right|^{n_k+1} \le \left|\frac{R}{z_k}\right|^{n_k+1}, \quad |z| \le R.$$

The convergence of the series $\sum_{|z_k|>2R} |\frac{R}{z_k}|^{n_k+1}$ implies the uniform convergence of the series $\sum_{|z_k|>2R} |\frac{z}{z_k}|^{n_k+1}$ in the closed disc $|z| \le R$ and hence on compact subsets of $\mathbb{C}$. Therefore by Theorem 3.3.10 the product

$$\prod_{|z_k|>2R} E_{n_k}\left(\frac{z}{z_k}\right)$$

converges uniformly on compact subsets of $\mathbb{C}$ and defines an entire function. Since we have omitted only a finite number of factors, the product F in (3.4.1) is entire. Clearly, F has the required zeros with the required multiplicities. □

In the paragraph following Corollary 3.4.2, we have factorized entire functions with finitely many zeros. We now use Theorem 3.4.4 to factorize entire functions with infinitely many zeros. The result is called the *Weierstrass factorization theorem*, which we will state and prove next.

Theorem 3.4.5. *Suppose that $f \not\equiv 0$ is an entire function having infinitely many nonzero zeros z_n, each zero being repeated according to its multiplicity. Then there exist an entire function g and a sequence $\{n_k\}$ of nonnegative integers such that*

$$f(z) = z^m e^{g(z)} \prod_{k=1}^{\infty} E_{n_k}\left(\frac{z}{z_k}\right),$$

where $m \in \mathbb{N} \cup \{0\}$ is the multiplicity of the zero of f at the origin.

Proof. We may choose the sequence $\{n_k\}$ such that the product F in (3.4.1) is entire and has precisely the same nonzero zeros as f. Hence $\frac{f(z)}{z^m F(z)}$ is entire and has no zeros. The statement now follows from Theorem 3.4.1. □

Theorem 3.4.1 was used in the proof of the Weiertrass factorization theorem and will also be used in the proofs of theorems in the next section. More generally, it can be mentioned that Theorem 3.4.1 has been utilized in the solutions of many concrete problems. As one of countless examples, Theorem 3.4.1 can be used to show that if f and g are entire functions that satisfy the Fermat-type equation $f^2 + g^2 = 1$, then $f = \cos w$ and $g = \sin w$ for some entire function w; see Theorem 12.4.1.

We now proceed to develop a well-known product representation of $\sin(\pi z)$. The next result will be used in the proof.

Lemma 3.4.6. *Suppose that g is an analytic function on $\mathbb{C} \setminus \mathbb{Z}$ that has the following properties:*
(a) *g has simple poles with residue 1 at the integers.*
(b) *g is an odd function, i. e., $g(-z) \equiv -g(z)$.*
(c) *$2g(2z) \equiv g(z) + g(z + 1/2)$.*

Then $g(z) \equiv \pi \cot(\pi z)$.

Proof. Set

$$H(z) = g(z) - \pi \cot(\pi z). \tag{3.4.3}$$

It is left as an exercise to prove that $\pi \cot(\pi z)$ has simple poles with residue 1 at the integers. Hence by (a) the principal parts in (3.4.3) cancel each other out, which means that H is entire. Since the cotangent is an odd function and g is an odd function by (b), we see that H is an odd function. Thus $H(0) = -H(0)$, which gives $H(0) = 0$. Using standard trigonometric identities, we obtain

$$2\cot(2\pi z) = \frac{\cot^2(\pi z) - 1}{\cot(\pi z)} = \cot(\pi z) - \tan(\pi z) = \cot(\pi z) + \cot(\pi z + \pi/2).$$

Consequently,

$$\begin{aligned} 2H(2z) &= g(z) + g(z + 1/2) - 2\pi \cot(2\pi z) \\ &= g(z) - \pi \cot(\pi z) + g(z + 1/2) - \pi \cot(\pi z + \pi/2) \\ &= H(z) + H(z + 1/2). \end{aligned} \tag{3.4.4}$$

Suppose that H is a nonconstant function, and consider the closed disc $\overline{D}(0,2)$. By the maximum modulus principle there exists a point $c \in \partial D(0,2)$ such that $|H(z)| < |H(c)|$ for all $z \in D(0,2)$. Since $\frac{c}{2}, \frac{1+c}{2} \in D(0,2)$, we have

$$\left|H\left(\frac{c}{2}\right) + H\left(\frac{1+c}{2}\right)\right| \le \left|H\left(\frac{c}{2}\right)\right| + \left|H\left(\frac{1+c}{2}\right)\right| < 2|H(c)|,$$

contradicting (3.4.4). Therefore H is a constant function. Since $H(0) = 0$, we obtain $H(z) \equiv 0$, and the statement follows from (3.4.3). □

Theorem 3.4.7. *We have*

$$\sin(\pi z) = \pi z \prod_{n=1}^{\infty} \left(1 - \frac{z^2}{n^2}\right), \quad z \in \mathbb{C}. \tag{3.4.5}$$

Proof. First note that $\sin(\pi z)$ has simple zeros at the integers and no other zeros. Since $\sum_{n\neq 0} \frac{1}{n^2} < \infty$, the sum $\sum_{n\neq 0} |\frac{z}{n}|^2$ converges uniformly on compact subsets of $\mathbb{C}$. Then by (3.4.2) and the Weierstrass factorization theorem, it follows that there exists an entire function g such that

$$\sin(\pi z) = ze^{g(z)} \prod_{n\neq 0} E_1\left(\frac{z}{n}\right) = ze^{g(z)} \prod_{n\neq 0}\left(1 - \frac{z}{n}\right) \exp\left(\frac{z}{n}\right).$$

In this convergent product, the corresponding exponential factors $\exp(\frac{z}{n})$ and $\exp(-\frac{z}{n})$ cancel out, whereas $(1 - \frac{z}{n}) \cdot (1 + \frac{z}{n}) = 1 - \frac{z^2}{n^2}$. This gives us

$$\sin(\pi z) = ze^{g(z)} \prod_{n=1}^{\infty}\left(1 - \frac{z^2}{n^2}\right).$$

We may write this as

$$f(z) = \sin(\pi z) = \pi z e^{h(z)} \prod_{n=1}^{\infty}\left(1 - \frac{z^2}{n^2}\right), \tag{3.4.6}$$

where $h(z)$ is entire and satisfies $\pi e^{h(z)} = e^{g(z)}$. Now

$$f'(z) = (\pi + \pi z h'(z))e^{h(z)} \prod_{n=1}^{\infty}\left(1 - \frac{z^2}{n^2}\right) + \pi z e^{h(z)} \sum_{n=1}^{\infty}\left(-\frac{2z}{n^2}\right)\prod_{k\neq n}\left(1 - \frac{z^2}{k^2}\right),$$

so that

$$\frac{f'(z)}{f(z)} = \frac{\pi\cos(\pi z)}{\sin(\pi z)} = \pi\cot(\pi z) = \frac{1}{z} + h'(z) + \sum_{n=1}^{\infty}\frac{2z}{z^2 - n^2}. \tag{3.4.7}$$

To determine h, we define

$$S(z) = \lim_{k\to\infty}\sum_{n=-k}^{k}\frac{1}{z+n} = \frac{1}{z} + \sum_{n=1}^{\infty}\frac{2z}{z^2 - n^2}.$$

The function S is analytic in $\mathbb{C}$ except for simple poles at $z = n$, $n \in \mathbb{Z}$, where we see that the residue = 1 at each of these poles. Clearly, S is an odd function. Further, define

$$s_k(z) = \frac{1}{z} + \sum_{n=1}^{k}\left(\frac{1}{z+n} + \frac{1}{z-n}\right).$$

Then

$$\begin{aligned}
&2s_{2k}(2z) - s_k(z) - s_k(z + 1/2)\\
&= \sum_{n=1}^{2k}\left(\frac{2}{2z+n} + \frac{2}{2z-n}\right) - \sum_{n=1}^{k}\left(\frac{1}{z+n} + \frac{1}{z-n}\right)\\
&\quad - \frac{2}{2z+1} - \sum_{n=1}^{k}\left(\frac{2}{2z+1+2n} + \frac{2}{2z+1-2n}\right).
\end{aligned}$$

This simplifies to

$$\begin{aligned}
&2s_{2k}(2z) - s_k(z) - s_k(z+1/2)\\
&= -\frac{2}{2z+1} + \sum_{n=1}^{2k}\frac{2}{2z+n} - \sum_{n=1}^{k}\frac{2}{2z+1+2n}\\
&\quad + \sum_{n=1}^{2k}\frac{2}{2z-n} - \sum_{n=1}^{k}\frac{2}{2z+1-2n} - \sum_{n=1}^{k}\left(\frac{1}{z+n}+\frac{1}{z-n}\right)\\
&= -\frac{2}{2z+1} + \frac{2}{2z+1} - \frac{2}{2z+2k+1} + \sum_{n=1}^{k}\frac{2}{2z+2n}\\
&\quad + \sum_{n=1}^{k}\frac{2}{2z-2n} - \sum_{n=1}^{k}\left(\frac{1}{z+n}+\frac{1}{z-n}\right)\\
&= -\frac{2}{2z+2k+1}.
\end{aligned}$$

Letting $k \to \infty$, we have $s_k(z) \to S(z)$ for $z \in \mathbb{C} \setminus Z$ (locally uniformly or uniformly on compact subsets), which gives

$$2S(2z) - S(z) - S(z+1/2) \equiv 0.$$

By Lemma 3.4.6 it follows that $S(z) = \pi\cot(\pi z)$. Equating the two expansions for $\pi\cot(\pi z)$, we have

$$\frac{1}{z} + \sum_{n=1}^{\infty}\frac{2z}{z^2-n^2} = \frac{1}{z} + h'(z) + \sum_{n=1}^{\infty}\frac{2z}{z^2-n^2}.$$

Thus $h'(z) \equiv 0$, and hence h is a constant function. Since

$$e^{h(0)} = \lim_{z\to 0}\frac{\sin(\pi z)}{\pi z} = 1, \tag{3.4.8}$$

we see that $e^{h(z)} \equiv 1$. This yields (3.4.5). □

In the above proof, we observe that $h'(z) \equiv 0$ and that (3.4.7) holds. This gives the following representation for the cotangent function.

Corollary 3.4.8. *We have*

$$\pi\cot(\pi z) = \frac{1}{z} + \sum_{n=1}^{\infty}\frac{2z}{z^2-n^2}, \quad z \neq 0, \pm1, \pm2, \pm3, \ldots. \tag{3.4.9}$$

Problems

Problem 3.4.9. Use the infinite product representation of $\sin(\pi z)$ to find the exact value of the product

$$\frac{5}{6}\cdot\frac{7}{6}\cdot\frac{11}{12}\cdot\frac{13}{12}\cdot\frac{17}{18}\cdot\frac{19}{18}\cdot\frac{23}{24}\cdot\frac{25}{24}\cdots.$$

Problem 3.4.10. Use the theory of entire functions to prove the following known identities from real analysis:
(a) The Wallis formula: $\prod_{n=1}^{\infty}\left(1-\frac{1}{(2n)^2}\right)=\frac{2}{\pi}$.
(b) The Basel problem: $\sum_{n=1}^{\infty}\frac{1}{n^2}=\frac{\pi^2}{6}$.

Problem 3.4.11. Prove that the function $\pi\cot(\pi z)$ has simple poles exactly at the integers $z=n\in\mathbb{Z}$, with residue 1 at each pole.

Problem 3.4.12. Find the exact values of the following expressions:
(a) $\prod_{n=1}^{\infty}\left(1-\frac{1}{4n}\right)\left(1+\frac{1}{4n}\right)$.
(b) $\sum_{n=1}^{\infty}\frac{1}{1+n^2}$.

3.5 Hadamard factorization theorem

The special case where $\{n_k\}$ in the Weierstrass factorization theorem is a constant sequence is of particular interest.

Definition 3.5.1. Let $\{z_n\}$ be an infinite sequence of nonzero complex numbers with no finite limit points and with finite exponent of convergence. Let $p\geq 0$ be the smallest integer that satisfies

$$\sum_{n=1}^{\infty}\frac{1}{|z_n|^{p+1}}<\infty.$$

Then the *canonical product* of $\{z_n\}$ is defined as

$$P(z)=\prod_{n=1}^{\infty}E_p\left(\frac{z}{z_n}\right). \tag{3.5.1}$$

The integer p is called the *genus* of the canonical product $P(z)$.

From the definition it is immediate that the genus p is unique and that the series $\sum_{n=1}^{\infty}|z_n|^{-p}$ diverges. It is possible that $p=0$ for an infinite sequence $\{z_n\}$. This happens, for example, when $z_n=n^2$.

The next result gives basic properties of the canonical product. Here we recall the definitions of the *floor function* $\lfloor x\rfloor$ and the *ceiling function* $\lceil x\rceil$ of a real number x, which are defined as the greatest integer $\leq x$ and the least integer $\geq x$, respectively. For example, $\lfloor\pi\rfloor=3$ and $\lceil\pi\rceil=4$.

Proposition 3.5.2. *Let $P(z)$ be the canonical product of $\{z_n\}$ with genus p as defined in* (3.5.1). *Then $P(z)$ is an entire function, and*

$$p \le \lambda(P) \le p+1. \tag{3.5.2}$$

Furthermore,

$$n(r, 1/P) = o(r^{p+1}), \quad r \to \infty. \tag{3.5.3}$$

Proof. Since

$$\sum_{n=1}^{\infty} \left|\frac{z}{z_n}\right|^{p+1} \le R^{p+1} \sum_{n=1}^{\infty} \frac{1}{|z_n|^{p+1}} = O(R^{p+1}), \quad |z| \le R,$$

it is clear from the proof of the Weierstrass factorization theorem that $P(z)$ is entire and has zeros precisely at the points z_n.

If $\lambda(P)$ is not an integer, then $p = \lfloor \lambda(P) \rfloor$, whereas if $\lambda(P)$ is an integer, then either $p = \lambda(P)$ or $p = \lambda(P) - 1$. It follows that (3.5.2) holds.

Finally, since the sum $\sum_{n=1}^{\infty} |z_n|^{-(p+1)}$ converges, the integral $\int_0^{\infty} \frac{n(t)}{t^{p+2}}\, dt$ converges by Lemma 3.1.4. Therefore,

$$\frac{n(r)}{(p+1)r^{p+1}} = n(r) \int_r^{\infty} \frac{dt}{t^{p+2}} \le \int_r^{\infty} \frac{n(t)}{t^{p+2}}\, dt = o(1), \quad r \to \infty. \tag{3.5.4}$$

This yields (3.5.3). □

We now prove the following useful result.

Theorem 3.5.3. *The canonical product P in* (3.5.1) *satisfies $\lambda(P) = \rho(P)$.*

The next lemma is used in the proof of Theorem 3.5.3.

Lemma 3.5.4. *If $\varepsilon > 0$ is a small constant and*

$$A = \max\left\{1, \frac{1+\varepsilon}{\varepsilon}\right\},$$

then

$$\log(1+x) \le Ax^{\varepsilon}, \quad x \ge 0. \tag{3.5.5}$$

Proof. Obviously, (3.5.5) holds for $x = 0$. Since $A \ge 1$, we obtain that for $0 < x < 1$,

$$Ax^{\varepsilon} - \log(1+x) \ge Ax^{\varepsilon} - x \ge x^{\varepsilon} - x > 0,$$

which gives (3.5.5). Finally, we see that $Ax^{\varepsilon} - \log(1+x)$ is strictly increasing and nonnegative for $x \ge 1$, which gives (3.5.5). □

Proof of Theorem 3.5.3. Since $\lambda(P) \leq \rho(P)$ by Theorem 3.2.4, it suffices to prove that $\rho(P) \leq \lambda(P)$. To this end, we need a suitable majorant of $M(r,P)$.

Fix z, $|z| = r > 0$, and $\varepsilon > 0$. Then

$$\log M(r,P) = \log \max_{|\zeta|=r} |P(\zeta)| = \max_{|\zeta|=r} \log|P(\zeta)|.$$

Clearly, using $\log^+ x = \max\{0, \log x\}$, $x \geq 0$, we have

$$\begin{aligned}\log|P(z)| &= \log \prod_{n=1}^{\infty} \left|E_p\left(\frac{z}{z_n}\right)\right| \\ &\leq \sum_{|z/z_n|\geq 1/2} \log^+\left|E_p\left(\frac{z}{z_n}\right)\right| + \sum_{|z/z_n|<1/2} \log^+\left|E_p\left(\frac{z}{z_n}\right)\right| =: S_1 + S_2.\end{aligned}$$

Note that S_1 is a finite sum by the uniqueness theorem of analytic functions because the zeros z_n cannot accumulate to a finite point in $\mathbb{C}$.

To estimate S_2, where $\left|\frac{z}{z_n}\right| < 1/2 < 1$, we recall Lemma 3.4.3, according to which

$$\left|E_p\left(\frac{z}{z_n}\right)\right| \leq 1 + \left|\frac{z}{z_n}\right|^{p+1}. \tag{3.5.6}$$

Since $\log^+(1+x) = \log(1+x) \leq x$ for all $x \geq 0$, we have

$$S_2 \leq \sum_{|z/z_n|<1/2} \left|\frac{z}{z_n}\right|^{p+1}. \tag{3.5.7}$$

The three cases in the definition of the genus p need to be checked. If $\lambda(P) \in \mathbb{N}$ and $p = \lambda(P) - 1$, then

$$S_2 \leq |z|^{p+1} \sum_{n=1}^{\infty} \frac{1}{|z_n|^{p+1}} = O(|z|^{p+1}) = O(|z|^{\lambda(P)}).$$

If $\lambda(P) \in \mathbb{N} \cup \{0\}$ and $p = \lambda(P)$ or if $\lambda(P) \notin \mathbb{N} \cup \{0\}$ and $p = \lfloor \lambda(P) \rfloor$, then we have $p + 1 > \lambda(P) + \varepsilon$ for $\varepsilon > 0$ small enough. Hence, by (3.5.7) and the definition of exponent of convergence,

$$\begin{aligned}S_2 &\leq |z|^{\lambda(P)+\varepsilon} \sum_{|z/z_n|<1/2} \left|\frac{z}{z_n}\right|^{p+1-\lambda(P)-\varepsilon} |z_n|^{-\lambda(P)-\varepsilon} \\ &\leq |z|^{\lambda(P)+\varepsilon} \left(\frac{1}{2}\right)^{p+1-\lambda(P)-\varepsilon} \sum_{n=1}^{\infty} |z_n|^{-\lambda(P)-\varepsilon} = O(|z|^{\lambda(P)+\varepsilon}).\end{aligned}$$

Thus in all possible cases, we have

$$S_2 = O(|z|^{\lambda(P)+\varepsilon}). \tag{3.5.8}$$

To estimate the finite sum S_1, we consider two separate cases.

Case 1. First, suppose that the genus $p = 0$. Then by Lemma 3.5.4 we obtain

$$
\begin{aligned}
S_1 &= \sum_{|z/z_n|\geq 1/2} \log^+ \left| E_0\left(\frac{z}{z_n}\right)\right| \leq \sum_{|z/z_n|\geq 1/2} \log\left(1 + \left|\frac{z}{z_n}\right|\right) \\
&\leq A \sum_{|z/z_n|\geq 1/2} \left|\frac{z}{z_n}\right|^{\varepsilon} = A|z|^{\varepsilon} \sum_{|z/z_n|\geq 1/2} |z_n|^{-\varepsilon}.
\end{aligned} \tag{3.5.9}
$$

Since $p = 0$, we must have $\lambda(P) \in [0, 1]$. If $\lambda(P) = 0$, then $\sum |z_n|^{-\varepsilon}$ converges, and by (3.5.9),

$$S_1 = O(|z|^{\varepsilon}) = O(|z|^{\lambda(P)+\varepsilon}).$$

If $\lambda(P) = 1$, then since $\sum |z_n|^{-1}$ converges, by (3.5.9) we get

$$S_1 \leq A|z| \sum_{|z/z_n|\geq 1/2} \left|\frac{z}{z_n}\right|^{\varepsilon-1} |z_n|^{-1} \leq 2^{1-\varepsilon} A|z| \sum_{n=1}^{\infty} |z_n|^{-1} = O(|z|) = O(|z|^{\lambda(P)}).$$

Assume next that $\lambda(P) \in (0, 1)$ and $\varepsilon \in (0, \lambda(P))$. Then by (3.5.9),

$$
\begin{aligned}
S_1 &\leq A|z|^{\lambda(P)+\varepsilon} \sum_{|z/z_n|\geq 1/2} \left|\frac{z}{z_n}\right|^{-\lambda(P)} |z_n|^{-\lambda(P)-\varepsilon} \\
&\leq 2^{\lambda(P)} A|z|^{\lambda(P)+\varepsilon} \sum_{|z/z_n|\geq 1/2} |z_n|^{-\lambda(P)-\varepsilon} = O(|z|^{\lambda(P)+\varepsilon}).
\end{aligned}
$$

Hence in all possible situations when $p = 0$, we have

$$S_1 = O(|z|^{\lambda(P)+\varepsilon}). \tag{3.5.10}$$

Case 2. Finally, suppose that $p \geq 1$. Then for each term in S_1,

$$
\begin{aligned}
\log^+ \left| E_p\left(\frac{z}{z_n}\right)\right| &\leq \log^+ \left|1 - \frac{z}{z_n}\right| + \left|\frac{z}{z_n}\right| + \cdots + \frac{1}{p}\left|\frac{z}{z_n}\right|^p \\
&\leq 2\left(\left|\frac{z}{z_n}\right| + \cdots + \left|\frac{z}{z_n}\right|^p\right) \\
&\leq 2\left|\frac{z}{z_n}\right|^p \left(1 + \left|\frac{z_n}{z}\right| + \cdots + \left|\frac{z_n}{z}\right|^{p-1}\right) \\
&\leq 2\left|\frac{z}{z_n}\right|^p (1 + 2 + \cdots + 2^{p-1}) \leq 2^{p+1}\left|\frac{z}{z_n}\right|^p.
\end{aligned} \tag{3.5.11}
$$

If now $p = \lambda(P) - 1$, then

$$\log^+ \left| E_p\left(\frac{z}{z_n}\right)\right| \leq 2^{p+1}\left|\frac{z}{z_n}\right|^{\lambda(P)-1} = 2^{p+1}\left|\frac{z}{z_n}\right|^{\lambda(P)} \left|\frac{z_n}{z}\right| \leq 2^{p+2}\left|\frac{z}{z_n}\right|^{\lambda(P)},$$

and, consequently,

$$S_1 \le 2^{p+2}|z|^{\lambda(P)} \sum_{|z/z_n|\ge 1/2} |z_n|^{-\lambda(P)} = O(|z|^{\lambda(P)}).$$

If $p \ne \lambda(P) - 1$ and $\varepsilon > 0$ is small enough, then $p < \lambda(P) + \varepsilon \le p + 1$, and by (3.5.11),

$$\begin{aligned}\log^+\left|E_p\left(\frac{z}{z_n}\right)\right| &\le 2^{p+1}\left|\frac{z}{z_n}\right|^p = 2^{p+1}\left|\frac{z}{z_n}\right|^{\lambda(P)+\varepsilon}\left|\frac{z_n}{z}\right|^{\lambda(P)+\varepsilon-p} \\ &\le 2^{p+1+\lambda(P)+\varepsilon-p}\left|\frac{z}{z_n}\right|^{\lambda(P)+\varepsilon} \le 2^{p+2}\left|\frac{z}{z_n}\right|^{\lambda(P)+\varepsilon},\end{aligned}$$

and, consequently,

$$S_1 \le 2^{p+2}|z|^{\lambda(P)+\varepsilon} \sum_{|z/z_n|\ge 1/2} |z_n|^{-\lambda(P)-\varepsilon} = O(|z|^{\lambda(P)+\varepsilon}).$$

Thus in all possible situations when $p \ge 1$, we have

$$S_1 = O(|z|^{\lambda(P)+\varepsilon}). \tag{3.5.12}$$

Putting together (3.5.8), (3.5.10), and (3.5.12), we obtain $S_1 + S_2 = O(|z|^{\lambda(P)+\varepsilon})$, which yields

$$\log|P(z)| = O(|z|^{\lambda(P)+\varepsilon}).$$

Since the upper bound is independent of the argument of z and $|z| = r$, we have

$$\log M(r, P) = O(r^{\lambda(P)+\varepsilon}).$$

Hence $\rho(P) \le \lambda(P) + \varepsilon$. Since ε is arbitrarily small, we get $\rho(P) \le \lambda(P)$, which proves the theorem. □

We use the following notation in the next example.

Definition 3.5.5. Suppose that two functions $F, G : (r_0, \infty) \to (0, \infty)$, where $r_0 \ge 0$, satisfy

$$C_1 F(r) \le G(r) \le C_2 F(r)$$

for all r large enough, where $C_1 > 0$ and $C_2 > 0$ are fixed constants. Then we say that $F(r)$ and $G(r)$ are *asymptotically comparable*, which is denoted by $F(r) \asymp G(r)$.

Example 3.5.6. Let $0 < a < 1$ and consider the function

$$f(z) = \prod_{n=1}^{\infty}(1 - a^n z). \tag{3.5.13}$$

The zeros of f are $z_n = 1/a^n$, $n \in \mathbb{N}$, and since

$$\sum_{n=1}^{\infty} \frac{1}{|z_n|^{\varepsilon}} = \sum_{n=1}^{\infty} (a^{\varepsilon})^n < \infty \tag{3.5.14}$$

for all $\varepsilon > 0$, we see that f in (3.5.13) is a canonical product with genus $p = 0$. Thus by Proposition 3.5.2, f is entire, and in view of (3.5.14), the exponent of convergence $\lambda(f) = 0$. Hence by Theorem 3.5.3, the order of f is $\rho(f) = 0$.

A more careful analysis gives the asymptotic growth for the frequency of zeros of f. To this end, let $r \geq 1/a$. Then there exists $k \in \mathbb{N}$ such that

$$a^{-k} \leq r < a^{-k-1}.$$

Then $n(r) = k$, and

$$k \log \frac{1}{a} \leq \log r < (k+1) \log \frac{1}{a},$$

from which we obtain $n(r) \asymp \log r$ and $N(r) \asymp \log^2 r$.

Entire functions of finite order have a unique factorization. This result is known as the *Hadamard factorization theorem* or *Hadamard's theorem*, the precise statement of which is given in Theorem 3.5.7. The uniqueness of the factorization carries over to entire functions of finite iterated k-order as well; see Section 11.3 and [150, Satz 12.4].

Theorem 3.5.7. *Let f be a nonconstant entire function of finite order ρ. Then*

$$f(z) = z^m P(z) e^{Q(z)}, \tag{3.5.15}$$

where $m \geq 0$ is the multiplicity of the zero of f at $z = 0$, $Q(z)$ is a polynomial of degree $\leq \rho$, and $P(z)$ has one of the following three forms:
(a) *$P(z) \equiv 1$ when f has no nonzero zeros.*
(b) *$P(z) = (z - z_1) \cdots (z - z_n)$ when f has $n \geq 1$ nonzero zeros $z_1, \ldots, z_n$.*
(c) *$P(z)$ is the canonical product of the nonzero zeros of f when f has an infinite number of zeros.*

The proof of Hadamard's theorem is postponed to Section 3.8, because we do not yet have the necessary tools to prove it.

Hadamard's theorem gives the product representation for the function $\sin(\pi z)$ with less effort than the Weierstrass theorem; see Theorem 3.4.7. Indeed, we know that the function h in (3.4.6) is a polynomial of degree at most one, and hence we may write $h(z) = az + b$. Since the sine function is odd, we find that $a = 0$. Similarly to (3.4.8), it follows that $e^b = 1$, which gives $e^{h(z)} \equiv 1$.

Jacques Salomon Hadamard (1865–1963) was a French mathematician, who made major contributions in number theory, complex analysis, differential geometry, and partial differential equations. He was elected to the French Academy of Sciences in 1916, in succession to Henri Poincaré.

Definition 3.5.8. Let f be an entire function of the form (3.5.15) with infinitely many zeros. Then the *genus* of f is the number $\max\{p, q\}$, where p is the genus of the canonical product $P(z)$, and $q = \deg Q$. If f is of the form (3.5.15) but has at most finitely many zeros, then we say that the genus of f is equal to q.

We now use Hadamard's theorem to prove some results about entire functions of finite order ρ when ρ is not a positive integer.

Corollary 3.5.9. *Let f be an entire function of finite noninteger order ρ. Then $\lambda(f) = \rho$.*

Proof. In general, $\lambda(f) \le \rho$ by Theorem 3.2.4, so we suppose to the contrary of the statement that $\lambda(f) < \rho$. By Hadamard's theorem, f has the representation (3.5.15), where $Q(z)$ is a polynomial of degree $n \le \rho$. Since n is an integer while ρ is not, we have $n < \rho$. Write

$$Q(z) = a_n z^n + \cdots + a_0, \quad a_n \neq 0.$$

By Lemma 2.1.4 there exists $R > 0$ such that

$$M(r, e^Q) \le e^{2|a_n| r^n}, \quad r \ge R,$$

so that $\rho(e^Q) \le n$. On the other hand, by the Borel–Carathéodory theorem and Lemma 2.1.4 (with $\varepsilon = 1/2$), we obtain

$$\begin{aligned} M(r, e^Q) &= \max_{|z|=r} |e^{Q(z)}| = \max_{|z|=r} e^{\operatorname{Re} Q(z)} = e^{A(r,Q)} \\ &\ge \exp(M(r/2, Q)/2 - 3|Q(0)|/2) \\ &\ge \exp(|a_n| r^n / 2^{n+2} - 3|Q(0)|/2) \end{aligned}$$

for all $r \ge R_0 \ge R$. This gives us $\rho(e^Q) \ge n$, and hence we have proved that $\rho(e^Q) = n < \rho$. By Theorems 2.1.10 and 3.5.3

$$\rho(f) \le \max\{\rho(z^m), \rho(P), \rho(e^Q)\} = \max\{0, \lambda(f), n\} < \rho = \rho(f),$$

a contradiction. □

Corollary 3.5.10. *If f is a transcendental entire function of finite order $\rho \notin \mathbb{N}$, then f has infinitely many zeros.*

Proof. We have $\rho = \lambda(f)$ by Corollary 3.5.9. So, if $\rho > 0$, then $\lambda(f) > 0$, and hence f must have infinitely many zeros. If $\rho = 0$, then Hadamard's theorem gives the representation

$$f(z) = cz^m P(z), \quad c \in \mathbb{C},\ m \geq 0.$$

Since f is transcendental, $P(z)$ cannot be a polynomial, which means that $P(z)$ is a canonical product with infinitely many zeros. Thus f has infinitely many zeros. □

Theorem 3.5.11. *If f is an entire function of finite positive order $\rho \notin \mathbb{N}$, then f is of finite type if and only if $n(r) = O(r^\rho)$ and of zero type if and only if $n(r) = o(r^\rho)$.*

We need an auxiliary result.

Lemma 3.5.12 ([37, p. 19]). *Let P be the canonical product in* (3.5.1) *with genus $p \geq 0$ and zeros $z_n \neq 0$. Let $n(r)$ denote the counting function of the zeros z_n. Then there exists a constant $C > 0$ such that*

$$\log|P(z)| \leq Cr^{p+1} \int_0^\infty \frac{n(t)}{t^{p+1}(r+t)}\, dt, \quad |z| = r \geq 0.$$

Proof. Denote $r_n = |z_n|$ and $|z| = r > 0$. Similarly as in the proof of Theorem 3.5.3,

$$\log|P(z)| \leq \sum_{r_n \leq 2r} \log^+\left|E_p\left(\frac{z}{z_n}\right)\right| + \sum_{r_n > 2r} \log^+\left|E_p\left(\frac{z}{z_n}\right)\right| =: S_1 + S_2,$$

where by (3.5.9) and (3.5.11),

$$S_1 \leq \begin{cases} 2^{p+1} r^p \sum_{r_n \leq 2r} r_n^{-p}, & p \geq 1, \\ \sum_{r_n \leq 2r} \log(1 + r/r_n), & p = 0, \end{cases} \tag{3.5.16}$$

and by (3.5.7),

$$S_2 \leq r^{p+1} \sum_{r_n > 2r} r_n^{-(p+1)}. \tag{3.5.17}$$

Now, if $p \geq 1$, then by Riemann–Stieltjes integration and integration by parts,

$$S_1 \leq 2^{p+1} r^p \int_0^{2r} \frac{dn(t)}{t^p} = 2n(2r) + p2^{p+1} r^p \int_0^{2r} \frac{n(t)}{t^{p+1}}\, dt.$$

Similarly, if $p = 0$, then

$$S_1 \leq \int_0^{2r} \log(1 + r/t)\, dn(t) \leq n(2r)\log(3/2) + \int_0^{2r} \frac{n(t)}{t}\, dt.$$

Thus for all values of $p \geq 0$,

$$S_1 \le 2n(2r) + (p+1)2^{p+1} r^p \int_0^{2r} \frac{n(t)}{t^{p+1}}\, dt.$$

Recall from (3.5.3) that $n(r) = o(r^{p+1})$. Therefore, we find that

$$\begin{aligned} S_2 &\le r^{p+1} \int_{2r}^{\infty} \frac{dn(t)}{t^{p+1}} = -\frac{n(2r)}{2^{p+1}} + (p+1) r^{p+1} \int_{2r}^{\infty} \frac{n(t)}{t^{p+2}}\, dt \\ &\le (p+1) r^{p+1} \int_{2r}^{\infty} \frac{n(t)}{t^{p+2}}\, dt. \end{aligned}$$

Hence for $B = (p+1)2^{p+1}$, it follows that

$$\log|P(z)| \le B\left(r^p \int_0^{2r} \frac{n(t)}{t^{p+1}}\, dt + r^{p+1} \int_{2r}^{\infty} \frac{n(t)}{t^{p+2}}\, dt \right) + 2n(2r),$$

where

$$2n(2r) = 2Br^{p+1} n(2r) \int_{2r}^{\infty} \frac{dt}{t^{p+2}} \le 2Br^{p+1} \int_{2r}^{\infty} \frac{n(t)}{t^{p+2}}\, dt.$$

We conclude that

$$\begin{aligned} \log|P(z)| &\le Br^p \int_0^{2r} \frac{n(t)}{t^{p+1}}\, dt + 3Br^{p+1} \int_{2r}^{\infty} \frac{n(t)}{t^{p+2}}\, dt \\ &\le 3Br^{p+1} \int_0^{2r} \frac{n(t)\, dt}{t^{p+1}(t+r)} + 5Br^{p+1} \int_{2r}^{\infty} \frac{n(t)\, dt}{t^{p+1}(t+r)} \\ &\le 5Br^{p+1} \int_0^{\infty} \frac{n(t)\, dt}{t^{p+1}(t+r)}, \qquad r > 0. \end{aligned}$$

Since $P(0) = 1$, the statement follows for all $r \ge 0$ by choosing $C = 5B$. □

Proof of Theorem 3.5.11. If f is of finite type, then the growth bounds for $n(r)$ follow by Theorem 3.2.7. Thus we assume the growth bounds for $n(r)$ and consider the type of f. By Hadamard's theorem

$$f(z) = z^m P(z) e^{Q(z)},$$

where $Q(z)$ is a polynomial with $q = \deg Q < \rho$ because $q \le \rho$ and $\rho \notin \mathbb{N}$. Thus $\log|e^{Q(z)}| = O(r^q) = o(r^\rho)$, where $|z| = r$. To estimate the type τ of f, we estimate the growth of $\log M(r, f)$, for which it suffices to estimate $\log|P(z)|$. Since $\rho(f) \notin \mathbb{N}$ and $\rho(f) < \infty$, it follows by Corollary 3.5.10 that $P(z)$ must be a canonical product. Let p denote the genus of $P(z)$. Since $\lambda(P) = \rho \notin \mathbb{N}$ by Theorem 3.5.3 and the assumption, we have $p < \rho < p+1$. On one hand, if $n(r) = O(r^\rho)$, then Lemma 3.5.12 yields (for some constant $C_0 > 0$)

$$\log|P(z)| \le C_0 r^{p+1} \int_0^\infty \frac{dt}{t^{p-\rho+1}(r+t)}$$
$$\le C_0 r^p \int_0^r \frac{dt}{t^{p-\rho+1}} + C_0 r^{p+1} \int_r^\infty \frac{dt}{t^{p-\rho+2}} = O(r^\rho).$$

Since the right-hand side does not depend on $\arg z$, we obtain $\log M(r,P) = O(r^\rho)$, and hence $\log M(r,f) = O(r^\rho)$. This implies that f is of finite type. On the other hand, if $n(r) = o(r^\rho)$, then for every $\varepsilon > 0$, there exists $R_0 > 0$ such that

$$\log|P(z)| \le O(r^\rho) + \varepsilon r^p \int_{R_0}^r \frac{dt}{t^{p-\rho+1}} + \varepsilon r^{p+1} \int_r^\infty \frac{dt}{t^{p-\rho+2}}, \qquad r > R_0.$$

This yields $\log M(r,P) = o(r^\rho)$, which in turn implies that $\tau(f) = 0$. □

Problems

Problem 3.5.13. Prove that the function $f(z) = \prod_{n=1}^\infty \left(1 - \frac{z^n}{n!}\right)$ is entire and determine the order of f.

Problem 3.5.14. Find the Hadamard product factorization of the functions

$$f(z) = e^z - 1 \quad \text{and} \quad g(z) = \cos(\pi/2 + z).$$

Problem 3.5.15. Find a transcendental entire function g satisfying $\rho(g) = 1/5$. Then find a transcendental entire function f with $\rho(f) = 4$ and $\lambda(f) = 1/5$.

Problem 3.5.16. Find the canonical product that has simple zeros at the points $\sqrt{n}\log n$, $n = 2, 3, 4, \ldots$.

Problem 3.5.17. Find the order of the entire functions:
(a) $f(z) = z^2 \prod_{n=1}^\infty \left(1 - \frac{z}{n^3}\right)$.
(b) $g(z) = e^{z^3} \prod_{n=1}^\infty \left(1 - \frac{z}{\sqrt{n}}\right) \exp\left(\frac{z}{\sqrt{n}} + \frac{1}{2}(\frac{z}{\sqrt{n}})^2\right)$.

Problem 3.5.18. Give examples of entire functions f, g, h that satisfy the following conditions:
(a) $\lambda(f) = 1$, and the genus of the canonical product of f is $p = 1$.
(b) $\lambda(g) = 1$, and the genus of the canonical product of g is $p = 0$.
(c) $\rho(h) = 2$, $\lambda(h) = 1$, and the genus of the canonical product of h is $p = 0$.

Problem 3.5.19. Prove that the differential equation

$$z^3 f'' + f' + f = 0$$

cannot possess a solution f that is an entire function of finite order with at most finitely many zeros.

Problem 3.5.20. Prove that the equation $\tan z = z$ has an infinite number of roots for $z \in \mathbb{C}$.

Hint: Use the entire function $\sin z - z \cos z$.

3.6 Theorems of Picard and Borel

In this section, we discuss two seminal results in the theory of entire functions, proved by Picard and Borel. Charles Émile Picard (1856–1941) was a French mathematician and a historian of science. He was a member of the French Science Academy and of the French Academy. Alongside his accomplishments in function theory, he made important contributions to the theory of differential equations, including Painlevé equations. Félix Édouard Justin Émile Borel (1871–1956) was a French mathematician and politician. In addition to his important work in entire functions, he was known for his founding work in measure theory and probability. Borel's lemma is of fundamental importance in Nevanlinna theory.

We now state *Picard's theorem*, which he proved in 1879.

Theorem 3.6.1 ([273, p. 277]). *If f is a transcendental entire function, then f assumes every finite value infinitely often, with at most one possible exception.*

An exceptional value is possible, since e^z never assumes the value 0. This is a special kind of exceptional value.

Definition 3.6.2. Let f be a nonconstant entire function. If for some $a \in \mathbb{C}$, we have $f(z) \neq a$ for all $z \in \mathbb{C}$, then a is called a *Picard value* of f.

Picard's eye-opening result was a cause for celebration. Indeed, when speaking about one of the forms of Picard's theorem, Hille said [139, p. 219]:

> This theorem was justly hailed as an outstanding achievement and for many years it stirred the imagination of many prominent mathematicians with results that were ultimately epoch making.

Picard used the elliptic modular function to prove Theorem 3.6.1. A complete proof of Theorem 3.6.1 that uses the elliptic modular function is given in [218]. In Section 4.6, Nevanlinna theory will be used to give a short proof of Theorem 3.6.1.

Recall that $\lambda(a,f) \leq \rho(f)$ for any entire function f and $a \in \mathbb{C}$. If a strict inequality $\lambda(a,f) < \rho(f)$ holds, then a is called a *Borel exceptional value* of f. The following result is known as *Borel's theorem.*

Theorem 3.6.3 ([142, Theorem 8.2]). *Let f be an entire function of positive integer order $\rho \in \mathbb{N}$. Then f has at most one finite Borel exceptional value.*

Proof. Suppose that f has two Borel exceptional values $a, b \in \mathbb{C}$. Then $\lambda(a,f) < \rho$ and $\lambda(b,f) < \rho$. By the Hadamard factorization theorem,

$$f(z) - a = z^{m_1} P_1(z) e^{Q_1(z)},$$
$$f(z) - b = z^{m_2} P_2(z) e^{Q_2(z)},$$

where $Q_1(z)$ and $Q_2(z)$ are polynomials of degree $= \rho$, and $P_1(z)$ and $P_2(z)$ are either canonical products or polynomials determined by the a-points and b-points of f, respectively. (If, for example, $a = 0$, then $P_1(z)$ is determined by the nonzero zeros of f.) In addition, $\rho_j = \rho(P_j) < \rho$ for $j = 1, 2$.

Subtracting and eliminating f, we get

$$b - a = z^{m_1} P_1(z) e^{Q_1(z)} - z^{m_2} P_2(z) e^{Q_2(z)}, \tag{3.6.1}$$

which can be rewritten as

$$z^{m_1} P_1(z) e^{Q_1(z) - Q_2(z)} = z^{m_2} P_2(z) + (b - a) e^{-Q_2(z)}.$$

Since the right-hand side of this equation is of order ρ, the left-hand side has order ρ as well. It follows that $\deg(Q_1(z) - Q_2(z)) = \rho$. Differentiating both sides of (3.6.1), we get

$$(z^{m_1} Q_1' P_1 + m_1 z^{m_1 - 1} P_1 + z^{m_1} P_1') e^{Q_1} = (z^{m_2} Q_2' P_2 + m_2 z^{m_2 - 1} P_2 + z^{m_2} P_2') e^{Q_2}.$$

Observe that this differentiated identity is valid even if one of m_1 and m_2 (or both) is zero. Recall that differentiation does not affect the order of growth by Theorem 2.3.1. Hence

$$\rho(P_j') = \rho_j < \rho, \quad j = 1, 2.$$

In particular, the coefficients of e^{Q_1} and e^{Q_2} are of order $< \rho$. Since the coefficients are entire functions, we may use Hadamard's theorem on both sides of the differentiated identity to obtain

$$z^{m_3} P_3(z) e^{Q_1(z) + Q_3(z)} = z^{m_4} P_4(z) e^{Q_2(z) + Q_4(z)}, \tag{3.6.2}$$

where $Q_3(z)$ and $Q_4(z)$ are polynomials of degree $\le \rho - 1$, and $P_3(z)$ and $P_4(z)$ are canonical products or polynomials. Since the zeros on both sides of (3.6.2) are the same, we must have $m_3 = m_4$ and $P_3(z) \equiv P_4(z)$. Therefore, for some integer n,

$$Q_1(z) - Q_2(z) = Q_4(z) - Q_3(z) + 2\pi i n.$$

This is a contradiction since $\deg(Q_1(z) - Q_2(z)) = \rho$, while $\deg(Q_4(z) - Q_3(z)) \le \rho - 1$. This proves that $\lambda(a, f) = \rho$ for all a with at most one exception. □

Complementing Theorem 3.6.3, we prove that entire functions of finite noninteger order have no finite Borel exceptional values.

Theorem 3.6.4 ([142, Theorem 8.4]). *Let f be an entire function of finite noninteger order ρ. Then f has no finite Borel exceptional values.*

Proof. Suppose on the contrary that $a \in \mathbb{C}$ is a Borel exceptional value of f. Without loss of generality (by considering $g(z) = f(z) - a$), we may suppose that $a = 0$. Hence $\lambda(f) < \rho$. However, by Corollary 3.5.9 we have $\lambda(f) = \rho$, a contradiction. □

In the 1920s, Rolf Nevanlinna developed his modern theory of meromorphic functions, which extended the theorems of Picard and Borel in major ways. Indeed, the theorems of Picard and Borel are immediate corollaries of the second fundamental theorem, which is the main result in his theory that is appropriately called *Nevanlinna theory*. The details can be found in Theorem 4.6.9. Function theory was elevated to a significantly higher level with the creation of Nevanlinna theory, which is a best possible theory in several ways. An introduction to Nevanlinna theory will be given in Section 4.6.

Problems

Problem 3.6.5. Suppose that f is an entire function that satisfies $|f(z) + 3| \geq 2$ for all $z \in \mathbb{C}$. Show that f must be a constant.

Problem 3.6.6. Show that an entire function f that assumes every value in the finite complex plane exactly once must have the form $f(z) = az + b$, where $a \neq 0$ and b are constants.

Problem 3.6.7. Suppose that f and g are entire functions that satisfy

$$e^{f(z)} + e^{g(z)} = 1, \quad z \in \mathbb{C}.$$

Prove that f and g are constant functions.

Problem 3.6.8. Construct an entire function f satisfying $\rho(f) = 6$ and $\lambda(1, f) = 8/3$. Then state the value of $\lambda(2, f)$ and justify why.

3.7 Lindelöf's theorem

The main result in this section is Lindelöf's theorem, which deals with entire functions of positive integer order. Ernst Leonard Lindelöf (1870–1946) was a Finnish mathematician, who founded an international school on function theory in Finland. He was the PhD supervisor of Rolf Nevanlinna. The Finnish Mathematical Society, founded in 1868 by Lorenz Lindelöf, the father of Ernst Lindelöf, gives an annual Ernst Lindelöf award for the best mathematical Master's thesis in Finland.

The next result is *Lindelöf's theorem*, where the assumption $\lambda(f) = \rho(f)$ is not necessary, but it leaves out the discussions of the trivial case $\lambda(f) < \rho(f)$.

Theorem 3.7.1 ([37, Theorem 2.10.1]). *Let f be an entire function of positive integer order $\rho = \rho(f)$ such that $\lambda(f) = \rho(f)$, and let $\{z_n\}$ denote the sequence of nonzero zeros of f, organized by increasing moduli. Then f is of finite type if and only if $n(r) = O(r^{\rho})$ and the expression*

$$S(r) = \left| \sum_{|z_n| \le r} z_n^{-\rho} \right| \tag{3.7.1}$$

is a bounded function of $r > 0$.

We need two lemmas for the proof.

Lemma 3.7.2. *Let* $\rho \in \mathbb{N}$, *and let* f *be an entire function of the form*

$$f(z) = z^m e^{Q(z)} P(z),$$

where $m \ge 0$ *is an integer,* $Q(z)$ *is a polynomial of degree* $\le \rho$, *and* $P(z)$ *is a canonical product formed by the nonzero zeros* z_n *of* f, *organized by increasing moduli, such that* $\lambda(P) = \rho$. *Suppose that the genus* p *of* $P(z)$ *satisfies* $p = \rho - 1$. *Then* $\rho(f) = \rho$, *and the following three statements hold:*

(a) *The function* $S(r)$ *in* (3.7.1) *is bounded.*
(b) $n(r, 1/P) = o(r^\rho)$.
(c) f *is of mean type if* $\deg Q = \rho$ *and of zero type if* $\deg Q < \rho$.

Proof. We begin by proving that $\rho(f) = \rho$. We have $\rho(P) = \lambda(P)$ by Theorem 3.5.3 and also $\rho(e^Q) \le \rho$. Since $\rho(z^m) = 0$, it follows from Theorem 2.1.10(b) that $\rho(f) \le \rho$. On the other hand, $\rho = \lambda(P) = \lambda(f) \le \rho(f)$ by Theorem 3.2.4. Hence $\rho(f) = \rho$.

We proceed to prove statements (a), (b), and (c). By the definition of the genus p and the fact that $p + 1 = \rho$, we have

$$\sum_n |z_n|^{-\rho} < \infty,$$

which gives (a) by Lemma 3.3.7. Since $\rho = p + 1$, statement (b) follows by (3.5.3). As for (c), we use Lemma 3.5.12, the last equation in (3.5.4), and statement (b) to find that there exists a constant $C > 0$ such that (with $n(t) = n(t, 1/P)$)

$$\begin{aligned} \log|P(z)| &\le Cr^\rho \int_0^\infty \frac{n(t)}{t^\rho (r+t)}\, dt \\ &\le Cr^{\rho-1} \int_0^r \frac{n(t)}{t^\rho}\, dt + Cr^\rho \int_r^\infty \frac{n(t)}{t^{\rho+1}}\, dt = o(r^\rho). \end{aligned}$$

Since $\rho(f) = \rho$, this shows that the type of f is determined by the leading coefficient a_ρ of $Q(z)$ if $a_\rho \ne 0$, and otherwise f is of zero type. □

Lemma 3.7.3 ([37, Lemma 2.10.5]). *Suppose that* f *is an entire function satisfying* $f(0) = 1$, *and let* $\sigma > 0$. *For* $r > 0$, *let* $2\pi r \cdot l(\sigma, r)$ *denote the linear measure of the part of the circle* $|z| = r$ *on which* $|f(re^{i\theta})| < M(r, f)^{-\sigma}$. *Then for all large* $r > 0$,

$$l(\sigma, r) \le (1 + \sigma)^{-1}.$$

Proof. Let $r > 0$ be large enough so that $\log M(r,f) > 0$. On the specified part of the circle $|z| = r$, we have

$$\frac{1}{2\pi}\int \log|f(re^{i\theta})|\,d\theta \leq -\sigma l(\sigma,r)\log M(r,f), \tag{3.7.2}$$

whereas on the rest of the circle $|z| = r$, we have $\log|f(re^{i\theta})| \leq \log M(r,f)$, and the corresponding contribution to the left-hand side of (3.7.2) is at most

$$(1 - l(\sigma,r))\log M(r,f).$$

Using Jensen's theorem together with $f(0) = 1$, we therefore obtain

$$\begin{aligned} 0 \leq N(r) &= \frac{1}{2\pi}\int_0^{2\pi} \log|f(re^{i\theta})|\,d\theta \\ &\leq (1 - l(\sigma,r))\log M(r,f) - \sigma l(\sigma,r)\log M(r,f), \end{aligned}$$

which yields the statement. □

Proof of Theorem 3.7.1. Using Hadamard's factorization, we may write

$$f(z) = z^m e^{Q(z)} P(z),$$

where $m \geq 0$ is an integer, $Q(z) = a_\rho z^\rho + a_{\rho-1}z^{\rho-1} + \cdots + a_0$, and $P(z)$ is the canonical product formed with the zeros $z_n \neq 0$ of f. More precisely,

$$P(z) = \prod_{n=1}^{\infty} E_p\left(\frac{z}{z_n}\right),$$

where $p \geq 0$ is the genus of $P(z)$, and $E_p(\zeta)$ is the Weierstrass primary factor. Since $\lambda(f) = \rho$ is an integer, we have either $p = \rho$ or $p = \rho - 1$. However, the case $p = \rho - 1$ is already covered in Lemma 3.7.2, so we may suppose that $p = \rho$.

(1) Suppose first that $n(r) = O(r^\rho)$ and that the function $S(r)$ in (3.7.1) is bounded. We proceed to prove that f is of finite type. Since $p \geq 1$, we may write $E_p(\zeta) = \exp(p^{-1}\zeta^p)E_{p-1}(\zeta)$, giving us

$$\begin{aligned} P(z) &= \prod_{n=1}^{N} E_p\left(\frac{z}{z_n}\right) \cdot \prod_{n=N+1}^{\infty} E_p\left(\frac{z}{z_n}\right) \\ &= \prod_{n=1}^{N} \exp\left(\frac{1}{p}\left(\frac{z}{z_n}\right)^p\right) E_{p-1}\left(\frac{z}{z_n}\right) \cdot \prod_{n=N+1}^{\infty} E_p\left(\frac{z}{z_n}\right), \end{aligned}$$

where the integer N will be specified later. Then by the Hadamard factorization,

$$
\begin{aligned}
D &:= \log|f(z)| - \operatorname{Re}\left(z^p\left(a_p + p^{-1}\sum_{n=1}^{N} z_n^{-p}\right)\right) \\
&= \operatorname{Re}(a_{p-1}z^{p-1} + \cdots + a_0) + m\log|z| \\
&\quad + \sum_{n=1}^{N}\log|E_{p-1}(z/z_n)| + \sum_{n=N+1}^{\infty}\log|E_p(z/z_n)|.
\end{aligned} \tag{3.7.3}
$$

Denoting $r_n = |z_n|$ and applying estimates (3.5.16) and (3.5.17), where $\sum_{r_n\le 2r}$ is replaced with $\sum_{n=1}^{N}$ and $\sum_{r_n>2r}$ is replaced with $\sum_{n=N+1}^{\infty}$, we obtain, for $|z| = r$,

$$
D \le \begin{cases} O\left(r^{p-1}\right) + O\left(r^{p-1}\sum_{n=1}^{N} r_n^{-p+1}\right) + O\left(r^{p+1}\sum_{n=N+1}^{\infty} r_n^{-p-1}\right), & p > 1, \\ O(\log r) + O\left(\sum_{n=1}^{N}\log(1 + r/r_n)\right) + O\left(r^2\sum_{n=N+1}^{\infty} r_n^{-2}\right), & p = 1. \end{cases}
$$

Since $n(r) = O(r^p)$, there exists a constant $A > 0$ such that

$$
n \le Ar_n^p, \quad n \in \mathbb{N}. \tag{3.7.4}
$$

Assume first that $p > 1$. Choosing $N = \lfloor r^p \rfloor$ and using (3.7.4), we have

$$
\begin{aligned}
D &\le O(r^{p-1}) + O\left(r^{p-1}\sum_{n=1}^{N} n^{-(p-1)/p}\right) + O\left(r^{p+1}\sum_{n=N+1}^{\infty} n^{-(p+1)/p}\right) \\
&\le O(r^{p-1}) + O(r^{p-1})\left(1 + \int_1^N t^{-(p-1)/p}\,dt\right) + O(r^{p+1})\int_N^{\infty} t^{-(p+1)/p}\,dt \\
&= O(r^{p-1}) + O(r^{p-1}N^{1/p}) + O(r^{p+1}N^{-1/p}) = O(r^p), \quad p > 1.
\end{aligned} \tag{3.7.5}
$$

Now suppose that $p = 1$. We have

$$
\begin{aligned}
\int_1^r \log(1 + Ar/t)\,dt &= \left[Ar\log\left(1 + \frac{t}{Ar}\right) + t\log\left(1 + \frac{Ar}{t}\right)\right]_1^r \\
&= Ar\log\left(\frac{(1+A)r}{1+Ar}\right) + r\log(1+A) - \log(1+Ar) = O(r),
\end{aligned}
$$

where A is the constant in (3.7.4). Choosing $N = \lfloor r \rfloor$ and using (3.7.4), we have

$$
\begin{aligned}
D &\le O(\log r) + O\left(\sum_{n=1}^{N}\log(1 + Ar/n)\right) + O\left(r^2\sum_{n=N+1}^{\infty} n^{-2}\right) \\
&\le O(\log r) + O(1)\left(\log(1+Ar) + \int_1^N \log(1 + Ar/t)\,dt\right) + O(r^2)\int_N^{\infty} t^{-2}\,dt \\
&= O(r) + O(r^2N^{-1}) = O(r), \quad p = 1.
\end{aligned} \tag{3.7.6}
$$

Since we assumed that the function $S(r)$ is bounded and since $p = \rho = \rho(f)$, it follows from (3.7.3), (3.7.5), and (3.7.6) that

$$\log|f(z)| = \operatorname{Re}\left(z^p \left(a_p + p^{-1} \sum_{n=1}^{N} z_n^{-p} \right) \right) + D = O(r^p).$$

This means that f is of finite type.

(2) Conversely, suppose that f is of finite type, still assuming that $p = \rho$. Then $n(r) = O(r^p)$ by Theorem 3.2.7. Hence it suffices to prove that the function $S(r)$ is bounded. To this end, applying (3.7.3), (3.7.5), and (3.7.6), we obtain

$$-\operatorname{Re}\left(z^p \left(a_p + p^{-1} \sum_{n=1}^{N} z_n^{-p} \right) \right) \leq -\log|f(z)| + O(r^p). \tag{3.7.7}$$

Note that $N = \lfloor r^p \rfloor$ in (3.7.7) because this choice for N is used in (3.7.5) and (3.7.6). For some constant $B > 0$, we have $\log M(r,f) \leq Br^p$ for all $r > 0$ large enough. Moreover, for $\sigma = 3$, Lemma 3.7.3 shows that

$$\log|f(z)| < -3 \log M(r,f)$$

on at most one-fourth of the circle $|z| = r$ when $r > 0$ is large enough. Thus

$$-3 \log M(r,f) \leq \log|f(z)| \leq \log M(r,f) \leq Br^p$$

on at least three-fourths of $|z| = r$ when $r > 0$ is large enough. In other words,

$$\big|\log|f(z)|\big| = O(r^p)$$

on at least three-fourths of $|z| = r$. Combining this with (3.7.7), it follows that

$$-\operatorname{Re}\left(z^p \left(a_p + p^{-1} \sum_{n=1}^{N} z_n^{-p} \right) \right) \leq O(r^p) \tag{3.7.8}$$

on at least three-fourths of $|z| = r$, where $N = \lfloor r^p \rfloor$.

After these preparations, we now suppose, on the contrary to the statement, that $S(r)$ is unbounded. The monotonicity of $S(r)$ as a function of r is not known. However, since $S(r)$ is unbounded, there exists infinitely many positive integers K such that

$$a_p + p^{-1} \sum_{n=1}^{K} z_n^{-p} = R_K e^{i\phi_K}, \quad R_K \to \infty.$$

For each such K, let $r_K > 0$ be a number satisfying $K = \lfloor r_K^p \rfloor$. For $z = r_K e^{i\theta}$, estimate (3.7.8) can be rewritten as

$$-\mathrm{Re}(z^p R_K \exp(i\phi_K)) \le O(r_K^p), \tag{3.7.9}$$

which is valid on at least three-fourths of $|z| = r_K$. On the other hand,

$$\mathrm{Re}(z^p R_K \exp(i\phi_K)) = r_K^p R_K \cos(p\theta + \phi_K) < -\frac{r_K^p R_K}{2}$$

on the part of the circle $|z| = r_K$ where $\cos(p\theta + \phi_K) < -1/2$. By the properties of cosine, this corresponds to one-third of $|z| = r_K$. Hence

$$-\mathrm{Re}(z^p R_K \exp(i\phi_K)) > \frac{r_K^p R_K}{2}$$

on one-third of $|z| = r_K$. Since $R_K \to \infty$, this contradicts the fact that (3.7.9) holds on at least three-fourths of $|z| = r_K$. This completes the proof. □

It is clear from Theorem 3.2.7 that if f is an entire function of finite positive order ρ and of finite type, then the number of zeros of f satisfies $n(r) = O(r^\rho)$. The converse is not true by Example 3.7.4. Indeed, if $n(r) = O(r^\rho)$ for an entire function f of order $\rho \in (0, \infty)$, then f might not be of finite type, even if ρ is any positive integer. This shows that the boundedness of the function $S(r)$ in Theorem 3.7.1 is a necessary assumption.

Example 3.7.4. For $m \in \mathbb{N}$, let $f(z) = 1/\Gamma(z^m)$, where $\Gamma(z)$ is the gamma function, to be discussed in more detail in Section 5.1. It is known that $\Gamma(z)$ has no zeros, but it has a simple pole at every nonpositive integer $n = 0, -1, -2, \ldots$. This leads to the facts that $1/\Gamma(z)$ is an entire function of order one [205, p. 307] and of maximal type [205, p. 319]. Hence f is an entire function of integer order $\rho = m$ and of maximal type. It is clear that f has an m-fold zero at the origin, m evenly spaced simple zeros on every circle $|z| = n^{1/m}$, where $n \in \mathbb{N}$, and no other zeros.

To estimate $n(r) = n(r, 1/\Gamma(z^m))$, let $r \ge 1$. Then there exists an integer $k \ge 2$ such that

$$(k-1)^{1/m} \le r < k^{1/m}.$$

It follows that $n(r) = km$. Then using $k - 1 \le r^m < k$, we obtain

$$mr^\rho = mr^m < n(r) \le m\left(r^m + 1\right) = m\left(r^\rho + 1\right).$$

Consequently, $n(r) \sim mr^\rho$, even though f is of maximal type.

Problems

For the first two problems below, recall that an entire function f is called an *even function* if $f(z) = f(-z)$ for all $z \in \mathbb{C}$ and an *odd function* if $f(z) = -f(-z)$ for all $z \in \mathbb{C}$.

Problem 3.7.5. Prove that an even entire function f of order one has infinitely many zeros. This proves, for example, that $z^2 + \cos z$ has infinitely many zeros.

Problem 3.7.6. Find the form of all odd entire functions of order two that are real on the real axis and have the same zeros as $\sin z$ counted by multiplicities.

Problem 3.7.7. Suppose that f is an entire function of order one that has the same zeros as $\cos z$ counted by multiplicities. If $f(0) = 1$ and $f'(0) = 0$, then show that $f(z) = \cos z$ for all $z \in \mathbb{C}$.

Problem 3.7.8. For $f(z) = \sin z$, find the function $S(r)$ in (3.7.1).

Problem 3.7.9. Let f be analytic in $|z| < 1$, and suppose that

$$f(z) = 2f(z/2)f'(z/2), \quad |z| < 1.$$

Show that f can be extended to an entire function, that is, show that there exists an entire function g such that $g(z) = f(z)$ for $|z| < 1$. State a transcendental entire function g that has this property.

3.8 Estimates for canonical products

In this section, we prove a lower bound for the modulus of canonical products, which will allow us to prove two earlier results, including the Hadamard factorization theorem. To begin with, we recall the following facts between the exponent of convergence λ and the genus p of a given canonical product P: If λ is not an integer, then $p = \lfloor\lambda\rfloor$, whereas if λ is an integer, then either $p = \lambda$ or $p = \lambda - 1$. Using these facts, a lower bound for the modulus of P can be found by restricting the variable z away from the zeros of P.

Theorem 3.8.1 ([278, Theorem V.19]). *Let P be a canonical product formed by an infinite sequence $\{z_n\}$ with finite exponent of convergence λ and genus p. If $p = \lambda - 1$, then let $\varepsilon = 0$, and if $p > \lambda - 1$, then let $0 < \varepsilon \leq 1 - (\lambda - \lfloor\lambda\rfloor)$. Then there exists a constant $K > 0$ such that*

$$\log|P(z)| \geq -Kr^{\lambda+\varepsilon}\log r \tag{3.8.1}$$

for $z = re^{i\theta}$ lying outside the closed discs $|z - z_n| \leq 1/|z_n|^{\lambda+\varepsilon}$ and $|z| \leq e$.

Proof. Suppose first that $p \geq 1$. We may write

$$P(z) = \prod_{n=1}^{\infty}\left(1 - \frac{z}{z_n}\right)\exp\left(\frac{z}{z_n} + \cdots + \frac{1}{p}\left(\frac{z}{z_n}\right)^p\right).$$

Using $\log w = \log|w| + i\arg w$, we get, for $z \neq z_n$,

$$\begin{aligned}
|P(z)| &= \prod_{n=1}^{\infty}\left|\frac{z_n - z}{z_n}\right|\exp\left(\operatorname{Re}\left(\frac{z}{z_n} + \cdots + \frac{1}{p}\left(\frac{z}{z_n}\right)^p\right)\right)\\
&= \prod_{n=1}^{\infty}\exp\left(\log\left|\frac{z_n - z}{z_n}\right| + \operatorname{Re}\left(\frac{z}{z_n} + \cdots + \frac{1}{p}\left(\frac{z}{z_n}\right)^p\right)\right)\\
&= \prod_{n=1}^{\infty}\exp\left(\operatorname{Re}\left(\log\left(\frac{z_n - z}{z_n}\right) + \frac{z}{z_n} + \cdots + \frac{1}{p}\left(\frac{z}{z_n}\right)^p\right)\right).
\end{aligned}$$

The branch for the logarithm is irrelevant in the above formula, because we only consider the real part of the logarithm in question.

In the reasoning below, we will consider the logarithms

$$\log \frac{1}{1-w}, \qquad w \neq 1,$$

in two different cases: (1) If $|w| \geq \frac{1}{2}$ and $w \neq 1$, then we do not need to be concerned about the branch of the logarithm because we will only be interested in the real part of $\log \frac{1}{1-w}$. (2) If $|w| < \frac{1}{2}$, then the principal branch with the negative real axis as the branch cut can be chosen for the logarithm, because in this case, we have

$$\operatorname{Re}\left(\frac{1}{1-w}\right) = \operatorname{Re}\left(\frac{1-\overline{w}}{|1-w|^2}\right) \geq \frac{1-|w|}{(1+|w|)^2} > \frac{2}{9}, \qquad |w| < \frac{1}{2}.$$

We denote

$$\Psi(z, z_n) = \log \frac{z_n}{z_n - z} - \left(\frac{z}{z_n} + \cdots + \frac{1}{p}\left(\frac{z}{z_n}\right)^p\right), \qquad z \neq z_n,$$

where the branch of the logarithm depends on whether $\left|\frac{z}{z_n}\right| \geq \frac{1}{2}$ or $\left|\frac{z}{z_n}\right| < \frac{1}{2}$, as described above. Note that this is indeed just a notation, and hence it does not define a unique function $\Psi(z, z_n)$. Clearly,

$$|P(z)| = \prod_{n=1}^{\infty} \exp(-\operatorname{Re}(\Psi(z, z_n))), \qquad z \neq z_n. \tag{3.8.2}$$

Using $\frac{z_n}{z_n - z} = \frac{1}{1 - z/z_n}$ together with

$$\log \frac{1}{1-w} = \sum_{n=1}^{\infty} \frac{w^n}{n}, \qquad |w| < 1,$$

we may rewrite $\Psi(z, z_n)$ as

$$\Psi(z, z_n) = \sum_{n=p+1}^{\infty} \frac{1}{n}\left(\frac{z}{z_n}\right)^n, \qquad \left|\frac{z}{z_n}\right| < \frac{1}{2}. \tag{3.8.3}$$

From (3.8.2) we obtain

$$\log^+ \frac{1}{|P(z)|} \leq \sum_{\frac{r}{r_n} \geq \frac{1}{2}} \operatorname{Re}^+(\Psi(z, z_n)) + \sum_{\frac{r}{r_n} < \frac{1}{2}} |\Psi(z, z_n)| =: \Sigma_1 + \Sigma_2, \tag{3.8.4}$$

where $r_n = |z_n|$ and $\operatorname{Re}^+ x = \max\{0, \operatorname{Re} x\}$.

In Σ_2, we have by (3.8.3) that

$$|\Psi(z, z_n)| \leq \sum_{j=p+1}^{\infty} \frac{1}{j}\left(\frac{r}{r_n}\right)^j \leq \left(\frac{r}{r_n}\right)^{p+1} \sum_{j=0}^{\infty} \left(\frac{1}{2}\right)^j = 2\left(\frac{r}{r_n}\right)^{p+1}.$$

If $p = \lambda - 1$, then the previous inequality is simply

$$|\Psi(z, z_n)| \le 2\left(\frac{r}{r_n}\right)^{\lambda}. \tag{3.8.5}$$

If $p > \lambda - 1$ and $0 < \varepsilon \le 1 - (\lambda - \lfloor \lambda \rfloor)$, then

$$0 \le p + 1 - (\lambda + \varepsilon),$$

and hence, for $\frac{r}{r_n} < \frac{1}{2}$,

$$\begin{aligned} |\Psi(z, z_n)| \le 2\left(\frac{r}{r_n}\right)^{p+1} &= \frac{2r^{\lambda+\varepsilon}}{r_n^{p+1}} \cdot r^{p+1-(\lambda+\varepsilon)} \\ &\le \frac{2r^{\lambda+\varepsilon}}{r_n^{p+1}} \cdot \left(\frac{r_n}{2}\right)^{p+1-(\lambda+\varepsilon)} \le 2\left(\frac{r}{r_n}\right)^{\lambda+\varepsilon}. \end{aligned} \tag{3.8.6}$$

From (3.8.5) and (3.8.6) we have

$$\Sigma_2 \le 2\sum_{n=1}^{\infty}\left(\frac{r}{r_n}\right)^{\lambda+\varepsilon} \le C_1 r^{\lambda+\varepsilon}, \tag{3.8.7}$$

where $\varepsilon = 0$ if $p = \lambda - 1$ and $0 < \varepsilon \le 1 - (\lambda - \lfloor \lambda \rfloor)$ if $p > \lambda - 1$.

In Σ_1, we have

$$\begin{aligned} \mathrm{Re}^+(\Psi(z, z_n)) &\le \log^+\left|\frac{z_n}{z_n - z}\right| + \sum_{j=1}^{p}\frac{1}{j}\left(\frac{r}{r_n}\right)^j \\ &\le \log^+\left|\frac{z_n}{z_n - z}\right| + \left(\frac{r}{r_n}\right)^p \sum_{j=1}^{p}\left(\frac{r}{r_n}\right)^{j-p} \\ &\le \log^+\left|\frac{z_n}{z_n - z}\right| + 2^p\left(\frac{r}{r_n}\right)^p. \end{aligned}$$

If z lies outside the discs $|z - z_n| \le 1/|z_n|^{\lambda+\varepsilon}$ and $|z| \le e$, where $\varepsilon \ge 0$ is as above, then it follows that

$$\log^+\left|\frac{z_n}{z_n - z}\right| \le \log(r_n^{1+\lambda+\varepsilon}) = O(\log r),$$

where we have used the fact that $r_n \le 2r$. Recall that $p \le \lambda$ is always true. Hence we may choose $x \ge 0$ such that $p + x = \lambda + \varepsilon$. Then

$$\left(\frac{r}{r_n}\right)^p = \left(\frac{r}{r_n}\right)^{p+x-x} = \left(\frac{r}{r_n}\right)^{\lambda+\varepsilon}\left(\frac{r_n}{r}\right)^x \le 2^x\left(\frac{r}{r_n}\right)^{\lambda+\varepsilon},$$

and thus

$$\operatorname{Re}^+(\Psi(z,z_n)) \le C_2\left(\frac{r}{r_n}\right)^{\lambda+\varepsilon}\log r.$$

This yields

$$\Sigma_1 \le C_2 r^{\lambda+\varepsilon}\log r\sum_{n=1}^{\infty}\frac{1}{r_n^{\lambda+\varepsilon}} = C_3 r^{\lambda+\varepsilon}\log r.$$

Combining this with (3.8.7), statement (3.8.1) follows for $K = C_1 + C_3$.

Finally, if $p = 0$, then P and Ψ reduce to the forms

$$P(z) = \prod_{n=1}^{\infty}\left(1 - \frac{z}{z_n}\right) \quad \text{and} \quad \Psi(z,z_n) = \log\frac{z_n}{z_n - z}.$$

In this case, estimate (3.8.4) still holds, where Σ_2 can be estimated as in (3.8.7) by keeping in mind that $p = 0$. For the sum Σ_1, if z lies outside the discs $|z-z_n| \le 1/|z_n|^{\lambda+\varepsilon}$ and $|z| \le e$, then

$$\operatorname{Re}^+(\Psi(z,z_n)) = \log^+\left|\frac{z_n}{z_n - z}\right| \le \log(r_n^{1+\lambda+\varepsilon}) = O(\log r),$$

because $r_n \le 2r$. Thus

$$\begin{aligned}\Sigma_1 &\le K_1\log r\sum_{\frac{r}{r_n}\ge\frac{1}{2}} 1 \le K_1\log r\sum_{\frac{r}{r_n}\ge\frac{1}{2}}\left(2\frac{r}{r_n}\right)^{\lambda+\varepsilon}\\ &\le (2r)^{\lambda+\varepsilon}K_1\log r\sum_{n=1}^{\infty}\frac{1}{r_n^{\lambda+\varepsilon}} \le K_3 r^{\lambda+\varepsilon}\log r\end{aligned}$$

when z lies outside the aforementioned discs. Finally, for such z,

$$\log^+\frac{1}{|P(z)|} \le \Sigma_1 + \Sigma_2 \le K_3 r^{\lambda+\varepsilon}\log r + C_1 r^{\lambda+\varepsilon} \le K r^{\lambda+\varepsilon}\log r,$$

where $K = K_3 + C_1$. This yields (3.8.1) when $p = 0$. □

By assuming that $\varepsilon > 0$ in Theorem 3.8.1, the lower estimate in (3.8.1) can be simplified. Moreover, the exceptional set can be expressed in alternative ways. We state this formally as a corollary.

Corollary 3.8.2. *Let P be a canonical product formed with an infinite sequence $\{z_n\}$ that has finite exponent of convergence λ, and let $\varepsilon > 0$. Then*

$$\log|P(z)| \ge -r^{\lambda+\varepsilon} \tag{3.8.8}$$

for all $z = re^{i\theta}$ *outside an R-set. Alternatively,* (3.8.8) *is valid*

(a) *for all* $|z| = r \notin E$*, where* $E \subset [0, \infty)$ *has finite linear measure, and*
(b) *on almost every ray* $\arg z = \theta$ *when* $|z| = r \geq R(\theta)$.

Proof. Since $\varepsilon > 0$, the factor $K \log r$ in (3.8.1) can be incorporated into $r^{\lambda+\varepsilon}$, provided that r is large enough, say $r \geq R$. This gives (3.8.8), provided that $r \geq R$ and that z lies outside the discs in Theorem 3.8.1. Rotating these discs around the origin, we find that the projection set F on the positive real axis consists of a union of intervals that has linear measure at most

$$e + \sum_{r_n \geq e} \frac{2}{r_n^{\lambda+\varepsilon}} < \infty.$$

Hence the collection of such discs forms an R-set by definition. Thus (3.8.8) holds outside an R-set and also for all $r \notin E := [0, R] \cup F$, where E has finite linear measure. Moreover, since an R-set is an $\mathcal{E}$-set, by Lemma 1.4.1 estimate (3.8.8) is valid on almost every ray $\arg z = \theta$ when $|z| \geq R(\theta)$. □

We are now ready to prove the Hadamard factorization theorem from Section 3.5.

Proof of Theorem 3.5.7. By the construction, $z^m P(z)$ has exactly the same zeros as f, with the same multiplicities. Therefore,

$$\frac{f(z)}{z^m P(z)}$$

is an entire function with no zeros. By Theorem 3.4.1 there exists an entire function $Q(z)$ such that

$$f(z) = z^m P(z) e^{Q(z)}.$$

It remains to show that $Q(z)$ is a polynomial of degree $\leq \rho$. For this, we suppose that $P(z)$ has infinitely many zeros.

Since f is of order ρ, for fixed $\varepsilon > 0$, there exists $R_1 > 1$ such that

$$M(r,f) \leq e^{r^{\rho+\varepsilon}}, \quad r \geq R_1.$$

By Theorems 3.2.4 and 3.5.3 we obtain $\lambda = \rho(P) = \lambda(P) = \lambda(f) \leq \rho(f) = \rho$. Let E be the exceptional set in Corollary 3.8.2. Then

$$\max_{|z|=r} |e^{Q(z)}| = \max_{|z|=r} e^{\operatorname{Re} Q(z)} \leq \frac{\max_{|z|=r} |f(z)|}{r^m \min_{|z|=r} |P(z)|} \leq \frac{e^{r^{\rho+\varepsilon}}}{e^{-r^{\lambda+\varepsilon}}} \leq e^{2r^{\rho+\varepsilon}}$$

for $r \geq R_2 \geq R_1$ such that $r \notin E$. Thus $A(r, Q) \leq 2r^{\rho+\varepsilon}$ for all such r. The Borel–Carathéodory theorem now gives

$$M(r/2, Q) \leq 2A(r, Q) + 3|Q(0)| = O(r^{\rho+\varepsilon})$$

for infinitely many values of r tending to infinity. By the maximum modulus principle, $Q(z)$ is a polynomial of degree $\leq \rho + \varepsilon$. Letting $\varepsilon \to 0^+$, we obtain $\deg Q \leq \rho$.

If f has finitely many zeros, then $P(z)$ is a polynomial, whereas if f has no zeros, then $P(z) \equiv 1$. In both cases the above reasoning simplifies because Corollary 3.8.2 is not needed in estimating $|P(z)|$. □

Regarding the remark that follows Theorem 2.1.10, we are now ready to prove the next result.

Theorem 3.8.3. *Let f and g be entire functions such that $\rho(f) > \rho(g)$. Then $\rho(fg) = \rho(f)$.*

Proof. Let $F = fg$. The inequality $\rho(F) \leq \rho(f)$ is already proved in the remark following Theorem 2.1.10. We suppose on the contrary to the statement that $\rho(F) < \rho(f)$. Let $\sigma \in (\max\{\rho(g), \rho(F)\}, \rho(f))$. Then there exists a constant $R > 1$ such that

$$|F(z)| \leq \exp(r^\sigma), \quad |z| = r \geq R.$$

By Hadamard's theorem the finite-order function g can be factorized as

$$g(z) = z^m P(z) e^{Q(z)},$$

where $m \geq 0$ is the multiplicity of the zero of g at $z = 0$, $P(z)$ is the Weierstrass product formed with the nonzero zeros of g, and $Q(z)$ is a polynomial of degree $\leq \sigma$. We may suppose that g has infinitely many zeros because the case of finitely many zeros is easy. Then $P(z)$ is a canonical product, and by Corollary 3.8.2 we find that

$$\log|P(z)| \geq -r^\sigma, \quad r \notin E,$$

where $E \subset [0, \infty)$ has finite linear measure. Putting all this together, there exists a constant $C > 0$ such that

$$|f(z)| = \left|\frac{F(z)}{g(z)}\right| \leq \frac{|F(z)|e^{|Q(z)|}}{r^m |P(z)|} \leq r^{-m} e^{Cr^\sigma} \leq e^{Cr^\sigma},$$

provided that $r \geq R$ and $r \notin E$. Since the upper bound in the above estimate is independent of $\theta = \arg z$, we obtain

$$M(r, f) \leq e^{Cr^\sigma}, \quad r \notin E_0,$$

where $E_0 = E \cup [0, R]$ has finite linear measure. Now $\rho(f) \leq \sigma < \rho(f)$ follows either by Lemma 1.2.8 with $\psi(r) = r$ or by Lemma 1.2.9 (since finite linear measure implies finite logarithmic measure). This contradiction proves that $\rho(F) = \rho(f)$. □

An alternative short proof of Theorem 3.8.3 in terms of Nevanlinna theory will be discussed in Section 4.6.

Problems

Problem 3.8.4. Suppose that an entire function f has a Hadamard factorization

$$f(z) = P(z)e^{Q(z)},$$

where $Q(z)$ is a polynomial of degree q, and $P(z)$ is a canonical product that has finite exponent of convergence λ and genus p. Let $\varepsilon > 0$. Prove that there exist a constant $K > 1$ and a set $F \subset [0, \infty)$ of finite linear measure such that

$$\log\left|f(z)\right| \geq -K \max\left\{r^q, r^{\lambda+\varepsilon}\right\}, \quad |z| = r \notin F.$$

3.9 Zeros of complex polynomials

We have already encountered the growth of complex polynomials in Lemma 2.1.4. In this section, we focus on the zeros of complex polynomials.

A natural starting point is the well-known *fundamental theorem of algebra* (Theorem 3.9.1), which students normally learn a long time before they see a proof. Usually, the first proofs that students see for this famous result are in complex analysis courses. We use Liouville's theorem in the proof.

Theorem 3.9.1. *Let $P(z)$ be a nonconstant polynomial of degree $n \geq 1$, i. e.,*

$$P(z) = a_n z^n + a_{n-1} z^{n-1} + \cdots + a_0, \quad a_n \neq 0,\ n \geq 1. \tag{3.9.1}$$

Then $P(z)$ has exactly n zeros $z_1, z_2, \ldots, z_n$ when counted according to their multiplicities. Moreover, $P(z)$ can be written in the factored form

$$P(z) = c(z - z_1)(z - z_2)\cdots(z - z_n), \tag{3.9.2}$$

where $c \neq 0$ is a complex constant.

Proof. First, we assume that $P(z)$ has no zeros. Then the function f defined by $f = 1/P$ is a nonconstant entire function. From Lemma 2.1.4 we see that $P(z) \to \infty$ as $z \to \infty$, which means that $f(z) \to 0$ as $z \to \infty$. Hence f is a bounded entire function. By Liouville's theorem, f is a constant, a contradiction. Thus our assumption is false, and $P(z)$ must have at least one zero z_1.

By factoring out $z - z_1$, we obtain

$$P(z) = (z - z_1)Q(z),$$

where $Q(z)$ is a polynomial of degree $n-1$. If $n = 1$, then $Q(z)$ is a nonzero constant, (3.9.2) holds, and the theorem is proved. If $n \geq 2$, then Q is a nonconstant polynomial, which means that Q has at least one zero z_2 by what we just proved. Therefore,

$$P(z) = (z - z_1)(z - z_2)R(z),$$

where $R(z)$ is a polynomial of degree $n-2$. Continuing in this manner, we obtain (3.9.2), which proves the theorem. □

In addition to Liouville's theorem, several other theorems from complex analysis can be used to prove Theorem 3.9.1. For instance, by considering the dominant term $a_n z^n$ in (3.9.1), Rouché's theorem or the argument principle can be used to prove Theorem 3.9.1.

Lemma 2.1.4 gives the growth of the modulus of a polynomial. From (3.9.2) we see that Lemma 3.9.2 gives a lower bound for the modulus of a polynomial outside of finitely many discs that contain its zeros. This is a classical result known as the *Boutroux–Cartan lemma* or simply as the *Cartan lemma.*

Lemma 3.9.2 ([280, Lemma 8.1]). *Let $z_1, \dots, z_n \in \mathbb{C}$ be fixed, and let $H > 0$. Then there exist closed discs $D_1, \dots, D_m$, $m \le n$, the sum of whose radii is $2H$, such that if $z \notin \bigcup_{j=1}^m D_j$, then*

$$|z - z_1| \cdots |z - z_n| > \left(\frac{H}{e}\right)^n. \tag{3.9.3}$$

Proof. Suppose there happens to be one closed disc of radius H that contains all the points $z_1, \dots, z_n$. Let D_1 be the concentric disc of radius $2H$. If $z \notin D_1$, then we have $|z - z_j| \ge H$ for all $j = 1, \dots, n$. Consequently,

$$|z - z_1| \cdots |z - z_n| \ge H^n > \left(\frac{H}{e}\right)^n.$$

In general, the points $z_1, \dots, z_n$ cannot be included in just one closed disc of radius H. However, there will be a largest integer k for which the following is true: There is a closed disc of radius kH/n that contains exactly k of the given points $z_1, \dots, z_n$. This integer k exists because the above statement is clearly true for $k = 1$. The assumption $k = n$ was made in the preceding paragraph. Set $k_1 = k$.

Choose a closed disc C_1 of radius $k_1 H/n$ containing exactly k_1 of the points $z_1, \dots, z_n$. The disc C_1 cannot contain more than k_1 points, for if it did, then the concentric disc of radius $(k_1 + 1)H/n$ would contain at least $k_1 + 1$ points, which would contradict the definition of k_1.

For the remaining $n - k_1$ points, we choose in the same way an integer $k_2 \in \{1, \dots, n - k_1\}$ and a closed disc C_2 of radius $k_2 H/n$ containing k_2 of the remaining points. By the definition of k_1 we have $k_2 \le k_1$.

This process is repeated for the remaining $n - k_1 - k_2$ points until after m steps, the points $z_1, \dots, z_n$ are all covered by discs $C_1, \dots, C_m$. The radius of C_j is $k_j H/n$ for each j, and $k_1 \ge k_2 \ge \cdots \ge k_m$. Since $k_1 + \cdots + k_m = n$ by the definition of the indices k_j, it follows that the sum of the radii of the discs C_j is H.

For $j \in \{1, \ldots, m\}$, let D_j denote the closed disc of radius $2k_jH/n$ that is concentric with C_j. Fix $z \notin \bigcup_{j=1}^{m} D_j$. We may renumber the finitely many points $z_1, \ldots, z_n$, if necessary, to ensure that

$$|z - z_1| \leq |z - z_2| \leq \cdots \leq |z - z_n|. \tag{3.9.4}$$

For $j \in \{1, \ldots, n\}$, we claim that

$$|z - z_j| > \frac{j}{n} H. \tag{3.9.5}$$

Suppose on the contrary to this claim that there exists an index $j_0 \in \{1, \ldots, n\}$ for which $|z - z_{j_0}| \leq j_0 H/n$. We find that $k_1 \geq j_0$, because the disc of radius $j_0 H/n$ about the fixed z contains $z_1, \ldots, z_{j_0}$ by (3.9.4). Since $k_1 \geq k_2 \geq \cdots \geq k_m$, this allows us to define the largest index $\ell \in \{1, \ldots, n\}$ for which $k_\ell \geq j_0$.

For $p \leq j_0$ and $q \leq \ell$, we proceed to prove that $z_p \notin C_q$. Suppose on the contrary to this claim that $z_p \in C_q$. Then, since $k_q \geq k_\ell \geq j_0$, by (3.9.4) we have

$$|z - z_1| \leq |z - z_2| \leq \cdots \leq |z - z_{j_0}| \leq \frac{j_0}{n} H \leq \frac{k_q}{n} H.$$

Since $p \leq j_0$, we deduce that $|z - z_p| \leq k_q H/n$. Then, since
(1) $z_p \in C_q$,
(2) the radius of C_q is $k_q H/n$ by definition,
(3) the disc D_q is concentric with C_q and has radius $2k_q H/n$,

it follows that $z \in D_q$, which contradicts the fact that $z \notin \bigcup_{j=1}^{m} D_j$. Thus $z_p \notin C_q$. Since $p \leq j_0$ and $q \leq \ell$ were arbitrarily chosen, it follows that $z_1, \ldots, z_{j_0} \notin \bigcup_{j=1}^{\ell} C_j$. Consequently,

$$z_1, \ldots, z_{j_0} \in \bigcup_{j=\ell+1}^{m} C_j. \tag{3.9.6}$$

Since $|z - z_{j_0}| \leq j_0 H/n$, it follows from (3.9.4) that there is a disc of radius $j_0 H/n$ containing the points $z_1, \ldots, z_{j_0}$. By the discussion above, this disc cannot be any of the discs $C_1, \ldots, C_\ell$. From (3.9.6) we see that one of the discs $C_{\ell+1}, \ldots, C_m$ contains the points $z_1, \ldots, z_{j_0}$, say C_{m_0}. By the construction, the disc C_{m_0} contains k_{m_0} points. Since the points $z_1, \ldots, z_{j_0}$ are already included in C_{m_0}, this means that $k_{m_0} \geq j_0$ for some $m_0 > \ell$, which contradicts the definition of ℓ.

Since we arrived at a contradiction, inequality (3.9.5) must be true for all $j \in \{1, \ldots, n\}$, as claimed. Inequality (3.9.5) then implies

$$|z - z_1| \cdots |z - z_n| > \frac{n!}{n^n} H^n.$$

The conclusion now follows from Lemma 2.2.5. □

The Boutroux–Cartan lemma will be applied in Section 11.1 to estimate the minimum modulus of entire functions of genus zero. The lemma is also applied in the literature to obtain, e. g., growth estimates for (1) logarithmic derivatives of meromorphic functions [53, 86] and for (2) logarithmic differences and logarithmic q-differences of meromorphic functions [49, 288].

Our next goal is to state and prove the Gauss–Lucas theorem, according to which, for a given polynomial P, the zeros of the derivative P' lie in the convex hull of the zeros of P. We begin with some definitions and auxiliary results.

Definition 3.9.3. A nonempty set $S \subset \mathbb{C}$ is called *convex* if for each pair of points $z_1, z_2 \in S$, the line segment $[z_1, z_2]$ joining z_1 and z_2 belongs to S. For every nonempty set $S \subset \mathbb{C}$, the intersection of all closed convex sets containing S is called the *convex hull* of S and is denoted by $\mathrm{co}(S)$; see Figure 3.2.

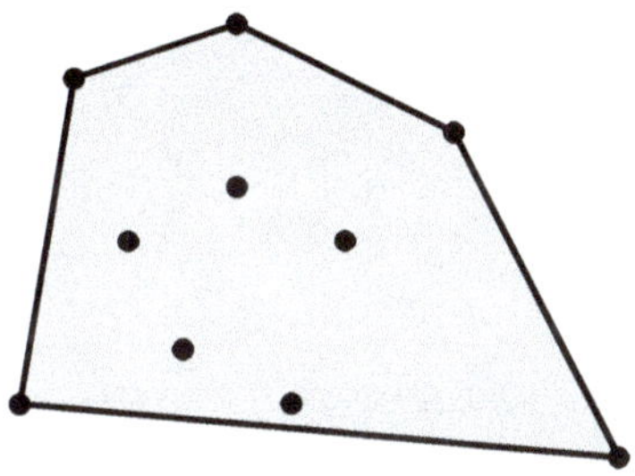

Figure 3.2: The convex hull of a set of finitely many points.

A singleton set is trivially convex. If S is nonempty, then the intersection $\mathrm{co}(S)$ must also be nonempty. In general, any nonempty intersection $\bigcap_n C_n$ of convex sets C_n is convex. Indeed, let $z_1, z_2 \in \bigcap_n C_n$. Then $z_1, z_2 \in C_n$ for every n, and since C_n is convex, we have $[z_1, z_2] \subset C_n$ for every n. Consequently, $[z_1, z_2] \subset \bigcap_n C_n$. In particular, $\mathrm{co}(S)$ is closed and convex.

The following lemma appears in many textbooks, but the proof is typically given as an exercise.

Lemma 3.9.4. *The convex hull* $\mathrm{co}(S)$ *of a nonempty set* $S \subset \mathbb{C}$ *is the intersection of all closed half-planes that contain* S*, provided that* $\mathrm{co}(S) \subsetneq \mathbb{C}$.

Proof. Since $\mathrm{co}(S)$ is convex and $\mathrm{co}(S) \subsetneq \mathbb{C}$, the set $\mathrm{co}(S)$ can be covered by a closed half-plane. Let T be the intersection of all the closed half-planes that contain $\mathrm{co}(S)$. Since closed half-planes are convex, we have $\mathrm{co}(S) \subset T$ because the intersection $\mathrm{co}(S)$ contains all the sets in the intersection T and more. The proof of $T \subset \mathrm{co}(S)$ is more complicated. The details can be found in [117, Lemma B.2], where the assumption $\mathrm{co}(S) \subsetneq \mathbb{C}$ is also deployed. The identity $\mathrm{co}(S) = T$ proves the claim. □

For the convenience of the reader, we prove a known result about convex hulls.

Lemma 3.9.5. *Let $S \subset \mathbb{C}$ be a set containing the points $z_1, \dots, z_n$. Let w be a linear combination of the form*

$$w = \alpha_1 z_1 + \cdots + \alpha_n z_n,$$

where $\alpha_j \geq 0$ and $\sum_{j=1}^{n} \alpha_j = 1$. Then $w \in \operatorname{co}(S)$.

Proof. The statement is trivial if $\operatorname{co}(S) = \mathbb{C}$, so we may suppose that $\operatorname{co}(S) \subsetneq \mathbb{C}$.

Let H be any closed half-plane that contains all the points $z_1, \dots, z_n$. Then there are constants $a, b, c \in \mathbb{R}$ such that

$$ax + by + c = 0$$

for all points $z = x + iy$ that lie on the boundary line of H. Denote $z_j = x_j + iy_j$, $j \in \{1, \dots, n\}$. Since $z_1, \dots, z_n \in H$, we have that either $ax_j + by_j + c \leq 0$ for all $j \in \{1, \dots, n\}$ or $ax_j + by_j + c \geq 0$ for all $j \in \{1, \dots, n\}$. Suppose that $ax_j + by_j + c \leq 0$ for all $j \in \{1, \dots, n\}$, and denote $w = \alpha + i\beta$. Then

$$\alpha = \alpha_1 x_1 + \cdots + \alpha_n x_n \quad \text{and} \quad \beta = \alpha_1 y_1 + \cdots + \alpha_n y_n,$$

and writing $c = (\alpha_1 + \cdots + \alpha_n)c$, we obtain

$$a\alpha + b\beta + c = \alpha_1(ax_1 + by_1 + c) + \cdots + \alpha_n(ax_n + by_n + c) \leq 0.$$

This shows that $w \in H$. If $ax_j + by_j + c \geq 0$ for all $j \in \{1, \dots, n\}$, then a similar proof shows that $w \in H$. Thus an arbitrary closed half-plane containing $z_1, \dots, z_n$ always contains w. This implies $w \in \operatorname{co}(S)$ by Lemma 3.9.4. □

An alternative proof for Lemma 3.9.5 that is independent of Lemma 3.9.4 can be found in [142, Section 5.8]. However, Lemma 3.9.4 will be used later.

We are now ready to state and prove the *Gauss–Lucas theorem.*

Theorem 3.9.6 ([253, p. 25]). *Let P be a nonconstant complex polynomial of degree n with zeros $z_1, \dots, z_n$, and set $S = \{z_1, \dots, z_n\}$. Then the zeros of the derivative P' lie in $\operatorname{co}(S)$.*

Proof. By the fundamental theorem of algebra, we may write P in the form (3.9.2). Then

$$P'(z) = c \sum_{j=1}^{n} \prod_{k \neq j} (z - z_k) \quad \text{and} \quad \frac{P'(z)}{P(z)} = \sum_{j=1}^{n} \frac{1}{z - z_j}.$$

Let ζ be a zero of P'. If ζ is also a zero of P, then the statement is clear. Otherwise, ζ is a zero of P'/P, and

$$\sum_{j=1}^{n} \frac{1}{\zeta - z_j} = 0.$$

Conjugating this equation and using the identity $1/\bar{a} = a/|a|^2$ for $a \in \mathbb{C} \setminus \{0\}$, we obtain

$$\sum_{j=1}^{n} \frac{\zeta - z_j}{|\zeta - z_j|^2} = 0.$$

Writing this as

$$\zeta \sum_{j=1}^{n} \frac{1}{|\zeta - z_j|^2} = \sum_{j=1}^{n} \frac{z_j}{|\zeta - z_j|^2}$$

gives

$$\zeta = \sum_{j=1}^{n} \alpha_j z_j, \quad \alpha_j = \frac{1}{|\zeta - z_j|^2} \cdot \left(\sum_{k=1}^{n} \frac{1}{|\zeta - z_k|^2} \right)^{-1}.$$

We see that $\alpha_j > 0$ and $\sum_{j=1}^{n} \alpha_j = 1$. Thus $\zeta \in \mathrm{co}(S)$ by Lemma 3.9.5. □

Problems

Problem 3.9.7. Suppose that all the zeros of a nonconstant polynomial P lie in a closed convex set $K \subset \mathbb{C}$. Prove that all the zeros of the derivative P' also lie in K.

Problem 3.9.8. Let $P(z) = a_n z^n + a_{n-1} z^{n-1} + \cdots + a_0$ be a polynomial with $a_n \neq 0$ and $n \geq 1$. Define

$$M = \max_{0 \leq k \leq n-1} \left| \frac{a_k}{a_n} \right|.$$

Prove that the zeros of P lie in the disc $D(0, r)$ of radius $r \leq 1 + M$.

Problem 3.9.9. Use the maximum modulus principle to prove the fundamental theorem of algebra.

Problem 3.9.10. Use Rouché's theorem to prove the fundamental theorem of algebra by considering $P(z)$ in (3.9.1) and $Q(z) = a_n z^n$.

4 Classical topics

We proceed to consider some well-known classical topics involving entire functions. To begin with, we find that the maximum modulus principle can be generalized from bounded to unbounded domains via Phragmén–Lindelöf theorems. Then we learn that it is convenient to express the radial growth of entire functions (i. e., growth along rays) by means of the Phragmén–Lindelöf indicator function. After this, the next two sections concern particular properties of entire functions of order less than one and a brief presentation of the Wiman–Valiron theory. Finally, the last three sections contain discussions of meromorphic functions, Nevanlinna theory, and complex interpolation theory, which cover material needed later in the book.

4.1 Phragmén–Lindelöf principle

The maximum modulus principle can be generalized from bounded to unbounded domains such as sectors or strips. The method also allows us to say something about the asymptotic values of analytic functions in such domains. All these results are known as Phragmén–Lindelöf theorems, and the theory itself is called the Phragmén–Lindelöf principle, named after Lars Edvard Phragmén (1863–1937, Sweden) and Ernst Leonard Lindelöf (1870–1946, Finland). Phragmén made contributions to complex analysis, voting theory, and actuarial science and was the President of the Board of the Mittag-Leffler Institute. Historical comments about Lindelöf were made earlier in Section 3.7.

Unless stated otherwise, throughout the book a *sector* is an open set of the form

$$S(a,b) = \{z \in \mathbb{C} : a < \arg z < b\},$$

where $a < b \le a + 2\pi$.

Theorem 4.1.1 ([142, p. 125]). *Suppose that f is analytic in a sector S of opening π/α where $\alpha > 1/2$. If $|f(z)| \le M$ on the boundary ∂S of S and if*

$$|f(z)| \le K e^{|z|^{\beta}}, \quad z \in S, \tag{4.1.1}$$

where $\beta < \alpha$, then $|f(z)| \le M$ for all $z \in \overline{S}$.

Proof. By appealing to a rotation we may suppose that

$$S = \{z : |\arg z| < \pi/2\alpha\}. \tag{4.1.2}$$

Let $\varepsilon > 0$ and $\gamma \in (\beta, \alpha)$. Define

$$F_{\varepsilon}(z) = e^{-\varepsilon z^{\gamma}} f(z), \tag{4.1.3}$$

https://doi.org/10.1515/9783112249130-004

where

$$z^\gamma = \exp(\gamma \log z) = \exp(\gamma(\log|z| + i \arg z))$$
$$= |z|^\gamma \exp(i\gamma \arg z), \quad -\pi < \arg z \leq \pi. \tag{4.1.4}$$

Denoting $z = re^{i\theta}$ and using Euler's formula $e^{ix} = \cos x + i \sin x$, we observe that

$$|F_\varepsilon(z)| = e^{\mathrm{Re}(-\varepsilon z^\gamma)}|f(z)| = e^{-\varepsilon r^\gamma \cos(\gamma\theta)}|f(z)|.$$

If $z \in \overline{S}$, then $|\gamma\theta| \leq \frac{\gamma\pi}{2\alpha} < \frac{\pi}{2}$. Consequently, $\cos(\gamma\theta) > 0$, and so

$$|F_\varepsilon(z)| \leq |f(z)|, \quad z \in \overline{S}.$$

In particular, $|F_\varepsilon(z)| \leq M$ on the boundary ∂S of S, whereas inside S, we have

$$|F_\varepsilon(re^{i\theta})| = e^{-\varepsilon r^\gamma \cos(\gamma\theta)}|f(z)| \leq Ke^{r^\beta - \varepsilon r^\gamma \cos(\gamma\theta)}, \quad z = re^{i\theta} \in S.$$

Since $\gamma > \beta$ and $\cos(\gamma\theta) > 0$, we have

$$\lim_{r\to\infty} Ke^{r^\beta - \varepsilon r^\gamma \cos(\gamma\theta)} = 0.$$

In particular, there exists $R > 0$ such that

$$|F_\varepsilon(z)| \leq M, \quad z \in \overline{S(R)},$$

where $S(R) = \{z : |\arg z| < \pi/2\alpha, \ |z| > R\}$. By the maximum modulus principle we deduce that

$$|F_\varepsilon(z)| \leq M, \quad z \in \overline{S}.$$

Since this is true for every $\varepsilon > 0$, we may let $\varepsilon \to 0^+$ and conclude from (4.1.3) that $|f(z)| \leq M$ for all $z \in \overline{S}$. This completes the proof. □

Remark. (a) Suppose that f is an entire function of finite order. If f tends to zero on almost every line through the origin, then f is bounded in the complex plane from Theorem 4.1.1. By Liouville's theorem, f is a constant function, and by the assumption, $f \equiv 0$. However, there exist nonconstant entire functions that do tend to zero on any line whatsoever; see [5] and the references therein. Clearly, such functions must be of infinite order.

(b) By using Theorem 4.1.1 and Liouville's theorem, we also observe that if f is an entire function of order < 1, that is bounded on the real axis (or the imaginary axis), then f is a constant.

The growth restriction (4.1.1) can be relaxed.

Theorem 4.1.2 ([142, p. 126]). *Suppose that f is analytic in a sector S of opening π/α, where $\alpha > 1/2$. If $|f(z)| \leq M$ on the boundary ∂S of S and if for every $\delta > 0$, there exist $K = K(\delta) > 0$ and $R = R(\delta) > 0$ such that*

$$|f(z)| \leq Ke^{\delta|z|^\alpha}, \quad z \in S,\ |z| \geq R, \tag{4.1.5}$$

then $|f(z)| \leq M$ for all $z \in \overline{S}$.

Proof. We may again assume that the sector S is given by (4.1.2). Let $\varepsilon > 0$. Define

$$F_\varepsilon(z) = e^{-\varepsilon z^\alpha} f(z),$$

where the principal branch for z^α is chosen similarly as in (4.1.4). In particular, $x^\alpha > 0$ for all $x > 0$. Choose $\delta \in (0, \varepsilon)$. Then on the real axis, we have

$$|F_\varepsilon(x)| = e^{-\varepsilon x^\alpha}|f(x)| \leq Ke^{(\delta-\varepsilon)x^\alpha} \to 0, \quad x \to \infty.$$

Since $|F_\varepsilon(x)|$ is a continuous function in x,

$$|F_\varepsilon(x)| \leq M_1 := \max_{t\geq 0}|F_\varepsilon(t)|, \quad x \geq 0.$$

Next, consider $F_\varepsilon(z)$ in the upper and lower half-sectors. If $z = re^{i\theta} \in \overline{S}$, then $-\pi/2 \leq \alpha\theta \leq \pi/2$, and so $\cos(\alpha\theta) \geq 0$, and

$$|e^{-\varepsilon z^\alpha}| = \exp(\operatorname{Re}(-\varepsilon z^\alpha)) = \exp(-\varepsilon r^\alpha \cos(\alpha\theta)) \leq 1.$$

Set $M_2 = \max\{M, M_1\}$. The reasoning above shows that $|F_\varepsilon(z)| \leq M_2$ on the boundaries of both half-sectors and that $|F_\varepsilon(z)| \leq Ke^{\delta r^\alpha}$ inside each half-sector. By Theorem 4.1.1,

$$|F_\varepsilon(z)| \leq M_2 \tag{4.1.6}$$

in both closed half-sectors, and hence in $\overline{S}$.

We proceed to prove that $M_2 = M$. Suppose on the contrary that $M_1 > M$, in which case $M_2 = M_1 > M$. Since $F_\varepsilon(x) \to 0$ as $x \to \infty$ and $|F_\varepsilon(0)| \leq M$, there exists a point $x_0 > 0$ such that

$$|F_\varepsilon(x_0)| = M_1 = M_2.$$

By (4.1.6) and the fact that the maximum modulus of $F_\varepsilon(z)$ is attained at an interior point of S, the maximum modulus principle yields $F_\varepsilon(z) \equiv M_2 e^{i\gamma}$ for some $\gamma \in \mathbb{R}$. But then $f(z) = M_2 e^{\varepsilon z^\alpha + i\gamma}$ would not satisfy the assumptions of the theorem, a contradiction. This contradiction means that $M_2 = M$.

It follows that (4.1.6) holds in $\overline{S}$ for M in place of M_2. Since this is true for every $\varepsilon > 0$, we may let $\varepsilon \to 0^+$ and conclude that $|f(z)| \leq M$ for all $z \in \overline{S}$. □

Corollary 4.1.3. *Let f be a nonconstant entire function of order $\rho < 1/2$. Then f is unbounded along every ray emanating from the origin.*

Proof. The statement is clear if f is a nonconstant polynomial, so we may suppose that f is a transcendental entire function of order $\rho < 1/2$.

Suppose on the contrary that there exists at least one such ray along which f is bounded. By appealing to a rotation we may suppose that f is bounded on the positive real axis. Define $g(z) = f(z^2)$, so that g is bounded on the whole real axis. By Lemma 2.1.8,

$$\rho(g) = 2\rho(f) = 2\rho < 1.$$

Next, we use Theorem 4.1.1 two times with $\alpha = 1$ and conclude that g is bounded in the upper and lower half-planes. Since g is now bounded in the entire plane, it must be a constant function by Liouville's theorem. Consequently, f is a constant function, a contradiction. □

We proceed to discuss entire functions that approach limits along rays.

Theorem 4.1.4 ([142, p. 127]). *Let L_1 and L_2 be two distinct rays emanating from the origin, and let f be analytic and bounded in one of the sectors S determined by L_1 and L_2. If $f(z) \to a \in \mathbb{C}$ as $z \to \infty$ along both L_1 and L_2, then $f(z) \to a$ uniformly as $z \to \infty$ in the closed sector $\overline{S}$.*

Proof. Considering $f-a$, if needed, we may suppose that $a = 0$. By appealing to a rotation, we may suppose that the positive real axis is the bisector of S, that is, $S = \{z : |\arg z| < \varphi\}$ for some $\varphi \in (0, \pi)$.

We first prove the case $\varphi \in (0, \pi/2)$. Let $\varepsilon > 0$. By the assumptions we have $|f(z)| \le M$ for all $z \in \overline{S}$ and some constant $M > 0$. Moreover, there exists $R = R(\varepsilon) > 1$ such that

$$|f(z)| < \varepsilon, \quad \arg z = \pm\varphi, \ |z| > R.$$

Denote $\lambda = \frac{MR}{\varepsilon} > 0$, and define

$$F_\lambda(z) = \frac{z}{z+\lambda} f(z).$$

Then for $z \in \overline{S}$, we have $\operatorname{Re} z > 0$ and

$$|F_\lambda(z)| = \frac{|z|}{(|z|^2 + 2\lambda \operatorname{Re} z + \lambda^2)^{1/2}} |f(z)| < \frac{|z|}{(|z|^2 + \lambda^2)^{1/2}} |f(z)|.$$

Moreover, if $z \in \overline{S}$ and $|z| \le R$, then

$$|F_\lambda(z)| \le \frac{M|z|}{\lambda} \le \frac{MR}{\lambda} = \varepsilon,$$

whereas if $\arg z = \pm\varphi$ and $|z| > R$, then

$$|F_\lambda(z)| \le |f(z)| < \varepsilon.$$

Clearly, if $z \in S$ and $|z| = r > 1$, then

$$|F_\lambda(z)| \le |f(z)| \le M \le Me^r \le Me^{r^\beta} \le Me^{r^\alpha}$$

for all α, β such that $1 < \beta < \alpha$. Choose $\alpha = \frac{\pi}{2\varphi}$. Since $\varphi \in (0, \pi/2)$, it follows that $\alpha > 1$. By Theorem 4.1.1 we deduce that

$$|F_\lambda(z)| \le \varepsilon, \quad z \in \overline{S}.$$

Consequently,

$$\begin{aligned} |f(z)| &= \left|1 + \frac{\lambda}{z}\right| |F_\lambda(z)| \\ &\le \left(1 + \frac{\lambda}{r}\right) |F_\lambda(z)| \le 2\varepsilon, \quad z \in \overline{S},\ |z| = r > \lambda. \end{aligned}$$

This proves that $f(z) \to 0$ uniformly as $z \to \infty$ in $\overline{S}$.

Suppose then that S is of the form $S = \{z : |\arg z| < \varphi\}$ for some $\varphi \in [\pi/2, \pi)$. Then, applying the first part of the proof to the function $g(z) = f(z^2)$, we find that $g(z) \to 0$ uniformly as $z \to \infty$ in $\overline{S}_0$, where $S_0 = \{z : |\arg z| < \varphi/2\}$. This implies, in particular, that $f(z) \to 0$ as $z \to \infty$ along the positive real axis. Now we may apply the first part of the proof to f in the sectors $S_1 = \{z : 0 < \arg z < \varphi\}$ and $S_2 = \{z : -\varphi < \arg z < 0\}$, both of which have an opening $< \pi$. □

Theorem 4.1.5 ([142, p. 128]). *Let L_1 and L_2 be two distinct rays emanating from the origin, and let f be analytic and bounded in one of the sectors S determined by L_1 and L_2. Suppose that $f(z) \to a$ as $z \to \infty$ along L_1 and that $f(z) \to b$ as $z \to \infty$ along L_2, where $a, b \in \mathbb{C}$. Then $a = b$ and $f(z) \to a$ as $z \to \infty$ uniformly in $\overline{S}$.*

Proof. Suppose that L_1 and L_2 correspond to $\arg z = \alpha$ and $\arg z = \beta$, respectively, where $\alpha < \beta$. The function $F(z) = (f(z) - (a+b)/2)^2$ is analytic and bounded in the sector S and has the asymptotic value $(a-b)^2/4$ along L_1 and along L_2. By Theorem 4.1.4, $F(z)$ tends to $(a-b)^2/4$ as $z \to \infty$ uniformly in $\overline{S}$. In other words,

$$\big(f(z) - (a+b)/2\big)^2 - (a-b)^2/4 = \big(f(z) - a\big)\big(f(z) - b\big) \tag{4.1.7}$$

tends uniformly to zero in $\overline{S}$. Thus for small $\varepsilon > 0$, the closed sector $\overline{S}$ contains a closed arc C of a circle $|z| = r$ with $r > 0$ large enough such that

$$\left|\big(f(z) - a\big)\big(f(z) - b\big)\right| \le \varepsilon, \quad z \in C,$$

and

$$|f(re^{i\alpha}) - a| < \sqrt{\varepsilon} \quad \text{and} \quad |f(re^{i\beta}) - b| < \sqrt{\varepsilon}.$$

At every point $z \in C$, we have

$$|f(z) - a| \le \sqrt{\varepsilon} \quad \text{or} \quad |f(z) - b| \le \sqrt{\varepsilon},$$

or both. By the continuity of f on C we can deduce that there exists a point $z_0 \in C$ that simultaneously satisfies

$$|f(z_0) - a| \le \sqrt{\varepsilon} \quad \text{and} \quad |f(z_0) - b| \le \sqrt{\varepsilon}.$$

Then

$$|a - b| = |(f(z_0) - a) - (f(z_0) - b)| \le |f(z_0) - a| + |f(z_0) - b| \le 2\sqrt{\varepsilon}.$$

Since ε can be arbitrarily small, we obtain $a = b$. Finally, it follows from (4.1.7) that $f(z) \to a$ uniformly in S. □

A version of the Phragmén–Lindelöf theorem is valid for strips; see [142, pp. 128–129]. A discussion parallel to that above can be found in the book of Titchmarsh [273]. Slightly different variations of the Phragmén–Lindelöf theorems discussed above can be found in the book of Markushevich [205].

Problems

Problem 4.1.6. Let f be an entire function given by

$$f(z) = \prod_{n=1}^{\infty} \left(1 - \frac{z}{n^2}\right).$$

Prove that f is unbounded on the real and imaginary axes.

Problem 4.1.7. Let f be an entire function of order $\rho = 1/2$ and type $\tau = 0$. Prove that f is unbounded along every ray emanating from the origin.

Problem 4.1.8. Give an example of an entire function f of order $\rho(f) = 1$ such that $f(z) \to \infty$ as $z \to \infty$ along any ray emanating from the origin.

4.2 Phragmén–Lindelöf indicator function

Let f be an entire function of order $0 < \rho < \infty$. Then the function

$$h_f(\theta) = \limsup_{r \to \infty} \frac{\log |f(re^{i\theta})|}{r^\rho}, \qquad \theta \in \mathbb{R},$$

is called the *Phragmén–Lindelöf indicator function* of f. It is clear that $h_f(\theta)$ is a 2π-periodic function. Moreover, if f is of bounded type with respect to its order ρ, we find that $h_f(\theta)$ is bounded from above. In this case, it is also true that $h_f(\theta)$ is bounded from below, but this will be proved later.

Example 4.2.1. If $f(z) = e^z$ and $z = re^{i\theta} = r\cos\theta + ir\sin\theta$, then $\rho = 1$ and $\log|f(re^{i\theta})| = r\cos\theta$, so that $h_f(\theta) = \cos\theta$.

The indicator function $h_f(\theta)$ describes the growth of f along the ray $\arg z = \theta$. If $h_f(\theta) > 0$, then f has exponential growth, whereas if $h_f(\theta) < 0$, then f decays to zero exponentially. If $h_f(\theta) = 0$, then f may have a polynomial type of growth, but this is not necessarily the case.

Lemma 4.2.2. *Let f and g be entire functions of the same order $0 < \rho < \infty$. Then $h_{fg}(\theta) \leq h_f(\theta) + h_g(\theta)$ and $h_{f+g}(\theta) \leq \max\{h_f(\theta), h_g(\theta)\}$.*

Proof. The second statement follows from

$$\begin{aligned} h_{f+g}(\theta) &= \limsup_{r\to\infty} \frac{\log|f(re^{i\theta}) + g(re^{i\theta})|}{r^\rho} \\ &\leq \limsup_{r\to\infty} \frac{\log(2\max\{|f(re^{i\theta})|, |g(re^{i\theta})|\})}{r^\rho} \\ &\leq \limsup_{r\to\infty} \frac{\log 2}{r^\rho} + \limsup_{r\to\infty} \frac{\log(\max\{|f(re^{i\theta})|, |g(re^{i\theta})|\})}{r^\rho} \\ &= \max\{h_f(\theta), h_g(\theta)\}. \end{aligned}$$

The first statement follows by writing $\log|fg| = \log|f| + \log|g|$. □

Example 4.2.3. Complex sine and cosine functions can be written as

$$\sin z = \frac{1}{2i}\left(e^{iz} - e^{-iz}\right) \quad \text{and} \quad \cos z = \frac{1}{2}\left(e^{iz} + e^{-iz}\right).$$

If $z = re^{i\theta}$, where θ is fixed and $r \to \infty$, then we have

$$\left|e^{iz} \pm e^{-iz}\right| = \begin{cases} (1 + o(1))e^{r\sin\theta}, & 0 < \theta < \pi, \\ (1 + o(1))e^{-r\sin\theta}, & \pi < \theta < 2\pi. \end{cases}$$

Moreover, $|\sin z| \leq 1$ and $|\cos z| \leq 1$ for $z \in \mathbb{R}$. Hence

$$h_{\sin z}(\theta) = h_{\cos z}(\theta) = \max\{-\sin\theta, \sin\theta\} = |\sin\theta| \geq 0.$$

Therefore, both $\sin z$ and $\cos z$ have exponential growth on every ray emanating from the origin except the positive and negative real axes, where both functions are bounded. The maximal radial growth is attained on the positive and negative imaginary axes, where the maximum modulus points are also found. Since $|\sin\theta|$ is not differentiable at the points $\theta = n\pi$, where $n \in \mathbb{Z}$, this example also illustrates the fact that the indicator function does not need to be differentiable everywhere.

We proceed toward proving that if $h_f(\theta)$ is bounded from above, then it is continuous. We first need to build some tools.

Theorem 4.2.4 ([142, Theorem 7.8.1], [205, Theorem 9.13]). *Let f be an entire function of order $0 < \rho < \infty$, and let $\theta_1, \theta_2 \in \mathbb{R}$ be such that $0 < \theta_2 - \theta_1 < \min\{2\pi, \pi/\rho\}$. Suppose that*

$$h_f(\theta_1) \le h_1 \quad \text{and} \quad h_f(\theta_2) \le h_2, \tag{4.2.1}$$

and define

$$H(\theta) = \frac{h_1 \sin(\rho(\theta_2 - \theta)) + h_2 \sin(\rho(\theta - \theta_1))}{\sin(\rho(\theta_2 - \theta_1))}.$$

Then

$$h_f(\theta) \le H(\theta), \quad \theta_1 \le \theta \le \theta_2. \tag{4.2.2}$$

Remark. We see that $H(\theta_1) = h_1$ and $H(\theta_2) = h_2$. Hence (4.2.2) is trivially true for $\theta = \theta_1, \theta_2$. Property (4.2.2) is known as *ρ-trigonometric convexity*.

Proof. First, fix any $\delta > 0$. By considering the sum formula for the sine we construct the function

$$\begin{aligned} H_\delta(\theta) &= \frac{(h_1 + \delta)\sin(\rho(\theta_2 - \theta)) + (h_2 + \delta)\sin(\rho(\theta - \theta_1))}{\sin(\rho(\theta_2 - \theta_1))} \\ &= a_\delta(\theta)\cos(\rho\theta) + b_\delta(\theta)\sin(\rho\theta), \quad a_\delta, b_\delta \in \mathbb{R}, \end{aligned}$$

where

$$a_\delta(\theta) = \frac{(h_1 + \delta)\sin(\rho\theta_2) - (h_2 + \delta)\sin(\rho\theta_1)}{\sin(\rho(\theta_2 - \theta_1))}$$

and

$$b_\delta(\theta) = \frac{(h_2 + \delta)\cos(\rho\theta_1) - (h_1 + \delta)\cos(\rho\theta_2)}{\sin(\rho(\theta_2 - \theta_1))}.$$

Then $H_\delta(\theta_1) = h_1 + \delta$ and $H_\delta(\theta_2) = h_2 + \delta$.

Next, define the function

$$\varphi_\delta(z) = f(z)e^{-(a_\delta - ib_\delta)z^\rho},$$

where $z = r^{i\theta}$, and

$$\begin{aligned} z^\rho &= \exp(\rho \log z) = \exp(\rho(\log|z| + i \arg z)) \\ &= |z|^\rho \exp(i\rho \arg z) = r^\rho e^{i\rho\theta} = r^\rho \cos(\rho\theta) + ir^\rho \sin(\rho\theta), \end{aligned}$$

where the branch cut for $\log z$ will be fixed as follows. To use Theorem 4.1.1, we want $\varphi_\delta(z)$ to be analytic in the sector $G = \{z : \theta_1 < \theta < \theta_2\}$. To achieve this, $\log z$ needs to be analytic in G. Hence the branch cut of $\log z$ must be chosen such that it lies outside $\overline{G}$. For example, we may choose $\arg z = \frac{\theta_1+\theta_2}{2} - \pi$ as the branch cut, and then $\log z$ will be analytic and well-defined in the slit plane

$$\frac{\theta_1 + \theta_2}{2} - \pi < \arg z < \frac{\theta_1 + \theta_2}{2} + \pi.$$

Indeed, the closed sector $\overline{G} = \{z : \theta_1 \le \theta \le \theta_2\}$ is contained in this slit plane by the assumption $0 < \theta_2 - \theta_1 < \min\{2\pi, \pi/\rho\}$. Then using this branch, $\varphi_\delta(z)$ will be analytic in G. For $z \in \overline{G}$,

$$\begin{aligned} |\varphi_\delta(z)| &= |f(z)| \exp(\operatorname{Re}(-(a_\delta(\theta) - ib_\delta(\theta))(r^\rho \cos(\rho\theta) + ir^\rho \sin(\rho\theta))) \\ &= |f(z)| \exp(-H_\delta(\theta) r^\rho), \quad |z| = r. \end{aligned} \tag{4.2.3}$$

Next, we use the following fact (Problem 1.1.12): If the limit $\lim_{r\to\infty} B(r)$ exists, then

$$\limsup_{r\to\infty}(A(r) + B(r)) = \limsup_{r\to\infty} A(r) + \lim_{r\to\infty} B(r).$$

Hence

$$\limsup_{r\to\infty} \frac{\log|\varphi_\delta(re^{i\theta})|}{r^\rho} = h_f(\theta) - H_\delta(\theta), \quad \theta_1 \le \theta \le \theta_2.$$

By assumptions (4.2.1) and the definition of $H_\delta(\theta)$, we obtain

$$\begin{aligned} h_f(\theta_1) - H_\delta(\theta_1) &\le h_1 - (h_1 + \delta) = -\delta < 0, \\ h_f(\theta_2) - H_\delta(\theta_2) &\le h_2 - (h_2 + \delta) = -\delta < 0, \end{aligned}$$

and it follows that $\varphi_\delta(z)$ tends to zero exponentially on the two rays $\arg z = \theta_1, \theta_2$. In particular, $|\varphi_\delta(z)|$ is bounded on the rays $\arg z = \theta_1, \theta_2$.

We proceed to prove that

$$|\varphi_\delta(z)| \le Ke^{|z|^\beta}, \quad z \in G, \tag{4.2.4}$$

for some $\beta < \frac{\pi}{\theta_2-\theta_1}$ and $K > 0$. Since $\frac{\pi}{\theta_2-\theta_1} > \rho$ by the hypothesis, we put $\frac{\pi}{\theta_2-\theta_1} = \rho + \alpha$, where $\alpha > 0$. Since f is an entire function of order ρ, we have

$$|f(z)| = O(e^{|z|^{\rho+\alpha/3}}).$$

Denoting $\mu_\delta = |a_\delta - ib_\delta|$, it follows that

$$|e^{-(a_\delta - ib_\delta)z^\rho}| \le e^{\mu_\delta |z|^\rho} = O(e^{|z|^{\rho+\alpha/3}}), \quad z \in G.$$

Hence by the definition of $\varphi_\delta(z)$, statement (4.2.4) follows for $\beta = \rho + 2\alpha/3$. Since, in addition to (4.2.4), we have proved that $|\varphi_\delta(z)|$ is bounded on the rays $\arg z = \theta_1, \theta_2$, we may now apply Theorem 4.1.1 to conclude that $|\varphi_\delta(z)|$ is bounded on $\overline{G}$. Thus (4.2.3) gives

$$h_f(\theta) \le H_\delta(\theta), \quad \theta_1 \le \theta \le \theta_2.$$

Since $\delta > 0$ can be as close to 0 as desired, we see that (4.2.2) holds. □

It follows from Theorem 4.2.4 that if the indicator $h_f(\theta)$ is finite for $\theta = \theta_1, \theta_2$, where $0 < \theta_2 - \theta_1 < \pi/\rho$, then $h_f(\theta)$ is bounded from above for all $\theta \in [\theta_1, \theta_2]$.

Theorem 4.2.5. *Let f be an entire function of order $0 < \rho < \infty$ such that $h_f(\theta) < \infty$ for all $0 \le \theta \le 2\pi$. Then*

$$h_f(\theta) > -\infty, \quad 0 \le \theta \le 2\pi.$$

Proof. Suppose on the contrary that there exists $\theta_0 \in [0, 2\pi]$ such that $h_f(\theta_0) = -\infty$. Choose an integer $m > \rho$, and then choose $\theta_1 = \theta_0$ and $\theta_2 = \theta_0 + \frac{\pi}{m}$ in Theorem 4.2.4. Then $0 < \theta_2 - \theta_1 = \frac{\pi}{m} < \min\{2\pi, \frac{\pi}{\rho}\}$, and by (4.2.2),

$$\begin{aligned} h_f\left(\theta_0 + \frac{\pi}{2m}\right) = h_f\left(\frac{\theta_1 + \theta_2}{2}\right) &\le H\left(\frac{\theta_1 + \theta_2}{2}\right) \\ &= \frac{(h_1 + h_2)\sin(\rho(\theta_2 - \theta_1)/2)}{\sin(\rho(\theta_2 - \theta_1))}. \end{aligned}$$

By the double angle formula $\sin(2x) = 2\sin x \cos x$ we obtain

$$h_f\left(\theta_0 + \frac{\pi}{2m}\right) \le \frac{h_1 + h_2}{2\cos(\rho(\theta_2 - \theta_1)/2)} = \frac{h_1 + h_2}{2\cos\left(\frac{\pi\rho}{2m}\right)},$$

where $0 < \frac{\pi\rho}{2m} < \frac{\pi}{2}$, $h_2 < \infty$, and where we may let $h_1 \to -\infty$. Thus

$$h_f\left(\theta_0 + \frac{\pi}{2m}\right) = -\infty.$$

Repeating this argument, we find that

$$h_f\left(\theta_0 + \frac{k\pi}{2m}\right) = -\infty, \quad k = 0, 1, \ldots, 4m - 1.$$

Hence, by the definition of the indicator function, f must be bounded on every ray

$$\arg z = \theta_0 + \frac{k\pi}{2m}, \quad k = 0, 1, \ldots, 4m - 1.$$

Then by Theorem 4.1.1, f is bounded in each closed sector determined by two adjacent rays. Consequently, f is bounded in the entire complex plane, and hence f is a constant function by Liouville's theorem. This contradicts the fact that $\rho(f) > 0$ and completes the proof. □

Theorem 4.2.6. *Let f be an entire function of order $0 < \rho < \infty$ such that $h_f(\theta) < \infty$ for all $0 \le \theta \le 2\pi$. Then $h_f(\theta)$ is continuous for all $\theta \in \mathbb{R}$.*

Proof. By the 2π-periodicity of the indicator function it suffices to prove the continuity on the interval $[0, 2\pi]$. By Theorem 4.2.5 we have $h_f(\theta) > -\infty$ for all $0 \le \theta \le 2\pi$. Hence we may rewrite (4.2.2) as

$$h_f(\theta) \le \frac{h_f(\theta_1)\sin(\rho(\theta_2 - \theta)) + h_f(\theta_2)\sin(\rho(\theta - \theta_1))}{\sin(\rho(\theta_2 - \theta_1))}, \quad \theta_1 \le \theta \le \theta_2, \tag{4.2.5}$$

for all $\theta_1, \theta_2 \in \mathbb{R}$ such that $0 < \theta_2 - \theta_1 < \min\{2\pi, \pi/\rho\}$.

Choose an arbitrary $\theta_0 \in [0, 2\pi]$, and set $\theta_1 = \theta_0$ in (4.2.5), where the right-hand side is a continuous function of θ. As $\theta \to \theta_0^+$, the right-hand side approaches $h_f(\theta_0)$. Taking the limit superiors on both sides of (4.2.5), we get

$$\limsup_{\theta \to \theta_0^+} h_f(\theta) \le h_f(\theta_0).$$

Then setting $\theta_2 = \theta_0$ and passing to the limit as $\theta \to \theta_0^-$, we find in the same way that

$$\limsup_{\theta \to \theta_0^-} h_f(\theta) \le h_f(\theta_0).$$

Putting the two results together, we have

$$\limsup_{\theta \to \theta_0} h_f(\theta) \le h_f(\theta_0). \tag{4.2.6}$$

Next, we choose $\theta = \theta_0 \in (\theta_1, \theta_2)$ in (4.2.5). Passing to the limit as $\theta_1 \to \theta_0$ and then to the limit as $\theta_2 \to \theta_0$, we obtain

$$h_f(\theta_0) \le \liminf_{\theta_1 \to \theta_0^-} h_f(\theta_1) \quad \text{and} \quad h_f(\theta_0) \le \liminf_{\theta_2 \to \theta_0^+} h_f(\theta_2),$$

that is,

$$\liminf_{\theta \to \theta_0} h_f(\theta) \ge h_f(\theta_0). \tag{4.2.7}$$

Combining (4.2.6) and (4.2.7), we conclude that

$$\lim_{\theta \to \theta_0} h_f(\theta) = h_f(\theta_0).$$

This shows, by definition, that h_f is continuous at θ_0 and hence on $[0, 2\pi]$. □

Corollary 4.2.7 ([205, Theorem 9.19]). *Let f be an entire function of order $0 < \rho < \infty$ such that $h_f(\theta) < \infty$ for all $0 \le \theta \le 2\pi$. Then $\max_{0\le\theta\le 2\pi} h_f(\theta)$ exists as a finite number and is equal to the type τ of f.*

Proof. The first statement is clear because $h_f(\theta)$ is continuous by Theorem 4.2.6 and because the interval $[0, 2\pi]$ is finite and closed. Thus there exists at least one $\theta_0 \in [0, 2\pi]$ such that

$$h_f(\theta_0) = \max_{0\le\theta\le 2\pi} h_f(\theta).$$

Since the inequality $h_f(\theta) \le \tau$ is clear for all $\theta \in [0, 2\pi]$, we suppose on the contrary that $h_f(\theta) \le \sigma < \tau$ for all $\theta \in [0, 2\pi]$. Let $\varepsilon > 0$ be a small constant. We choose $h_1 = h_2 = \sigma$ in (4.2.1) and conclude by Lemma 4.2.4 that

$$\log|f(re^{i\theta})| \le \left(\frac{\sin(\rho(\theta_2 - \theta)) + \sin(\rho(\theta - \theta_1))}{\sin(\rho(\theta_2 - \theta_1))}\sigma + \varepsilon\right) r^\rho, \quad \theta_1 \le \theta \le \theta_2,$$

whenever $r \ge R(\varepsilon)$. Using the trigonometric identities

$$\sin(2x) = 2\sin x \cos x,$$
$$\sin A + \sin B = 2\cos\left(\frac{A - B}{2}\right)\sin\left(\frac{A + B}{2}\right)$$

with $2x = \rho(\theta_2 - \theta_1)$, $A = \rho(\theta_2 - \theta)$, and $B = \rho(\theta - \theta_1)$, we obtain

$$\log|f(re^{i\theta})| \le \left(\frac{\cos\left(\rho\left(\frac{\theta_1+\theta_2}{2} - \theta\right)\right)}{\cos\left(\rho\left(\frac{\theta_2-\theta_1}{2}\right)\right)}\sigma + \varepsilon\right) r^\rho, \quad \theta_1 \le \theta \le \theta_2,$$

whenever $r \ge R(\varepsilon)$. Setting

$$\frac{\theta_1 + \theta_2}{2} = \theta_0 \quad \text{and} \quad \frac{\theta_2 - \theta_1}{2} = \delta,$$

it follows that

$$\log|f(re^{i\theta})| \le \left(\frac{\cos(\rho(\theta_0 - \theta))}{\cos(\rho\delta)}\sigma + \varepsilon\right) r^\rho, \quad \theta_0 - \delta \le \theta \le \theta_0 + \delta,$$

whenever $r \ge R(\varepsilon)$. Let $\sigma_0 \in (\sigma, \tau)$. For $\varepsilon > 0$ and $\delta > 0$ small enough, we deduce that

$$\log|f(re^{i\theta})| \le \sigma_0 r^\rho \tag{4.2.8}$$

whenever $\theta_0 - \delta \le \theta \le \theta_0 + \delta$ and $r \ge R(\varepsilon)$. We can choose $\delta = \frac{\pi}{2m}$, where m is a large positive integer, and conclude that inequality (4.2.8) holds in all closed intervals

$$\frac{k\pi}{m} \le \theta \le \frac{(k+1)\pi}{m}, \quad k = 0, \ldots, 2m - 1,$$

whenever r is large enough. Consequently, (4.2.8) holds for all θ whenever r is large enough. This leads to $\tau \le \sigma_0 < \tau$, a contradiction. This contradiction proves the statement. □

Regarding lower bounds for $h_f(\theta)$ in Corollary 4.2.7, we have the following result.

Theorem 4.2.8 ([205, pp. 276–277]). *Let f be an entire function of order $0 < \rho < \infty$ such that $h_f(\theta) < \infty$ for all $0 \le \theta \le 2\pi$. Then f has finite type $\tau(f)$, and we have the following properties:*
(a) *If $1/2 < \rho < \infty$, then*

$$-\tau(f) \le h_f(\theta) \le \tau(f), \quad 0 \le \theta \le 2\pi.$$

(b) *If $0 < \rho \le 1/2$, then*

$$0 \le h_f(\theta) \le \tau(f), \quad 0 \le \theta \le 2\pi.$$

Concerning the sharpness of Theorem 4.2.8, Corollary 4.2.7 shows that there always exists $\theta_1 \in [0, 2\pi]$ such that $h_f(\theta_1) = \tau(f)$. On the other hand, the function $f(z) = e^z$ in Example 4.2.1 satisfies $h_f(\pi) = -1 = -\tau(f)$. However, it is not in general guaranteed that there exists $\theta_2 \in [0, 2\pi]$ such that $h_f(\theta_2) = -\tau(f)$; see Examples 4.2.3 and 4.2.10.

We finish this section with a discussion about the length of intervals of θ where the indicator function $h_f(\theta)$ can be negative or positive. Suppose first that f is an entire function of order $\rho(f) \in (0, \infty)$ such that $h_f(\theta) < 0$ for all $\theta \in \mathbb{R}$. Then $|f(z)|$ tends to zero on every ray emanating from the origin, which means that f is bounded on each ray. It follows from the Phragmén–Lindelöf principle that f is bounded in the whole complex plane, and therefore it is a constant function by Liouville's theorem. Since this is a contradiction, we obtain that it is impossible to have $h_f(\theta) < 0$ for all $\theta \in \mathbb{R}$.

Theorem 4.2.9 ([43, p. 47], [205, p. 275]). *Let f be an entire function of order $\rho \in [1/2, \infty)$ such that $h_f(\theta) < \infty$ for all $\theta \in [0, 2\pi]$. Then we have the following properties:*
(a) *If $h_f(\theta) < 0$ for $a < \theta < b$, then $b - a \le \pi/\rho$.*
(b) *If $h_f(\theta) > 0$ for $a < \theta < b$, then $b - a \ge \pi/\rho$.*

Proof. To prove part (a), we suppose on the contrary that $h_f(\theta) < 0$ for $a < \theta < b$, where $b - a > \pi/\rho$. Then there exists a real constant c such that for every $\varepsilon > 0$ satisfying

$$0 < \varepsilon < \frac{\pi}{2\rho},$$

we have

$$a < c < c + \varepsilon < c + \frac{\pi}{\rho} - \varepsilon < c + \frac{\pi}{\rho} < b.$$

Set $\theta_1 = c + \varepsilon$ and $\theta_2 = c + \pi/\rho - \varepsilon$. Then $0 < \theta_2 - \theta_1 = \pi/\rho - 2\varepsilon < \pi/\rho$. Now applying Theorem 4.2.4, we obtain

$$h_f(\theta) \leq \frac{\alpha \sin(\rho(\theta_2 - \theta)) + \beta \sin(\rho(\theta - \theta_1))}{\sin(\rho(\theta_2 - \theta_1))}, \quad \theta_1 \leq \theta \leq \theta_2, \tag{4.2.9}$$

where $\alpha = h_f(\theta_1)$ and $\beta = h_f(\theta_2)$. Since $\rho(\theta_2 - \theta_1) = \pi - 2\varepsilon\rho$, we get

$$\sin(\rho(\theta_2 - \theta_1)) = \sin(\pi - 2\varepsilon\rho) = \sin(2\varepsilon\rho).$$

Choosing θ as the midpoint of the interval $[\theta_1, \theta_2]$, that is, $\theta = (\theta_1 + \theta_2)/2$, from (4.2.9) we obtain that

$$h_f\left(\frac{\theta_1 + \theta_2}{2}\right) \leq \frac{\alpha \sin(\frac{\rho(\theta_2 - \theta_1)}{2}) + \beta \sin(\frac{\rho(\theta_2 - \theta_1)}{2})}{\sin(2\varepsilon\rho)}.$$

Since

$$\sin\left(\frac{\rho(\theta_2 - \theta_1)}{2}\right) = \sin\left(\frac{\pi}{2} - \varepsilon\rho\right) = \cos(\varepsilon\rho),$$

we get

$$h_f\left(\frac{\theta_1 + \theta_2}{2}\right) \leq (\alpha + \beta)\frac{\cos(\varepsilon\rho)}{\sin(2\varepsilon\rho)} = \frac{\alpha + \beta}{2\sin(\varepsilon\rho)}.$$

Letting $\varepsilon \to 0^+$ and noting that $\alpha + \beta < 0$, we obtain

$$h_f\left(\frac{\theta_1 + \theta_2}{2}\right) = -\infty. \tag{4.2.10}$$

But our hypothesis is that $h_f(\theta) < \infty$ for all $\theta \in [0, 2\pi]$, which means that $h_f(\theta) > -\infty$ for all $\theta \in \mathbb{R}$ from Theorem 4.2.5. This contradicts (4.2.10), which completes the proof of part (a).

To prove part (b), we suppose on the contrary that $h_f(\theta) > 0$ for $a < \theta < b$ and $h_f(a) = h_f(b) = 0$, where $b - a < \pi/\rho$. Then applying Theorem 4.2.4 with $\theta_1 = a$, $\theta_2 = b$, and $h_1 = h_2 = 0$, we obtain

$$h_f(\theta) \leq 0, \quad a \leq \theta \leq b,$$

which contradicts our assumption and thus proves part (b). □

Remark. (a) The simple function $f(z) = e^z$ with indicator $h_f(\theta) = \cos\theta$ illustrates the sharpness of statements (a) and (b) in Theorem 4.2.9.

(b) The restriction $\rho(f) \geq 1/2$ in Theorem 4.2.9 is required so that $\pi/\rho(f) \leq 2\pi$. Moreover, if $0 < \rho(f) < 1/2$, then $h_f(\theta) \geq 0$ for all $\theta \in \mathbb{R}$; see Problem 4.2.13.

(c) From the proof of part (a) we can see that a little more can be said, namely, that there cannot exist two real numbers φ and ψ that satisfy $\psi - \varphi = \pi/\rho$ such that $h_f(\varphi) < 0$ and $h_f(\psi) < 0$ [205, Theorem 9.16].

It is possible to have an interval where the indicator function is identically zero.

Example 4.2.10. Let f be the function defined as follows:

$$f(z) = \begin{cases} \frac{e^z-1}{z} & \text{if } z \neq 0, \\ 1 & \text{if } z = 0. \end{cases}$$

Using the definition of derivative, we obtain that $f'(0) = 1/2$. Hence f is an entire function. It can be shown that $\rho(f) = 1$ and $\tau(f) = 1$ and that the indicator function is given as follows:

$$h_f(\theta) = \begin{cases} \cos\theta & \text{for } \theta \in [-\pi/2, \pi/2], \\ 0 & \text{for } \theta \in [\pi/2, 3\pi/2]. \end{cases}$$

Hence, in fact, $h_f(\theta) = \max\{0, \cos\theta\}$ for all $\theta \in \mathbb{R}$.

Problems

Problem 4.2.11. Find $h_f(\theta)$ when $f(z) = e^{3z^2} + e^z$.

Problem 4.2.12. Let f be an entire function of finite positive order ρ, and let $P \not\equiv 0$ be a polynomial. What can be said about the indicator of the function $f + P$?

Problem 4.2.13. Let f be an entire function of order $0 < \rho < 1/2$. Prove that $h_f(\theta) \geq 0$ for all $\theta \in \mathbb{R}$.

4.3 Entire functions of order less than one

In this section, we consider entire functions f of orders $\leq 1/2$ and < 1. We have already encountered such functions in the previous chapters, but this time we will compare the growth of the minimum modulus $L(r,f) := \min_{|z|=r} |f(z)|$ to the growth of the maximum modulus $M(r,f)$. Obviously, $L(r,f) = 0$ whenever f has a zero on $|z| = r$, but it turns out that for some values of r, the function $L(r,f)$ can be large or even close to $M(r,f)$. We will also give an analogue of the Gauss–Lucas theorem for transcendental entire functions of order < 1.

Following Boas [37, p. 8], we say that an entire function f is *of growth* (ρ, τ), where $0 < \rho < \infty$, if either (a) $\rho(f) < \rho$ or (b) $\rho(f) = \rho$ and $\tau(f) \leq \tau$.

According to Hadamard's theorem (Section 3.5), a transcendental entire function f of order $\rho(f) < 1$ is essentially a canonical product. By Corollaries 3.5.9 and 3.8.2, for every $\varepsilon > 0$, we have

$$L(r,f) \geq \exp(-r^{\rho+\varepsilon}), \quad r \notin E,$$

where the exceptional set E has finite linear measure and is a union of circular projections of discs in which f has its zeros. Thus $L(r,f)$ cannot tend to zero arbitrarily fast outside E. We can say even more about functions of growth $(1/2, 0)$.

Theorem 4.3.1 ([37, Theorem 3.1.1]). *If f is a nonconstant entire function of growth $(1/2, 0)$, then*

$$\limsup_{r\to\infty} L(r,f) = \infty.$$

Proof. We may assume that $f(0) = 1$ without affecting the statement. If $\{z_n\}$ denotes the nonzero zeros of f, counting multiplicities, then by Hadamard's theorem,

$$f(z) = \prod_{n=1}^{\infty}(1 - z/z_n).$$

Denote $r = |z|$ and $r_n = |z_n|$, and define

$$g(z) = \prod_{n=1}^{\infty}(1 + z/r_n). \tag{4.3.1}$$

The zeros of g are negative real numbers. In fact, if f has k zeros on a certain circle $|z| = r$, where k includes the multiplicity of each zero, then g has a zero of multiplicity k at the point $z = -r$. Thus $n(r, 1/f) = n(r, 1/g)$.

Since $|1 - r/r_n| \leq |1 - z/z_n|$, we have

$$|g(-r)| = L(r,g) \leq L(r,f) \leq M(r,f) \leq g(r) = M(r,g). \tag{4.3.2}$$

By Theorem 3.1.6 we know that

$$\lambda(f) = \limsup_{r\to\infty} \frac{\log n(r, 1/f)}{\log r}.$$

Moreover, since the order ρ of f is not a positive integer, we have $\lambda(f) = \rho$ by Corollary 3.5.9. Since $n(r, 1/f) = n(r, 1/g)$, it therefore follows that f and g have the same order ρ. Further, if $\rho = 1/2$, then $\tau(f) = 0$ by the assumption, in which case $\tau(g) = 0$ by Theorem 3.5.11. In particular, g is of growth $(1/2, 0)$, same as f.

By Corollary 4.1.3 and Problem 4.1.7, an entire function of growth $(1/2, 0)$ is unbounded along every ray emanating from the origin. Thus g is unbounded on the negative real axis. The statement now follows from (4.3.2). □

A slightly weaker version of Theorem 4.3.1 for entire functions of order $< 1/2$ can be found in the book of Holland [142, p. 134].

Theorem 4.3.2 ([37, Theorem 3.1.6]). *If f is a nonconstant entire function of growth $(\rho, 0)$, where $0 < \rho < 1/2$, then there exists a strictly increasing and unbounded sequence $\{r_n\}$ of positive real numbers such that*

$$\log L(r_n, f) > \cos \pi\rho \cdot \log M(r_n, f).$$

To prove Theorem 4.3.2, we need some preparations. Let $-\pi \le \alpha < \beta \le \pi$, and suppose that f is analytic in the open sector $S(\alpha, \beta)$ and continuous in the corresponding closed sector $\overline{S}(\alpha, \beta)$. We define

$$M(r, \alpha, \beta, f) := \max_{\alpha \le \theta \le \beta} |f(re^{i\theta})|.$$

Clearly, for an entire function f, we have $M(r, -\pi, \pi, f) = M(r, f)$.

We proceed to state a result known as *Beurling's theorem*, where a function f that is analytic in the right half-plane is dominated on the imaginary axis by a function $\phi(r)$. This is in contrast to the Phragmén–Lindelöf theorems, where a function is analytic in a sector and is bounded on the sides of this sector.

Theorem 4.3.3 ([37, p. 4]). *Let f be analytic in the right half-plane $\operatorname{Re} z > 0$ and continuous in $\operatorname{Re} z \ge 0$. Suppose that $|f(iy)| \le \phi(|y|)$, where $\phi : (0, \infty) \to (0, \infty)$ is continuous and satisfies*

$$\limsup_{r \to \infty} \frac{\log \phi(r)}{r^\rho} = 0$$

for some $0 < \rho < 1$. Suppose further that for each $\varepsilon > 0$, there exists a sequence of positive values $t_n \to \infty$ such that $|f(z)| = O(e^{\varepsilon |z|})$ whenever $|z| = t_n$ and $\operatorname{Re} z \ge 0$ (as $t_n \to \infty$). Then either f is bounded in $\operatorname{Re} z \ge 0$, or there exists a strictly increasing and unbounded sequence of positive numbers $\{s_n\}$ such that

$$\log M(s_n, -\pi/2, \pi/2, f) \cdot \cos(\pi\rho/2) < \log \phi(s_n).$$

Proof of Theorem 4.3.2. Let g be the entire function in (4.3.1) associated with f and having negative real zeros. If we can prove that there exists a strictly increasing and unbounded sequence $\{r_n\}$ of positive real numbers such that

$$\log L(r_n, g) > \cos(\pi\rho) \cdot \log M(r_n, g), \tag{4.3.3}$$

then the statement for f will follow from (4.3.2).

The function $G(z) = g(z^2)$ is entire, of growth $(2\rho, 0)$, unbounded along rays emanating from the origin, and $|G(iy)| = |g(-y^2)| = L(y^2, g)$. Choose $\phi(|y|) = L(y^2, g)$. Then

$$\limsup_{r \to \infty} \frac{\log \phi(r)}{r^{2\rho}} \le \limsup_{r \to \infty} \frac{\log M(r, G)}{r^{2\rho}} = 0.$$

We now apply Beurling's theorem to G with $2\rho < 1$ in place of ρ. It follows that there exists a strictly increasing and unbounded sequence of positive real numbers $\{s_n\}$ such that

$$\log M(s_n, -\pi/2, \pi/2, G) \cdot \cos(\pi\rho) < \log \phi(s_n) = \log L(s_n^2, g).$$

Since $G(z) = g(z^2)$, this implies that (4.3.3) holds for $r_n = s_n^2$. □

The function $f(z) = e^z$ satisfies $L(r,f) = e^{-r}$, and therefore it is an example of an entire function having bounded minimum modulus. In 1948, Heins [114] proved that if f is entire satisfying $L(r,f) = O(1)$ and

$$\liminf_{r\to\infty} r^{-1/2} \log M(r,f) < \infty,$$

then the limit

$$\lim_{r\to\infty} r^{-1/2} \log M(r,f)$$

exists on $(0, \infty]$. If this limit is finite, then the function f in question is of perfectly regular growth; see Section 7.4.

Replacing the condition $0 < \rho < 1/2$ in Theorem 4.3.2 with $0 < \rho < 1$, we obtain a result that is commonly known as the $\cos \pi\rho$ *theorem.*

Theorem 4.3.4 ([37, Theorem 3.2.1], [142, Theorem 7.9.3]). *If f is a nonconstant entire function of growth $(\rho, 0)$, where $0 < \rho < 1$, then*

$$\limsup_{r\to\infty} \frac{\log L(r,f)}{\log M(r,f)} \geq \cos(\pi\rho). \tag{4.3.4}$$

Sketch of Proof. Let g be the function defined in (4.3.1). By (4.3.2),

$$\frac{\log L(r,f)}{\log M(r,f)} \geq \frac{\log L(r,g)}{\log M(r,g)}. \tag{4.3.5}$$

Hence if $0 < \rho < 1/2$, then $\cos(\pi\rho) > 0$, and (4.3.4) follows from (4.3.5) by applying Theorem 4.3.2 to g. If $1/2 \leq \rho < 1$, then we may suppose as before that $f(0) = 1$. Also, let $\{z_n\}$ denote the nonzero zeros of f, and set $r_n = |z_n|$. Choose a point ζ such that $|\zeta| = r$ and $|f(\zeta)| = L(r,f)$. Since

$$\left|\left(1 + \frac{r}{r_n}\right)\left(1 - \frac{r}{r_n}\right)\right| = \left|1 - \frac{r^2}{r_n^2}\right| \leq \left|1 - \frac{\zeta^2}{z_n^2}\right| = \left|\left(1 + \frac{\zeta}{z_n}\right)\left(1 - \frac{\zeta}{z_n}\right)\right|,$$

it follows that

$$|g(r)g(-r)| \leq |f(\zeta)f(-\zeta)| \leq L(r,f)M(r,f). \tag{4.3.6}$$

If we can prove that

$$\limsup_{r\to\infty} \frac{\log L(r,g)}{\log M(r,g)} \geq \cos(\pi\rho), \tag{4.3.7}$$

then for every $\varepsilon > 0$, there exist arbitrarily large values of r for which

$$|g(-r)| = L(r,g) \geq M(r,g)^{\cos(\pi\rho)-\varepsilon} = |g(r)|^{\cos(\pi\rho)-\varepsilon}.$$

For these values of r, using (4.3.2) and (4.3.6), we have

$$L(r,f)M(r,f) \geq |g(r)g(-r)| \geq |g(r)|^{1+\cos(\pi\rho)-\varepsilon} \geq M(r,f)^{1+\cos(\pi\rho)-\varepsilon},$$

which implies (4.3.4). Thus it remains to prove (4.3.7) for $1/2 \leq \rho < 1$. This part involves complex calculations of integrals. The details can be found in [37, Section 3.2] or [142, Section 7.9], where the latter reference contains more details than the former. □

The following example shows that inequality (4.3.4) is sharp.

Example 4.3.5. Let $0 < \rho < 1/2$, let $\{z_n\}$ be the sequence defined by $z_n = -n^{1/\rho}$, and then set

$$f(z) = \prod_{n=1}^{\infty}(1 - z/z_n).$$

Then f is an entire function satisfying $\rho(f) = \lambda(f) = \rho$. The discussion in Example 7.4.7 gives us $n(r) \sim r^\rho$. Therefore, from the proof of Theorem 7.4.6 we obtain

$$\log f(re^{i\theta}) \sim \frac{\pi r^\rho e^{i\theta\rho}}{\sin(\pi\rho)}, \qquad -\pi < \theta < \pi.$$

Taking the real parts, we get

$$\log|f(re^{i\theta})| \sim \frac{\pi r^\rho \cos(\theta\rho)}{\sin(\pi\rho)}, \qquad -\pi < \theta < \pi,$$

from which

$$\log M(r,f) \sim \frac{\pi r^\rho}{\sin(\pi\rho)} \quad \text{and} \quad \log L(r,f) \sim \frac{\pi r^\rho \cos(\pi\rho)}{\sin(\pi\rho)}.$$

This gives an equality in (4.3.4).

Even though the $\cos\pi\rho$ theorem is for entire functions of order $\rho \in (0,1)$, we point out that if $f(z) = e^z$, then $\log M(r,f) = r$, $\log L(r,f) = -r$, and $\cos\pi\rho = -1$.

A generalization of the $\cos\pi\rho$ theorem was given by Baernstein II [6] in 1964. In 1963, Kjellberg [157] proved that if f is a nonconstant entire function such that

$$\log L(r,f) < \cos(\pi\rho) \cdot \log M(r,f) + O(1), \quad r \to \infty,$$

where $0 < \rho < 1$, then the limit

$$\lim_{r\to\infty} \frac{\log M(r,f)}{r^\rho}$$

exists on $(0, \infty]$. Similarly, as in the aforementioned result by Heins, if this limit is finite, then f is of perfectly regular growth. In 1975, Hayman [110] proved an analogue of Kjellberg's result for the case $\rho = 1$.

Let α be a constant satisfying $\rho < \alpha < 1$. Inequality (4.3.4) guarantees the existence of a strictly increasing and unbounded sequence $\{r_n\}$ of positive real numbers such that

$$\frac{\log L(r_n,f)}{\log M(r_n,f)} \geq \cos(\pi\alpha), \quad n \in \mathbb{N}.$$

In 1970, Barry [16] showed this inequality holds on a much larger set of points than just a sequence. The size of this set is measured in terms of strong logarithmic density in [16].

To close this section, we give an analogue of the Gauss-Lucas theorem for transcendental entire functions of order < 1. It turns out that the proof of the Gauss–Lucas theorem works here as well, with some slight modifications. The original proof is by Marden [203], who used a slightly different approach.

Theorem 4.3.6 ([203]). *Let f be a transcendental entire function of order < 1. If the zeros of f lie in a set $S \subset \mathbb{C}$, then the zeros of the derivative f' lie in the convex hull* $\mathrm{co}(S)$.

Proof. Clearly, we may suppose that $\mathrm{co}(S) \subsetneq \mathbb{C}$. Let $\{z_n\}$ denote the sequence of nonzero zeros of f counting multiplicities and organized with respect to increasing moduli. Then by the Hadamard factorization,

$$f(z) = Cz^k \prod_{n=1}^{\infty}(1 - z/z_n)$$

for some constants $k \in \mathbb{N} \cup \{0\}$ and $C \in \mathbb{C} \setminus \{0\}$. From Problem 5.5.7 with $p = 0$ we have

$$\frac{f'(z)}{f(z)} = \frac{k}{z} + \sum_{n=1}^{\infty} \frac{1}{z - z_n}.$$

If $k > 0$, then we may redefine the sequence $\{z_n\}$ so that the first k terms of the new sequence are zero and denote the new sequence again by $\{z_n\}$. Therefore, we may write

$$\frac{f'(z)}{f(z)} = \sum_{n=1}^{\infty} \frac{1}{z - z_n}$$

independently of whether $k = 0$ or $k > 0$.

Let ζ be a zero of f'. If ζ is also a zero of f, then the statement is clear. Otherwise, ζ is a zero of f'/f, and

$$\sum_{n=1}^{\infty} \frac{1}{\zeta - z_n} = 0.$$

Conjugating this equation and using the identity $1/\bar{a} = a/|a|^2$ for $a \in \mathbb{C} \setminus \{0\}$, we obtain

$$\sum_{n=1}^{\infty} \frac{\zeta - z_n}{|\zeta - z_n|^2} = 0.$$

We may rewrite this as

$$\zeta \sum_{n=1}^{\infty} \frac{1}{|\zeta - z_n|^2} = \sum_{n=1}^{\infty} \frac{z_n}{|\zeta - z_n|^2}, \tag{4.3.8}$$

where the convergence of the series on either side needs to be addressed. However, since $|z_n| \to \infty$ as $n \to \infty$, there exists $N \in \mathbb{N}$ such that

$$|z_n| \geq 2|\zeta| + 1, \quad n \geq N,$$

and hence

$$\frac{1}{|\zeta - z_n|^2} \leq \frac{1}{(|z_n| - |\zeta|)^2} \leq \frac{1}{|z_n|^2 - 2|z_n||\zeta|} \leq \frac{1}{|z_n|}, \quad n \geq N.$$

Since $\sum_{n \geq k+1} |z_n|^{-1} < \infty$, the first series in (4.3.8) converges by the majorant principle. Moreover, since

$$\lim_{n\to\infty} \frac{|z_n|^2}{|\zeta - z_n|^2} = 1 \in (0, \infty),$$

the second series in (4.3.8) is absolutely convergent by the limit comparison test, and therefore it is convergent by Lemma 3.3.7. Thus (4.3.8) gives

$$\zeta = \sum_{n=1}^{\infty} a_n z_n, \quad a_n = \frac{1}{|\zeta - z_n|^2} \cdot \left(\sum_{k=1}^{\infty} \frac{1}{|\zeta - z_k|^2} \right)^{-1},$$

where $a_n > 0$ for all n, and $\sum_{n=1}^{\infty} a_n = 1$. In particular, the point ζ can be represented as an infinite convergent series involving the zeros z_n of f.

Since $\mathrm{co}(S) \subsetneq \mathbb{C}$ by the assumption, it follows that $\mathrm{co}(S)$ is the intersection of all closed half-planes including S by Lemma 3.9.4. Hence there exists at least one closed half-plane containing S and therefore, all the zeros z_n of f. Let H be such a closed half-plane. Then $\zeta \in H$, and, consequently, $\zeta \in \mathrm{co}(S)$, by using reasoning similar to the proof of Lemma 3.9.5. □

The example

$$f(z) = (1 + z)e^{z^2}, \quad f'(z) = (2z^2 + 2z + 1)e^{z^2}, \tag{4.3.9}$$

shows that the conclusion of Theorem 4.3.6 is not valid for all transcendental entire functions (of arbitrary order). A similar example is given in [203, p. 829]. We also mention that [203, Theorem (3.1)] is a higher-order version of Theorem 4.3.6.

Problems

Problem 4.3.7. Let f be the entire function

$$f(z) = \prod_{n=1}^{\infty}\left(1 - \frac{z}{n^{3/2}}\right).$$

Prove that all the zeros of the derivative f' lie on the real interval $[1, \infty)$.

4.4 Wiman–Valiron theory

Wiman–Valiron theory describes the local behavior of a transcendental entire function f near a point z where $|f(z)|$ is large in terms of the Taylor series of f. We use Langley's Postgraduate Notes [177] as the main reference, which in turn is based on Hayman's paper [109] and Valiron's book [279]. Other references relevant to this section are Fenton's paper [68] and the book [150] by Jank and Volkmann.

Anders Wiman (1865–1959) was a Swedish mathematician, who worked in algebraic geometry, group theory, and entire functions. One evidence of his impact on mathematics is the large number of papers with the word "Wiman" in the title, which is used to describe various mathematical entities. Georges Valiron (1884–1955) was a French mathematician, who was known for his work in entire functions and Tauberian theorems. His classic book "Lectures on the theory of entire functions" [279], first published in 1923, is still often cited.

Let f be a transcendental entire function given as a Taylor series

$$f(z) = \sum_{n=0}^{\infty} a_n z^n. \tag{4.4.1}$$

As this series converges uniformly on compact subsets of $\mathbb{C}$, it is clear that $|a_n|r^n \to 0$ as $n \to \infty$ when $r = |z|$ is fixed. This allows us to define the *maximal term* of f by

$$\mu(r,f) = \max\{|a_n|r^n : n = 0, 1, \ldots\}.$$

Clearly, there exists $n \geq 1$ such that $a_n \neq 0$, giving us $\mu(r,f) \geq |a_n|r^n$. Consequently, $\mu(r,f) \to \infty$ as $r \to \infty$. The largest of the integers k for which $\mu(r,f) = |a_k|r^k$ is called the *central index* and is denoted by $\nu(r,f)$. In other words,

$$\nu(r,f) = \max\{k \geq 0 : \mu(r,f) = |a_k|r^k\}.$$

We make two simple observations.

(1) Clearly, $\mu(0,f) = |a_0|$. Hence, if $a_0 = 0$, then $\nu(0,f)$ is not defined, whereas if $a_0 \neq 0$, then $\nu(0,f) = 0$.

(2) If $r > 0$, then $\mu(r,f) > 0$, and if $n \neq k$ are two indices for which $a_n a_k \neq 0$, then the equation $|a_n|r^n = |a_k|r^k$ holds for exactly one positive value of r. Conversely, if $\mu(r,f) = |a_n|r^n$ for some n and $r > 0$, then there are at most finitely many indices k such that $|a_k|r^k = |a_n|r^n$ because $|a_m|r^m \to 0$ as $m \to \infty$. Therefore, since there are only countably many Taylor coefficients, this shows that there are at most countably many values of r for which there does not exist a unique n such that $|a_n|r^n = \mu(r,f)$.

Example 4.4.1. Let $f(z) = e^z = \sum_{n=0}^{\infty} \frac{z^n}{n!}$. Set $T_n(r) = \frac{r^n}{n!}$ for $r > 0$. Then

$$\mu(r,f) = \max\{T_n(r) : n \geq 0\}, \quad \text{and} \quad \frac{T_{n+1}(r)}{T_n(r)} = \frac{r}{n+1}.$$

Fix $r > 0$. If $\frac{r}{n+1} > 1$, then the sequence $\{T_n(r)\}$ is strictly increasing in n, whereas if $\frac{r}{n+1} < 1$, then $\{T_n(r)\}$ is strictly decreasing in n. Since the central index $\nu(r,f)$ is defined as the largest of the integers n for which $T_n(r)$ takes the maximal value, we obtain

$$\nu(r,f) \sim r.$$

Since it is known that the gamma function $\Gamma(z)$ is a generalization of the factorial (Section 5.1), we obtain $\mu(r,f) \sim \frac{r^r}{\Gamma(r+1)}$. Using the Stirling approximation $\Gamma(r+1) \sim r^r e^{-r}\sqrt{2\pi r}$ (Theorem 5.1.4), we obtain

$$\mu(r,f) \sim \frac{e^r}{\sqrt{2\pi r}}.$$

Both the maximal term and the central index are key concepts in Wiman–Valiron theory. We proceed to discuss some of their fundamental properties in a sequence of lemmas. In these lemmas, f denotes the transcendental entire function in (4.4.1) and $M(r,f)$ denotes its maximum modulus.

Lemma 4.4.2. *We have*

$$\mu(r,f) \leq M(r,f) \leq 2\mu(2r,f), \quad r > 0. \tag{4.4.2}$$

Further, the orders of the functions $\log M(r,f)$ *and* $\log \mu(r,f)$ *are equal:*

$$\rho(f) = \limsup_{r\to\infty} \frac{\log\log M(r,f)}{\log r} = \limsup_{r\to\infty} \frac{\log\log \mu(r,f)}{\log r}. \tag{4.4.3}$$

Proof. The first inequality in (4.4.2) comes from Cauchy's estimate (2.1.6). The second inequality in (4.4.2) is proved as follows. For $n \geq 0$, we have

$$|a_n|(2r)^n \leq \mu(2r,f), \quad |a_n|r^n \leq 2^{-n}\mu(2r,f),$$

and so

$$M(r,f) \leq \sum_{n=0}^{\infty} |a_n|r^n \leq \sum_{n=0}^{\infty} 2^{-n}\mu(2r,f) = 2\mu(2r,f).$$

The first equality in (4.4.3) is simply the definition of the order $\rho(f)$, whereas the second equality in (4.4.3) follows easily from (4.4.2). □

Lemma 4.4.3. *The maximal term $\mu(r,f)$ is continuous and nondecreasing on $[0,\infty)$, and there exists $R \geq 0$ such that $\mu(r,f)$ is strictly increasing on $[R,\infty)$.*

Proof. By the definition of $\mu(r,f)$ and by Lemma 4.4.2,

$$\mu(0,f) = |a_0| \leq \mu(r,f) \leq M(r,f) \to |a_0|, \quad r \to 0^+,$$

and so $\mu(r,f)$ is continuous as $r \to 0^+$. Choose $m > 0$ such that $a_m \neq 0$. If $r_0 > 0$ is arbitrarily fixed, then there exists $n_0 > m$ such that

$$|a_n|(2r_0)^n < |a_m|r_0^m, \quad n > n_0. \tag{4.4.4}$$

This follows from the fact that $|a_n|(2r_0)^n \to 0$ as $n \to \infty$. For $r_0 \leq r \leq 2r_0$, we have

$$\mu(r,f) \geq |a_m|r^m \geq |a_m|r_0^m. \tag{4.4.5}$$

Suppose that $\mu(r,f) = |a_k|r^k$ for some $k > n_0$ and $r_0 \leq r \leq 2r_0$. Then $\mu(r,f) \leq |a_k|(2r_0)^k$, which contradicts (4.4.4) and (4.4.5). Hence it follows that

$$\mu(r,f) = \max\{|a_n|r^n : 0 \leq n \leq n_0\}, \quad r_0 \leq r \leq 2r_0.$$

We conclude that $\mu(r,f)$ is continuous on $[r_0, 2r_0]$ as the maximum of finitely many continuous functions. Since $r_0 > 0$ is arbitrary, $\mu(r,f)$ is continuous on $[0,\infty)$.

Let $0 \leq r < s < \infty$, and choose n such that $\mu(r,f) = |a_n|r^n$. Then

$$\mu(s,f) \geq |a_n|s^n \geq |a_n|r^n = \mu(r,f), \tag{4.4.6}$$

which shows that $\mu(r,f)$ is nondecreasing. Next, choose $R \geq 0$ large enough so that $|a_m|R^m \geq |a_0|$ for some $m > 0$ with $a_m \neq 0$. Then for $R \leq r < s < \infty$, we have

$$|a_m|r^m \geq |a_0|,$$

and, consequently, $\mu(r,f) = |a_n|r^n$ for some $n > 0$ with $a_n \neq 0$. This gives a strict inequality in (4.4.6), proving that $\mu(r,f)$ is strictly increasing on $[R,\infty)$. □

Lemma 4.4.4. *The central index $\nu(r,f)$ is nondecreasing, and $\nu(r,f) \to \infty$ as $r \to \infty$. Moreover, $\nu(r,f)$ is continuous from the right at each $s > 0$, that is,*

$$\lim_{r\to s^+} \nu(r,f) = \nu(s,f).$$

Proof. Let $1 \leq r < s$ and $\nu(r,f) = N$. If $N = 0$, then clearly $\nu(s,f) \geq \nu(r,f)$. Suppose that $N > M$, where $M \in \mathbb{N} \cup \{0\}$. Since $\nu(r,f) = N$, we have

$$|a_N|r^N \geq |a_M|r^M,$$

and since $r^{M-N} \geq s^{M-N}$ for $1 \leq r < s$,

$$|a_N|s^N = |a_N|r^N\left(\frac{s}{r}\right)^N \geq |a_M|r^M\left(\frac{s}{r}\right)^N \geq |a_M|s^M,$$

from which it follows that $\nu(s,f) \geq N = \nu(r,f)$. Hence $\nu(r,f)$ is nondecreasing on $[1,\infty)$.

Let $L > 0$ be a large constant, and choose an integer $k > L$ such that $a_k \neq 0$. If $n < k$, then $|a_n|r^n < |a_k|r^k$ for all $r > 0$ large enough. Consequently, $\nu(r,f) \geq k > L$ for all $r > 0$ large enough. This proves that $\nu(r,f) \to \infty$ as $r \to \infty$.

It remains to prove that $\nu(r,f)$ is continuous from the right. Let $s > 0$, and denote $N = \nu(s,f)$. Since $|a_n|(2s)^n \to 0$ as $n \to \infty$, we may choose $n_0 > N$ such that

$$|a_n|(2s)^n < \mu(s,f), \quad n > n_0. \tag{4.4.7}$$

Then by Lemma 4.4.3 and (4.4.7)

$$\mu(s,f) \leq \mu(r,f) = \max\{|a_n|r^n : 0 \leq n \leq n_0\}, \quad s \leq r \leq 2s.$$

Since N is the largest integer n for which $|a_n|s^n = \mu(s,f)$, we have

$$|a_n|s^n < \mu(s,f), \quad n > N.$$

By the continuity of $\mu(r,f)$ there exists $\delta \in (0,s)$ such that

$$|a_n|r^n < |a_N|s^N = \mu(s,f), \quad s \leq r \leq s+\delta,\ N < n \leq n_0.$$

By the choice of n_0 we now have

$$|a_n|r^n < |a_N|r^N, \quad s \leq r \leq s+\delta,\ n > N.$$

Hence $\nu(r,f) = N$ for $s \leq r \leq s+\delta$, which proves that $\nu(r,f)$ is continuous from the right at $s > 0$. A similar argument shows that $\nu(r,f)$ is continuous as $r \to 0^+$, provided that $a_0 \neq 0$. □

The quantities $\mu(r,f)$ and $\nu(r,f)$ are related to each other by means of the following lemma.

Lemma 4.4.5. *There exists a strictly increasing and unbounded sequence $\{r_n\}_{n=0}^{\infty}$ of non-negative real numbers, with $r_0 = 0$, such that $\nu(r,f)$ is constant on (r_0, r_1) and $[r_n, r_{n+1})$ for all $n \geq 1$. Also, if $0 < s < r$, then*

$$\log\mu(r,f) = \log\mu(s,f) + \int_s^r \frac{\nu(t,f)}{t}\,dt. \tag{4.4.8}$$

For all $r > 0$ *large enough, we have*

$$\log \mu(r,f) \le \nu(r,f)\log r + O(1) \tag{4.4.9}$$

and

$$\nu(r,f)\log 2 \le \mu(2r,f), \quad \nu(r,f)\log r \le \log \mu(r^2,f). \tag{4.4.10}$$

Proof. Recall that $\nu(r,f)$ is a step function. If $r > 0$ and $\nu(r,f) = N$, then the function ν cannot have more than N discontinuities in $(0,r)$. Let $\{r_n\}$ be the sequence of discontinuities of $\nu(r,f)$ on $(0,\infty)$ such that $0 < r_1 < r_2 < \cdots$. Since $\nu(r,f)$ is integer-valued and continuous from the right by Lemma 4.4.4, it is a constant on (r_0, r_1) and $[r_n, r_{n+1})$ for all $n \ge 1$. This proves the first statement.

We proceed to prove (4.4.8). By the previous paragraph we may suppose that $\nu(r,f) = N$ for $r_n \le r < r_{n+1}$. Then on this interval, we have $\mu(r,f) = |a_N|r^N$. Consequently,

$$\frac{d\log \mu(r,f)}{d\log r} = N = \nu(r,f). \tag{4.4.11}$$

Hence by the continuity of $\mu(r,f)$

$$\log \mu(b,f) - \log \mu(a,f) = \int_a^b \frac{\nu(t,f)}{t}\,dt, \quad r_n \le a \le b \le r_{n+1}.$$

Adding suitably many of these identities and using a telescoping argument, we deduce (4.4.8).

To prove (4.4.9) and (4.4.10), choose $s \ge 1$ large enough so that $\mu(s,f) \ge 1$. Recalling that $\nu(r,f)$ is nondecreasing by Lemma 4.4.4, from (4.4.8) we find that

$$\begin{aligned}\log \mu(r,f) &\le \log \mu(s,f) + \nu(r,f)\int_s^r \frac{dt}{t}\\ &\le \log \mu(s,f) + \nu(r,f)\log r, \quad r \ge s.\end{aligned}$$

This implies (4.4.9). Using again (4.4.8), we deduce

$$\log \mu(2r,f) \ge \int_r^{2r} \frac{\nu(t,f)}{t}\,dt \ge \nu(r,f)\int_r^{2r}\frac{dt}{t} = \nu(r,f)\log 2,$$

$$\log \mu(r^2,f) \ge \int_r^{r^2} \frac{\nu(t,f)}{t}\,dt \ge \nu(r,f)\int_r^{r^2}\frac{dt}{t} = \nu(r,f)\log r,$$

which give (4.4.10). □

From (4.4.3), (4.4.9), and (4.4.10), we obtain the following important result.

Theorem 4.4.6. *We have*

$$\rho(f) = \limsup_{r\to\infty} \frac{\log\log\mu(r,f)}{\log r} = \limsup_{r\to\infty} \frac{\log\nu(r,f)}{\log r}. \tag{4.4.12}$$

Lemma 4.4.7. *Let $\varepsilon > 0$. Then there exists a set $E \subset (1,\infty)$ of finite logarithmic measure such that*

$$\nu(r,f) \le (\log\mu(r,f))^{1+\varepsilon} \le (\log M(r,f))^{1+\varepsilon}, \quad r \notin E \cup [0,1]. \tag{4.4.13}$$

Proof. Choose $s > 1$ such that $\mu(s,f) > 1$, and let F be the set of points $r \ge s$ for which the first inequality in (4.4.13) fails, that is,

$$F := \{r \ge s : \nu(r,f) > (\log\mu(r,f))^{1+\varepsilon}\}.$$

From (4.4.11) we have

$$\begin{aligned}\int_{F\cap[1,R]} \frac{dt}{t} &\le \int_s^R \frac{\nu(t,f)\,dt}{t(\log\mu(t,f))^{1+\varepsilon}} = \int_s^R \frac{d\log\mu(t,f)}{(\log\mu(t,f))^{1+\varepsilon}} \\ &= \frac{1}{\varepsilon}\left(\frac{1}{(\log\mu(s,f))^{\varepsilon}} - \frac{1}{(\log\mu(R,f))^{\varepsilon}}\right), \quad R > s.\end{aligned}$$

Letting $R \to \infty$ and using the fact that $\mu(r,f) \to \infty$ as $r \to \infty$, we find that the set F has finite logarithmic measure. Define $E := F \cup (1,s)$. Then E has finite logarithmic measure, and the first inequality in (4.4.13) is valid for all $r \notin E \cup [0,1]$. The second inequality in (4.4.13) is valid for all $r \ge s$ by (4.4.2). □

Finally, we state the main result of Wiman–Valiron theory, of which there are many versions in the literature. The version below is from Fenton's paper [68], but analogous results can be found in [150, 177, 279].

Theorem 4.4.8. *Suppose that f is a transcendental entire function. Let $\gamma \in (0,1/2)$, and let z be such that $|z| = r$ and*

$$|f(z)| \ge \nu(r,f)^{-\gamma} M(r,f). \tag{4.4.14}$$

Then for all integers $n \ge 0$,

$$f^{(n)}(z) = (1+o(1))\left(\frac{\nu(r,f)}{z}\right)^n f(z),$$

uniformly as $r \to \infty$ outside a set of finite logarithmic measure.

Remark. The restriction for the values of z in (4.4.14) leads us to discuss the measure of the set

$$E(r) = \{\theta \in [0, 2\pi] : |f(re^{i\theta})| \geq \nu(r,f)^{-\gamma} M(r,f)\}$$

for large values of r. From (4.4.10) and (4.4.2) we find that

$$\nu(r,f) \leq \frac{\log \mu(r^2,f)}{\log r} \leq \log \mu(r^2,f) \leq \log M(r^2,f), \quad r \geq e.$$

Hence $E(r)$ is a subset of the set

$$E_0(r) = \left\{\theta \in [0, 2\pi] : |f(re^{i\theta})| \geq \frac{M(r,f)}{\sqrt{\log M(r^2,f)}}\right\},$$

where $r \geq e$ is such that $M(r^2,f) > 1$. For example, if $f(z) = e^z$, then

$$E_0(r) = \left\{\theta \in [0, 2\pi] : e^{r\cos\theta} \geq \frac{e^r}{r}\right\} = \left\{\theta \in [0, 2\pi] : \frac{\log r}{r} \geq 1 - \cos\theta\right\}.$$

Using $1 - \cos\theta \sim \frac{\theta^2}{2}$ for $\theta \sim 0$, we find that the measure of the set $E_0(r)$ for $f(z) = e^z$ is asymptotically at most $2\sqrt{\frac{2\log r}{r}}$, because θ must asymptotically satisfy $-\sqrt{\frac{2\log r}{r}} \leq \theta \leq \sqrt{\frac{2\log r}{r}}$.

Wiman–Valiron theory is frequently used in calculations with differential equations, especially when there are polynomial coefficients.

Example 4.4.9. By Corollary 1.7.2, all the solutions of the linear differential equation

$$f''' - zf'' - z^2 f' + z^3 f = 0 \tag{4.4.15}$$

are entire functions. Let f be a nontrivial solution. It is easy to see that equation (4.4.15) cannot possess a nontrivial polynomial solution. Hence f is transcendental.

Now divide (4.4.15) by f and use Theorem 4.4.8 to get

$$(1 + o(1))\left(\frac{\nu(r,f)}{z}\right)^3 - (1 + o(1))z\left(\frac{\nu(r,f)}{z}\right)^2 - (1 + o(1))z^2\left(\frac{\nu(r,f)}{z}\right) + z^3 = 0$$

with the aforementioned restrictions for z and r. Multiplying by z^3, we get

$$(1 + o(1))\nu(r,f)^3 - (1 + o(1))z^2\nu(r,f)^2 - (1 + o(1))z^4\nu(r,f) + z^6 = 0.$$

Since the product and the ratio of two terms of the form $1 + o(1)$ are both of the form $1 + o(1)$, we may factor the previous equation into the following form:

$$((1 + o(1))\nu(r,f) - z^2)^2((1 + o(1))\nu(r,f) + z^2) = 0.$$

From this, we find that $\nu(r,f) = (1 + o(1))r^2$. Thus $\rho(f) = 2$ by (4.4.12).

Problems

Problem 4.4.10. If f is a nontrivial entire solution of the differential equation

$$f'' + z^6 f = 0,$$

then show that $\rho(f) = 4$.

Problem 4.4.11. If f is a nontrivial entire solution of the differential equation

$$f''' + zf'' + z^4 f' + z^5 f = 0,$$

then show that either $\rho(f) = 2$ or $\rho(f) = 3$.

4.5 Meromorphic functions

A function f is called *meromorphic* at z_0 if it is either analytic at z_0 or has a pole at z_0. We will consider meromorphic functions in the complex plane $\mathbb{C}$, although it makes sense to consider them in other types of plane domains as well. Throughout this book, a *meromorphic function* will mean a meromorphic function in $\mathbb{C}$, unless otherwise stated. A meromorphic function that is not a rational function is called *transcendental.*

An important subclass of meromorphic functions is the collection of entire functions because an entire function has no poles. The simplest examples of nonentire meromorphic functions are rational functions that are not polynomials, for instance, $f(z) = 1/z$.

Theorem 4.5.1. *A meromorphic function f has at most finitely many poles in any bounded domain $G \subset \mathbb{C}$.*

Proof. Suppose on the contrary that f has infinitely many poles $\{\zeta_n\}$ in some bounded domain G. Then the sequence $\{\zeta_n\}$ must have at least one finite limit point ζ_0 in the finite plane. In other words, any disc $D(\zeta_0, r)$ centered at ζ_0 and having an arbitrarily small radius $r > 0$ contains infinitely many points ζ_n. Therefore, f cannot be analytic at ζ_0.

If ζ_0 is a pole of f, then it has a multiplicity of $k \in \mathbb{N}$. Then $g(z) = (z - \zeta_0)^k f(z)$ is analytic at ζ_0, and as such, it has a Taylor series expansion in a (small) neighborhood $D(\zeta_0, r)$ of ζ_0. This is a contradiction because the poles ζ_n of f accumulate to ζ_0. Hence ζ_0 cannot be a pole either.

Since f is not analytic at ζ_0 and since ζ_0 is not a pole of f, it follows that f is not meromorphic at ζ_0, a contradiction. □

Corollary 4.5.2. *If f is meromorphic, then the only possible accumulation point for the poles of f is the point at infinity.*

If f_1 and f_2 are entire functions with $f_1 f_2 \not\equiv 0$, then the only possible singularities of $f = f_1/f_2$ are at the zeros of f_2, and each such point is either a pole of f or a removable singularity. Technically, the zeros of f_1 and f_2 may cancel out, and such a zero is a removable singularity for f. This is the case with $f(z) = \frac{\sin z}{z}$ at the origin. By the uniqueness theorem of analytic functions, the (possible) zeros of f_2 can only accumulate to the point at infinity and nowhere else. Hence the ratio $f = f_1/f_2$ is meromorphic. In particular, $\tan z = \frac{\sin z}{\cos z}$ is meromorphic.

In the converse direction, if f is meromorphic, then we may multiply out the poles of f with a suitable entire function f_2. Indeed, if f has finitely many poles, then we choose f_2 to be a polynomial with zeros precisely at the poles of f, whereas if f has infinitely many poles, then we choose f_2 to be a corresponding Weierstrass product. Then the product $f_1 := ff_2$ is entire, from which $f = f_1/f_2$. Thus every meromorphic function can be written as a ratio of two entire functions. Theorem 4.6.8 gives a more detailed account of this reasoning.

Next, we proceed to discuss a metric that works particularly well with meromorphic functions, as meromorphic functions are continuous with respect to this metric; see Theorem 4.5.9. Following the presentation in [244, Section 1.2], let Σ denote the *Riemann sphere* in $\mathbb{R}^3$ defined as

$$\Sigma = \{(x_1, x_2, x_3) \in \mathbb{R}^3 : x_1^2 + x_2^2 + (x_3 - 1/2)^2 = 1/4\},$$

that is, Σ is the sphere of radius 1/2 centered at $(0, 0, 1/2)$. Obviously, Σ is tangent to the xy-plane (= $\mathbb{C}$) at the origin $(0, 0, 0)$. The north pole $(0, 0, 1)$ of Σ is denoted by N, whereas the origin is the south pole of Σ.

A one-to-one correspondence between the points on $\Sigma \setminus N$ and the complex plane $\mathbb{C}$ can be obtained by defining as the image of $z \in \mathbb{C}$ the point $\Sigma(z)$ on $\Sigma \setminus N$, where the line segment from N to z intersects Σ; see Figure 4.1.

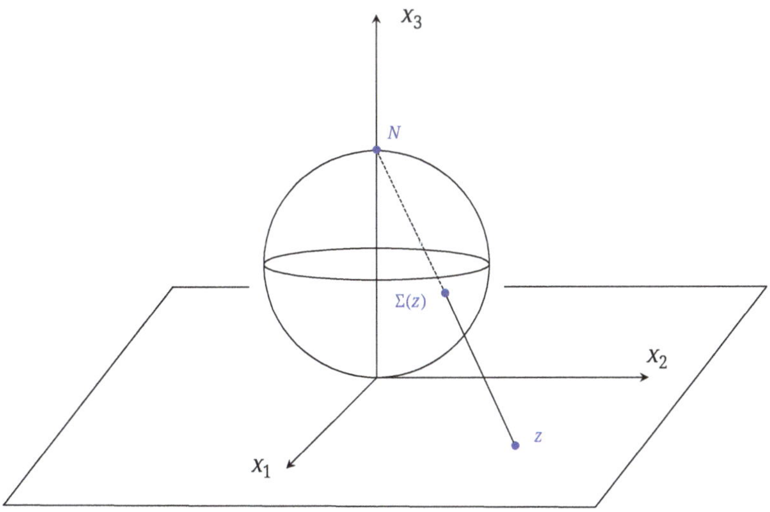

Figure 4.1: The Riemann sphere and the stereographic projection.

Setting N as the image of ∞, we obtain a one-to-one correspondence between Σ and the *extended complex plane* $\widehat{\mathbb{C}} = \mathbb{C} \cup \{\infty\}$. This correspondence is known as the *stereographic projection*.

Note that the point at infinity becomes unique in the stereographic projection. Hence the equations

$$1/\infty = 0 \quad \text{and} \quad 1/0 = \infty$$

make sense and are often agreed to in complex analysis textbooks.

Choose an arbitrary point $z = (x, y, 0) \in \mathbb{C}$, and let $\Sigma(z)$ denote the projection of z onto Σ. The line segment from N to z in $\mathbb{R}^3$ can be expressed as the vector

$$\bar{v} = \bar{k} + t(x\bar{i} + y\bar{j} - \bar{k}) = xt\bar{i} + yt\bar{j} + (1 - t)\bar{k}, \quad t \in [0, 1],$$

where $\bar{i}, \bar{j}, \bar{k}$ denote the standard unit vectors. The parameter value $t = 0$ corresponds to N, and $t = 1$ corresponds to z.

Lemma 4.5.3. *The length of* $\Sigma(z)$ *as a vector from the origin* $(0, 0, 0)$ *is*

$$|\Sigma(z)| = \frac{|z|}{\sqrt{1 + |z|^2}}.$$

Proof. Since the point $\Sigma(z)$ is on the sphere Σ, we have

$$(x^2 + y^2)t^2 + ((1 - t) - 1/2)^2 = 1/4.$$

This simplifies to

$$(x^2 + y^2 + 1)t^2 - t = 0.$$

Since $t \neq 0$ for the point $\Sigma(z)$, the equation above has a unique solution

$$t = \frac{1}{1 + x^2 + y^2} = \frac{1}{1 + |z|^2}.$$

Hence $\Sigma(z)$ has the vectorial representation

$$\Sigma(z) = \left(\frac{x}{1 + |z|^2}, \frac{y}{1 + |z|^2}, \frac{|z|^2}{1 + |z|^2}\right). \tag{4.5.1}$$

The distance from $\Sigma(z)$ to the origin $(0, 0, 0)$ is

$$|\Sigma(z)| = \sqrt{\left(\frac{x}{1 + |z|^2}\right)^2 + \left(\frac{y}{1 + |z|^2}\right)^2 + \left(\frac{|z|^2}{1 + |z|^2}\right)^2} = \frac{|z|}{\sqrt{1 + |z|^2}},$$

which completes the proof. □

Definition 4.5.4. The *chordal distance* $\chi(z_1, z_2)$ of two points $z_1, z_2 \in \widehat{\mathbb{C}}$ is the $\mathbb{R}^3$-distance between the points $\Sigma(z_1)$ and $\Sigma(z_2)$.

It is clear that $0 \le \chi(z_1, z_2) \le 1$ for all $z_1, z_2 \in \widehat{\mathbb{C}}$.

Lemma 4.5.5. *If $z_1, z_2, z \in \mathbb{C}$, then*

$$\chi(z_1, z_2) = \frac{|z_1 - z_2|}{\sqrt{1+|z_1|^2}\sqrt{1+|z_2|^2}} \quad \textit{and} \quad \chi(z, \infty) = \frac{1}{\sqrt{1+|z|^2}}.$$

Proof. Denote $z_j = (x_j, y_j)$, $j = 1, 2$, and $z = (x, y)$. Using representation (4.5.1) for $\Sigma(z_1)$ and $\Sigma(z_2)$ along with the $\mathbb{R}^3$-distance formula, we obtain

$$\begin{aligned}\chi(z_1, z_2)^2 &= \left(\frac{x_2}{1+|z_2|^2} - \frac{x_1}{1+|z_1|^2}\right)^2 + \left(\frac{y_2}{1+|z_2|^2} - \frac{y_1}{1+|z_1|^2}\right)^2 + \left(\frac{|z_2|^2}{1+|z_2|^2} - \frac{|z_1|^2}{1+|z_1|^2}\right)^2 \\ &= \frac{x_2^2 + y_2^2}{(1+|z_2|^2)^2} + \frac{x_1^2 + y_1^2}{(1+|z_1|^2)^2} - 2\frac{x_1x_2 + y_1y_2}{(1+|z_1|^2)(1+|z_2|^2)} + \left(\frac{1}{1+|z_1|^2} - \frac{1}{1+|z_2|^2}\right)^2.\end{aligned}$$

This simplifies to

$$\begin{aligned}\chi(z_1, z_2)^2 &= \frac{x_2^2 + y_2^2 + 1}{(1+|z_2|^2)^2} + \frac{x_1^2 + y_1^2 + 1}{(1+|z_1|^2)^2} - 2\frac{x_1x_2 + y_1y_2 + 1}{(1+|z_1|^2)(1+|z_2|^2)} \\ &= \frac{1}{1+|z_1|^2} + \frac{1}{1+|z_2|^2} - 2\frac{x_1x_2 + y_1y_2 + 1}{(1+|z_1|^2)(1+|z_2|^2)} \\ &= \frac{x_1^2 + y_1^2 + x_2^2 + y_2^2 - 2x_1x_2 - 2y_1y_2}{(1+|z_1|^2)(1+|z_2|^2)} \\ &= \frac{(x_1 - x_2)^2 + (y_1 - y_2)^2}{(1+|z_1|^2)(1+|z_2|^2)} = \frac{|z_1 - z_2|^2}{(1+|z_1|^2)(1+|z_2|^2)}.\end{aligned}$$

The first statement follows by taking square roots on this. Moreover, using (4.5.1) and $N = (0, 0, 1)$ along with the $\mathbb{R}^3$-distance formula, we obtain

$$\begin{aligned}\chi(z, \infty)^2 &= \frac{x^2}{(1+|z|^2)^2} + \frac{y^2}{(1+|z|^2)^2} + \left(1 - \frac{|z|^2}{1+|z|^2}\right)^2 \\ &= \frac{x^2 + y^2 + 1}{(1+|z|^2)^2} = \frac{1}{1+|z|^2},\end{aligned}$$

which implies the second statement. □

Lemma 4.5.6. *For all $z_1, z_2 \in \widehat{\mathbb{C}}$, we have $\chi(\frac{1}{z_1}, \frac{1}{z_2}) = \chi(z_1, z_2)$. Moreover, if $|z_1| \le |z_2| \le \infty$, then $\chi(0, z_1) \le \chi(0, z_2)$.*

Proof. The first statement follows immediately from Lemma 4.5.5. The second statement is trivial if $z_1 = 0$ or $z_2 = \infty$, and hence we may suppose that $z_1, z_2 \in \mathbb{C} \setminus \{0\}$. Writing

$$\chi(0, z_j) = \frac{1}{\sqrt{1 + 1/|z_j|^2}},$$

the second statement becomes trivial. □

Lemma 4.5.7. *The chordal distance χ is a metric on $\widehat{\mathbb{C}}$. Moreover,*

$$\chi(z_1, z_2) \leq |z_1 - z_2|$$

for all $z_1, z_2 \in \mathbb{C}$.

Proof. The stated inequality is trivial by Lemma 4.5.5. The fact that χ defines a metric is an immediate consequence of the definition of the chordal distance and the fact that the $\mathbb{R}^3$-distance is a metric. □

Definition 4.5.8. Let $D \subset \mathbb{C}$ be a simply connected domain. A function $f : D \to \widehat{\mathbb{C}}$ is *spherically continuous* at $z_0 \in D$ if for every $\varepsilon > 0$, there exists $\delta > 0$ such that

$$\chi(f(z), f(z_0)) < \varepsilon$$

whenever $|z - z_0| < \delta$ and $z \in D$. If f is spherically continuous at every $z_0 \in D$, then we say that f is spherically continuous in D.

Theorem 4.5.9. *If f is meromorphic in a simply connected domain $D \subset \mathbb{C}$, then f is spherically continuous in D.*

Proof. If f is analytic at z_0, then f is continuous at z_0, and the spherical continuity follows immediately from the inequality

$$\chi(f(z), f(z_0)) \leq |f(z) - f(z_0)|$$

proved in Lemma 4.5.7. If z_0 is a pole of f, then $1/f$ is analytic at z_0 and satisfies $1/f(z_0) = 0$. Moreover, by Lemma 4.5.6,

$$\chi(f(z), f(z_0)) = \chi\left(\frac{1}{f(z)}, \frac{1}{f(z_0)}\right) \leq \left|\frac{1}{f(z)} - \frac{1}{f(z_0)}\right|,$$

which implies the spherical continuity. □

An analogue of Definition 1.6.1 can now be formulated for the uniform convergence of meromorphic functions. This analogue is given for completeness only as it will not be applied in this book. See [244] for more details.

Definition 4.5.10. A sequence $\{f_n(z)\}$ of meromorphic functions *converges spherically uniformly on compact subsets* of a domain $G \subset \mathbb{C}$ to a function $f(z)$ if for any compact set $K \subset G$ and any $\varepsilon > 0$, there exists a number $N = N(\varepsilon, K) \in \mathbb{N}$ such that

$$\chi(f_n(z), f(z)) < \varepsilon, \quad z \in K,\ n \geq N.$$

Definition 4.5.11. Suppose that f is meromorphic in a simply connected domain $D \subset \mathbb{C}$. If z is not a pole of f, then the *spherical derivative* of f at z is defined as

$$f^{\#}(z) = \lim_{\zeta \to z} \frac{\chi(f(z), f(\zeta))}{|z - \zeta|}.$$

If z is a pole of f, we define

$$f^{\#}(z) = \lim_{\zeta \to z} \frac{|f'(\zeta)|}{1 + |f(\zeta)|^2}.$$

If z is not a pole of f, then using Lemma 4.5.5 and the continuity of f at z, we may write

$$f^{\#}(z) = \lim_{\zeta \to z} \frac{|f(z) - f(\zeta)|}{|z - \zeta|} \frac{1}{\sqrt{1 + |f(z)|^2}} \frac{1}{\sqrt{1 + |f(\zeta)|^2}} = \frac{|f'(z)|}{1 + |f(z)|^2}.$$

This explains the definition of the spherical derivative at a pole. From Lemma 4.5.6, it is clear that $f^{\#}(z) = (1/f)^{\#}(z)$. The spherical derivative is used, for example, in the theories of normal functions and normal families [244].

In the next section, we discuss Nevanlinna theory, which is a powerful tool to deal with the value distribution of both meromorphic and entire functions. In layman's terms, the value distribution of a meromorphic f has to do with the quantity and location of a-points of f, where $a \in \widehat{\mathbb{C}}$. Last, meromorphic functions will be needed in dealing with complex interpolation problems in Section 4.7.

Problems

Problem 4.5.12. Construct a meromorphic function f with simple zeros at the points $1, \sqrt{3}, \sqrt{5}, \ldots$ and simple poles at $2, 4, 8, \ldots$, such that f has no other zeros or poles.

Problem 4.5.13. Find the multiplicity of the pole of the function

$$f(z) = \frac{z^2 \tan z}{1 - \cos(z^4)}$$

at $z_0 = 0$.

Problem 4.5.14. Suppose that $w(z)$ is a meromorphic function such that

$$w(z) = \sum_{n=0}^{\infty} (-1)^n \frac{z^{3n+1}}{8^n} \quad \text{for } |z| < 2.$$

Find $w(3)$.

Problem 4.5.15. Prove that the meromorphic function $f(z) = \frac{\cos z}{z}$ has infinitely many 1-points.

Problem 4.5.16. A meromorphic function f is called a *Yoshida function* if

$$\sup_{z\in\mathbb{C}} f^{\#}(z) < \infty.$$

Prove that $f(z) = e^z$ is a Yoshida function, whereas $g(z) = e^{z^2}$ is not.

4.6 Nevanlinna theory

In a series of papers from 1922–1925, Rolf Nevanlinna developed his celebrated theory of meromorphic functions called *Nevanlinna theory*. The most important of these papers is [216], about which Weyl made this comment in 1943 [289, p. 8]:

> The appearance of this paper has been one of the few great mathematical events in our century.

Although Nevanlinna theory was created for meromorphic functions, its usefulness for entire functions cannot be overestimated, because it has been used countless times in the last one hundred years to prove theorems and solve problems concerning entire functions. More specifically, as Hayman observed [106, p. 21], Nevanlinna used his theory to obtain conclusions about entire functions that went further than what had been known before, where some of these conclusions are essentially best possible. As a technique, Nevanlinna theory has been thought of as the best method to use for many problems concerning entire functions, even though Nevanlinna theory is set up for the more general class of meromorphic functions. This kind of technique reminds us of some less comprehensive situations, such as where the evaluation of certain real integrals is accomplished by using the calculus of residues, which is in the more general framework of the complex plane.

In this section, we discuss the basic notation and results in Nevanlinna theory, some of which are used later in the book. Books such as [82, 106, 166, 278, 299, 300] are often given as standard references for Nevanlinna theory.

Rolf Nevanlinna (1895–1980) was born in Joensuu, Finland. His family moved to Helsinki in 1903, where during his early years, he developed an interest in music, the classics, and mathematics. He especially enjoyed the music of Jean Sibelius and the mathematical work of Ernst Lindelöf, who later became his PhD advisor at the University of Helsinki. Nevanlinna held prestigious positions at the University of Helsinki, the Academy of Finland, and the University of Zurich, where he conducted his groundbreaking research and had thirty-four PhD students, including Lars Ahlfors and Olli Lehto. Nevanlinna theory is also called the value distribution theory of meromorphic functions, and this theory has had a profound and global impact on function theory in the last one hundred years.

The *Nevanlinna characteristic* $T(r,f)$ of a meromorphic function $f \not\equiv 0$ is defined as

$$T(r,f) = m(r,f) + N(r,f), \quad 0 \le r < \infty,$$

where $m(r,f)$ and $N(r,f)$ are defined as follows. The *proximity function* $m(r,f)$ is defined as

$$m(r,f) = \frac{1}{2\pi} \int_0^{2\pi} \log^+ |f(re^{i\theta})| \, d\theta,$$

where $\log^+ x = \max\{0, \log x\}$ for $x \ge 0$. Following the definitions in Section 3.2, the *integrated counting function* $N(r,f)$ of the poles of f is given as

$$N(r,f) = \int_0^r \frac{n(t,f)}{t} \, dt$$

when $f(0) \neq \infty$ and as

$$N(r,f) = \int_0^r \frac{n(t,f) - n(0,f)}{t} \, dt + n(0,f) \log r$$

when $f(0) = \infty$, where the *nonintegrated counting function* $n(r,f)$ counts the number of poles of f in $|z| \le r$ (counted by multiplicities).

We also write

$$T(r,a,f) = \begin{cases} T(r, \frac{1}{f-a}), & a \in \mathbb{C}, \\ T(r,f), & a = \infty, \end{cases}$$

when f is nonconstant, and define $m(r,a,f)$ and $N(r,a,f)$ similarly. Since for all $x, y \ge 0$, we have

$$\log^+ (xy) \le \log^+ x + \log^+ y,$$
$$\log^+ (x+y) \le \log^+ x + \log^+ y + \log 2,$$

and since

$$\max\{N(r,fg), N(r,f+g)\} \le N(r,f) + N(r,g),$$

it follows that

$$T(r,fg) \le T(r,f) + T(r,g),$$
$$T(r,f+g) \le T(r,f) + T(r,g) + \log 2.$$

Moreover, if $c \in \mathbb{C}$, then

$$T(r,f) = T(r,f+c-c) \le T(r,f+c) + O(1) \le T(r,f) + O(1),$$

that is, $T(r,f+c) = T(r,f) + O(1)$.

In the following examples, we will calculate the characteristic T of some elementary functions by using the above properties of T.

Example 4.6.1. Let $a > 0$ and $n \in \mathbb{N}$. Denote $z = re^{i\theta}$. Then by 2π-periodicity,

$$T\left(r, e^{az^n}\right) = \frac{1}{2\pi} \int_{-\pi/2n}^{2\pi-\pi/2n} \log^+ e^{\mathrm{Re}\left(ar^n e^{in\theta}\right)}\, d\theta = \frac{1}{2\pi} \int_{-\pi/2n}^{2\pi-\pi/2n} \log^+ e^{ar^n \cos(n\theta)}\, d\theta$$
$$= \frac{anr^n}{2\pi} \int_{-\pi/2n}^{\pi/2n} \cos(n\theta)\, d\theta = \frac{ar^n}{\pi}.$$

In general, if $a \in \mathbb{C} \setminus \{0\}$, then by a change of variable and the above calculation we have

$$T\left(r, e^{az^n}\right) = \frac{|a|r^n}{\pi}.$$

Next, let P be a polynomial $P(z) = az^n + \cdots$, and denote $g(z) = P(z) - az^n$. Then

$$\begin{aligned}\frac{|a|r^n}{\pi} &= T\left(r, e^{az^n}\right) = T\left(r, e^{P-g}\right) \le T\left(r, e^P\right) + T\left(r, e^{-g}\right)\\ &\le T\left(r, e^P\right) + O\left(r^{n-1}\right) = T\left(r, e^{az^n + g(z)}\right) + O\left(r^{n-1}\right)\\ &\le T\left(r, e^{az^n}\right) + O\left(r^{n-1}\right) = \frac{|a|r^n}{\pi} + O\left(r^{n-1}\right),\end{aligned}$$

so that

$$T\left(r, e^{P(z)}\right) = \frac{|a|r^n}{\pi} + O\left(r^{n-1}\right).$$

Example 4.6.2. Let f be the infinite-order entire function given by $f(z) = \exp(e^z)$. Our goal is to prove that

$$T(r,f) \sim \frac{e^r}{\sqrt{2\pi^3 r}} \tag{4.6.1}$$

by following the proof in [106, Lemma 4.3]. Define $\phi(t) = \max\{0, \cos t\}$, where $t \in \mathbb{R}$. Since ϕ is continuous and even, we have

$$T(r,f) = \frac{1}{2\pi} \int_{-\pi}^{\pi} \log^+ e^{\mathrm{Re}\left(e^{re^{i\theta}}\right)}\, d\theta = \frac{1}{2\pi} \int_{-\pi}^{\pi} \log^+ e^{e^{r\cos\theta}\cos(r\sin\theta)}\, d\theta$$
$$= \frac{1}{2\pi} \int_{-\pi}^{\pi} e^{r\cos\theta} \phi(r\sin\theta)\, d\theta = \frac{1}{\pi} \int_{0}^{\pi} e^{r\cos\theta} \phi(r\sin\theta)\, d\theta.$$

Clearly,

$$\left|\frac{1}{\pi}\int_{\pi/6}^{\pi} e^{r\cos\theta}\phi(r\sin\theta)\,d\theta\right| \le \frac{1}{\pi}\int_{\pi/6}^{\pi} e^{r\cos(\pi/6)}\,d\theta = O\big(e^{\sqrt{3}r/2}\big),$$

and hence

$$T(r,f) = \frac{1}{\pi}\int_{0}^{\pi/6} e^{r\cos\theta}\phi(r\sin\theta)\,d\theta + O\big(e^{\sqrt{3}r/2}\big).$$

Making a change of variable $y = r\sin\theta$, we obtain

$$\begin{aligned} r\cos\theta &= \sqrt{r^2\cos^2\theta} = \sqrt{r^2 - y^2}, \\ dy &= r\cos\theta\,d\theta = \sqrt{r^2 - y^2}\,d\theta, \end{aligned}$$

so that

$$T(r,f) = \frac{1}{\pi}\int_{0}^{r/2} \frac{e^{\sqrt{r^2-y^2}}\phi(y)}{\sqrt{r^2-y^2}}\,dy + O\big(e^{\sqrt{3}r/2}\big).$$

Using the auxiliary functions

$$\psi(y) = \int_{0}^{y}\phi(t)\,dt \quad \text{and} \quad \chi(y) = \frac{e^{\sqrt{r^2-y^2}}}{\sqrt{r^2-y^2}},$$

and integrating by parts, it follows that

$$\begin{aligned} T(r,f) &= \frac{1}{\pi}\int_{0}^{r/2}\chi(y)\psi'(y)\,dy + O\big(e^{\sqrt{3}r/2}\big) \\ &= -\frac{1}{\pi}\int_{0}^{r/2}\chi'(y)\psi(y)\,dy + O\big(e^{\sqrt{3}r/2}\big). \end{aligned}$$

Regarding $\psi(y)$, we see that

$$\begin{aligned} \lim_{y\to\infty}\frac{1}{y}\psi(y) &= \lim_{m\to\infty}\frac{1}{\frac{5\pi}{2}+2\pi m}\left(\int_{0}^{\pi/2}\cos t\,dt + \sum_{n=0}^{m}\int_{3\pi/2+2\pi n}^{5\pi/2+2\pi n}\cos t\,dt\right) \\ &= \lim_{m\to\infty}\frac{2m+3}{\frac{5\pi}{2}+2\pi m} = \frac{1}{\pi}. \end{aligned}$$

Since $\psi(y) \le y$ and

$$-\chi'(y) = \frac{ye^{\sqrt{r^2-y^2}}}{r^2-y^2}\left(1 - \frac{1}{\sqrt{r^2-y^2}}\right) = O\big(ye^r/r^2\big) = O\big(e^r/r\big), \quad 0 < y < r^{1/6},$$

we conclude that

$$-\int_0^{r^{1/6}} \chi'(y)\psi(y)\,dy = O\Big(\frac{e^r}{r}\Big)\int_0^{r^{1/6}} y\,dy = O\Big(\frac{r^{1/3}e^r}{r}\Big) = o\Big(\frac{e^r}{r^{2/3}}\Big).$$

Moreover, integrating by parts, we have

$$\begin{aligned}
-\int_{r^{1/6}}^{r/2} \chi'(y)\psi(y)\,dy &= -\frac{1+o(1)}{\pi}\int_{r^{1/6}}^{r/2} \chi'(y)y\,dy \\
&= -\frac{1+o(1)}{\pi}\int_0^{r/2} \chi'(y)y\,dy + O\Big(\frac{e^r}{r^{2/3}}\Big) \\
&= \frac{1+o(1)}{\pi}\int_0^{r/2} \chi(y)\,dy + O\Big(\frac{e^r}{r^{2/3}}\Big).
\end{aligned}$$

Putting everything together, we obtain

$$T(r,f) = \frac{1+o(1)}{\pi^2}\int_0^{r/2} \chi(y)\,dy + o\Big(\frac{e^r}{\sqrt{r}}\Big). \tag{4.6.2}$$

Thus it remains to calculate this integral of $\chi(y)$. To this end, set $a = 1 - \sqrt{3}/2$ and make the change of variable

$$\sqrt{r^2 - y^2} = r - u, \quad du = \frac{y\,dy}{r-u} = \frac{\sqrt{2ru - u^2}}{r-u}\,dy.$$

Then for $r > 1/a^2$,

$$\begin{aligned}
\int_0^{r/2} \chi(y)\,dy &= e^r\int_0^{ar} \frac{e^{-u}\,du}{\sqrt{2ru-u^2}} = e^r\int_0^{r^{1/2}} \frac{e^{-u}\,du}{\sqrt{2ru-u^2}} + e^r\int_{r^{1/2}}^{ar} \frac{e^{-u}\,du}{\sqrt{2ru-u^2}} \\
&= e^r\int_0^{r^{1/2}} \frac{e^{-u}}{\sqrt{2ru}}\cdot\frac{\sqrt{2r}}{\sqrt{2r-u}}\,du + O\Big(\frac{r-\sqrt{r}}{\sqrt{2r^{3/2}-r}}\cdot e^{r-\sqrt{r}}\Big) \\
&= e^r\big(1+o(1)\big)\int_0^{r^{1/2}} \frac{e^{-u}}{\sqrt{2ru}}\,du + O\big(r^{1/4}e^{r-\sqrt{r}}\big).
\end{aligned}$$

Since $e^x \ge x^2/2!$, we have $r^{1/4}e^{r-\sqrt{r}} = O\big(\frac{e^r}{r^{3/4}}\big)$. Moreover, since

$$e^r\int_{r^{1/2}}^{\infty} \frac{e^{-u}}{\sqrt{2ru}}\,du \le \frac{e^r}{\sqrt{2r^{3/2}}}\int_{r^{1/2}}^{\infty} e^{-u}\,du = \frac{e^{r-\sqrt{r}}}{\sqrt{2r^{3/2}}} = O\Big(\frac{e^r}{r^{3/4}}\Big),$$

it follows that

$$\int_0^{r/2} \chi(y)\,dy = \frac{e^r(1+o(1))}{\sqrt{2r}}\int_0^{\infty} e^{-u}u^{-1/2}\,du + O\Big(\frac{e^r}{r^{3/4}}\Big).$$

Since

$$\int_0^\infty e^{-u}u^{-1/2}\,du = \sqrt{\pi}$$

(see Example 5.1.8), we have

$$\int_0^{r/2} \chi(y)\,dy = \frac{\sqrt{\pi}e^r}{\sqrt{2r}} + o\left(\frac{e^r}{\sqrt{r}}\right). \tag{4.6.3}$$

Then from (4.6.3) and (4.6.2) we obtain (4.6.1).

Theorem 4.6.3 ([166, Theorem 2.2.3]). *Let $f \not\equiv 0$ be a meromorphic function. Then f is a rational function if and only if $T(r,f) = O(\log r)$.*

If f is a transcendental meromorphic function, then by Theorem 4.6.3 there exists a strictly increasing sequence $\{r_n\}$ of real numbers on (e,∞) tending to infinity such that

$$\lim_{n\to\infty} \frac{T(r_n,f)}{\log r_n} = \infty.$$

However, the following lemma shows that even more can be said. This lemma is a meromorphic counterpart to Lemma 2.1.5.

Lemma 4.6.4. *Let $f \not\equiv 0$ be a meromorphic function. Then f is transcendental if and only if*

$$\lim_{r\to\infty} \frac{T(r,f)}{\log r} = \infty. \tag{4.6.4}$$

Proof. Suppose that (4.6.4) holds. If f is a rational function, then Theorem 4.6.3 gives $T(r,f) = O(\log r)$, which contradicts (4.6.4). Hence f must be transcendental.

Now suppose that (4.6.4) does not hold. Then there exist $C > 1$ and a strictly increasing sequence $\{r_n\}$ of real numbers on (e,∞) tending to infinity such that

$$T(r_n,f) \le C\log r_n, \quad n \in \mathbb{N}.$$

Denoting $s_n = \log r_n$, we obtain

$$T(e^{s_n},f) \le Cs_n, \quad n \in \mathbb{N}.$$

Let $s \in [s_n, s_{n+1}]$, where $n \in \mathbb{N}$ is arbitrary but fixed. Then there exists $t \in [0,1]$ such that $s = ts_n + (1-t)s_{n+1}$. It is well known (and will be proved in Section 7.1) that $T(r,f)$ is convex in $\log r$. This means that $U(r) := T(e^r,f)$ is convex. Consequently,

$$\begin{aligned} T(e^s,f) = U(s) = U(ts_n + (1-t)s_{n+1}) &\le tU(s_n) + (1-t)U(s_{n+1}) \\ &\le tCs_n + (1-t)Cs_{n+1} = C(ts_n + (1-t)s_{n+1}) = Cs. \end{aligned}$$

By the change of variable $s = \log r$, we obtain

$$T(r,f) \leq C \log r, \quad r \geq r_1.$$

Then f is a rational function by Theorem 4.6.3. This completes the proof. □

An alternative (but more laborious) proof of Lemma 4.6.4 that does not use convexity can be found in [299, Theorem 1.5].

Nevanlinna theory begins with a generalization of Jensen's theorem (Theorem 3.2.2), which is often called the *Poisson–Jensen formula.*

Theorem 4.6.5 ([166, Proposition 2.1.7]). *If f is a meromorphic function with $f(0) \neq 0, \infty$, then*

$$\frac{1}{2\pi}\int_0^{2\pi} \log|f(re^{i\theta})|\,d\theta + N(r,f) - N\left(r,\frac{1}{f}\right) = O(1).$$

It is easy to see that

$$\log x = \log^+ x - \log^+ \frac{1}{x}$$

for all $x > 0$. Hence, when $f \not\equiv 0$, Theorem 4.6.5 yields

$$T(r,f) = T\left(r,\frac{1}{f}\right) + O(1).$$

Writing this in its general form gives the *first fundamental theorem* of Nevanlinna.

Theorem 4.6.6 ([166, p. 28]). *If f is a nonconstant meromorphic function, then*

$$T(r,f) = T(r,a,f) + O(1), \quad a \in \mathbb{C}.$$

Recall from Definition 2.1.6 that the order of growth of an entire function f involves $\log M(r,f)$, which is not practical for meromorphic functions. It is more practical to use the Nevanlinna characteristic.

Definition 4.6.7. The *T-order of growth* of a meromorphic function f is defined as

$$\rho_T(f) = \limsup_{r\to\infty} \frac{\log T(r,f)}{\log r} \in [0,\infty] \tag{4.6.5}$$

when f is nonconstant and $\rho_T(f) = 0$ when f is a constant.

It will be proved in Section 7.1 that $\rho(f) = \rho_T(f)$ whenever f is entire. We now use Nevanlinna theory to give an alternative short proof of Theorem 3.8.3.

Proof of Theorem 3.8.3. We are given that f and g are entire functions such that $\rho(f) > \rho(g)$. Since $\rho(fg) \leq \max\{\rho(f),\rho(g)\} = \rho(f)$, we suppose on the contrary to the statement $\rho(fg) = \rho(f)$ that $\rho(fg) < \rho(f)$. Let $\sigma \in (\max\{\rho(g),\rho(fg)\},\rho(f))$. We may assume that $g \not\equiv 0$.

By the first fundamental theorem and the properties of the characteristic function, we get

$$T(r,f) = T\left(r, \frac{fg}{g}\right) \leq T(r,fg) + T\left(r, \frac{1}{g}\right)$$
$$= T(r,fg) + T(r,g) + O(1) \leq r^{\sigma}, \quad r \geq R.$$

Hence $\rho(f) = \rho_T(f) \leq \sigma < \rho(f)$, a contradiction. □

Theorem 4.6.8. *If f is a nonconstant meromorphic function of finite order $\rho_T(f)$, then f can be written in the form*

$$f = f_1/f_2,$$

where f_1 and f_2 are entire functions of order $\leq \rho_T(f)$.

Proof. If f has no poles, then it is entire, and we may choose $f_1 \equiv f$ and $f_2 \equiv 1$. Since $\rho(f) = \rho_T(f)$, the statement follows. If f has finitely many poles $\zeta_1, \ldots, \zeta_k$ where each $\zeta_n \neq 0$, then let P be the polynomial

$$P(z) = (z - \zeta_1) \cdots (z - \zeta_k),$$

whereas if f has infinitely many non-zero poles ζ_n, then let P be the canonical product formed with the points ζ_n. If f does not have a pole at $z = 0$, then we set $f_2(z) = P(z)$. If f has a pole of multiplicity m at $z = 0$, then we set $f_2(z) = z^m P(z)$. In both cases, f_2 is an entire function such that $\rho(f_2) = \lambda(\infty, f)$ by Theorem 3.5.3. On the other hand, $\lambda(\infty, f) \leq \rho_T(f)$ by Theorem 4.6.16. Consequently, $\rho(f_2) \leq \rho_T(f)$. Then we put $f_1(z) = f(z)f_2(z)$, which is clearly entire and satisfies

$$T(r,f_1) \leq T(r,f) + T(r,f_2) \leq 2\max\{T(r,f), T(r,f_2)\},$$

from which $\rho(f_1) = \rho_T(f_1) \leq \max\{\rho_T(f), \rho_T(f_2)\} = \rho_T(f)$. Then $f = f_1/f_2$ is the desired decomposition. □

Remark. For a nonconstant meromorphic function f, the *Ahlfors–Shimizu characteristic function* is defined as

$$T_0(r,f) = \int_0^r \frac{A(t,f)}{t}\, dt,$$

where

$$A(t,f) = \frac{1}{\pi} \int_0^{2\pi} \int_0^t f^{\#}\left(se^{i\theta}\right)^2 s\, ds\, d\theta,$$

and where

$$f^{\#}(z) = \frac{|f'(z)|}{1 + |f(z)|^2}$$

is the *spherical derivative* of f, as discussed in Section 4.5. The Nevanlinna characteristic $T(r,f)$ satisfies

$$T(r,f) = T_0(r,f) + O(1).$$

The nontrivial proof of this identity can be found, for example, in Hayman's book [106]. Hence the characteristic functions T and T_0 can be used interchangeably.

We are now ready to state the seminal result in Nevanlinna theory, the *second fundamental theorem* of Nevanlinna.

Theorem 4.6.9 ([106, pp. 31–32]). *If f is a nonconstant meromorphic function and $a_1, a_2, \ldots, a_q$ are $q \geq 2$ distinct constants, then*

$$m(r,f) + \sum_{j=1}^{q} m\left(r, \frac{1}{f - a_j}\right) \leq 2T(r,f) - N_1(r,f) + S(r,f), \tag{4.6.6}$$

where

$$N_1(r,f) = N\left(r, \frac{1}{f'}\right) + 2N(r,f) - N(r,f'),$$

and where $S(r,f)$ is a term that satisfies $S(r,f) = o(T(r,f))$ as $r \to \infty$ outside a possible exceptional set of finite linear measure.

The quantity $S(r,f)$ is often called the *error term*. Observe that the term $N_1(r,f)$ satisfies $N_1(r,f) \geq 0$ for all $r \geq 0$.

An important tool for proving the second fundamental theorem is the following result, which is often called the *lemma on the logarithmic derivative* or the fundamental estimate of the logarithmic derivative. Throughout this section (and the book), by $S(r,f)$ we will denote any quantity that satisfies $S(r,f) = o(T(r,f))$ as $r \to \infty$ outside a possible exceptional set of finite linear measure.

Lemma 4.6.10 ([106, Theorem 3.1]). *If f is a nonconstant meromorphic function, then for all $k \in \mathbb{N}$, we have*

$$m\left(r, \frac{f^{(k)}}{f}\right) = S(r,f).$$

Moreover, if f is of finite order, then

$$m\left(r, \frac{f^{(k)}}{f}\right) = O(\log r), \quad r \to \infty.$$

Adding

$$N(r,f) + \sum_{j=1}^{q} N\left(r, \frac{1}{f - a_j}\right)$$

to both sides of (4.6.6) and using the first fundamental theorem, we obtain alternate formulations of the second fundamental theorem; see [106, p. 43]

Theorem 4.6.11. *If f is a nonconstant meromorphic function and $a_1, a_2, \ldots, a_q$ are $q \geq 2$ distinct constants, then the following statements hold:*

$$(q-1)T(r,f) \leq N(r,f) + \sum_{j=1}^{q} N\left(r, \frac{1}{f - a_j}\right) - N_1(r,f) + S(r,f). \tag{4.6.7}$$

$$(q-2)T(r,f) \leq \sum_{j=1}^{q} N\left(r, \frac{1}{f - a_j}\right) - N_1(r,f) + S(r,f). \tag{4.6.8}$$

We now turn our attention to distinct a-points of meromorphic functions. For a nonconstant meromorphic function f, let $\overline{n}(r,f)$ denote the number of poles of f in $|z| \leq r$, counted once each. If $f(0) \neq \infty$, then we set

$$\overline{N}(r,f) = \int_0^r \frac{\overline{n}(t,f)}{t}\, dt,$$

and if $f(0) = \infty$, then we set

$$\overline{N}(r,f) = \int_0^r \frac{\overline{n}(t,f) - \overline{n}(0,f)}{t}\, dt + \overline{n}(0,f) \log r.$$

Observe that the term $N_1(r,f)$ in Theorem 4.6.11 provides a measure of the frequency of the multiple a-points of f for $a \in \widehat{\mathbb{C}}$. Using the property that $N_1(r,f) \geq 0$ in Theorem 4.6.11, it is easy to obtain the following form of the second fundamental theorem in terms of the distinct a-points of f; see [106, p. 43].

Theorem 4.6.12. *If f is a nonconstant meromorphic function and $a_1, a_2, \ldots, a_q$ are $q \geq 2$ distinct constants, then the following statements hold:*

$$(q-1)T(r,f) \leq \overline{N}(r,f) + \sum_{j=1}^{q} \overline{N}\left(r, \frac{1}{f - a_j}\right) + S(r,f). \tag{4.6.9}$$

$$(q-2)T(r,f) \leq \sum_{j=1}^{q} \overline{N}\left(r, \frac{1}{f - a_j}\right) + S(r,f). \tag{4.6.10}$$

The next result is Picard's theorem for meromorphic functions, which is sometimes called *Picard's great theorem*. Picard's theorem for entire functions (Theorem 3.6.1) is a corollary of this result.

Theorem 4.6.13. *If f is a transcendental meromorphic function, then f takes every value in the extended plane an infinite number of times with at most two possible exceptional values.*

Proof. Suppose on the contrary that there exist a transcendental meromorphic function f and three distinct values a, b, c in the extended plane such that f takes a, b, c only a finite number of times. Then $N(r,a,f) = O(\log r)$, $N(r,b,f) = O(\log r)$, and $N(r,c,f) = O(\log r)$ as $r \to \infty$. If a, b, c are all finite, then we use (4.6.10) with $q = 3$, whereas if one of a, b, c equals ∞, then we use (4.6.9) with $q = 2$ and $\overline{N}(r,f) = O(\log r)$. In both cases, we obtain

$$T(r,f) \leq O(\log r) + S(r,f). \tag{4.6.11}$$

Since f is transcendental, from Lemma 4.6.4 we have that

$$\lim_{r\to\infty} \frac{T(r,f)}{\log r} = \infty,$$

which gives a contradiction in (4.6.11). This proves the theorem. □

Analogously to the definition for entire functions, we define a *Picard value* for a meromorphic function f as a value $a \in \widehat{\mathbb{C}}$ that satisfies the following property: $f(z) \neq a$ for all $z \in \mathbb{C}$.

A meromorphic function may have two exceptional values in Theorem 4.6.13, since, for example, e^z has the Picard values 0 and ∞, and

$$f(z) = \frac{e^z}{e^z + 1}$$

has the Picard values 0 and 1.

Analogously to the definition for entire functions, we define the *exponent of convergence* $\lambda(a,f)$ of the a-points ($a \in \widehat{\mathbb{C}}$) of a nonconstant meromorphic function f as the exponent of convergence of the sequence of nonzero a-points of f.

We use the following lemma.

Lemma 4.6.14. *Let f be a nonconstant meromorphic function, and let $a \in \widehat{\mathbb{C}}$. If $\lambda(a,f) < \infty$ and α is a constant satisfying $\alpha > \lambda(a,f)$, then*

$$N(r,a,f) = O(r^\alpha), \quad r \to \infty. \tag{4.6.12}$$

Proof. Without loss of generality, we may suppose that $f(0) \neq a$. First observe that (4.6.12) is clearly true when f has only a finite number of a-points. Hence we assume that f has an infinite sequence $\{z_n\}$ of a-points. Since $\lambda(a,f)$ and $n(r,a,f)$ only involve the sequence of a-points of f, from Lemma 3.1.4 we obtain that if $\beta > 0$ is a constant, then the series $\sum_{n=1}^{\infty} |z_n|^{-\beta}$ converges if and only if the integral $\int_0^{\infty} \frac{n(t,a,f)}{t^{\beta+1}}$ converges. Hence, since $\alpha > \lambda(a,f)$, there exists a constant $K > 0$ such that

$$\int_0^r \frac{n(t,a,f)}{t^{\alpha+1}}\,dt \le \int_0^{\infty} \frac{n(t,a,f)}{t^{\alpha+1}}\,dt = K.$$

Therefore,

$$\begin{aligned} N(r,a,f) &= \int_0^r \frac{n(t,a,f)}{t}\,dt = \int_0^r t^{\alpha} \cdot \frac{n(t,a,f)}{t^{\alpha+1}}\,dt \\ &\le r^{\alpha} \int_0^r \frac{n(t,a,f)}{t^{\alpha+1}}\,dt \le Kr^{\alpha}, \end{aligned}$$

which gives (4.6.12). □

Since $\lambda(a,f)$ and $n(r,a,f)$ only involve the sequence of a-points of a meromorphic function f, it is easy to see that the proof of Theorem 3.1.6 can be used to obtain the following result.

Theorem 4.6.15. *If f is a nonconstant meromorphic function and $a \in \widehat{\mathbb{C}}$, then*

$$\lambda(a,f) = \limsup_{r\to\infty} \frac{\log n(r,a,f)}{\log r}. \tag{4.6.13}$$

The next theorem generalizes the analogous result for entire functions.

Theorem 4.6.16. *If f is a nonconstant meromorphic function and $a \in \widehat{\mathbb{C}}$, then*

$$\lambda(a,f) \le \rho_T(f).$$

Proof. Without loss of generality, we assume that $f(0) \ne 0$. We argue as in [166, p. 25]. The first fundamental theorem gives

$$N(r,a,f) \le T(r,a,f) = T(r,f) + O(1). \tag{4.6.14}$$

For $r \ge 1$, we have

$$N(3r,a,f) = \int_0^{3r} \frac{n(t,a,f)}{t}\,dt \ge \int_r^{3r} \frac{n(t,a,f)}{t}\,dt \ge n(r,a,f)\log 3 \ge n(r,a,f).$$

Hence from (4.6.14) we obtain

$$n(r,a,f) \le N(3r,a,f) \le T(3r,f) + O(1).$$

Then by formula (4.6.13),

$$\lambda(a,f) = \limsup_{r\to\infty} \frac{\log n(r,a,f)}{\log r} \le \limsup_{r\to\infty} \frac{T(3r,f)}{\log(3r) - \log 3} = \rho_T(f),$$

which proves the statement. □

Analogously to the definition for entire functions, we define a *Borel exceptional value* of a meromorphic function f as a value $c \in \widehat{\mathbb{C}}$ such that $\lambda(c,f) < \rho_T(f)$. The next result is *Borel's theorem* for meromorphic functions.

Theorem 4.6.17. *If f is a transcendental meromorphic function, then f can have at most two Borel exceptional values in $\widehat{\mathbb{C}}$.*

Proof. Suppose on the contrary that there exist a transcendental meromorphic function f and three distinct values a, b, c in the extended plane that are Borel exceptional values of f. Then by (4.6.12), $N(r,a,f) = O(r^\alpha)$, $N(r,b,f) = O(r^\alpha)$, and $N(r,c,f) = O(r^\alpha)$, where α is a constant that satisfies

$$\max\{\lambda(a,f), \lambda(b,f), \lambda(c,f)\} < \alpha < \rho_T(f).$$

If a, b, c are all finite, then we use (4.6.10) with $q = 3$, whereas if one of a, b, c equals ∞, then we use (4.6.9) with $q = 2$ and $\overline{N}(r,f) = O(r^\alpha)$. In both cases, we obtain

$$T(r,f) \le O(r^\alpha) + S(r,f). \tag{4.6.15}$$

Then applying Lemma 1.2.9 to (4.6.15), we obtain

$$T(r,f) \le O(r^\alpha).$$

Hence $\rho_T(f) \le \alpha < \rho_T(f)$, a contradiction. This proves the theorem. □

Example 4.6.18. We can see that $a = 0$ is a Borel exceptional value of the function

$$f(z) = e^{z^2} \sin z,$$

but $a = 0$ is not an exceptional value in the sense of Theorem 4.6.13. Here we have $\lambda(0,f) = 1$ and $\rho_T(f) = 2$.

The next well-known result is useful.

Lemma 4.6.19. *If f is a nonconstant meromorphic function, then*

$$T(r,f') \le 2T(r,f) + S(r,f). \tag{4.6.16}$$

If f is a nonconstant entire function, then

$$T(r,f') \le T(r,f) + S(r,f). \tag{4.6.17}$$

Proof. If f is a nonconstant meromorphic function, then by the lemma on the logarithmic derivative we obtain

$$m(r,f') = m\left(r,\frac{f'f}{f}\right) \le m(r,f) + m\left(r,\frac{f'}{f}\right) = m(r,f) + S(r,f).$$

Also, since a pole of f of multiplicity $k \ge 1$ is a pole of f' of multiplicity $k+1$, we get $N(r,f') \le 2N(r,f)$. Hence

$$T(r,f') \le m(r,f) + 2N(r,f) + S(r,f) \le 2T(r,f) + S(r,f),$$

which gives (4.6.16).

If f is a nonconstant entire function, then $N(r,f) = 0$, and the above proof then gives (4.6.17). □

We finish this section with a discussion of two quantities that measure the frequency of a-points of meromorphic functions. First, the *Nevanlinna deficiency* of an a-point of a nonconstant meromorphic function f is defined as

$$\delta(a,f) = \liminf_{r\to\infty} \frac{m(r,a,f)}{T(r,f)} = 1 - \limsup_{r\to\infty} \frac{N(r,a,f)}{T(r,f)}, \quad a \in \widehat{\mathbb{C}}.$$

Clearly, $0 \le \delta(a,f) \le 1$ for $a \in \widehat{\mathbb{C}}$. If $\delta(a,f) > 0$, then a is called a *deficient value* for f. As illustrations, it can be shown by Theorems 6.3.7 and 6.3.9 that the functions

$$f(z) = e^z - 1, \quad g(z) = (e^z - 1)e^z, \quad h(z) = (e^z - 1)e^{2z},$$

satisfy $\delta(0,f) = 0$, $\delta(0,g) = 1/2$, $\delta(0,h) = 2/3$.

We now recall the well-known *deficiency relation.*

Theorem 4.6.20 ([300, Theorem 1.9]). *The number of deficient values for any transcendental meromorphic function f is at most countable, and*

$$\sum_{a\in\widehat{\mathbb{C}}} \delta(a,f) \le 2. \tag{4.6.18}$$

Proof. Since $N_1(r,f) \ge 0$, estimate (4.6.6) gives

$$\sum_{j=1}^{p} m(r,b_j,f) \le 2T(r,f) + S(r,f),$$

where $b_1,\dots,b_p \in \widehat{\mathbb{C}}$ are distinct, and $p \ge 3$. Then

$$\sum_{j=1}^{p} \frac{m(r, b_j, f)}{T(r,f)} \le 2 + \frac{S(r,f)}{T(r,f)},$$

and

$$\liminf_{r\to\infty} \sum_{j=1}^{p} \frac{m(r, b_j, f)}{T(r,f)} \le \liminf_{r\to\infty}\left(2 + \frac{S(r,f)}{T(r,f)}\right) = 2 + \liminf_{r\to\infty} \frac{S(r,f)}{T(r,f)}.$$

Consequently,

$$\sum_{j=1}^{p} \delta(b_j, f) \le 2. \tag{4.6.19}$$

For each positive integer k, the number of deficient values a for which $\delta(a,f) > 1/k$ is $\le 2k$ by (4.6.19). Moreover, since

$$\{a \in \widehat{\mathbb{C}} : \delta(a,f) > 0\} = \bigcup_{k=1}^{\infty} \{a \in \widehat{\mathbb{C}} : \delta(a,f) > 1/k\},$$

the number of deficient values is at most countable. Further, as inequality (4.6.19) holds for all finite $p \ge 3$, the sum of the deficiencies does not exceed 2. □

Inequality (4.6.18) is sharp since $\delta(0, e^z) = \delta(\infty, e^z) = 1$. By Theorem 4.6.20, when a is a deficient value, i. e., when $\delta(a,f) > 0, f$ has less than the "usual frequency" of a-points. A deficient value is also called an exceptional value in the sense of Nevanlinna. A proof for the existence of an entire function with prescribed finite deficient values was given by Fuchs and Hayman; see [106, Theorem 4.1].

The next quantity measures the frequency of distinct a-points of meromorphic functions. For a nonconstant meromorphic function f, we define

$$\Theta(a,f) = 1 - \limsup_{r\to\infty} \frac{\overline{N}(r,a,f)}{T(r,f)}, \quad a \in \widehat{\mathbb{C}}. \tag{4.6.20}$$

Obviously, $0 \le \delta(a,f) \le \Theta(a,f) \le 1$. If $\Theta(a,f) > 0$, then f has less than the "usual frequency" of distinct a-points.

Dividing (4.6.9) by $T(r,f)$ and reorganizing terms, we obtain

$$\left(1 - \frac{\overline{N}(r,f)}{T(r,f)}\right) + \sum_{j=1}^{q}\left(1 - \frac{\overline{N}(r, \frac{1}{f-a_j})}{T(r,f)}\right) \le 2 + \frac{S(r,f)}{T(r,f)}.$$

Then using reasoning similar to the proof of Theorem 4.6.20, we obtain the following analogous result.

Theorem 4.6.21 ([300, Theorem 1.10]). *For any transcendental meromorphic function f, the number of values a where $\Theta(a,f) > 0$ is at most countable, and*

$$\sum_{a\in\widehat{\mathbb{C}}} \Theta(a,f) \le 2. \tag{4.6.21}$$

Inequality (4.6.21) is also sharp, since $\Theta(0, e^z) = \Theta(\infty, e^z) = 1$. For applications of Theorem 4.6.21 and connections between multiple a-points and Theorem 4.6.21, see Section 12.5.

We close this section with the comment that when Nevanlinna theory has been used in the literature to solve problems about entire functions, the proof has sometimes involved having to address questions about meromorphic functions with poles. Therefore, in order to solve problems about entire functions, it is often a benefit to be familiar with both (a) the basic results of Nevanlinna theory and (b) the algebraic manipulations of the Nevanlinna functionals $T(r,f)$, $m(r,f)$, $N(r,f)$, etc.

Problems

Problem 4.6.22. Write down entire functions f, g, h that satisfy the following conditions. Reasoning is not needed.

(a) $\delta(0,f) = 1/6$.

(b) $\delta(0,g) = 1 - 1/n$ for fixed $n \in \mathbb{N}$.

(c) $\Theta(0,h) = 2/3$ and $\delta(0,h) = 0$.

Problem 4.6.23. Show that $m(r,f^k) = km(r,f)$, $k \in \mathbb{N}$, for every nonconstant meromorphic function f.

Problem 4.6.24. Show that $T(r, \sin z) = 2r/\pi + O(1)$ as $r \to \infty$.

Problem 4.6.25. Let f be a transcendental entire function, and set

$$g = (f - a)(f - b),$$

where a and b are distinct constants. Prove that $\rho(g) = \rho(f)$.

Problem 4.6.26. Show that

$$T\left(r, e^{z^2} + \cos z\right) = \frac{r^2}{\pi} + o\left(r^2\right), \quad r \to \infty.$$

Problem 4.6.27. Let f be a nonconstant meromorphic function, and set

$$h = \frac{f'(f' - f)}{f^2 - 1}.$$

Prove that $m(r,h) = S(r,f)$.

Problem 4.6.28. Let f be a transcendental entire solution of the differential equation

$$f'(z) = \frac{a(z)f(z) + b(z)}{c(z)f(z) + d(z)},$$

where $a(z)$, $b(z)$, $c(z)$, $d(z)$ are nonconstant polynomials. Prove that $T(r,f) = T(r,f') + S(r,f)$.

Problem 4.6.29. Prove that Yoshida functions f satisfy $\rho_T(f) \leq 2$. See Problem 4.5.16 for the definition of Yoshida functions.

Problem 4.6.30. Let f be a nonconstant meromorphic function. Concerning the Ahlfors–Shimizu characteristic function $T_0(r,f) = \int_0^r \frac{A(t,f)}{t}\,dt$, prove the following four statements:

(a) $T_0(r,f) \le A(r,f)\log r + O(1)$, $r > 1$.

(b) $A(r,f)\log 2 \le T_0(2r,f)$, $r > 0$.

(c) $A(r,f)\log r \le T_0(r^2,f)$, $r > 1$.

(d) $A(r,f) \asymp \log r$ if and only if $T_0(r,f) \asymp \log^2 r$.

Problem 4.6.31. Show that $\tan z$ has the two Picard values $\pm i$.

Problem 4.6.32.

(a) Give an example of an entire function $w(z)$ that has two distinct finite values α and β such that $\Theta(\alpha, w) = \Theta(\beta, w) = 1/2$. Briefly state why $\Theta(\alpha, w) = \Theta(\beta, w) = 1/2$ in your example.

(b) Explain why $\delta(c, w) = 0$ for all finite values of c satisfying $c \ne \alpha, \beta$.

4.7 Complex interpolation

Let $\{z_n\}$ be a sequence of distinct points in $\mathbb{C}$ having no finite accumulation points. By the Weierstrass theorem there exists an entire function f such that $f(z_n) = 0$. Instead of sending each point z_n to zero, we may ask if each z_n could be sent to any prescribed point ζ_n. This is known as an *interpolation problem*. We first consider this problem for finite sequences.

Example 4.7.1. Let $z_1, \dots, z_k \in \mathbb{C}$ be distinct points, and let $\zeta_1, \dots, \zeta_k \in \mathbb{C}$ be any points (repetition allowed). Define

$$L(z) = \sum_{n=1}^{k} \frac{\zeta_n}{P'(z_n)} \cdot \frac{P(z)}{z - z_n},$$

where $P(z) = (z - z_1)\cdots(z - z_k)$ is a polynomial having zeros precisely at the points z_n. As a finite sum of polynomials, $L(z)$ is a polynomial.

Since $P'(z_n) = \prod_{j \ne n}(z_n - z_j)$ and $P(z_n) = 0$, it follows that

$$L(z_n) = \zeta_n + \sum_{j \ne n} \frac{\zeta_j}{P'(z_j)} \cdot \frac{P(z_n)}{z_n - z_j} = \zeta_n, \quad n \in \{1, \dots, k\}.$$

The polynomial $L(z)$ is known as *Lagrange's interpolation polynomial* of degree $k - 1$ [205, p. 71], [242, p. 192].

We now consider infinite sequences and state a solution to the interpolation problem formally as a theorem.

Theorem 4.7.2 ([99, Corollary 7.2.3]). *Let $\{z_n\}$ be an infinite sequence of distinct points in $\mathbb{C}$ having no finite accumulation points, and let $\{\zeta_n\}$ be any infinite sequence in $\mathbb{C}$, not necessarily distinct. Then there exists an entire function f such that*

$$f(z_n) = \zeta_n, \quad n \in \mathbb{N}.$$

For the proof, we need Mittag-Leffler's theorem on meromorphic functions with prescribed singular parts. The theorem is named after Gösta Mittag-Leffler (1846–1927, Sweden), whose surname as a child was Leffler according to his father. As an adult, he added his mother's maiden name Mittag to his surname.

Recall that a meromorphic function f has a Laurent series expansion at $a \in \mathbb{C}$ of the form

$$f(z) = \sum_{n=-m}^{\infty} a_n(z-a)^n, \quad m \in \mathbb{Z}.$$

If $m \le 0$, then f is analytic at a, whereas if $m > 0$, then f has a pole of multiplicity m at a, and the finite sum

$$\sum_{n=-m}^{-1} a_n(z-a)^n$$

is called the *singular part* of f at $z = a$.

We now state and prove *Mittag-Leffler's theorem.*

Theorem 4.7.3 ([99, Theorem 7.1.2]). *Let $\{z_n\}$ be an infinite sequence of distinct points in $\mathbb{C}$ having no finite accumulation points, and let $\{P_n(z)\}$ be an infinite sequence of nonconstant polynomials such that $P_n(0) = 0$ for all $n \in \mathbb{N}$. Then there exists a meromorphic function f in $\mathbb{C}$ with poles at z_n and no other poles, where $P_n(\frac{1}{z-z_n})$ is the singular part of f at each $z = z_n$.*

Proof. We may assume that $|z_1| \le |z_2| \le |z_3| \le \cdots$. Suppose first that $z_1 \ne 0$. Since $P_n(z)$ is a polynomial, the function $P_n(\frac{1}{z-z_n})$ must be analytic in the disc $D(0, |z_n|)$. Therefore it has a Taylor expansion

$$P_n\left(\frac{1}{z-z_n}\right) = \sum_{j=0}^{\infty} a_{j,n} z^j, \quad z \in D(0, |z_n|).$$

This power series converges absolutely and uniformly on the closed disc $\overline{D}(0, a)$, where $|z_n|/2 < a < |z_n|$. Let $\{c_n\}$ be a sequence of positive numbers such that $\sum_{n=1}^{\infty} c_n$ converges. Denotc

$$Q_n(z) = \sum_{j=0}^{k_n} a_{j,n} z^j,$$

where the indices $k_n \in \mathbb{N}$ are chosen large enough such that

$$\sup_{z \in \overline{D}(0,|z_n|/2)} \left| P_n\left(\frac{1}{z-z_n}\right) - Q_n(z) \right| < c_n. \tag{4.7.1}$$

For arbitrary $R > 0$, we consider the sum

$$S(z) = \sum_{n=0}^{\infty}\left(P_n\left(\frac{1}{z-z_n}\right) - Q_n(z)\right)$$
$$= \sum_{|z_n|\le 2R}\left(P_n\left(\frac{1}{z-z_n}\right) - Q_n(z)\right) + \sum_{|z_n|>2R}\left(P_n\left(\frac{1}{z-z_n}\right) - Q_n(z)\right).$$

The sum $\sum_{|z_n|>2R}$ has no poles in $D(0,R)$, and since $R < |z_n|/2$, we have by (4.7.1) that

$$\sup_{z\in\overline{D}(0,R)}\left|P_n\left(\frac{1}{z-z_n}\right) - Q_n(z)\right| < c_n.$$

By the majorant principle, the sum $\sum_{|z_n|>2R}$ converges absolutely and uniformly in $\overline{D}(0,R)$. This means that the partial sums of $\sum_{|z_n|>2R}$ converge to an analytic function in $D(0,R)$ by Theorem 1.6.6, and hence $\sum_{|z_n|>2R}$ is analytic in $D(0,R)$. Meanwhile, the finite sum $\sum_{|z_n|\le 2R}$ is a rational function with prescribed singular parts at every $z_n \in D(0,R)$.

Since $R > 0$ was chosen arbitrarily, the sum $S(z)$ converges uniformly on compact subsets of $\mathbb{C} \setminus \cup_{n=1}^{\infty}\{z_n\}$ and has prescribed singular parts at the points $z = z_n$, except perhaps at the origin. Adding one singular part to $S(z)$, say

$$S_0(z) = S(z) + P_0(1/z),$$

if necessary, a prescribed singular part at the origin is also covered. The function f in the statement is one of $S(z)$ or $S_0(z)$. □

Proof of Theorem 4.7.2. By the Weierstrass theorem we may construct an entire function $F(z)$ having simple zeros at the points $z = z_n$ and no other zeros. Then $F'(z_n) \neq 0$ for all $n \in \mathbb{N}$. By the Mittag-Leffler theorem, whenever $\zeta_n \neq 0$, there exists a meromorphic function $G(z)$ on $\mathbb{C}$ with simple poles of residue $\zeta_n/F'(z_n)$ at the points z_n and no other poles. If $\zeta_n = 0$, then $G(z)$ is analytic at z_n.

Define $f(z) = F(z)G(z)$, which is meromorphic in $\mathbb{C}$. Indeed, f is analytic in $\mathbb{C}$ except perhaps at the points z_n corresponding to $\zeta_n \neq 0$. If $\zeta_n = 0$, then $f(z_n) = 0$. Near each $z = z_n$ corresponding to $\zeta_n \neq 0$,

$$F(z) = F'(z_n)(z-z_n) + \frac{F''(z_n)}{2!}(z-z_n)^2 + \cdots =: (z-z_n)F_n(z),$$
$$G(z) = \frac{\zeta_n}{F'(z_n)}\frac{1}{z-z_n} + G(z_n) + G'(z_n)(z-z_n) + \cdots =: \frac{G_n(z)}{z-z_n},$$

where $F_n(z)$ and $G_n(z)$ are analytic in a (possibly small) disc around $z = z_n$ such that $F_n(z_n) = F'(z_n)$ and $G_n(z_n) = \frac{\zeta_n}{F'(z_n)}$. Thus $f(z) = F_n(z)G_n(z)$ near $z = z_n$, and hence f is analytic at each z_n, independently of whether the corresponding target point ζ_n satisfies $\zeta_n = 0$ or $\zeta_n \neq 0$. Consequently, f is entire, and

$$f(z_n) = F_n(z_n)G_n(z_n) = F'(z_n)\cdot\frac{\zeta_n}{F'(z_n)} = \zeta_n, \quad n \in \mathbb{N},$$

where $\zeta_n = 0$ is allowed. This completes the proof. □

The interpolation problem can be extended to derivatives.

Theorem 4.7.4. *Let $\{z_n\}$ be a sequence of distinct points in $\mathbb{C}$ having no finite accumulation points, let $N \in \mathbb{N}$, and let $\{\zeta_{jn}\}$ be any sequences in $\mathbb{C}$ for $j = 0, \ldots, N-1$, in each of which repetition of points is allowed. Then there exists an entire function f such that*

$$f^{(j)}(z_n) = \zeta_{jn}, \quad n \in \mathbb{N}, \ j = 0, \ldots, N-1.$$

Proof. The case $N = 1$ is Theorem 4.7.2. We prove the general case by induction on j. Suppose that for $K \in \mathbb{N}$, there exists an entire function g such that

$$g^{(j)}(z_n) = \zeta_{jn}, \quad n \in \mathbb{N}, \ j = 0, \ldots, K-1.$$

By Weierstrass' theorem there exists an entire function P that has simple zeros at the points z_n and nowhere else. Let h be an entire function, to be defined later on, and define the entire function f by

$$f(z) = g(z) + P(z)^K h(z). \tag{4.7.2}$$

By Leibniz' rule for derivatives

$$f^{(j)}(z) = g^{(j)}(z) + \sum_{m=0}^{j} \binom{j}{m} (P(z)^K)^{(j-m)} h^{(m)}(z), \quad j = 0, \ldots, K.$$

Since the function $(P(z)^K)^{(j-m)}$ has a zero of multiplicity $K - j + m$ at every z_n, we deduce that

$$f^{(j)}(z_n) = \begin{cases} \zeta_{jn}, & j = 0, \ldots, K-1, \\ g^{(K)}(z_n) + (P^K)^{(K)}(z_n) h(z_n), & j = K. \end{cases}$$

Therefore the function f in (4.7.2) is the function we are looking for, provided that h satisfies

$$h(z_n) = \frac{\zeta_{Kn} - g^{(K)}(z_n)}{(P^K)^{(K)}(z_n)} =: w_n, \quad n \in \mathbb{N}. \tag{4.7.3}$$

Here $w_n \in \mathbb{C}$ because $(P^K)^{(K)}(z_n) \neq 0$. However, by Theorem 4.7.2 we can find an entire function h satisfying (4.7.3), and the proof is now complete. □

Problems

Problem 4.7.5. Find Lagrange's interpolation polynomial $L(z)$ of degree two that satisfies $L(0) = 1$, $L(1) = i$, and $L(-1) = 0$.

Problem 4.7.6. Let $\{z_n\}$ be an infinite sequence of distinct points in $\mathbb{C}$ having no finite accumulation points, and let $\{\zeta_n\}$ be any infinite sequence in $\mathbb{C}$, not necessarily distinct. Further, let $P(z)$ be an entire function with simple zeros at the points z_n and no other zeros. Prove that the Lagrange interpolation series

$$f(z) = \sum_{k=1}^{\infty} \frac{\zeta_k}{P'(z_k)} \cdot \frac{P(z)}{z - z_k}$$

is an entire function and satisfies $f(z_n) = \zeta_n$ for all $n \in \mathbb{N}$, provided that the series $\sum_{k=1}^{\infty} \left| \frac{\zeta_k}{P'(z_k)} \right|$ is convergent.

Remark. The Lagrange interpolation series is a more simple solution of the interpolation problem than the function coming from the Mittag-Leffler theorem. However, the assumption that $\sum_{k=1}^{\infty} \left| \frac{\zeta_k}{P'(z_k)} \right|$ converges is rather harsh and allows the Lagrange series to solve only certain types of interpolation problems.

5 Entire functions with real zeros

The main focus of this chapter is the special class of entire functions whose zeros are all real numbers. These functions often have other interesting properties. Some of these special functions are solutions of certain linear differential equations, which will be discussed informally. General discussions on the theory of linear differential equations will be postponed to Chapters 8 and 9.

5.1 Reciprocal of the gamma function

The gamma function $\Gamma(z)$ was first introduced by Euler. Leonhard Euler (1707–1783, Switzerland) may have been the most prolific mathematical author of all time, and even though he was totally blind during the last seventeen years of his life, his mind was still powerful, and with the aid of helpers, he actually increased his extraordinary production of work. He made important contributions to many areas of mathematics, and “he preferred concrete special problems to the general theories.” See [260, pp. 107–111].

The gamma function has several important properties and satisfies interesting functional equations. Although $\Gamma(z)$ is a meromorphic function with poles, the reciprocal function $1/\Gamma(z)$ is an entire function with all real zeros and has maximal type, which are useful features in both theoretical considerations and in providing interesting examples. We proceed to define and discuss $\Gamma(z)$, relying mostly on the presentation in [57, Section 7.7].

By the Weierstrass theorem, the product

$$f(z) = \prod_{n=1}^{\infty}\left(1+\frac{z}{n}\right)e^{-z/n}$$

is an entire function having simple zeros precisely at the negative integers. Hence $1/f(z)$ is a meromorphic function that has simple poles precisely at the negative integers. Set

$$c = \frac{1}{f(1)} = \prod_{n=1}^{\infty}\left(1+\frac{1}{n}\right)^{-1} e^{1/n} > 0.$$

The constant $\gamma = \log c$, known as *Euler’s constant*, satisfies

$$\begin{aligned}\gamma &= \sum_{k=1}^{\infty}\log\left(\left(1+\frac{1}{k}\right)^{-1} e^{1/k}\right) = \lim_{n\to\infty}\sum_{k=1}^{n}\left(\frac{1}{k} - \log(k+1) + \log k\right)\\ &= \lim_{n\to\infty}\left(\left(1+\frac{1}{2}+\frac{1}{3}+\cdots+\frac{1}{n}\right) - \log(n+1)\right).\end{aligned}$$

Since $\log(\frac{n+1}{n}) \to 0$ as $n \to \infty$, we obtain the simplified form

https://doi.org/10.1515/9783112249130-005

$$\gamma = \lim_{n\to\infty}\left(\left(1+\frac{1}{2}+\frac{1}{3}+\cdots+\frac{1}{n}\right)-\log n\right). \tag{5.1.1}$$

This form can be used to approximate γ, and it follows that $\gamma \approx 0.57722$.

Definition 5.1.1. The *gamma function* $\Gamma(z)$ is the meromorphic function with simple poles at the integers $n = 0, -1, -2, \ldots$ defined by

$$\Gamma(z) = \frac{e^{-\gamma z}}{z}\prod_{n=1}^{\infty}\left(1+\frac{z}{n}\right)^{-1} e^{z/n}. \tag{5.1.2}$$

The purpose of Euler's constant is to induce the normalization $\Gamma(1) = 1$. Moreover, since

$$\overline{e^{z}} = \overline{e^{x+iy}} = \overline{e^{x}(\cos y + i\sin y)} = e^{x}(\cos y - i\sin y) = e^{x-iy} = e^{\overline{z}},$$

we find that $\Gamma(\overline{z}) = \overline{\Gamma(z)}$ for $z \neq 0, -1, -2, \ldots$.

Observe from (5.1.2) that $\Gamma(z) \neq 0$ for all $z \in \mathbb{C}$. Hence $1/\Gamma(z)$ is an entire function with all real zeros, which are simple zeros at the points $z = 0, -1, -2, \ldots$, such that

$$\frac{1}{\Gamma(z)} = ze^{\gamma z}\prod_{n=1}^{\infty}\left(1+\frac{z}{n}\right)e^{-z/n}, \quad z \in \mathbb{C}. \tag{5.1.3}$$

The following representation for the gamma function is known as the *Gauss formula*.

Theorem 5.1.2. *For $z \neq 0, -1, -2, \ldots$, we have*

$$\Gamma(z) = \lim_{n\to\infty}\frac{n!n^{z}}{z(z+1)\cdots(z+n)}.$$

Proof. We use the definition of the gamma function together with (5.1.1) to get

$$\begin{aligned}\Gamma(z) &= \frac{e^{-\gamma z}}{z}\lim_{n\to\infty}\prod_{k=1}^{n}\left(1+\frac{z}{k}\right)^{-1} e^{z/k} = \frac{e^{-\gamma z}}{z}\lim_{n\to\infty}\prod_{k=1}^{n}\frac{ke^{z/k}}{z+k}\\ &= \lim_{n\to\infty}\frac{n!e^{-\gamma z}}{z(z+1)\cdots(z+n)}\exp\left(z\left(1+\frac{1}{2}+\cdots+\frac{1}{n}\right)\right)\\ &= \lim_{n\to\infty}\frac{n!e^{z\log n}}{z(z+1)\cdots(z+n)},\end{aligned}$$

which yields the statement. □

Using the Gauss formula, we can quickly verify the well-known *functional equation of the gamma function*

$$\Gamma(z+1) = z\Gamma(z), \quad z \neq 0, -1, -2, \ldots, \tag{5.1.4}$$

since

$$\Gamma(z+1) = \lim_{n\to\infty} \frac{n!n^{z+1}}{(z+1)(z+2)\cdots(z+n+1)}$$
$$= z \lim_{n\to\infty} \frac{n!n^{z}}{z(z+1)\cdots(z+n)} \cdot \frac{n}{n+z+1} = z\Gamma(z).$$

Then by (5.1.4),

$$\Gamma(z+2) = \Gamma\big((z+1)+1\big) = (z+1)\Gamma(z+1) = z(z+1)\Gamma(z),$$

and more generally

$$\Gamma(z+n) = z(z+1)\cdots(z+n-1)\Gamma(z).$$

In particular, using $\Gamma(1) = 1$, we obtain the famous formula

$$\Gamma(n+1) = n!, \quad n = 0, 1, 2, \ldots. \tag{5.1.5}$$

From (5.1.5) we see that $\Gamma(z)$ is a generalization of the factorial.

The next result gives a useful relationship between the gamma and sine functions.

Lemma 5.1.3. *We have*

$$\Gamma(z)\Gamma(1-z) = \frac{\pi}{\sin(\pi z)}, \quad z \in \mathbb{C} \setminus \mathbb{Z}. \tag{5.1.6}$$

Proof. Recall that by Theorem 3.4.7,

$$\sin(\pi z) = \pi z \prod_{n=1}^{\infty} \left(1 - \frac{z^2}{n^2}\right).$$

Note that by (5.1.3),

$$\frac{1}{\Gamma(z)\Gamma(-z)} = -z^2 \prod_{n=1}^{\infty} \left(1 - \frac{z^2}{n^2}\right) = -\frac{z\sin(\pi z)}{\pi}.$$

Since $z\Gamma(z) = \Gamma(z+1)$, we have $-z\Gamma(-z) = \Gamma(1-z)$. Thus (5.1.6) follows. □

The following result, known as *Stirling's formula*, is important and useful in many applications.

Theorem 5.1.4 ([218, p. 284]). *Along the positive real axis $x > 0$, we have*

$$\Gamma(x+1) = (1+o(1))x^x e^{-x}\sqrt{2\pi x}$$

as $x \to \infty$. In the important case of integer values $x = n$, we have

$$\Gamma(n+1) = n! = (1+o(1))n^n e^{-n}\sqrt{2\pi n}$$

as $n \to \infty$.

Remark. We mention that the entire function $1/\Gamma(z)$ has order 1 and type $\tau = \infty$, i. e., *maximum type*; see [290].

Lemma 5.1.5. *We have* $\Gamma(x) > 0$ *for all* $x > 0$ *and* $\log\Gamma(x)$ *is convex on* $(0,\infty)$.

Proof. By the definition,

$$\Gamma(x) = \frac{e^{-\gamma x}}{x}\prod_{n=1}^{\infty}\left(1+\frac{x}{n}\right)^{-1} e^{x/n}, \quad x > 0. \tag{5.1.7}$$

Since each factor in (5.1.7) is positive for $x > 0$, trivially, $\Gamma(x) > 0$ for all $x > 0$. Consequently, $\log\Gamma(x)$ is well-defined for $x > 0$. Moreover, since each factor in (5.1.7) is differentiable, $\log\Gamma(x)$ is differentiable.

Furthermore, since $\Gamma(z)$ is analytic and has no zeros in the right half-plane, $\log\Gamma(z)$ is analytic in the right half-plane. By the uniqueness theorem for analytic functions the derivative of $\log\Gamma(x)$ agrees with that of $\log\Gamma(z)$ when z is restricted to the positive real axis. Hence we do not have to worry about possible branches of $\log\Gamma(x)$ – we can treat it as we would treat an ordinary logarithm in real analysis. Therefore,

$$\log\Gamma(x) = -\gamma x - \log x + \sum_{n=1}^{\infty}\left(\log\left(\frac{n}{n+x}\right) + \frac{x}{n}\right)$$

implies

$$(\log\Gamma(x))' = \frac{\Gamma'(x)}{\Gamma(x)} = -\gamma - \frac{1}{x} + \sum_{n=1}^{\infty}\frac{x}{n(n+x)},$$

which in turn yields

$$(\log\Gamma(x))'' = \left(\frac{\Gamma'(x)}{\Gamma(x)}\right)' = \frac{1}{x^2} + \sum_{n=1}^{\infty}\frac{1}{(n+x)^2}.$$

Since $(\log\Gamma(x))'' > 0$ for $x > 0$, we find that $\log\Gamma(x)$ is convex on $(0,\infty)$. □

Next, we state that if a function f satisfies the previously described properties of the gamma function, then f must be equal to the gamma function. This result is known as the *Bohr–Mollerup theorem*.

Theorem 5.1.6. *Let* $f : (0,\infty) \to (0,\infty)$ *be a function that has the following properties:*
(a) $\log f(x)$ *is a convex function.*
(b) $f(x+1) = xf(x)$ *for all* $x > 0$.

(c) $f(1) = 1$.

Then $f(x) = \Gamma(x)$ *for all* $x > 0$.

A proof for the Bohr–Mollerup theorem can be found in Conway's book [57]. The result itself is used to prove the following more familiar representation for the gamma function, which is valid in the right half-plane and, in particular, on $(0, \infty)$.

Theorem 5.1.7 ([57, Theorem 7.15]). *If* $\operatorname{Re} z > 0$, *then*

$$\Gamma(z) = \int_0^\infty e^{-t} t^{z-1}\, dt. \tag{5.1.8}$$

Example 5.1.8. Substituting $z = 1/2$ into (5.1.6), we obtain $\Gamma(1/2)^2 = \pi$. Since $\Gamma(x) > 0$ for all $x > 0$ by Lemma 5.1.5, this gives us $\Gamma(1/2) = \sqrt{\pi}$. This is the initial step for proving the half-integer formulas

$$\Gamma\left(\frac{1}{2} + n\right) = \frac{(2n)!}{4^n n!}\sqrt{\pi} \quad \text{and} \quad \Gamma\left(\frac{1}{2} - n\right) = \frac{(-4)^n n!}{(2n)!}\sqrt{\pi}, \tag{5.1.9}$$

where $n \geq 0$ is an integer. Suppose that the first identity in (5.1.9) is valid for some integer $k \geq 0$. Then by (5.1.4),

$$\begin{aligned}\Gamma\left(\frac{1}{2} + k + 1\right) &= \left(\frac{1}{2} + k\right) \cdot \Gamma\left(\frac{1}{2} + k\right) = \left(\frac{1}{2} + k\right) \cdot \frac{(2k)!}{4^k k!}\sqrt{\pi} \\ &= \frac{2(2k+1)!}{4^{k+1} k!}\sqrt{\pi} = \frac{(2(k+1))!}{4^{k+1}(k+1)!}\sqrt{\pi}.\end{aligned}$$

The first equality in (5.1.9) follows by induction. We leave the proof of the second equality as an exercise.

Problems

Problem 5.1.9. Prove the second equality in (5.1.9).

Problem 5.1.10. What is the exact value of $\Gamma(-3/2)$?

Problem 5.1.11. Verify the following:

(a) $\int_{-\infty}^{\infty} e^{-x^2}\, dx = \sqrt{\pi}$.

(b) $\int_{-\infty}^{\infty} x^2 e^{-x^2}\, dx = \sqrt{\pi}/2$.

(c) $\int_0^{\infty} \sqrt{x} e^{-x^3}\, dx = \sqrt{\pi}/3$.

Problem 5.1.12. Prove that $\lim_{x\to+\infty} \frac{\Gamma(x)}{e^x} = \infty$. Does this result change if e^x is replaced with e^{2x}?

Problem 5.1.13. Prove that $\lim_{y\to+\infty} \Gamma\left(\frac{1}{2} + iy\right) = 0$.

Problem 5.1.14. Set

$$\psi(z) = \frac{\Gamma'(z)}{\Gamma(z)}.$$

Prove the following:
(a) $\psi(z+1) - \psi(z) = 1/z$.
(b) $\psi(1-z) - \psi(z) = \pi\cot(\pi z)$.

5.2 Mittag-Leffler functions

For $\alpha > 0$, we define *Mittag-Leffler's function* as the Taylor series

$$E_\alpha(z) = \sum_{n=0}^{\infty} a_n z^n \quad \text{with } a_n = \frac{1}{\Gamma(\alpha n + 1)}.$$

The notation should not be confused with that of Weierstrass primary factors.

Theorem 5.2.1. *For each $\alpha > 0$, the Mittag-Leffler function $E_\alpha(z)$ is an entire function of order $\rho = 1/\alpha$ and type $\tau = 1$.*

Proof. By Stirling's formula,

$$\Gamma(\alpha n + 1) = (1 + o(1))\sqrt{2\pi}(\alpha n)^{\alpha n + 1/2} e^{-\alpha n}, \quad n \to \infty, \tag{5.2.1}$$

which gives

$$\limsup_{n\to\infty} \sqrt[n]{|a_n|} = \limsup_{n\to\infty}(1 + o(1))\frac{e^\alpha}{(\alpha n)^\alpha} = 0.$$

It follows that the radius of convergence R satisfies $R = \infty$ by the Cauchy–Hadamard theorem, which ensures that $E_\alpha(z)$ is entire for each $\alpha > 0$.

Moreover, by using (5.2.1), we get

$$\rho = \limsup_{n\to\infty} \frac{n\log n}{-\log|a_n|} = \limsup_{n\to\infty} \frac{n\log n}{\log\Gamma(\alpha n + 1)} = \limsup_{n\to\infty} \frac{n\log n}{O(n) + \alpha n\log n} = \frac{1}{\alpha}$$

by Theorem 2.2.2, and then

$$\tau = \frac{1}{e\rho}\limsup_{n\to\infty}(n|a_n|^{\rho/n}) = \frac{\alpha}{e}\limsup_{n\to\infty} \frac{n}{\Gamma(\alpha n + 1)^{1/\alpha n}} = \frac{\alpha}{e}\limsup_{n\to\infty} \frac{(1 + o(1))en}{(\alpha n)^{1+1/2\alpha n}} = 1$$

by Theorem 2.2.8. This completes the proof. □

Remark. Some Mittag-Leffler functions reduce to known functions:

$$E_1(z) = \sum_{n=0}^{\infty} \frac{z^n}{\Gamma(n+1)} = \sum_{n=0}^{\infty} \frac{z^n}{n!} = e^z,$$

$$E_2(-z^2) = \sum_{n=0}^{\infty} \frac{(-1)^n z^{2n}}{\Gamma(2n+1)} = \sum_{n=0}^{\infty} \frac{(-1)^n z^{2n}}{(2n)!} = \cos z,$$

$$E_2(z^2) = \sum_{n=0}^{\infty} \frac{z^{2n}}{\Gamma(2n+1)} = \sum_{n=0}^{\infty} \frac{z^{2n}}{(2n)!} = \cosh z.$$

Obviously, $E_1(z) = e^z$ has no zeros. As the next two theorems state, all the other functions $E_\alpha(z)$, $\alpha \neq 1$, possess an infinite number of zeros. Furthermore, $E_\alpha(z)$ has all real zeros when $\alpha \geq 2$.

Theorem 5.2.2 ([83, p. 31]). *The function $E_\alpha(z)$, $\alpha \geq 2$, possesses an infinite number of zeros, all the zeros are negative real numbers, and all the zeros are simple. Assuming that the sequence $\{x_n\}$ of zeros is ordered by increasing modulus with $n = 1, 2, \ldots$, we have the inequalities*

$$-\left(\frac{\pi n}{\sin\frac{\pi}{\alpha}}\right)^{\alpha} < x_n < -\left(\frac{\pi(n-1)}{\sin\frac{\pi}{\alpha}}\right)^{\alpha}.$$

Theorem 5.2.3 ([83, p. 32]). *The function $E_\alpha(z)$, $0 < \alpha < 2$, $\alpha \neq 1$, possesses an infinite number of zeros, and the zeros with sufficiently large modulus are simple. We have the following asymptotic formula for the zeros z_n:*

$$z_n^{\pm} = e^{\pm i\frac{\alpha\pi}{2}}(2\pi n)^{\alpha}\left(1 + O\left(\frac{\log n}{n}\right)\right), \quad n \to \infty.$$

Mittag-Leffler functions of order > 1 have specific asymptotic growth properties in sectors, which have often been used to give illustrative examples in the literature. We give an example after the proof of the next theorem.

Theorem 5.2.4 ([43, p. 50]). *If $0 < \alpha < 1$, then*

$$E_\alpha(z) = \frac{\exp(z^{1/\alpha})}{\alpha} - \frac{1}{z\Gamma(1-\alpha)} + O\left(\frac{1}{(\alpha|z|)^2}\right)$$

as $z \to \infty$ in $|\arg z| \leq \frac{3}{4}\alpha\pi$, and

$$E_\alpha(z) = -\frac{1}{z\Gamma(1-\alpha)} + O\left(\frac{1}{(\alpha|z|)^2}\right)$$

as $z \to \infty$ in $\frac{3}{4}\alpha\pi \leq \arg z \leq 2\pi - \frac{3}{4}\alpha\pi$.

The proof of Theorem 5.2.4 in [43, p. 50] uses the following integral representation:

$$E_\alpha(z) = \frac{1}{2\pi i\alpha}\int_C \frac{\exp(w^{1/\alpha})}{w-z}\,dw, \quad z \in \mathbb{C}, \quad 0 < \alpha < 2, \tag{5.2.2}$$

where C is the contour that goes from ∞ to $r_0 e^{i\beta}$ along $\arg w = \beta$, then from $r_0 e^{i\beta}$ to $r_0 e^{-i\beta}$ along the arc of the circle $|w| = r_0$ that lies in the right half-plane, and finally from $r_0 e^{-i\beta}$ to ∞ along $\arg w = -\beta$; see Figure 5.1. Here $r_0 > 0$ and $\beta \in (\frac{1}{2}\alpha\pi, \alpha\pi]$ are any pregiven numbers, and $\exp(w^{1/\alpha})$ has the principal branch of the logarithm, with the negative real axis being the branch cut. Note that $|\exp(w^{1/\alpha})|$ tends to zero exponentially as $w \to \infty$ along the rays $\arg w = \pm\beta$. This property of tending to zero is needed for the convergence of the integral in (5.2.2). Moreover, Cauchy's integral formula can be used to prove that the integral in (5.2.2) is independent of the choice of $\beta \in (\frac{1}{2}\alpha\pi, \alpha\pi]$ or $r_0 > 0$.

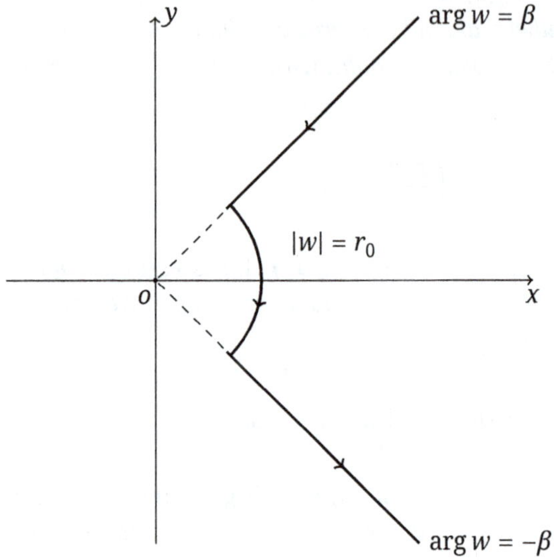

Figure 5.1: Contour C for $E_\alpha(z)$.

The proof of (5.2.2) in turn is based on the following integral representation for the reciprocal of the gamma function [43, p. 51], which we know is an entire function:

$$\frac{1}{\Gamma(z)} = \frac{1}{2\pi i}\int_{C_1} e^w w^{-z}\, dw, \quad z \in \mathbb{C}, \tag{5.2.3}$$

where C_1 is a contour similar to C in (5.2.2), but along the rays $\arg w = \pm\gamma$ and a circular arc of $|w| = s_0$ joining them, where $\gamma \in (\frac{\pi}{2}, \pi)$ and $s_0 > 0$. Similarly as above, Cauchy's integral formula can be used to prove that the integral in (5.2.3) is independent of the choice of $\gamma \in (\frac{\pi}{2}, \pi)$ or $s_0 > 0$.

Full proofs of both (5.2.2) and (5.2.3) along with alternatives for the integral paths and branch cuts are given by Saenko [241].

Proof of Theorem 5.2.4. We use the integral representations (5.2.2) and (5.2.3), and rely on the reasoning in [43], but with added details.

To prove the first statement, we suppose that $|\arg z| \le \frac{3}{4}\alpha\pi < \frac{3}{4}\pi$ and $|z| > 2$. Suppose further, for a moment, that $\beta \in (\frac{1}{2}\alpha\pi, \alpha\pi)$ and write

$$\frac{1}{w-z} = -\frac{1}{z} + \frac{w}{z(w-z)}.$$

Then by (5.2.2),

$$E_\alpha(z) = -\frac{1}{2\pi i\alpha z}\int_C \exp(w^{1/\alpha})\,dw + \frac{1}{2\pi i\alpha z}\int_C \exp(w^{1/\alpha})\frac{w\,dw}{w-z}.$$

We make a change of variable $w = \zeta^\alpha$, $dw = \alpha\zeta^{\alpha-1}\,d\zeta$ in the first integral and denote the second integral by $I_1(C)$. By (5.2.3) this gives us

$$E_\alpha(z) = -\frac{1}{2\pi iz}\int_{C_1} e^\zeta \zeta^{-(1-\alpha)}\,d\zeta + I_1(C) = -\frac{1}{z\Gamma(1-\alpha)} + I_1(C). \tag{5.2.4}$$

From now on we assume that $\beta = \alpha\pi$ on C. Let C_2 be a contour similar to C but formed by using an arc of the circle $|w| = 1$ instead of the arc of $|w| = r_0$; see Figure 5.2. Assuming $r_0 > 0$ to be sufficiently large and keeping in mind that $|\arg z| \le \frac{3}{4}\alpha\pi$ and $\beta = \alpha\pi$, we see that the variable z lies inside the simple closed contour $C - C_2$, traversed in the clockwise direction. Then by Cauchy's integral formula,

$$I_1(C) = \alpha^{-1}\exp(z^{1/\alpha}) + I_1(C_2). \tag{5.2.5}$$

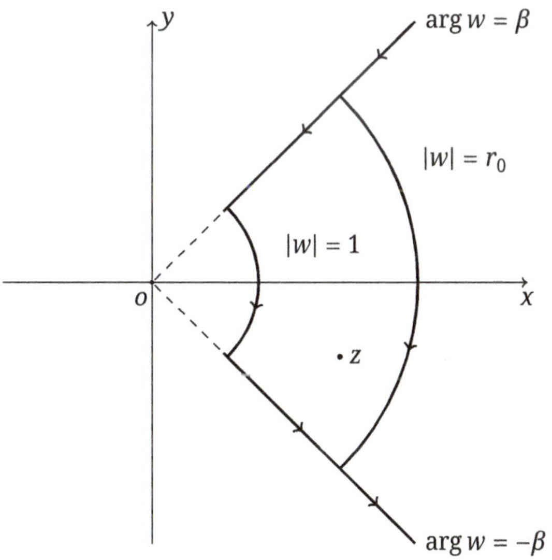

Figure 5.2: Contours C and C_2. The boundary of the annular sector is $C - C_2$, traversed in the clockwise direction.

We proceed to estimate $I_1(C_2)$. On $\arg w = \pm\alpha\pi = \pm\beta$, $|w| \geq 1$, and for the restrictions on z, the expression $|w - z|$ takes the minimal value in either of the following two cases:

(1) $\arg w = \alpha\pi$, $\arg z = \frac{3}{4}\alpha\pi$, and $|w| = |z|$.

(2) $\arg w = -\alpha\pi$, $\arg z = -\frac{3}{4}\alpha\pi$, and $|w| = |z|$.

Since $|e^{i\alpha\pi/4} - 1| = |e^{-i\alpha\pi/4} - 1|$ by symmetry, in both cases, we obtain

$$|w - z| = |z| \cdot |e^{i\alpha\pi/4} - 1| = 2|z| \cdot \sqrt{\frac{1 - \cos\frac{\alpha\pi}{4}}{2}} = 2|z| \cdot \sin\frac{\alpha\pi}{8} > \frac{1}{2}|z|\alpha,$$

where we have used the standard estimate $\frac{\sin x}{x} > \frac{2}{\pi}$ for $0 < x < \pi/2$. Still on $\arg w = \pm\alpha\pi = \pm\beta$, $|w| \geq 1$, it is clear that

$$\exp(w^{1/\alpha}) = \exp(-|w|^{1/\alpha}) \leq \exp(-|w|),$$

because $0 < \alpha < 1$. We move on to consider the circular part of C_2, that is, we now assume that $|w| = 1$ and $|\arg w| \leq \alpha\pi$. It follows that

$$\begin{aligned} |\exp(w^{1/\alpha})| &\leq \exp(|w|^{1/\alpha}) = e, \\ |w - z| &\geq |z| - |w| = |z| - 1 > \frac{|z|}{2}, \end{aligned} \tag{5.2.6}$$

because $|z| > 2$. We now obtain

$$\begin{aligned} |I_1(C_2)| &< \frac{2}{\pi\alpha^2|z|^2} \int_1^\infty te^{-t}\,dt + \frac{e}{\pi\alpha|z|^2} \int_{-\alpha\pi}^{\alpha\pi} dt \\ &= \frac{4}{e\pi\alpha^2|z|^2} + \frac{2e}{|z|^2} = \frac{K}{(\alpha|z|)^2}, \end{aligned}$$

where $K = \frac{4+2e^2\pi\alpha^2}{e\pi}$. Thus the first statement follows from (5.2.4) and (5.2.5).

To prove the second statement, we suppose that $\frac{3}{4}\alpha\pi \leq \arg z \leq 2\pi - \frac{3}{4}\alpha\pi$ and $|z| > 2$. Choose $\beta = \frac{5}{8}\alpha\pi$ in (5.2.2), and denote the resulting contour by C_3. Analogously as in (5.2.4), but this time directly for the aforementioned choice of β, we obtain

$$E_\alpha(z) = -\frac{1}{z\Gamma(1-\alpha)} + I_1(C_3). \tag{5.2.7}$$

We proceed to estimate $I_1(C_3)$. On $\arg w = \pm\beta$, $|w| \geq 1$, and for the restrictions on z, the expression $|w - z|$ takes the minimal value in either of the following two cases:

(1) $\arg w = \frac{5}{8}\alpha\pi$, $\arg z = \frac{3}{4}\alpha\pi$, and $|w| = |z|$.

(2) $\arg w = -\frac{5}{8}\alpha\pi$, $\arg z = 2\pi - \frac{3}{4}\alpha\pi$, and $|w| = |z|$.

In both cases, similarly as above, we have

$$|w-z| = |z| \cdot \left|e^{i\alpha\pi/8} - 1\right| = 2|z| \cdot \sqrt{\frac{1-\cos\frac{\alpha\pi}{8}}{2}} = 2|z| \cdot \sin\frac{\alpha\pi}{16} > \frac{1}{4}|z|\alpha.$$

Still on $\arg w = \pm\beta$, $|w| \geq 1$, it is clear that

$$\left|\exp(w^{1/\alpha})\right| = \exp\left(|w|^{1/\alpha}\cos\frac{5\pi}{8}\right) \leq \exp\left(-|w|\sin\frac{\pi}{8}\right).$$

On the arc of $|w| = 1$, we have (5.2.6) as before. We now obtain

$$\begin{aligned}|I_1(C_3)| &< \frac{4}{\pi\alpha^2|z|^2}\int_1^\infty t\exp\left(-t\sin\frac{\pi}{8}\right)dt + \frac{e}{\pi\alpha|z|^2}\int_{-5\alpha\pi/8}^{5\alpha\pi/8} dt \\ &< \frac{28}{\pi\alpha^2|z|^2} + \frac{5e}{4|z|^2} = \frac{K}{(\alpha|z|)^2},\end{aligned}$$

where $K = \frac{112+5e\pi\alpha^2}{4\pi}$. Thus the second statement follows from (5.2.7). □

Remark. In 1988, Shen [256] proved that if f is an entire function of finite order $\rho \geq 1$, then

$$\limsup_{r\to\infty}\frac{A(r)}{r^2} \geq \frac{\pi}{2\rho}, \tag{5.2.8}$$

where $A(r)$ is the area of the region $\{z \in \mathbb{C} : |f(z)| \geq 1,\ |z| \leq r\}$. Further, he observed that this inequality is sharp for Mittag-Leffler functions. To see this, we first recall that $E_\alpha(z)$ has order $\rho = 1/\alpha$. Moreover, if $0 < \alpha < 1$, then by Theorem 5.2.4 we find that $|E_\alpha(z)|$ is radially unbounded in the sector $|\arg z| < \frac{\alpha\pi}{2}$ and tends to zero in the rest of the plane. This gives us

$$A(r) = \left(1 + o(1)\right)\frac{\alpha\pi r^2}{2}, \quad r \to \infty,$$

and, consequently, (5.2.8) becomes an equality for $E_\alpha(z)$. Further, if $f(z) = e^z$, then $\rho = 1$ and $|f(z)| \geq 1$ in the closed right half-plane. Thus $A(r) = \frac{\pi r^2}{2}$, and (5.2.8) becomes an equality for $f(z)$. Finally, if $f(z) = e^z + z$, then we see that $A(r) \sim \pi r^2$, and (5.2.8) is a strict inequality.

Additional properties and more general Mittag-Leffler functions can be found in the book [83].

Problems

Problem 5.2.5. Find the exponent of convergence λ of the zeros of the Mittag-Leffler function $E_\alpha(z)$ when $\alpha \geq 2$.

Problem 5.2.6. Let

$$w(z) = \sum_{n=0}^{\infty}\frac{(az)^n}{\Gamma(cn+1)},$$

where $a > 0$ and $c > 0$ are fixed constants. Show that w is entire and that the order and type of w are

$$\rho(w) = \frac{1}{c} \quad \text{and} \quad \tau(w) = a^{\frac{1}{c}}.$$

The function $w(z)$ is a generalization of the Mittag-Leffler function $E_a(z)$.

Problem 5.2.7. Prove the following identity:

$$2E_{2a}(z^2) = E_a(z) + E_a(-z), \quad z \in \mathbb{C}, \quad a > 0.$$

When $a = 1$, this identity is $2\cosh z = e^z + e^{-z}$.

5.3 Airy integral

If $P(z) \not\equiv 0$ is a polynomial of degree n, then the nontrivial solutions f of the linear differential equation

$$f'' + P(z)f = 0 \tag{5.3.1}$$

are known to be entire functions of order $\rho(f) = (n+2)/2$ [166]. This is easy to see by applying the Wiman–Valiron theory, as in Example 4.4.9. Further results of this nature will be discussed in Chapter 8. Thus if n is odd, then $\lambda(f) = (n+2)/2$ by Corollary 3.5.9. On the other hand, if $Q(z)$ is a polynomial of degree $k \geq 1$, then $f(z) = e^{Q(z)}$ solves (5.3.1) with

$$P(z) = -Q'(z)^2 - Q''(z).$$

In this case, $n = 2(k-1)$ is even. This shows that if n is even, then the solutions of (5.3.1) might not have zeros. An easy proof by contradiction shows that the zeros of nontrivial solutions of (5.3.1) are simple. Indeed, the polynomial $P(z) = -f''/f$ would have a double pole at any multiple zero of f, which is not possible. More discussion on this topic, including for more general linear differential equations, can be found in Section 9.1.

An important special case of (5.3.1) is the *Airy differential equation*

$$f'' - zf = 0. \tag{5.3.2}$$

Sir George Biddell Airy (1801–1892) [260, p. 158] was the Astronomer Royal of England for many years. He did some early work in the theory of light and developed a keen interest in elaborate numerical computations.

For nontrivial solutions of the Airy equation, the order and exponent of convergence of the zeros are both equal to 3/2.

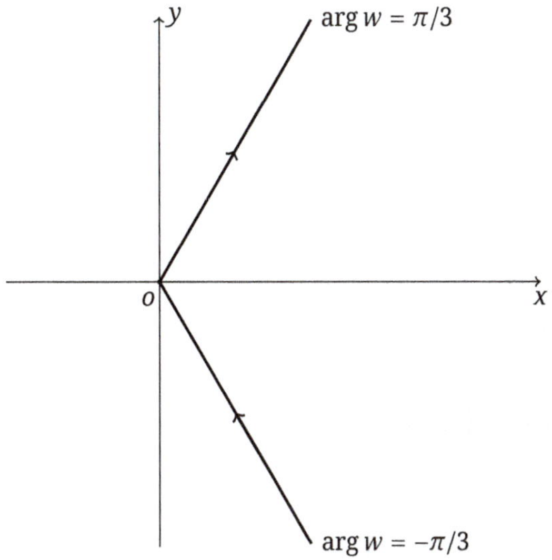

Figure 5.3: Contour C for Ai(z).

Definition 5.3.1. The *Airy integral* is the function Ai(z) that is defined for all complex numbers z by

$$\mathrm{Ai}(z) = \frac{1}{2\pi i}\int_C \exp\left(\frac{1}{3}w^3 - zw\right)dw, \tag{5.3.3}$$

where the contour C runs from ∞ to 0 along the ray $\arg w = -\pi/3$ and then from 0 to ∞ along the ray $\arg w = \pi/3$; see Figure 5.3.

Theorem 5.3.2. *The Airy integral* Ai(z) *is a nontrivial entire solution of the Airy differential equation* (5.3.2).

Proof. By the definition,

$$\begin{aligned}\mathrm{Ai}(z) &= \frac{1}{2\pi i}\int_{\infty e^{-\frac{\pi}{3}i}}^{0} \exp\left(\frac{1}{3}w^3 - zw\right)dw + \frac{1}{2\pi i}\int_0^{\infty e^{\frac{\pi}{3}i}} \exp\left(\frac{1}{3}w^3 - zw\right)dw \\ &= \frac{e^{-\frac{\pi}{3}i}}{2\pi i}\int_{\infty}^{0} \exp\left(-\frac{1}{3}t^3 - zte^{-\frac{\pi}{3}i}\right)dt + \frac{e^{\frac{\pi}{3}i}}{2\pi i}\int_0^{\infty} \exp\left(-\frac{1}{3}t^3 - zte^{\frac{\pi}{3}i}\right)dt.\end{aligned}$$

Therefore, we may write

$$\mathrm{Ai}(z) = \frac{e^{-\frac{\pi}{3}i}}{2\pi i}f_1(z) + \frac{e^{\frac{\pi}{3}i}}{2\pi i}f_2(z), \tag{5.3.4}$$

where

$$f_1(z) = -\int_0^\infty \exp\left(-\frac{1}{3}t^3 - zte^{-\frac{\pi}{3}i}\right) dt$$

and

$$f_2(z) = \int_0^\infty \exp\left(-\frac{1}{3}t^3 - zte^{\frac{\pi}{3}i}\right) dt.$$

Suppose that $|z| \leq R$, and let $\varepsilon > 0$. Then there exists a constant $T = T(\varepsilon, R) > 0$ such that

$$\begin{aligned}\left|\int_T^\infty \exp\left(-\frac{1}{3}t^3 - zte^{-\frac{\pi}{3}i}\right) dt\right| &\leq \int_T^\infty \exp\left(-\frac{1}{3}t^3 + |z|t\right) dt\\ &\leq \int_T^\infty \exp\left(-\frac{1}{3}t^3\left(1 - \frac{3R}{t^2}\right)\right) dt\\ &\leq \int_T^\infty \exp\left(-\frac{1}{6}t^3\right) dt < \varepsilon.\end{aligned}$$

Hence $f_1(z)$ converges uniformly on $|z| \leq R$ at the upper limit, where $R > 0$ can be chosen to be arbitrarily large. This conclusion holds for $f_2(z)$ as well. On the other hand, there exists a constant $\delta = \delta(\varepsilon, R) > 0$ such that

$$\begin{aligned}\left|\int_0^\delta \exp\left(-\frac{1}{3}t^3 - zte^{-\frac{\pi}{3}i}\right) dt\right| &\leq \int_0^\delta \exp\left(-\frac{1}{3}t^3 + |z|t\right) dt\\ &\leq \int_0^\delta \exp(Rt)\, dt \leq \delta e^{R\delta} < \varepsilon.\end{aligned}$$

Hence $f_1(z)$ converges uniformly on $|z| \leq R$ at the lower limit. Similarly, $f_2(z)$ converges uniformly on $|z| \leq R$ at the lower limit.

It follows from Theorem 1.6.12 that $f_1(z)$ and $f_2(z)$ are entire functions, and their derivatives of all orders can be found by differentiating under the integral signs. Thus by (5.3.4), $\mathrm{Ai}(z)$ is entire. Differentiating under the integral sign in (5.3.3), we obtain

$$\begin{aligned}\mathrm{Ai}''(z) - z\mathrm{Ai}(z) &= \frac{1}{2\pi i}\int_C (w^2 - z)\exp\left(\frac{1}{3}w^3 - zw\right) dw\\ &= \frac{1}{2\pi i}\left[\exp\left(\frac{1}{3}w^3 - zw\right)\right]_C = \frac{1}{2\pi i} - \frac{1}{2\pi i} = 0,\end{aligned}$$

so that $\mathrm{Ai}(z)$ solves (5.3.2).

Finally, $\mathrm{Ai}(z) \not\equiv 0$ follows from the next result. □

Theorem 5.3.3. *The Airy integral* Ai(*z*) *satisfies the initial conditions*

$$\operatorname{Ai}(0) = \frac{1}{3^{2/3}\Gamma(2/3)} \quad \text{and} \quad \operatorname{Ai}'(0) = -\frac{1}{3^{1/3}\Gamma(1/3)}.$$

Proof. We use the expanded form of Ai(*z*) in (5.3.4). Making the substitution $r = t^3/3$ and using representation (5.1.8), we obtain

$$\begin{aligned} f_1(0) &= -\int_0^\infty \exp\left(-\frac{1}{3}t^3\right) dt \\ &= -3^{-2/3} \int_0^\infty e^{-r} r^{1/3-1}\, dr = -3^{-2/3}\Gamma(1/3). \end{aligned}$$

Similarly, $f_2(0) = 3^{-2/3}\Gamma(1/3)$. Then by (5.3.4),

$$\begin{aligned} \operatorname{Ai}(0) &= \frac{e^{-\frac{\pi}{3}i}}{2\pi i} f_1(0) + \frac{e^{\frac{\pi}{3}i}}{2\pi i} f_2(0) \\ &= \frac{3^{-2/3}}{\pi}\Gamma(1/3)\frac{e^{\frac{\pi}{3}i} - e^{-\frac{\pi}{3}i}}{2i} = \frac{3^{-2/3}}{\pi}\Gamma(1/3)\sin(\pi/3). \end{aligned}$$

We now use identity (5.1.6), namely

$$\Gamma(z)\Gamma(1-z) = \frac{\pi}{\sin(\pi z)}, \quad z \in \mathbb{C} \setminus \mathbb{Z}, \tag{5.3.5}$$

and obtain $\operatorname{Ai}(0) = \frac{1}{3^{2/3}\Gamma(2/3)}$.

Next, we differentiate under the integral sign and again use the substitution $r = t^3/3$ and (5.1.8) to obtain

$$\begin{aligned} f_1'(0) &= e^{-\frac{\pi}{3}i} \int_0^\infty t \exp\left(-\frac{1}{3}t^3\right) dt \\ &= 3^{-1/3} e^{-\frac{\pi}{3}i} \int_0^\infty e^{-r} r^{2/3-1}\, dr = 3^{-1/3} e^{-\frac{\pi}{3}i}\Gamma(2/3), \end{aligned}$$

and, similarly, $f_2'(0) = -3^{-1/3} e^{\frac{\pi}{3}i}\Gamma(2/3)$. Now

$$\begin{aligned} \operatorname{Ai}'(0) &= \frac{e^{-\frac{\pi}{3}i}}{2\pi i} f_1'(0) + \frac{e^{\frac{\pi}{3}i}}{2\pi i} f_2'(0) \\ &= \frac{3^{-1/3}}{\pi}\Gamma(2/3)\frac{e^{-\frac{2\pi}{3}i} - e^{\frac{2\pi}{3}i}}{2i} = -\frac{3^{-1/3}}{\pi}\Gamma(2/3)\sin(2\pi/3). \end{aligned}$$

Using again (5.3.5), we obtain $\operatorname{Ai}'(0) = -\frac{1}{3^{1/3}\Gamma(1/3)}$. □

We now state two important results that exhibit special properties of the Airy integral.

Theorem 5.3.4 ([221, p. 415]). *All the zeros of the Airy integral* $\mathrm{Ai}(z)$ *are real and negative.*

Theorem 5.3.5 ([181, p. 138]). *The function* $|\mathrm{Ai}(z)|$ *has exponential decay of order* $\exp(-|z|^{3/2})$ *on each ray* $\arg z$ *in the sector* $-\pi/3 < \arg z < \pi/3$.

The properties of $\mathrm{Ai}(z)$ in Theorems 5.3.4 and 5.3.5 are unusual properties for solutions of equations of the form (5.3.1) to possess. This is explained in detail in Section 8.3, where such special solutions are called shortage solutions. Section 8.4 contains generalizations of the Airy integral, which includes a detailed statement and proof of a more general result (Theorem 8.4.6) than Theorem 5.3.5.

Finally, we mention a representation for the Airy integral on the real axis.

Theorem 5.3.6 ([181, p. 138], [221, p. 102]). *On the real line, the Airy integral has the representation*

$$\mathrm{Ai}(x) = \frac{1}{\pi}\int_0^\infty \cos\left(\frac{1}{3}t^3 + xt\right)dt, \quad x \in \mathbb{R}.$$

Problems

Problem 5.3.7. Suppose that a_1, a_2, a_3 are the three distinct cube roots of unity, that is, the distinct solutions of $a^3 = 1$. Prove that the functions

$$f_j(z) = \mathrm{Ai}(a_j z), \quad j = 1, 2, 3,$$

are three pairwise linearly independent solutions of Airy's equation (5.3.2). This shows that every solution of (5.3.2) can be expressed as a contour integral function.

Problem 5.3.8. By Theorem 5.3.6, $\mathrm{Ai}(x) \in \mathbb{R}$ whenever $x \in \mathbb{R}$. Show that $\mathrm{Ai}(x) > 0$ whenever $x \geq 0$.

Problem 5.3.9. Suppose that f is an entire function that satisfies the differential equation

$$f'' - e^z f = 0$$

and the conditions $f(0) = 1$ and $f'(0) = 2$. Show that $f(x) \in \mathbb{R}$ whenever $x \in \mathbb{R}$.

5.4 Solutions of $f'' + P(z)f = 0$ with real zeros

In this section, we discuss the history of results concerning equations of the form

$$f'' + P(z)f = 0, \tag{5.4.1}$$

where $P(z) \not\equiv 0$ is a polynomial that possesses a nontrivial solution f that has one of the following two properties:

(a) f has an infinite number of real zeros and no other zeros.

(b) f has an infinite number of real zeros and at most finitely many nonreal zeros.

For convenience in this section, we will call a nontrivial solution f of (5.4.1) that satisfies either (a) or (b) a *special solution.* Recall from Corollary 1.7.2 that all solutions of equations of the form (5.4.1) are entire functions.

In the previous section, we discussed the classical Airy integral Ai(z), which is a solution of Airy's equation $f'' - zf = 0$ that has an infinite number of negative real zeros and no other zeros. Here $\deg P = 1$ in (5.4.1).

Obviously, $\sin z$ and $\cos z$ are solutions of $f'' + f = 0$, where these solutions have an infinite number of real zeros and no other zeros. Here $\deg P = 0$ in (5.4.1).

The next example comes from the solution of a singular Sturm–Liouville problem.

Example 5.4.1. It was shown by Titchmarsh (see [274, pp. 172–173], and [85]) that there exists a sequence of positive constants β_k satisfying $0 < \beta_0 < \beta_1 < \beta_2 < \cdots < \beta_k < \cdots$ where $\beta_k \to +\infty$ as $k \to +\infty$, such that for each k, the equation

$$f'' + (z^4 - \beta_k)f = 0 \tag{5.4.2}$$

possesses a solution $f = f_k(z)$ that has an infinite number of real positive zeros, an infinite number of real negative zeros, and at most finitely many nonreal zeros. Any nonreal zero of f_k must lie on the imaginary axis, and f_k has exactly k zeros that lie on the imaginary axis. When k is even, f_k is an even function, and when k is odd, f_k is an odd function. It follows that all the zeros of f_0 and f_1 are real. Here $\deg P = 4$ in (5.4.1).

The above examples show that there exist such special solutions for certain polynomials $P(z)$ of degree $0, 1, 4$. On the other hand, such special solutions do not exist when $\deg P = 2 \pmod 4$.

Theorem 5.4.2 ([85, 131]). *Let f be a nontrivial solution of equation* (5.4.1), *where* $P(z)$ *satisfies* $\deg P = 2 + 4k$ *for some nonnegative integer k. Then either f has only finitely many zeros, or the exponent of convergence of the sequence of nonreal zeros of f is equal to* $2 + 2k$.

Example 5.4.3 ([260, 274]). It is well known and classical that the *Hermite function*

$$\psi_n(z) = H_n(z)\exp\left(-\frac{z^2}{2}\right), \quad n = 0, 1, 2, \ldots,$$

where, from Rodrigues' formula,

$$H_n(z) = (-1)^n \exp(z^2)\left(\frac{d}{dz}\right)^n \exp(-z^2)$$

is the *Hermite polynomial of degree n*, is a solution of the differential equation

$$f'' + (2n + 1 - z^2)f = 0. \tag{5.4.3}$$

Thus $\psi_n(z)$ has only a finite number of zeros, which are the zeros of $H_n(z)$. It is known that $H_n(z)$ has n distinct real zeros and no other zeros. On the other hand, if $f_n(z)$ is any solution of (5.4.3) that is linearly independent with $\psi_n(z)$, then it follows from Theorems 5.4.2 and 9.3.5 that the exponent of convergence of the nonreal zeros of $f_n(z)$ is equal to two.

Example 5.4.3 gives the solution of a singular Sturm–Liouville problem for $L^2(\mathbb{R})$, where $\{\psi_n(z)\}_{n=0}^{\infty}$ are the eigenfunctions, and $\{\lambda_n = 2n+1\}_{n=0}^{\infty}$ are the eigenvalues for the associated harmonic oscillator equation $f'' + (\lambda - z^2)f = 0$; see [274]. This example shows that a solution of an equation of the form (5.4.1) can have any finite number of real zeros and no other zeros. When a nontrivial solution of an equation of the form (5.4.1) possesses an infinite number of real zeros, then certain restrictions must hold.

Theorem 5.4.4 ([85]). *Suppose that f is a nontrivial solution of an equation of the form* (5.4.1) *such that either* (a) *f has an infinite number of real zeros or* (b) *f has k real zeros where k satisfies $k \geq 2 + \deg P$. Then $P(z)$ is a real polynomial, and $f \equiv Cg$, where g is a real entire function, and C is a constant.*

Theorem 5.4.4 is a immediate corollary of Theorem 5.4.5 and Lemma 5.4.8.

Theorem 5.4.5 ([85, Theorem 3]). *Let $P(z)$ be a nonreal polynomial of degree n, let $Q(z)$ be the polynomial defined by*

$$Q(z) = \frac{P(z) - \overline{P(\overline{z})}}{2i}, \tag{5.4.4}$$

and let q denote the number of distinct zeros of $Q(z)$. Then for any nontrivial solution f of (5.4.1), *the number of real zeros k of f is finite, and*

$$k \leq q + 1 \leq n + 1. \tag{5.4.5}$$

Proof. Suppose that x_1 and x_2 are two consecutive real zeros of f, where $x_1 < x_2$. Taking the imaginary parts of the Green's transform [144, p. 509] of equation (5.4.1), we obtain

$$\int_{x_1}^{x_2} \operatorname{Im}(P(x))|f(x)|^2\,dx = 0. \tag{5.4.6}$$

First, we observe that if it were true that $\operatorname{Im}(P(x)) \equiv 0$ on the segment $[x_1, x_2]$, then it could be deduced that $\operatorname{Im}(P(x)) \equiv 0$ on the whole real line, which would contradict the hypothesis that $P(z)$ is a nonreal polynomial. Hence $\operatorname{Im}(P(x)) \not\equiv 0$ on $[x_1, x_2]$. Since $f(x_1) = f(x_2) = 0$, it follows from (5.4.6) that $\operatorname{Im}(P(x))$ must change sign on the open interval (x_1, x_2). Since, by (5.4.4), $\operatorname{Im}(P(x)) = Q(x)$ for all $x \in \mathbb{R}$, we obtain that the polynomial $Q(z)$ has a real zero in (x_1, x_2). The double inequality (5.4.5) now follows. □

The following simple example shows that it is possible to have two equalities in the double inequality (5.4.5).

Example 5.4.6. If c is a constant satisfying $c^2 = -i$, then $f(z) = \sinh(cz)$ satisfies the equation

$$f'' + if = 0,$$

where f has exactly one real zero at $z = 0$, and $Q(z) = 1$ in (5.4.4). Hence, for this example, we have $k = 1$, $q = 0$, $n = 0$, and (5.4.5) becomes the double equality $1 = 1 = 1$.

The above proof of Theorem 5.4.5 can be used to prove the next general result.

Lemma 5.4.7 ([85, Lemma 1]). *Let $A(z)$ be a nonreal entire function, and set*

$$F(z) = \frac{A(z) - \overline{A(\overline{z})}}{2i}.$$

If f is a nontrivial solution of $f'' + A(z)f = 0$, then between every two consecutive real zeros of f, there must exist a real zero of F.

Lemma 5.4.8 ([85, Lemma 5]). *Let f be a solution of $f'' + A(z)f = 0$, where $A(z)$ is a real entire function. If f possesses at least one real zero, then $f \equiv Cg$, where g is a real entire function, and C is a constant.*

Proof. Since $A(z)$ is real, it follows that there exist linearly independent real entire solutions f_1 and f_2 of $f'' + A(z)f = 0$ such that $f(z) = c_1 f_1(z) + c_2 f_2(z)$, where c_1 and c_2 are constants. If $c_1 c_2 = 0$, then the statement is immediate. Hence we suppose that $c_1 c_2 \neq 0$. The hypothesis states that f has a real zero x_0, which means that $c_1 f_1(x_0) + c_2 f_2(x_0) = 0$. Since f_1 and f_2 are real, we obtain $\overline{c_1} f_1(x_0) + \overline{c_2} f_2(x_0) = 0$. It follows that $c_1 = bc_2$, where b is real. Then $f = c_2(bf_1 + f_2)$, which proves the statement. □

Rossi and Wang proved the following result.

Theorem 5.4.9 ([235]). *If an equation of the form* (5.4.1) *admits a solution that has an infinity of real zeros and no other zeros, then the number of nonreal zeros m of $P(z)$ counting multiplicities satisfies*

$$m < 1 + \frac{1}{2} \deg P. \tag{5.4.7}$$

For the functions f_0 and f_1 in Example 5.4.1, we have $m = 2$ and $\deg P = 4$, which gives $2 < 3$ in inequality (5.4.7). Also, for the equation $f'' + (z^3 - \beta_k)f = 0$ in Example 5.4.12, we have $m = 2$ and $\deg P = 3$, which gives $2 < 5/2$ in (5.4.7).

On the other hand, for the equation $f'' + (z^3 + z)^3 f = 0$, we have $m = 6$ and $\deg P = 9$, which means that this equation cannot possess a solution f that has an infinity of real zeros and no other zeros (from Theorem 5.4.9).

Hellerstein and Rossi posed the following question.

Question 5.4.10 ([112, Problem 2.71]). Can all the polynomials $P(z)$ be characterized such that equation (5.4.1) possesses a solution with an infinite number of real zeros and no other zeros?

The example in [88] extended Example 5.4.1, and the next example of Shin extended the example in [88].

Example 5.4.11 ([258]). If $a > 0$ and $b \in \mathbb{R}$, then there exists an infinite sequence of distinct real constants β_k satisfying $\beta_0 < \beta_1 < \beta_2 < \cdots < \beta_k < \cdots$, where $\beta_k \to \infty$ as $k \to \infty$, such that for each k, the equation

$$f'' + (az^4 + bz^2 - \beta_k)f = 0$$

possesses a solution $f = f_k(z)$ that has an infinite number of real zeros and at most finitely many nonreal zeros.

Shin found the next example.

Example 5.4.12 ([258]). Let a, b, c be real numbers satisfying $a \neq 0$ and $ac \leq 0$. Then there exists an infinite sequence of positive numbers β_k where $\beta_k \to \infty$ as $k \to \infty$, such that for each k, the equation

$$f'' + (az^3 + bz^2 + cz - \beta_k)f = 0$$

possesses a solution $f = f_k(z)$ that has infinitely many real zeros and at most finitely many nonreal zeros. Here $\deg P = 3$ in (5.4.1).

Up until 2002, the above examples and Theorem 5.4.2 showed that it was known that these special solutions of (5.4.1) exist for certain polynomials $P(z)$ satisfying $\deg P = 0, 1, 3, 4$, and that there cannot exist such special solutions when $\deg P = 2 \pmod 4$. In 2005, Eremenko and Merenkov proved the following two theorems, which showed that such special solutions exist for all the remaining degrees of $P(z)$.

Theorem 5.4.13 ([67]). *Let $d = 4k$ for some nonnegative integer k. Then there exists an equation of the form* (5.4.1), *where* $\deg P = d$, *which possesses a solution that has an infinite number of real zeros and no other zeros, such that the zeros are unbounded from above and below (as a subset of the real axis).*

Theorem 5.4.14 ([67]). *Let d be an odd positive integer. Then there exists an equation of the form* (5.4.1), *where* $\deg P = d$, *which possesses a solution that has an infinite number of real zeros and no other zeros, such that the zeros all lie on one ray.*

As pointed out in [67], Theorems 5.4.13 and 5.4.14 give the strong condition that all the zeros of the solution are real, but on the other hand, less can be seen about the form of the polynomial $P(z)$.

Problems

Problem 5.4.15. Show that the equation

$$f'' + \left(z^6 + z^5 + z^4 - z^2 - z - 1\right)f = 0$$

cannot possess a solution f that has an infinite number of real zeros and no other zeros.

Problem 5.4.16. Suppose that f is an entire solution of $f'' + A(z)f = 0$, where $A(z)$ is a real entire function. If f is an odd function and has no zeros in the first quadrant $\{z = re^{i\theta} : r > 0,\ 0 < \theta < \pi/2\}$, then show that all the zeros of f must lie on the real or imaginary axis.

Problem 5.4.17. Let $B(z)$ and $C(z)$ be entire functions, and let $D(z)$ be a primitive function of $-B(z)/2$, that is, $D'(z) = -B(z)/2$. Show that $g = fe^D$ transforms the equation

$$g'' + B(z)g' + C(z)g = 0$$

into an equation of the form $f'' + A(z)f = 0$. Find a representation for $A(z)$ in terms of $B(z)$ and $C(z)$.

5.5 Canonical products with real zeros

Lindelöf's functions serve as a nice starting point for the topic of this section, because they are canonical products of any order $\rho \in (0, \infty)$ with straightforward negative real zeros and no other zeros.

Example 5.5.1. For $\rho \in (0, \infty)$, the *Lindelöf function* $L_\rho(z)$ is the canonical product with simple zeros precisely at the points $z_n = -n^{1/\rho}$, $n \in \mathbb{N}$, on the negative real axis. It is defined by

$$L_\rho(z) = \prod_{n=1}^{\infty} E_p\left(-\frac{z}{n^a}\right),$$

where $a = 1/\rho$ and $p = \lfloor \rho \rfloor$ (= integer part of ρ). Indeed, since

$$\sum_{n=1}^{\infty} \frac{1}{|z_n|^{p+1}} < \infty \quad \text{and} \quad \sum_{n=1}^{\infty} \frac{1}{|z_n|^{p}} = \infty,$$

$L_\rho(z)$ is entire by Hadamard's theorem, and p is its genus. Moreover,

$$\sum_{n=1}^{\infty} \frac{1}{|z_n|^{\rho+\varepsilon}} < \infty \quad \text{and} \quad \sum_{n=1}^{\infty} \frac{1}{|z_n|^{\rho-\varepsilon}} = \infty$$

for every $\varepsilon > 0$, so that the exponent of convergence of the zeros z_n of $L_\rho(z)$ is ρ. Thus the order of growth of $L_\rho(z)$ is ρ by Theorem 3.5.3.

For a real differentiable function of a real variable, Rolle's theorem states that the derivative vanishes at least once between any two consecutive distinct zeros of the function. This theorem of course applies to entire functions that are real on the real axis, i. e., *real entire functions*. (For example, $\sin z$ and $\cos z$ are real entire functions.)

If f is a polynomial with real zeros only, then Rolle's theorem implies that the same is true for f'. The example in (4.3.9) shows that this conclusion does not hold for arbitrary entire functions. Next, we show that this conclusion does hold for canonical products with real zeros; see [139].

Theorem 5.5.2. *Let f be a canonical product of finite genus $p \geq 0$ with real zeros. Then the following statements hold:*

(a) *The zeros of f' are real. The point $z = 0$ is a zero of f' of multiplicity p or $p + 1$. If this multiplicity is $p + 1$, then p must be even.*

(b) *If r_1 and r_2 (s_1 and s_2) are consecutive distinct positive (negative) zeros of f, then there exists one and only one zero μ of f' satisfying $0 < r_1 < \mu < r_2$ ($s_1 < \mu < s_2 < 0$), and μ is a simple zero of f'.*

Proof. Let $\{x_n\}$ denote the sequence of (nonzero) real zeros of $f(z)$ listed according to multiplicities and organized with respect to increasing absolute value. Then

$$f(z) = \prod_{n=1}^{\infty} E_p\left(\frac{z}{x_n}\right).$$

By Problem 5.5.7,

$$F(z) := \frac{f'(z)}{z^p f(z)} = \sum_{n=1}^{\infty} \frac{1}{x_n^p (z - x_n)}. \tag{5.5.1}$$

Suppose to the contrary of the first sentence in (a) that $f'(c) = 0$ at some $c = a + ib$, where $a, b \in \mathbb{R}$ with $b \neq 0$. Then $c \neq 0$ and $f(c) \neq 0$. Hence by (5.5.1), $F(c) = 0$. Noting that $x_n \in \mathbb{R}$, we may write

$$\begin{aligned} F(c) &= \sum_{n=1}^{\infty} \frac{\overline{c} - x_n}{x_n^p (c - x_n)(\overline{c} - x_n)} = \sum_{n=1}^{\infty} \frac{\overline{c} - x_n}{x_n^p |c - x_n|^2} \\ &= \sum_{n=1}^{\infty} \frac{a - x_n}{x_n^p |c - x_n|^2} - i \sum_{n=1}^{\infty} \frac{b}{x_n^p |c - x_n|^2} = 0. \end{aligned} \tag{5.5.2}$$

The real and imaginary parts of $F(c)$ must both vanish. However, if p is even, then we see that the imaginary part could not vanish, a contradiction. Suppose next that p is odd. Then

$$\sum_{n=1}^{\infty} \frac{a - x_n}{x_n^p |c - x_n|^2} = 0 = \sum_{n=1}^{\infty} \frac{1}{x_n^p |c - x_n|^2}$$

shows that

$$0 = a \cdot 0 = a \sum_{n=1}^{\infty} \frac{1}{x_n^p |c - x_n|^2} = \sum_{n=1}^{\infty} \frac{1}{x_n^{p-1} |c - x_n|^2},$$

which is a contradiction because $p-1$ is even. This shows that the zeros of f' are real, which proves the first sentence in (a).

To prove the second part of (a), we now consider the point $z = 0$. Since the right-hand side of (5.5.1) does not have a pole at the origin, we see at once that f' must have a zero of multiplicity at least p at the origin.

From (5.5.1),

$$F'(x) = -\sum_{n=1}^{\infty} \frac{1}{x_n^p (x - x_n)^2}, \qquad x \in \mathbb{R},\ x \neq x_n, \tag{5.5.3}$$

and

$$F(0) = -\sum_{n=1}^{\infty} \frac{1}{x_n^{p+1}} \quad \text{and} \quad F'(0) = -\sum_{n=1}^{\infty} \frac{1}{x_n^{p+2}}. \tag{5.5.4}$$

If p is odd, then $F(0) < 0$ because $p+1$ is even, which implies that $z = 0$ is a zero of exact multiplicity p of f'. Now suppose that p is even. Then $F'(0) < 0$ because $p + 2$ is even. Thus $z = 0$ is a zero of f' of multiplicity p or $p + 1$. This proves the second part of (a), which completes the proof of (a).

We now proceed to prove part (b). Let r_1 and r_2 be consecutive distinct positive zeros of f. Since f is a real entire function, we can conclude by Rolle's theorem that f' has at least one zero that lies strictly between r_1 and r_2. We prove part (b) in two steps.

(1) First, we will show that there exists exactly one distinct positive zero μ of f' that lies in the interval (r_1, r_2).

Suppose that p is even. Then from (5.5.3) we see that

$$F'(x) < 0 \quad \text{for } r_1 < x < r_2.$$

Therefore, $F(x)$ strictly decreases as x increases between r_1 and r_2. By (5.5.1),

$$F(x) = \frac{f'(x)}{x^p f(x)} = \sum_{n=1}^{\infty} \frac{1}{x_n^p (x - x_n)}, \qquad x \in \mathbb{R},\ x \neq x_n, \tag{5.5.5}$$

from which we can conclude that $F(x)$ strictly decreases from $+\infty$ to $-\infty$ as x increases between r_1 and r_2. From the intermediate value theorem for continuous functions, it follows that $F(x)$ has exactly one zero in the interval (r_1, r_2). Since $x^p f(x) \neq 0$ when $x \in (r_1, r_2)$, we obtain from (5.5.5) that f' has exactly one distinct zero in (r_1, r_2).

Now suppose that p is odd. We assume that $f'(x)$ has two distinct positive zeros α and β satisfying $\alpha, \beta \in (r_1, r_2)$ and aim for a contradiction. Now

$$F(\alpha) + F(\beta) = \sum_{n=1}^{\infty} \frac{\alpha + \beta}{x_n^p (\alpha - x_n)(\beta - x_n)} - \sum_{n=1}^{\infty} \frac{2}{x_n^{p-1} (\alpha - x_n)(\beta - x_n)} = 0,$$

$$F(\alpha) - F(\beta) = \sum_{n=1}^{\infty} \frac{\beta - \alpha}{x_n^p (\alpha - x_n)(\beta - x_n)} = 0.$$

From the latter equation, since $\alpha \neq \beta$, we get

$$\sum_{n=1}^{\infty} \frac{1}{x_n^p(\alpha - x_n)(\beta - x_n)} = 0.$$

The former equation then gives

$$\sum_{n=1}^{\infty} \frac{2}{x_n^{p-1}(\alpha - x_n)(\beta - x_n)} = 0.$$

This is a contradiction since $p-1$ is even and $(\alpha - x_n)(\beta - x_n) > 0$ for all $n \in \mathbb{N}$. Therefore we have shown that there exists exactly one distinct zero μ of f' that lies in (r_1, r_2).

(2) Finally, we will show that the one distinct zero μ of f' that lies in (r_1, r_2) must be a simple zero. Suppose on the contrary that there exists a multiple positive zero ζ of f' that lies in (r_1, r_2). Then $f'(\zeta) = f''(\zeta) = 0$ and $f(\zeta) \neq 0$. Since

$$F'(z) = \frac{f''(z)}{z^p f(z)} - \frac{f'(z)(f'(z) + pf(z)/z)}{z^p f(z)^2},$$

we have $F'(\zeta) = 0$. From (5.5.3) with $x = \zeta \in \mathbb{R}$, we obtain

$$\sum_{n=1}^{\infty} \frac{1}{x_n^p(\zeta - x_n)^2} = 0. \tag{5.5.6}$$

If p is even, then this gives an immediate contradiction. Suppose that p is odd. Now

$$0 = F(\zeta) + F'(\zeta) = \sum_{n=1}^{\infty} \frac{\zeta - 1}{x_n^p(\zeta - x_n)^2} - \sum_{n=1}^{\infty} \frac{1}{x_n^{p-1}(\zeta - x_n)^2}.$$

By (5.5.6) it follows that

$$\sum_{n=1}^{\infty} \frac{1}{x_n^{p-1}(\zeta - x_n)^2} = 0,$$

which is again a contradiction since $p - 1$ is even. These contradictions prove that the one distinct zero μ of f' in (r_1, r_2) must be a simple zero.

Steps (1) and (2) prove part (b) for consecutive distinct positive zeros r_1 and r_2 of f. The proof of part (b) for consecutive distinct negative zeros s_1 and s_2 of f is similar. □

Example 5.5.3. We will discuss the second part of Theorem 5.5.2(a). Let $\{x_n\}_{n=1}^{\infty}$ be the sequence of real numbers defined by

$$x_n = \begin{cases} \sqrt{k}, & n = 2k-1, \\ -\sqrt{k}, & n = 2k. \end{cases}$$

Then the canonical product

$$f(z) = \prod_{n=1}^{\infty}\left(1 - \frac{z}{x_n}\right)\exp\left(\frac{z}{x_n} + \frac{z^2}{2x_n^2}\right)$$

represents an entire function and has genus $p = 2$. By Problem 5.5.7,

$$F(z) := \frac{f'(z)}{z^2 f(z)} = \sum_{n=1}^{\infty} \frac{1}{x_n^2(z - x_n)}.$$

By the definition of the sequence $\{x_n\}$,

$$F(0) = -\sum_{n=1}^{\infty} \frac{1}{x_n^3} = 0,$$

from which it follows that the point $z = 0$ is a zero of f' of multiplicity $p + 1 = 3$. For $z = 0$ to be a zero of f of multiplicity $p + 1$, the condition $F(0) = 0$ must hold, and p must be even. However, it is clear that for most canonical products f of genus $p \geq 0$, we have $F(0) \neq 0$. This is particularly true if the genus p is odd. Hence both cases can happen in the second part of Theorem 5.5.2(a).

We next discuss a useful classical result known as *Laguerre's theorem*.

Theorem 5.5.4 ([37, 273]). *Let f be a nonconstant entire function of order $\rho(f) < 2$, which is real on the real axis and has only real zeros. Then the zeros of f' are also real and have the following interlacing property: if x_1 and x_2 are any two consecutive distinct real zeros of f, then there exists one and only one zero μ of f' satisfying $x_1 < \mu < x_2$, and μ is a simple zero of f'.*

Proof. First, assume that f has an infinite number of zeros such that the canonical product of the nonzero zeros has genus $p = 1$. By the hypothesis,

$$f(z) = Cz^k e^{az} \prod_{n=1}^{\infty}\left(1 - \frac{z}{b_n}\right)\exp\left(\frac{z}{b_n}\right),$$

where k is a nonnegative integer, $C, a \in \mathbb{R}$, and $\{b_n\}$ is the sequence of nonzero zeros of f, which are real numbers. Taking the logarithmic derivative of both sides, we get

$$\frac{f'(z)}{f(z)} = \frac{k}{z} + a + \sum_{n=1}^{\infty}\left(\frac{1}{z - b_n} + \frac{1}{b_n}\right). \tag{5.5.7}$$

By setting $z = x + iy$, where $x, y \in \mathbb{R}$, we obtain

$$\operatorname{Im}\left(\frac{f'(z)}{f(z)}\right) = -y\left(\frac{k}{x^2 + y^2} + \sum_{n=1}^{\infty} \frac{1}{(x - b_n)^2 + y^2}\right), \tag{5.5.8}$$

which cannot be zero when $y \neq 0$. Therefore $f'(z) \neq 0$ whenever $z \notin \mathbb{R}$. Hence the zeros of f' are real (if there are any).

Next, differentiate (5.5.7) to get

$$\frac{d}{dz}\left(\frac{f'(z)}{f(z)}\right) = -\frac{k}{z^2} - \sum_{n=1}^{\infty} \frac{1}{(z-b_n)^2}, \tag{5.5.9}$$

which is real and negative whenever z is real and $f(z) \neq 0$. It follows that $f'(z)/f(z)$ is strictly decreasing in any interval that does not contain a zero of f.

Suppose that b_m and b_{m+1} are consecutive distinct real zeros of f. Then from (5.5.9) we see that f'/f strictly decreases from $+\infty$ to $-\infty$ in the interval (b_m, b_{m+1}) and thus assumes zero exactly once at some point μ satisfying $b_m < \mu < b_{m+1}$, where μ is a zero of f'. By (5.5.9), μ can be the only zero of f' that lies in (b_m, b_{m+1}), and μ must be a simple zero of f'. This proves the theorem when f has an infinite number of zeros and $p = 1$.

Assume then that f has an infinite number of zeros such that the canonical product of the nonzero zeros has genus $p = 0$. Then f is of the form

$$f(z) = Cz^k e^{az} \prod_{n=1}^{\infty}\left(1 - \frac{z}{b_n}\right),$$

and hence (5.5.7) takes the form

$$\frac{f'(z)}{f(z)} = \frac{k}{z} + a + \sum_{n=1}^{\infty}\left(\frac{1}{z-b_n}\right).$$

The expression $\operatorname{Im}(\frac{f'(z)}{f(z)})$ is the same as in (5.5.8), so the rest of the proof remains the same as in the case $p = 1$. The case where f has only finitely many zeros follows the same line of reasoning. The proof is now complete. □

Example 5.5.5. Obviously, $\sin z$ and $\cos z$ are examples of Laguerre's theorem. There exist some examples of Laguerre's theorem, where in addition to the interlacing property, f and f' have zeros at the same points. For instance, consider $\sin^k z$ and $\cos^k z$ for $k \geq 2$.

Example 5.5.6. The condition that $\rho(f) < 2$ in Theorem 5.5.4 is sharp. Indeed, the theorem is not true for entire functions of order two, since [273, p. 267] for the example

$$f(z) = ze^{z^2}, \quad f'(z) = (2z^2+1)e^{z^2},$$

we have that f' has complex zeros, whereas for the example

$$g(z) = (z^2-4)e^{z^2/3}, \quad g'(z) = \frac{2}{3}z(z^2-1)e^{z^2/3},$$

$g'(z)$ has three real zeros between consecutive real zeros of g.

Titchmarsh [273, pp. 267–268] used Laguerre's theorem to show that the differential equation

$$f(z)f''(z) = -\sin^2 z$$

cannot possess a real finite-order entire solution other than $f(z) = \pm \sin z$. In 2019, Li, Lü, and Yang used Laguerre's theorem to improve and generalize this result; see Theorem 10.2.1.

Problems

Problem 5.5.7. Let f be a canonical product of finite genus $p \geq 0$ having zeros z_n listed according to multiplicities and organized with respect to increasing moduli. Prove that

$$\frac{f'(z)}{f(z)} = \sum_{n=1}^{\infty} \frac{z^p}{z_n^p (z - z_n)}.$$

Problem 5.5.8. Find all possible constants $A, B \in \mathbb{C}$ such that

$$f(z) = A \sin z + B \cos z$$

solves the differential equation $2f(z)f'(z) = \sin(2z)$.

Problem 5.5.9. Find all possible constants $A, B \in \mathbb{C}$ such that

$$f(z) = A \sin z + B \cos z$$

solves the differential equation $f(z)^2 - f'(z)^2 = \cos(2z)$.

5.6 Laguerre–Pólya class

The *Laguerre–Pólya class* of functions, which we denote by $\mathcal{LP}$, consists of entire functions f of the form

$$f(z) = z^m e^{a+bz-cz^2} \prod_{n=1}^{\infty} \left(1 - \frac{z}{z_n}\right) \exp\left(\frac{z}{z_n}\right) = z^m e^{a+bz-cz^2} P(z), \tag{5.6.1}$$

where the nonzero zeros $z_n \in \mathbb{R}$ of f satisfy $\sum_{n=1}^{\infty} 1/z_n^2 < \infty$, $m \in \mathbb{N} \cup \{0\}$, $a \in \mathbb{C}$, $b \in \mathbb{R}$, and $c \geq 0$. Thus the functions in the $\mathcal{LP}$-class are of order ≤ 2 and are constant multiples of real entire functions with all real zeros. We allow the product in (5.6.1) to have only finitely many factors or even to reduce to the constant 1. To include these cases without additional notation, we agree that from a certain index on, we could have $z_n = \infty$.

Known examples of functions belonging to $\mathcal{LP}$:

$$1, \quad 1 - z, \quad e^z, \quad e^{-z^2}, \quad \sin(\pi z), \quad 1/\Gamma(z).$$

The function e^{z^2} does not belong to $\mathcal{LP}$. Also note that the product of two functions in $\mathcal{LP}$ is again in $\mathcal{LP}$.

The class $\mathcal{LP}$ was introduced by Edmond Laguerre (1834–1886) and George Pólya (1887–1985). Laguerre was a French mathematician known for work in geometry, complex analysis, and orthogonal polynomials, and Pólya was a Hungarian mathematician known for work in combinatorics, number theory, mathematical analysis, and geometry. Pólya said:

> It is better to prove one problem five different ways than to prove five different problems one way.

In 1914, Pólya [226, 227] posed the question of classifying the entire functions that together with all their derivatives, have all real zeros. He conjectured that such a function f must either belong to the $\mathcal{LP}$-class or have one of the forms $f(z) = Ae^{Bz}$ or $f(z) = A(e^{icz} - e^{id})$, where $A, B \in \mathbb{C}$ and $c, d \in \mathbb{R}$. In a series of papers, Hellerstein, Shen, and Williamson settled Pólya's question and conjecture.

Theorem 5.6.1 ([134–136]). *Suppose that f is an entire function such that f, f', f'' have only real zeros. Then f has one of the following forms:*

$$f \in \mathcal{LP}, \tag{5.6.2}$$
$$f(z) = Ae^{Bz}, \tag{5.6.3}$$
$$f(z) = A(e^{icz} - e^{id}), \tag{5.6.4}$$
$$f(z) = A\exp(e^{i(cz+d)}), \tag{5.6.5}$$
$$f(z) = A\exp(K(i(cz+d) - e^{i(cz+d)})), \tag{5.6.6}$$

where A, B, K, c, d are constants, $Ac \neq 0$, B is nonreal, and $K, c, d \in \mathbb{R}$ with $K \leq -1/4$.

Observe that (5.6.3)–(5.6.6) are strictly nonreal entire functions, that is, they are not a constant multiple of a real entire function.

When the third derivative is considered, we have the following strong affirmation of Pólya's conjecture.

Theorem 5.6.2 ([134–136]). *If f is an entire function such that f, f', f'', f''' have only real zeros, then f has one of the forms* (5.6.2), (5.6.3), (5.6.4).

Levin and Ostrovskii showed that entire functions f where ff'' has all real zeros cannot grow too fast.

Theorem 5.6.3 ([184]). *If f is an entire function such that all the zeros of ff'' are real, then*

$$\log^+ \log^+ |f(z)| = O(|z| \log |z|), \quad z \to \infty.$$

We now define a collection of classes of entire functions with real zeros. For every integer $p \geq 0$, let V_{2p} denote the set of entire functions of the form

$$f(z) = e^{-cz^2} g(z),$$

where $c \geq 0$ is a constant, and g is the constant multiple of a real entire function with only real zeros of genus at most $2p+1$. Set $U_0 = V_0$ and $U_{2p} = V_{2p} \setminus V_{2p-2}$ for $p \geq 1$. Observe that $U_0 = \mathcal{LP}$. If f is an entire function of finite order with all real zeros, then $f \in U_{2p}$ for some unique p.

Sheil-Small proved the following result, which confirmed a conjecture of Wiman.

Theorem 5.6.4 ([252]). *If $f \in U_{2p}$, then f'' has at least $2p$ nonreal zeros.*

For more results in the spirit of Theorem 5.6.4, see Edwards and Hellerstein [64]. For infinite-order functions, Bergweiler, Eremenko, and Langley proved the following result.

Theorem 5.6.5 ([30]). *For every real entire function f of infinite order, ff'' has infinitely many nonreal zeros.*

It was noted in [30] that Theorems 5.6.4 and 5.6.5 yield the following result.

Theorem 5.6.6. *If f is a real entire function and ff'' has only real zeros, then $f \in \mathcal{LP}$.*

Some observations in [30] are as follows. The function $f(z) = \exp(\sin z)$ shows that the condition for ff'' in Theorem 5.6.6 cannot be replaced by the same condition for ff'. Also in Theorem 5.6.6, the condition that f is real is necessary; consider $f(z) = \exp(e^{iz})$, which is an example of Edrei [62]. Theorem 5.6.6 confirmed a conjecture, which goes back to Wiman (1911).

For the history of Pólya's question and conjecture, as well as for more about the collection $U_0, U_2, U_4, \ldots$, including the contributions of others and related results, see the above papers and references therein.

An important feature of the $\mathcal{LP}$ class is the fact that polynomials with only real zeros converge to these functions uniformly on compact subsets of $\mathbb{C}$ and only to these functions. This result is known as the *Laguerre–Pólya theorem.*

Theorem 5.6.7 ([141, pp. 42–43]). *The following two statements hold:*

(a) *If $f \in \mathcal{LP}$, then there exists a sequence of polynomials $\{P_n\}$, each with real zeros only, such that*

$$\lim_{n\to\infty} P_n(z) = f(z) \tag{5.6.7}$$

uniformly on compact subsets of $\mathbb{C}$.

(b) *Conversely, suppose that for any $n \in \mathbb{N}$, the points $z_k = z_k(n)$, $k = 1, \ldots, n$, are real and uniformly bounded away from zero (where $z_k = \infty$ is allowed), such that $\sum_{k=1}^{\infty} 1/z_k^2 < \infty$. If*

$$P_n(z) := \prod_{k=1}^{n} \left(1 - \frac{z}{z_k}\right) \tag{5.6.8}$$

and (5.6.7) holds uniformly on compact subsets of $\mathbb{C}$, then $f \in \mathcal{LP}$.

The proof of Theorem 5.6.7 is lengthy, where part (b) takes longer to prove than part (a). We need an elementary technical lemma.

Lemma 5.6.8 ([141, p. 42]). *Let $R > 0$. If $|z| \le R$ and if n is an integer satisfying $n \ge 2R$, then*

$$\left|z + n\log\left(1 - \frac{z}{n}\right)\right| \le \frac{|z|^2}{n} \le \frac{R^2}{n}.$$

Here the logarithm is the principal branch for which the negative real axis is the branch cut and for which $\log 1 = 0$.

Proof. Since $\operatorname{Re}(1 - \frac{z}{n}) \ge 1 - \frac{|z|}{n} \ge \frac{1}{2}$, it follows that the logarithmic term in the statement is well-defined. By using the Taylor series expansion we can see that

$$\begin{aligned}\left|z + n\log\left(1 - \frac{z}{n}\right)\right| &\le n\left(\frac{1}{2}\left|\frac{z}{n}\right|^2 + \frac{1}{3}\left|\frac{z}{n}\right|^3 + \frac{1}{4}\left|\frac{z}{n}\right|^4 + \cdots\right) \\ &\le \frac{n}{2}\left|\frac{z}{n}\right|^2\left(1 + \left|\frac{z}{n}\right| + \left|\frac{z}{n}\right|^2 + \cdots\right) \\ &= \frac{n}{2}\left|\frac{z}{n}\right|^2 \frac{1}{1 - |z/n|} \le \frac{|z|^2}{n} \le \frac{R^2}{n},\end{aligned}$$

proving the statement. □

Proof of Theorem 5.6.7(a). We suppose that $f \in \mathcal{LP}$. Then f has the representation (5.6.1). Let $R > 0$ be a large but fixed constant.

Consider the function e^{bz} in (5.6.1) with $b \in \mathbb{R} \setminus \{0\}$. Replacing z with $-bz$ in Lemma 5.6.8, we obtain that for $|z| \le R/|b|$ and $n \ge 2R/|b|$,

$$\left|-bz + n\log\left(1 + \frac{bz}{n}\right)\right| \le \frac{|bz|^2}{n} \le \frac{R^2}{n}.$$

Using

$$|e^w - 1| = \left|\sum_{n=1}^{\infty} \frac{w^n}{n!}\right| \le \sum_{n=1}^{\infty} \frac{|w|^n}{n!} = e^{|w|} - 1, \tag{5.6.9}$$

we deduce

$$\begin{aligned}\left|\left(1 + \frac{bz}{n}\right)^n e^{-bz} - 1\right| &= \left|\exp\left(-bz + n\log\left(1 + \frac{bz}{n}\right)\right) - 1\right| \\ &\le e^{R^2/n} - 1, \quad n \ge 2R/|b|,\end{aligned}$$

from which

$$\lim_{n\to\infty}\left(1+\frac{bz}{n}\right)^n = e^{bz}$$

uniformly on $|z| \le R/|b|$. This shows that e^{bz} with $b \in \mathbb{R}\setminus\{0\}$ can be uniformly approximated on compact subsets of $\mathbb{C}$ by polynomials $(1+\frac{bz}{n})^n$ with real zeros only. As for e^{-cz^2} in (5.6.1) with $c > 0$, replacing $-bz$ with cz^2 in the reasoning above, we obtain

$$\left|cz^2 + n\log\left(1-\frac{cz^2}{n}\right)\right| \le \frac{|cz^2|^2}{n} \le \frac{R^2}{n}, \quad |z| \le \sqrt{R/c},\ n \ge 2\sqrt{\frac{R}{c}},$$

and, consequently,

$$\lim_{n\to\infty}\left(1-\frac{cz^2}{n}\right)^n = e^{-cz^2}$$

uniformly on $|z| \le \sqrt{R/c}$.

By the assumption $\sum_{n=1}^{\infty} 1/z_n^2 < \infty$ the product $P(z)$ in (5.6.1) converges uniformly on compact subsets of $\mathbb{C}$. By definition this means that the sequence $\{T_n\}$ of partial products of the form

$$T_n(z) = \prod_{k=1}^{n}\left(1-\frac{z}{z_k}\right)\exp\left(\frac{z}{z_k}\right)$$

converges uniformly on compact subsets of $\mathbb{C}$ to $P(z)$. Denote $d_n = \sum_{k=1}^{n}\frac{1}{z_k}$. Then $d_n \in \mathbb{R}$, and

$$T_n(z) = e^{d_n z}\prod_{k=1}^{n}\left(1-\frac{z}{z_k}\right). \tag{5.6.10}$$

In general, the sequence $\{d_n\}$ may diverge. Hence, despite the fact that $\{T_n(z)\}$ converges, the exponential factor $e^{d_n z}$ and the finite product in (5.6.10) can both diverge as $n \to \infty$ for every $z \ne 0$. To resolve this problem, for $n \in \mathbb{N}$, let p_n be the smallest positive integer satisfying

$$p_n \ge \max\{2R|d_n|,\ nR|d_n|^2\},$$

and define a sequence of polynomials $\{S_n\}$ by

$$S_n(z) = \left(1+\frac{d_n z}{p_n}\right)^{p_n}\prod_{k=1}^{n}\left(1-\frac{z}{z_k}\right).$$

Note that the zeros of each polynomial $S_n(z)$ are real. We will prove that the sequence $\{S_n(z)\}$ converges uniformly on compact subsets of $\mathbb{C}$ to $P(z)$. Since

$$\left|S_n(z) - P(z)\right| \le \left|T_n(z) - P(z)\right| + \left|S_n(z) - T_n(z)\right|,$$

it suffices to prove that $|S_n(z)-T_n(z)| \to 0$ as $n \to \infty$ uniformly on compact subsets of $\mathbb{C}$. To see this, we find that

$$\mathrm{Re}\left(1+\frac{d_n z}{p_n}\right) \geq 1-\frac{|d_n z|}{p_n} \geq 1-\frac{|d_n|R}{p_n} \geq \frac{1}{2}, \quad |z| \leq R,$$

and hence, by (5.6.9) and Lemma 5.6.8,

$$\begin{aligned}
|S_n(z)-T_n(z)| &= |T_n(z)| \cdot \left|\left(1+\frac{d_n z}{p_n}\right)^{p_n} e^{-d_n z} - 1\right| \\
&= |T_n(z)| \cdot \left|\exp\left(-d_n z + p_n \log\left(1+\frac{d_n z}{p_n}\right)\right) - 1\right| \\
&\leq |T_n(z)| \cdot \left(\exp\left(\left|-d_n z + p_n \log\left(1+\frac{d_n z}{p_n}\right)\right|\right) - 1\right) \\
&= |T_n(z)| \cdot \left(\exp\left(\left|-d_n z + p_n \log\left(1-\frac{(-d_n z)}{p_n}\right)\right|\right) - 1\right) \\
&\leq |T_n(z)| \cdot \left(\exp\left(\frac{|d_n|^2 R^2}{p_n}\right) - 1\right) \\
&\leq |T_n(z)| \cdot \left(\exp\left(\frac{R}{n}\right) - 1\right), \quad |z| \leq R.
\end{aligned}$$

Using

$$|T_n(z)| \leq |T_n(z) - P(z)| + |P(z)|$$

together with the facts that (i) $T_n(z) \to P(z)$ uniformly in $|z| \leq R$ and (ii) $|P(z)|$ is bounded in $|z| \leq R$, we find that $|T_n(z)|$ is uniformly bounded on $|z| \leq R$. Thus $|S_n(z) - T_n(z)| \to 0$ as $n \to \infty$ uniformly in $|z| \leq R$.

Finally, the polynomials P_n in the statement are of the form

$$P_n(z) = z^m e^a \left(1+\frac{bz}{n}\right)^n \left(1-\frac{cz^2}{n}\right)^n S_n(z),$$

and the proof of Theorem 5.6.7(a) is now complete. □

The proof of Theorem 5.6.7(b) is more involved. Further, observe that the notation $z_k = z_k(n)$ means that each polynomial P_n could have a unique set of zeros.

Proof of Theorem 5.6.7(b). We prove the statement in the converse direction, supposing that $\{P_n\}$ is the sequence of polynomials defined in (5.6.8) such that (5.6.7) holds. We aim to prove that $f \in \mathcal{LP}$.

Taking into account our convention on the points z_k, it could be that $z_k = \infty$ for all k large enough. Hence the polynomials P_n may have degree $= n$ or degree $< n$, or they may even reduce to the constant function 1. Therefore we suppose that $z_k \neq \infty$ for all k. Consequently, we consider a sequence $\{P_n(z)\}$ of polynomials with real zeros that converges uniformly to f on compact subsets of $\mathbb{C}$ and satisfies $P_n(0) = 1$.

By Weierstrass' theorem (Theorem 1.6.6), the function f is analytic at the origin, and

$$\lim_{n\to\infty} P_n'(0) = f'(0) \quad \text{and} \quad \lim_{n\to\infty} P_n''(0) = f''(0), \tag{5.6.11}$$

where

$$P_n'(0) = -\sum_{k=1}^{n} \frac{1}{z_k} \quad \text{and} \quad P_n''(0) = \sum_{k=1}^{n} \frac{1}{z_k} \sum_{\substack{\ell=1\\ \ell\neq k}}^{n} \frac{1}{z_\ell}.$$

Define

$$p_n = \sum_{k=1}^{n} \frac{1}{z_k(n)} \quad \text{and} \quad q_n = \sum_{k=1}^{n} \frac{1}{z_k(n)^2}.$$

Clearly, $p_n = -P_n'(0)$, and by using the calculation rule

$$\begin{aligned}\left(\sum_{k=1}^{n} a_k\right)^2 &= \left(\sum_{k=1}^{n} a_k\right)\left(\sum_{k=1}^{n} a_k\right) = \sum_{k=1}^{n}\sum_{\ell=1}^{n} a_k a_\ell \\ &= \sum_{k=1}^{n} a_k^2 + \sum_{k=1}^{n} \sum_{\substack{\ell=1\\ \ell\neq k}}^{n} a_k a_\ell, \quad a_k \in \mathbb{C},\end{aligned}$$

we obtain

$$P_n'(0)^2 - P_n''(0) = \left(\sum_{k=1}^{n} \frac{1}{z_k}\right)^2 - \sum_{k=1}^{n} \frac{1}{z_k} \sum_{\substack{\ell=1\\ \ell\neq k}}^{n} \frac{1}{z_\ell} = \sum_{k=1}^{n} \frac{1}{z_k^2} = q_n.$$

Thus the corresponding equations in (5.6.11) are equivalent to

$$\lim_{n\to\infty} p_n = -f'(0) =: p \quad \text{and} \quad \lim_{n\to\infty} q_n = f'(0)^2 - f''(0) =: q.$$

Since the sequences $\{p_n\}$ and $\{q_n\}$ are convergent, they are bounded as well, and so there exists $M > 0$ such that

$$|p_n| < M \quad \text{and} \quad 0 \le q_n < M, \quad n \in \mathbb{N}. \tag{5.6.12}$$

We need a number of inequalities. First, by Lemma 5.6.8 with $n = 1$ we have

$$\left|z + \log(1-z)\right| \le |z|^2, \quad |z| \le 1/2.$$

This gives rise to

$$|(1-z)e^z| = |\exp(z + \log(1-z))| \le e^{|z|^2} \le e^{4|z|^2}, \quad |z| \le 1/2.$$

Moreover, by elementary estimates,

$$|(1-z)e^z| \le (1+|z|)e^{|z|} \le e^{2|z|} \le e^{4|z|^2}, \quad |z| \ge 1/2.$$

Hence for all $z \in \mathbb{C}$,

$$|P_n(z)e^{p_n z}| = \left|\prod_{k=1}^{n}\left(1 - \frac{z}{z_k}\right)e^{z/z_k}\right| \le e^{4q_n|z|^2},$$

from which

$$|P_n(z)| \le e^{|p_n z| + 4q_n|z|^2} \le e^{M|z| + 4M|z|^2}, \quad z \in \mathbb{C}.$$

Thus the sequence $\{P_n(z)\}$ is uniformly bounded on compact subsets of $\mathbb{C}$. By the Vitali–Porter theorem (Theorem 1.6.11) we have $P_n \to f$ uniformly on compact subsets of $\mathbb{C}$, and f is entire.

Suppose that there exists a point $z_0 \in \mathbb{C} \setminus \mathbb{R}$ such that $f(z_0) = 0$. Choose $r > 0$ small enough such that the disc $D(z_0, r)$ does not intersect with the real line. By Hurwitz's theorem (Theorem 1.6.8), the polynomials P_n should have a zero in $D(z_0, r)$ for all n large enough. But since the polynomials P_n have real zeros only, this is clearly a contradiction. Thus all possible zeros of f must be real. By Hurwitz's theorem and the assumption on the points $z_k = z_k(n)$, it follows that $f(0) \neq 0$.

Let $\{a_k\}$ denote the sequence of (real) zeros of f, listed according to their multiplicities and arranged with respect to increasing absolute value, bearing in mind that the whole sequence could be empty. For example, if $f(z) = e^z$, then f has no zeros even though the polynomials $Q_n(z) = (1 + \frac{z}{n})^n$ that converge to f do have real zeros.

For the time being, we assume that the sequence $\{a_k\}$ is not empty. Since by assumption the points $z_k = z_k(n)$ are uniformly bounded away from zero and $f(0) \neq 0$, there exists a constant $\delta > 0$ such that

$$\begin{aligned} |a_k| &\ge \delta, \quad k \in \mathbb{N}, \\ |z_k(n)| &\ge \delta, \quad n \in \mathbb{N},\ k \in \{1, \dots, n\}. \end{aligned} \tag{5.6.13}$$

By Hurwitz' theorem again, if a is a zero of f of order ℓ, then for a sufficiently small radius $r > 0$, the polynomials $P_n(z)$ have precisely ℓ zeros in the disc $D(a, r)$ for all n large enough. Thus we may assume that the points $z_k(n)$ are arranged so that

$$\lim_{n\to\infty} z_k(n) = a_k, \quad k \in \mathbb{N}. \tag{5.6.14}$$

Then for every $\varepsilon \in (0, \delta)$, there exists $N(\varepsilon) \in \mathbb{N}$ such that

$$|z_k(n) - a_k| < \varepsilon, \quad n \ge N(\varepsilon),\ k \in \mathbb{N}. \tag{5.6.15}$$

We have

$$\sum_{k=1}^{\infty}\frac{1}{a_k^2}\le\sum_{k=1}^{\infty}\frac{1}{(|z_k(n)|-\varepsilon)^2}=:C<\infty,\qquad n\ge N(\varepsilon),$$

by using the limit comparison test with $A:=\sum_{k=1}^{\infty}1/z_k(n)^2<\infty$. Let m be an integer satisfying $1\le m\le n$. Then by (5.6.13) and (5.6.15),

$$\begin{aligned}\left|\sum_{k=1}^{m}\frac{1}{z_k(n)^2}-\sum_{k=1}^{m}\frac{1}{a_k^2}\right|&\le\sum_{k=1}^{m}\frac{|a_k^2-z_k(n)^2|}{z_k(n)^2a_k^2}\le\varepsilon\sum_{k=1}^{m}\frac{|a_k|+|z_k(n)|}{z_k(n)^2a_k^2}\\&\le\frac{\varepsilon}{\delta}\sum_{k=1}^{m}\left(\frac{1}{z_k(n)^2}+\frac{1}{a_k^2}\right)\le\frac{\varepsilon}{\delta}(A+C),\qquad n\ge N(\varepsilon).\end{aligned}$$

Therefore, since

$$\sum_{k=1}^{m}\frac{1}{z_k(n)^2}\le\sum_{k=1}^{n}\frac{1}{z_k(n)^2}=q_n\le q,$$

and since $\varepsilon>0$ can be made arbitrarily small, we can conclude that

$$\sum_{k=1}^{m}\frac{1}{a_k^2}\le q.$$

Since the upper bound q is independent of m, letting $m\to\infty$, we conclude that

$$a:=\sum_{k=1}^{\infty}\frac{1}{a_k^2}\le q. \tag{5.6.16}$$

Next, we define

$$\begin{aligned}g(z)&=az+\sum_{k=1}^{\infty}\left(\frac{1}{z-a_k}+\frac{1}{a_k}\right)=\sum_{k=1}^{\infty}\frac{z^2}{a_k^2(z-a_k)},\\g_n(z)&=q_nz+\sum_{k=1}^{n}\left(\frac{1}{z-z_k}+\frac{1}{z_k}\right)=\sum_{k=1}^{n}\frac{z^2}{z_k^2(z-z_k)}.\end{aligned}$$

It is clear that $g_n(z)$ is a rational function in $\mathbb{C}$ for every n. Meanwhile, writing

$$g(z)=\sum_{|a_k|<2|z|}\frac{z^2}{a_k^2(z-a_k)}+\sum_{|a_k|\ge2|z|}\frac{z^2}{a_k^2(z-a_k)}=:S_1+S_2$$

and observing that

$$|S_2|\le\sum_{|a_k|\ge2|z|}\frac{|z|^2}{a_k^2|a_k-z|}\le\sum_{|a_k|\ge2|z|}\frac{|z|}{a_k^2}\le q|z|,$$

we see that S_2 converges absolutely uniformly on compact subsets of $\mathbb{C}$. Since S_1 is a rational function (or S_1 vanishes identically), we find that g is a meromorphic function on $\mathbb{C}$. Let D be an arbitrary pathwise connected domain in $|\zeta| \le R$ containing the origin but no points a_k.[1] We will show that $g_n(z) \to g(z)$ uniformly on D as $n \to \infty$.

Let $T > R > 0$, and let N denote the number of points a_k in the disc $|\zeta| < T$. Recall our assumption that the sequence $\{a_k\}$ is not empty. Hence, assuming that T is large enough, we may suppose that $N \ge 1$. By the Hurwitz theorem there are precisely N points $z_k(n)$ in $|\zeta| < T$ for all n large enough, say $n \ge K$. In other words, for $n \ge K$,

$$\begin{aligned} &|a_k| < T, \quad |z_k(n)| < T, \quad k = 1, 2, \dots, N, \\ &|a_k| \ge T, \quad |z_k(n)| \ge T, \quad k = N+1, N+2, \dots. \end{aligned}$$

Then for $|z| \le R$ and $n \ge K$,

$$\begin{aligned} &\left| \sum_{k=N+1}^{\infty} \frac{z^2}{a_k^2(z - a_k)} \right| \le \frac{R^2}{T - R} \sum_{k=N+1}^{\infty} \frac{1}{a_k^2} \le \frac{qR^2}{T - R}, \\ &\left| \sum_{k=N+1}^{\infty} \frac{z^2}{z_k^2(z - z_k)} \right| \le \frac{R^2}{T - R} \sum_{k=N+1}^{\infty} \frac{1}{z_k(n)^2} \le \frac{qR^2}{T - R}. \end{aligned}$$

This gives rise to

$$|g(z) - g_n(z)| \le \left| \sum_{k=1}^{N} \left(\frac{z^2}{a_k^2(z - a_k)} - \frac{z^2}{z_k^2(z - z_k)} \right) \right| + \frac{2qR^2}{T - R}. \tag{5.6.17}$$

By (5.6.14) the first term in the right-hand side of (5.6.17) approaches zero pointwise in D as $n \to \infty$. Hence the right-hand side of (5.6.17) can be made smaller than an arbitrary $\varepsilon > 0$, first by the choice of T, and then by the choice of n. This proves that $g_n \to g$ pointwise in D. The functions $g_n(z)$ are locally bounded in D, for if $\zeta \in D$ such that $\zeta \ne z_k$, then

$$\gamma := \min\left\{ \operatorname{dist}(\zeta, \partial D), \min_k |\zeta - z_k| \right\} > 0,$$

where $\operatorname{dist}(\zeta, \partial D)$ denotes the shortest distance from the point ζ to the boundary ∂D. Hence

$$|g_n(z)| \le \frac{2|z|^2}{\gamma} \sum_{k=1}^{\infty} \frac{1}{z_k^2} = C_0 |z|^2, \quad z \in D(\zeta, \gamma/2).$$

1 A domain $D \subset \mathbb{C}$ is called *pathwise connected* if for any two distinct points $w_1, w_2 \in D$, there exists a piecewise continuously differentiable path $\alpha : [0, 1] \to D$ such that $\alpha(0) = w_1$ and $\alpha(1) = w_2$; see [57, pp. 229–230] and [240, p. 200].

Then by the Vitali–Porter theorem, $g_n \to g$ uniformly in D.

An easy calculation shows that

$$g_n(z) = \frac{P_n'(z)}{P_n(z)} + p_n + q_n z,$$

where $p_n \to p$ and $q_n \to q$ as $n \to \infty$. Hence by the Weierstrass theorem,

$$\frac{f'(z)}{f(z)} + p + qz = az + \sum_{k=1}^{\infty}\left(\frac{1}{z-a_k} + \frac{1}{a_k}\right) = az + \sum_{k=1}^{\infty}\frac{1}{a_k}\left(1 - \frac{1}{1-z/a_k}\right).$$

The convergence of the above formula is uniform in D. Hence we can replace the variable z with ζ and then integrate term by term from 0 to z along a path in D to obtain

$$f(z)e^{pz+qz^2/2} = e^{az^2/2}\prod_{k=1}^{\infty}\left(1 - \frac{z}{a_k}\right)e^{z/a_k}, \quad z \in D.$$

Since $a \le q$ from (5.6.16), we find that f has the desired form in D and hence in $\mathbb{C}$ by the uniqueness theorem. This completes the proof that $f \in \mathcal{LP}$ under the assumption that f has zeros (that is, $\{a_k\}$ is not empty).

Finally, we still have to address the case that the (entire) limit function f has no zeros, in which case f has the form $f = e^g$ for some entire function g. According to our interpretation, this means that $a_k = \infty$ for all $k \in \mathbb{N}$. Then by (5.6.14),

$$\lim_{n\to\infty} z_k(n) = \infty, \quad k \in \mathbb{N}. \tag{5.6.18}$$

Let $R > 1$ be a large constant, and suppose that $|z| \le R$. Then there exists $N_1 = N_1(R) \in \mathbb{N}$ such that

$$|z_k(n)| \ge R^3 + R, \quad k \in \mathbb{N},\ n \ge N_1.$$

Consequently, by (5.6.12),

$$|g_n(z)| \le \sum_{k=1}^{n}\frac{|z|^2}{z_k(n)^2(|z_k| - |z|)} \le \frac{1}{R}\sum_{k=1}^{n}\frac{1}{z_k(n)^2} \overset{\text{def}}{=} \frac{q_n}{R} < \frac{M}{R}$$

for $n \ge N_1$ and $|z| \le R$. We have

$$g_n(z) = \frac{P_n'(z)}{P_n(z)} + p_n + q_n z \to \frac{f'(z)}{f(z)} + p + qz = g'(z) + p + qz$$

as $n \to \infty$ uniformly for $|z| \le R$. For all $\varepsilon > 0$, this gives

$$|g'(z) + p + qz - g_n(z)| < \frac{\varepsilon}{2}, \quad n \ge N_3 \ge N_2,\ |z| \le R,$$

which in turn gives

$$|g'(z) + p + qz| < |g_n(z)| + \frac{\varepsilon}{2} < \frac{M}{R} + \frac{\varepsilon}{2}, \qquad n \geq N_3, \ |z| \leq R.$$

By requiring that $R > \max\{1, \frac{2M}{\varepsilon}\}$ we have

$$|g'(z) + p + qz| < \varepsilon, \qquad n \geq N_4 \geq N_3, \ |z| \leq R.$$

Hence $g'(z) = -p - qz$. Then $f(z) = e^{b-pz-qz^2/2}$, where $b \in \mathbb{C}$, $p \in \mathbb{R}$, and $q \geq 0$. Thus $f \in \mathcal{LP}$. □

The case $q = 0$ at the end of the previous proof is possible, as shown by the polynomials $P_n(z) = (1 + \frac{z}{n})^n$ with real zeros only and converging to $f(z) = e^z$.

6 Exponential polynomials

An *exponential polynomial* f is a finite sum of terms of the form $P(z)e^{Q(z)}$, where $P(z)$ and $Q(z)$ are polynomials. A polynomial is a special case of an exponential polynomial, whereas transcendental exponential polynomials are entire functions of finite positive integer order. If $\deg Q \leq 1$ for each $Q(z)$ in the sum, then f is called an *exponential sum*.

The survey [117] gives an overview of the theory of exponential polynomials from its initial stages in the early 1920s to research papers published just recently. We discuss some of these results in this chapter, with an emphasis on keeping all results and proofs as simple as possible. This means that some of the results are not in their most general or best possible form. The advantage of doing this is that we avoid tens of pages of tedious computations. An interested reader can easily find the original results and original references (containing the proofs) in [117].

Since it is impossible to cover over a century of research on exponential polynomials in this chapter, even if the proofs would be skipped, we have to focus on certain topics. In this chapter, we discuss the location and frequency of zeros, growth properties, value distribution theory, and dual exponential polynomials, which are found in various research works. We also (1) give a proof that every exponential polynomial solves a linear differential equation with polynomial coefficients and (2) discuss the local separation of zeros of exponential polynomials, which are, to the best of our knowledge, topics covered in the original paper [282] only. Regarding topics that are not discussed in this chapter, we refer the reader to [117] and the references therein.

6.1 Critical rays and orthogonal rays

An *exponential sum* is an entire function of the form

$$f(z) = \sum_{j=1}^{m} P_j(z)e^{w_j z}, \tag{6.1.1}$$

where the *multipliers* $P_j(z) \not\equiv 0$ are polynomials, and the *frequencies* $w_j \in \mathbb{C}$ are distinct. For convenience, we do not consider the trivial function $f(z) \equiv 0$. At most one of the constants w_j is allowed to vanish, and in case one of them does vanish, we may suppose that $w_1 = 0$. The trigonometric functions

$$\sin z = \frac{1}{2i}(e^{iz} - e^{-iz}) \quad \text{and} \quad \cos z = \frac{1}{2}(e^{iz} + e^{-iz})$$

and finite partial sums

$$\sum_{n=1}^{m} \frac{1}{n^z} = \sum_{n=1}^{m} e^{(-\log n)z}$$

of the *Riemann zeta-function* $\zeta(z)$ are typical examples of exponential sums.

https://doi.org/10.1515/9783112249130-006

Example 6.1.1. Suppose that $f(z)$ is either $\sin z$ or $\cos z$. Then the exponential term e^{-iz} dominates the growth of f in the upper half-plane, whereas e^{iz} dominates the growth in the lower half-plane. The positive and negative real axes are the turning points for the Phragmén–Lindelöf indicator function

$$h_f(\theta) = |\sin\theta| = \max\{\mathrm{Re}(ie^{i\theta}), \mathrm{Re}(-ie^{i\theta})\}, \quad \theta \in \mathbb{R};$$

see Example 4.2.3. Note that the zeros of f lie on the real axis.

We proceed to express the Phragmén–Lindelöf indicator of f in (6.1.1).

Lemma 6.1.2. *A transcendental exponential sum f in* (6.1.1) *is of order one and of finite positive type. Moreover, the indicator function of f is*

$$h_f(\theta) = \max_{1\le j\le m}\{\mathrm{Re}(w_j e^{i\theta})\}, \quad \theta \in \mathbb{R}. \tag{6.1.2}$$

Proof. It is clear that f has order one. Each term $g_j(z) = P_j(z)e^{w_j z}$ has the indicator

$$h_{g_j}(\theta) = \mathrm{Re}(w_j e^{i\theta}), \quad \theta \in \mathbb{R},$$

so that by Lemma 4.2.2,

$$h_f(\theta) \le \max_{1\le j\le m}\{\mathrm{Re}(w_j e^{i\theta})\}, \quad \theta \in \mathbb{R}. \tag{6.1.3}$$

Hence $h_f(\theta) < \infty$ for $\theta \in \mathbb{R}$. Then by Lemma 4.2.6, $h_f(\theta)$ is continuous on $\mathbb{R}$.

Suppose that for some $\theta_0 \in \mathbb{R}$, there is a unique integer $j_0 \in \{1, \dots, m\}$ for which the maximum in (6.1.3) is attained, that is,

$$\mathrm{Re}(w_{j_0}e^{i\theta_0}) = \max_{1\le j\le m}\{\mathrm{Re}(w_j e^{i\theta_0})\} \quad \text{and} \quad \mathrm{Re}(w_j e^{i\theta_0}) < \mathrm{Re}(w_{j_0}e^{i\theta_0}),\ j \ne j_0.$$

Then from

$$|f(re^{i\theta_0})| \ge |P_{j_0}(re^{i\theta_0})e^{w_{j_0}re^{i\theta_0}}| - \sum_{j\ne j_0}|P_j(re^{i\theta_0})e^{w_j re^{i\theta_0}}|,$$

we see that $h_f(\theta_0) \ge \mathrm{Re}(w_{j_0}e^{i\theta_0})$. Therefore, $h_f(\theta_0) = \mathrm{Re}(w_{j_0}e^{i\theta_0})$ for every such θ_0. This gives an equality in (6.1.3) for $\theta = \theta_0$.

On the other hand, if $h_{g_j}(\theta) = h_{g_k}(\theta)$ for $j \ne k$, then

$$\mathrm{Re}((w_j - w_k)e^{i\theta}) = 0.$$

Denoting $w_j - w_k = a + ib$, we obtain

$$a\cos\theta - b\sin\theta = 0.$$

However, since $w_j \neq w_k$, at least one of a or b is nonzero, and consequently this equation has at most two solutions θ on $[0, 2\pi)$. We deduce that the equality in (6.1.3) holds for $\theta \in [0, 2\pi) \setminus I$, where I is a finite set. By the continuity and 2π-periodicity, we see that the equality in (6.1.3) holds for all $\theta \in \mathbb{R}$, that is, (6.1.2) holds.

Since $h_f(\theta) < \infty$ for $\theta \in \mathbb{R}$, from Corollary 4.2.7 it follows that

$$\max_{0 \le \theta \le 2\pi} h_f(\theta) = \tau < \infty, \tag{6.1.4}$$

where τ is the type of f. To show that $\tau > 0$, observe that for each j where $w_j \neq 0$, it can be deduced that $\mathrm{Re}(w_j e^{i\theta}) > 0$ for some values of θ. Then from (6.1.2) and (6.1.4), we obtain $\tau > 0$. This completes the proof. □

Lemma 6.1.2 and its proof give rise to the following definition.

Definition 6.1.3 ([115]). Let f be a transcendental exponential sum of the form (6.1.1). Suppose that $\theta^* \in \mathbb{R}$ satisfies

$$h_f(\theta^*) = \mathrm{Re}(w_j e^{i\theta^*}) = \mathrm{Re}(w_k e^{i\theta^*}) \tag{6.1.5}$$

for a pair of indices $j \neq k$. Then the ray $\arg z = \theta^*$ is called a *critical ray* for f, and θ^* is called a *cusp*. For $\varepsilon > 0$, the set

$$S(\varepsilon, \theta^*) = \{z \in \mathbb{C} : |\arg z - \theta^*| < \varepsilon\}$$

is called a *critical sector* for f.

In view of Lemma 6.1.2, the equalities in (6.1.5) mean that at least two terms $\mathrm{Re}(w_j e^{i\theta^*})$ and $\mathrm{Re}(w_k e^{i\theta^*})$ are at the same time maximal and equal. Moreover, from Lemma 6.1.2 we see that $h_f(\theta)$ is differentiable everywhere except at cusps, of which there are at most finitely many on any closed interval of length 2π. To avoid ambiguity, we assume that $\varepsilon > 0$ is chosen small enough so that the critical sectors do not intersect.

Let $W = \{\overline{w}_1, \dots, \overline{w}_m\}$ denote the set of *conjugate frequencies* of f, and let $\mathrm{co}(W)$ be the convex hull of W; see Definition 3.9.3. Then $\mathrm{co}(W)$ is the intersection of all closed half-planes that contain W by Lemma 3.9.4. As W has only finitely many points, $\mathrm{co}(W)$ is either a convex polygon or a line segment. The vertices of $\mathrm{co}(W)$ are formed by some but not necessarily by all of the points $\overline{w}_1, \dots, \overline{w}_m$, and hence the number of vertices of $\mathrm{co}(W)$ is $\le m$. Some of the points $\overline{w}_\ell$ can be boundary points or interior points of $\mathrm{co}(W)$, but none of them is an exterior point. This follows trivially from the definition of convex hulls.

There is a geometric way of detecting critical rays without having to calculate the indicator function. This can be done by means of orthogonal rays.

Definition 6.1.4 ([115, 117]). Let f be a transcendental exponential sum of the form (6.1.1), and let $\overline{w}_j$, $\overline{w}_k$ be two consecutive vertices of $\mathrm{co}(W)$. Then the ray $\arg z = \theta^\perp$ that is parallel to the outer normal vector of $\mathrm{co}(W)$ determined by the line through $\overline{w}_j$, $\overline{w}_k$ is called an *orthogonal ray* for f.

Lemma 6.1.5 ([115, 117]). *For a transcendental exponential sum f of the form* (6.1.1)*, the critical rays and orthogonal rays coincide.*

Proof. The proof relies on the concept of a supporting function of $\mathrm{co}(W)$, which is defined by

$$k(\theta) = \sup_{z\in\mathrm{co}(W)} \mathrm{Re}(ze^{-i\theta}).$$

It follows from Lemma 6.1.2 and [117, Lemma B.7] that $k(\theta) = h_f(\theta)$ for every $\theta \in \mathbb{R}$. We take this property for granted.

Suppose first that $\arg z = \theta^\perp$ is an orthogonal ray for f. We may suppose that it is parallel to the outer normal of $\mathrm{co}(W)$ determined by the line through $\overline{w}_j$, $\overline{w}_k$, where $\overline{w}_j$, $\overline{w}_k$ are two consecutive vertices of $\mathrm{co}(W)$. Define two vectors

$$\boldsymbol{\alpha} = \overline{w}_k - \overline{w}_j = (\mathrm{Re}(w_k - w_j), -\mathrm{Im}(w_k - w_j)) \quad \text{and} \quad \boldsymbol{\beta} = (\cos\theta^\perp, \sin\theta^\perp).$$

Then $\boldsymbol{\alpha}$ and $\boldsymbol{\beta}$ are orthogonal, and hence

$$\boldsymbol{\alpha}\cdot\boldsymbol{\beta} = \mathrm{Re}(w_k - w_j)\cos\theta^\perp - \mathrm{Im}(w_k - w_j)\sin\theta^\perp = 0,$$

which in turn is equivalent to

$$\mathrm{Re}(w_j e^{i\theta^\perp}) = \mathrm{Re}(w_k e^{i\theta^\perp}).$$

Taking the complex conjugates will not affect the real parts:

$$\mathrm{Re}(\overline{w}_j e^{-i\theta^\perp}) = \mathrm{Re}(\overline{w}_k e^{-i\theta^\perp}).$$

Note that whenever $\theta = \theta^\perp$, the quantity $|k(\theta)|$ equals the distance from the origin to the supporting line of $\mathrm{co}(W)$ [117, Lemma 6.B]. In this case the supporting line is the line through the points $\overline{w}_j$, $\overline{w}_k$; see Figure 6.1.

Keeping (6.1.2) and $k(\theta^\perp) = h_f(\theta^\perp)$ in mind, we conclude that

$$k(\theta^\perp) = \mathrm{Re}(w_j e^{i\theta^\perp}) = \mathrm{Re}(w_k e^{i\theta^\perp}) \geq \max_{\ell\neq j,k}\{\mathrm{Re}(w_\ell e^{i\theta^\perp})\}.$$

Thus the two terms $\mathrm{Re}(w_j e^{i\theta^\perp})$ and $\mathrm{Re}(w_k e^{i\theta^\perp})$ are at the same time maximal and equal, which makes $\arg z = \theta^\perp$ a critical ray for f by the definition.

Conversely, suppose that $\arg z = \theta^*$ is a critical ray for f. Then (6.1.5) holds for a pair of indices $j \neq k$. Similarly as above, we define two vectors

$$\boldsymbol{\alpha} = \overline{w}_k - \overline{w}_j = (\mathrm{Re}(w_k - w_j), -\mathrm{Im}(w_k - w_j)) \quad \text{and} \quad \boldsymbol{\beta} = (\cos\theta^*, \sin\theta^*).$$

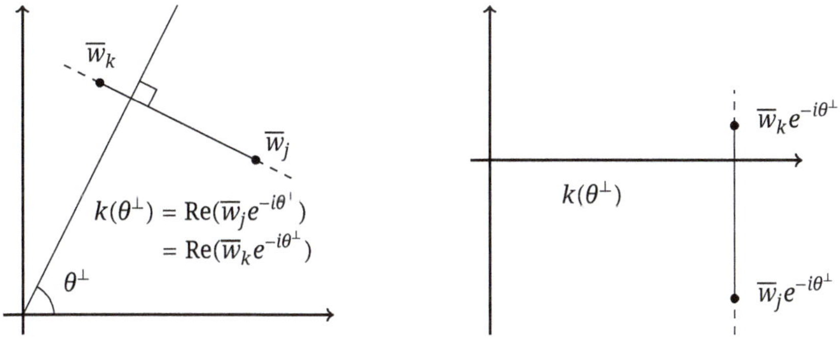

Figure 6.1: The distance from the origin to the supporting line through $\overline{w}_j, \overline{w}_k$.

From (6.1.5) it follows that $\boldsymbol{\alpha} \cdot \boldsymbol{\beta} = 0$, which means that $\boldsymbol{\alpha}$ and $\boldsymbol{\beta}$ are orthogonal. Consequently, $\arg z = \theta^*$ is an orthogonal ray for f. □

The number of outer normals of $\mathrm{co}(W)$ equals the number of vertices of $\mathrm{co}(W)$. This is true even if $\mathrm{co}(W)$ is a line segment because a line segment has two outer normals and two vertices.

The next result is a quick consequence of Lemma 6.1.5.

Corollary 6.1.6. *A transcendental exponential sum f of the form* (6.1.1) *has $s \leq m$ critical/orthogonal rays, where s is the number of vertices of* $\mathrm{co}(W)$.

Originally, the critical rays were defined by means of orthogonal rays, but this approach requires some knowledge of convex sets, which is not the focus of this book. As we have seen in the proof of Lemma 6.1.5, we might as well use the indicator function for detecting the critical/orthogonal rays. In Section 6.2, we will discuss Pólya's theorem, according to which the zeros of exponential sums lie in the critical sectors with at most finitely many exceptional zeros. The asymptotic number of zeros in each individual critical sector will also be discussed.

Example 6.1.7. Consider the exponential sum

$$f(z) = e^{2z} + z^3 e^{iz} + e^{2iz} + (1+z)e^{-2z}.$$

The statements below follow from the definitions and results in this section. The frequencies of f are $\{2, i, 2i, -2\}$, and the conjugate frequencies of f are $W = \{2, -i, -2i, -2\}$. Thus the convex hull $\mathrm{co}(W)$ is the triangular region with vertices at $2, -2i, -2$. The orthogonal and critical rays of f are the same, and they are $\arg z = \pi/2, 5\pi/4, 7\pi/4$.

Since $p(f) = 1$ and

$$\mathrm{Re}\big(2e^{i\theta}\big) = 2\cos\theta, \quad \mathrm{Re}\big(ie^{i\theta}\big) = -\sin\theta,$$
$$\mathrm{Re}\big(2ie^{e^{i\theta}}\big) = -2\sin\theta, \quad \mathrm{Re}\big(-2e^{i\theta}\big) = -2\cos\theta,$$

we obtain that the Phragmén–Lindelöf indicator function is given by

$$h_f(\theta) = \begin{cases} 2\cos\theta, & -\pi/2 \le \theta \le \pi/4, \\ 2\sin\theta, & \pi/4 \le \theta \le 3\pi/4, \\ -2\cos\theta, & 3\pi/4 \le \theta \le 3\pi/2. \end{cases} \quad (6.1.6)$$

From (6.1.6) we obtain

$$\max_{0\le\theta\le 2\pi} \{h_f(\theta)\} = 2.$$

Hence by Corollary 4.2.7, the type is $\tau(f) = 2$.

6.2 Zeros in critical sectors

We adopt the notation from Section 6.1 regarding exponential sums

$$f(z) = \sum_{j=1}^{m} P_j(z)e^{w_j z}. \quad (6.2.1)$$

The main result in this section is the following theorem by Pólya.

Theorem 6.2.1 ([228]). *Let f be a transcendental exponential sum of the form* (6.2.1), *and let $\varepsilon > 0$. Then all the zeros of f lie in the critical sectors $S(\varepsilon, \theta^*)$, except for at most finitely many zeros. Moreover, for each cusp θ^*, the number $n(r, S, f)$ of zeros of f in $S(\varepsilon, \theta^*) \cap \{|z| \le r\}$ has the asymptotic growth*

$$n(r, S, f) = \frac{|w_j - w_k|}{2\pi} r + o(r),$$

where w_j, w_k are the frequencies of f that determine θ^* in the sense of (6.1.5).

Pólya gives only an outline of the proof of Theorem 6.2.1 in [228]. Our proof is based on this outline and on the ideas in [263], which are developed further in [115, 126]. Three auxiliary results are needed for the proof.

The following lemma does not seem to exist in the literature, but the result itself is known at least to experts. The lemma is straightforward and is given as a replacement of a technical auxiliary result [263, Lemma 3] (see also [115, Lemma 5.3]). This is done at the expense of weakening the result but at the same time for the benefit of keeping things simple.

Lemma 6.2.2. *Let f be a transcendental exponential sum of the form* (6.2.1), *and let $\varepsilon > 0$. Then there exists $\delta = \delta(\varepsilon) > 0$ with the following property: If $S(\varepsilon, \theta_1^*)$ and $S(\varepsilon, \theta_2^*)$ are two consecutive critical sectors of f with $\theta_1^* < \theta_2^*$, then there exists a unique index j such that*

$$\operatorname{Re}(w_j e^{i\theta}) - \max_{k\ne j}\{\operatorname{Re}(w_k e^{i\theta})\} \ge \delta, \quad \theta_1^* + \varepsilon < \theta < \theta_2^* - \varepsilon.$$

Proof. The fact that there exists a unique index j such that

$$\operatorname{Re}(w_j e^{i\theta}) - \max_{k \neq j}\{\operatorname{Re}(w_k e^{i\theta})\} > 0, \quad \theta_1^* < \theta < \theta_2^*, \tag{6.2.2}$$

follows by the proof of Lemma 6.1.2. We still need to bring ε and δ into these inequalities. To this end, consider an auxiliary function

$$g(\theta) = A\cos\theta + B\sin\theta$$

on an interval $[a,b]$, where $a,b,A,B \in \mathbb{R}$ are chosen such that at least one of A, B is nonzero and $g(\theta) > 0$ for $\theta \in (a,b)$. The function g may or may not vanish at the endpoints of the interval $[a,b]$. Since $g''(\theta) = -g(\theta) < 0$ for $\theta \in (a,b)$, the function g is concave. By concavity and continuity, if $\varepsilon > 0$ is arbitrarily small such that $a+\varepsilon < b-\varepsilon$, then there exists $\delta = \delta(\varepsilon) > 0$ such that

$$g(\theta) \geq \delta, \quad a+\varepsilon < \theta < b-\varepsilon.$$

Set $w_\ell = a_\ell + ib_\ell$ for $\ell \in \{1,\dots,m\}$. Since $\operatorname{Re}(w_\ell e^{i\theta}) = a_\ell\cos\theta - b_\ell\sin\theta$, we may rewrite (6.2.2) for each index $k \neq j$ as

$$(a_j - a_k)\cos\theta + (b_k - b_j)\sin\theta > 0, \quad \theta_1^* < \theta < \theta_2^*.$$

Since $w_j \neq w_k$, at least one of $a_j - a_k$ or $b_k - b_j$ is nonzero. By the discussion above, for every $\varepsilon_k > 0$, there exists $\delta_k > 0$ such that

$$(a_j - a_k)\cos\theta + (b_k - b_j)\sin\theta \geq \delta_k, \quad \theta_1^* + \varepsilon_k < \theta < \theta_2^* - \varepsilon_k.$$

Finally, choose $\varepsilon = \max_k\{\varepsilon_k\}$ and $\delta = \min_k\{\delta_k\}$, and the statement follows. □

The next lemma combines Lemmas 7.2 and 7.3 from [126], but it is tailored for exponential sums instead of exponential polynomials. This simplifies the proof quite remarkably.

Lemma 6.2.3. *Let f be a transcendental exponential sum of the form* (6.2.1), *and let* $\varepsilon > 0$. *Then there exists a constant* $\delta = \delta(\varepsilon) > 0$ *with the following properties:*

(a) *If S denotes the region between two consecutive critical sectors $S(\varepsilon,\theta_1^*)$ and $S(\varepsilon,\theta_2^*)$ of f and if $z = re^{i\theta} \in S$, then there exists a unique index j_0 such that*

$$f(z) = P_{j_0}(z)e^{w_{j_0}z}(1+\varepsilon(\delta,r)), \tag{6.2.3}$$

where $|\varepsilon(\delta,r)| \leq \exp(-\delta r)$ *for* $r \geq R$ *for some* $R > 0$.

(b) *Denote*

$$G_j(z) = P_j'(z) + w_jP_j(z).$$

For the same values of z and j_0 as in (a), *we have*

$$f'(z) = \begin{cases} G_{j_0}(z)e^{w_{j_0}z}(1 + \varepsilon(\delta, r)) & \text{if } w_{j_0} \neq 0, \\ P_1'(z) + \varepsilon(\delta, r) & \text{if } w_{j_0} = 0, \end{cases} \tag{6.2.4}$$

and

$$f'(z)/f(z) = w_{j_0} + O(1/r), \tag{6.2.5}$$

regardless of whether $w_{j_0} \neq 0$ *or* $w_{j_0} = 0$.

Proof. (a) Since the multipliers $P_j(z)$ are all polynomials, there exist constants $d, r_0 \geq 1$ such that

$$M(r, P_k) \leq r^d, \quad k \in \{0, \dots, m\}, \tag{6.2.6}$$

whenever $r \geq r_0$. By Lemma 6.2.2, if $z = re^{i\theta} \in S$ and if r is large enough, then there exist a constant $\delta > 0$ and a unique index j_0 such that

$$\mathrm{Re}(w_{j_0}z) \geq \mathrm{Re}(w_k z) + 2\delta r \tag{6.2.7}$$

for all $k \neq j_0$. Then estimates (6.2.6) and (6.2.7) applied to (6.2.1) yield

$$\begin{aligned} |f(z)e^{-w_{j_0}z}| &\geq |P_{j_0}(z)| - \sum_{k \neq j_0} |P_k(z)||e^{(w_k - w_{j_0})z}| \\ &\geq |P_{j_0}(z)| - \exp(\log m + d \log r - 2\delta r), \end{aligned}$$

from which

$$\frac{|f(z)|}{|P_{j_0}(z)e^{w_{j_0}z}|} \geq 1 - \exp(-\delta r)$$

for all $r \geq r_0$ large enough. On the other hand,

$$\frac{|f(z)|}{|P_{j_0}(z)e^{w_{j_0}z}|} \leq 1 + \sum_{k \neq j_0} \frac{|P_k(z)|}{|P_{j_0}(z)|}|e^{(w_k - w_{j_0})z}| \leq 1 + \exp(-\delta r)$$

whenever $z \in S$ and r is large enough. This proves (6.2.3).

(b) To prove (6.2.4), we first notice that

$$f'(z) = G_1(z)e^{w_1 z} + \cdots + G_m(z)e^{w_m z},$$

where $G_j(z) = P_j'(z) + w_j P_j(z)$ for $j \geq 1$. If $G_j(z) \equiv 0$ and $w_j \neq 0$, then either $P_j(z) \equiv 0$ or $P_j(z) = Ce^{-w_j z}$ for some $C \in \mathbb{C} \setminus \{0\}$. Since neither conclusion is possible, we must have $w_j = 0$. We have agreed earlier that w_1 is the only frequency that can vanish. Thus

$G_j(z) \not\equiv 0$ for $j \geq 2$, and if $G_1(z) \equiv 0$, then $w_1 = 0$. Hence f' is an exponential sum and shares the frequencies with f, except possibly in the case where $w_1 = 0$ and $P_1(z)$ is a constant function. Thus by considering separately the cases $w_{j_0} \neq 0$ and $w_{j_0} = w_1 = 0$, the first part of the proof applies to f', and we obtain (6.2.4) for the same index j_0 that appears in (6.2.3).

Statement (6.2.5) is a simple consequence of (6.2.3), (6.2.4), and the fact that $P'_{j_0}(z)/P_{j_0}(z) = O(1/r)$ for a polynomial $P_{j_0}(z)$. □

The following lemma originates from ideas that Backlund used to study the zeros of Riemann's zeta-function in 1914. In his PhD thesis [246] published in 1925, Schwengeler called the result modestly as the Backlund lemma even though he also had a strong contribution to it. The version below together with a proof and original references can be found in [127].

Lemma 6.2.4 ([127]). *Suppose that F is an entire function and that $z_1, z_2 \in \mathbb{C}$ are distinct points such that F has no zeros on the line segment that connects z_1 and z_2. Then*

$$\left| \mathrm{Re}\left\{ \frac{1}{2\pi i} \int_{z_1}^{z_2} \frac{F'(z)}{F(z)}\, dz \right\} \right| \leq \frac{\max_{|\zeta| \leq R} \log |F(z_1 + \zeta)| - \log |F(z_1)|}{2 \log(R/T)} + \frac{1}{2},$$

where $R > T = |z_2 - z_1|$.

Proof of Theorem 6.2.1. The first statement that all the zeros of f lie in the critical sectors $S(\varepsilon, \theta^*)$, with at most finitely many exceptions, follows from Lemma 6.2.3. Thus it remains to estimate the number of zeros in each critical sector.

To simplify the situation, we may suppose that the positive real axis is a critical ray for f by appealing to the exponential polynomial $g(z) = f(e^{-i\theta^*} z)$. Thus we assume that $\theta^* = 0$ is a cusp of f, determined by its leading coefficients w_j, w_k as in (6.1.5). It follows that w_j and w_k are the unique dominant leading coefficients of f around the positive real axis. Without loss of generality, we may assume that w_j determines the dominant term of f below the x-axis and that w_k determines the dominant term of f above the x-axis. The uniqueness of the frequencies w_j, w_k carries over to the application of Lemma 6.2.3.

For a large enough $R_0 > 0$, define the closed path $\Gamma = \Gamma_1 + \Gamma_2 + \Gamma_3 + \Gamma_4$ with

$$\begin{aligned}
&\Gamma_1: \quad z = te^{-i\varepsilon}, \quad \text{where } t \text{ goes from } R_0 \text{ to } r,\\
&\Gamma_2: \quad z = r\cos\varepsilon + iy, \quad \text{where } y \text{ goes from } -r\sin\varepsilon \text{ to } r\sin\varepsilon,\\
&\Gamma_3: \quad z = te^{i\varepsilon}, \quad \text{where } t \text{ goes from } r \text{ to } R_0,\\
&\Gamma_4: \quad z = R_0 e^{i\theta}, \quad \text{where } \theta \text{ goes from } \varepsilon \text{ to } -\varepsilon.
\end{aligned}$$

The idea is to keep R_0 fixed and eventually to let r increase. Let Γ^s be the path obtained from Γ by replacing r with a constant s satisfying $\cos\varepsilon = \frac{r}{s}$. Then Γ_2 and Γ_2^s are respectively a cord and a tangent to the circle $|z| = r$; see Figure 6.2.

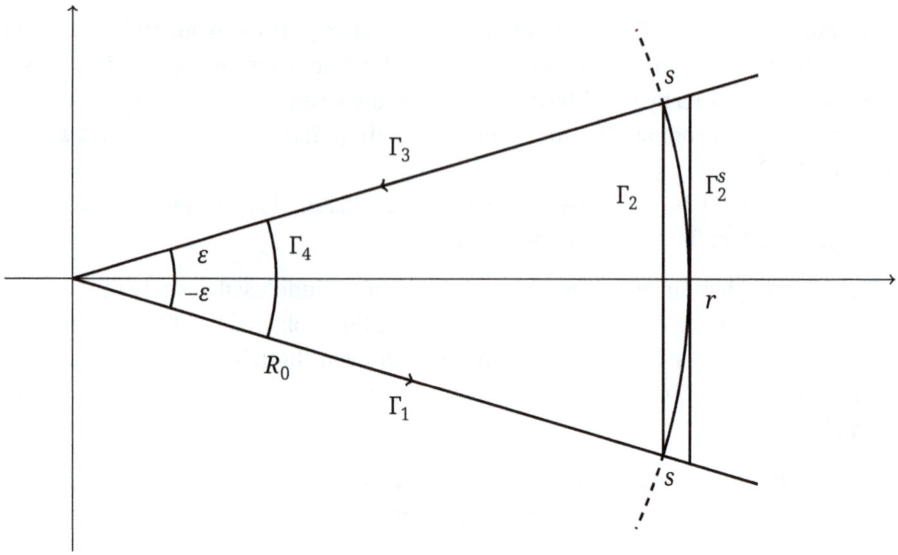

Figure 6.2: The critical sector $S(\varepsilon, 0)$ and the domains bounded by Γ and Γ^s.

Since f has only countably many zeros and there are uncountably many choices for ε, r, s, we may suppose that neither Γ nor Γ^s contains any zeros of f. This minor inconvenience will be addressed later in the proof.

Set

$$I := \int_{\Gamma_1} w_j \, dz = \int_{R_0}^{r} w_j e^{-i\varepsilon} \, dt.$$

Writing $w_j = a_j + ib_j$, we have

$$\operatorname{Im} I = \int_{R_0}^{r} (b_j \cos\varepsilon - a_j \sin\varepsilon) \, dt.$$

Using the standard estimates $1 - x^2/2 \le \cos x \le 1$ and $x - x^3/6 \le \sin x \le x$, we get

$$\operatorname{Im} I = b_j r + \varepsilon(r),$$

where $|\varepsilon(r)| = O(\varepsilon r)$. Using (6.2.5) for $j_0 = j$, it follows that

$$\operatorname{Im} \int_{\Gamma_1} \frac{f'(z)}{f(z)} \, dz = b_j r + \varepsilon(r).$$

Analogously, if $w_k = a_k + ib_k$, then

$$\operatorname{Im} \int_{\Gamma_3} \frac{f'(z)}{f(z)} \, dz = -b_k r + \varepsilon(r).$$

Here the minus sign is a result of integrating in the opposite direction. Since f'/f is bounded on the path Γ_4 of finite length (R_0 is fixed), it is clear that

$$\left|\frac{1}{2\pi i}\int_{\Gamma_4}\frac{f'(z)}{f(z)}\,dz\right| = O(1).$$

Next, we will integrate f'/f over Γ_2. Backlund's lemma will be used for this with $R = 2|z_2 - z_1|$, where $z_1 = re^{-i\varepsilon}$ and $z_2 = re^{i\varepsilon}$. Then since $P_j(z)e^{w_j z}$ is the dominant term of f on Γ_1, we have by Lemma 6.2.3 that there exists $\delta = \delta(\varepsilon) > 0$ such that

$$f(z_1) = P_j(z_1)e^{w_j z_1}(1 + \varepsilon(\delta, r)), \tag{6.2.8}$$

where $|\varepsilon(\delta, r)| \le \exp(-\delta r)$. By Lemma 2.1.4 we may write

$$P_\nu(z) = p_\nu z^{d_\nu}(1 + o(1)),$$

where $d_\nu = \deg(P_\nu)$. Using the obvious

$$\frac{R}{4r} = \frac{|z_2 - z_1|}{2r} = \sin\varepsilon \le \varepsilon, \tag{6.2.9}$$

it follows that

$$\begin{aligned}\max_{1\le\nu\le m}\max_{|\zeta|\le R}\left|\frac{P_\nu(z_1+\zeta)}{P_\nu(z_1)}\right| &= \max_{1\le\nu\le m}\max_{|\zeta|\le R}\left|\left(1+\frac{\zeta}{z_1}\right)^{d_\nu}(1+o(1))\right|\\ &\le 2\max_{1\le\nu\le m}\left(1+\frac{R}{r}\right)^{d_\nu} \le 2\max_{1\le\nu\le m}(1+4\varepsilon)^{d_\nu} =: C,\end{aligned}$$

where the $o(1)$-term refers to $z_1 \to \infty$ along Γ_1 or, equivalently, to $r \to \infty$. Denoting $w = \max_{1\le\nu\le m}|w_j|$, assuming that $|\zeta| \le R$, and using (6.2.8), we obtain the estimate

$$\begin{aligned}|f(z_1+\zeta)| &\le \left(|P_1(z_1+\zeta)e^{w_1 z_1}| + \cdots + |P_m(z_1+\zeta)e^{w_m z_1}|\right)\cdot e^{wR}\\ &\le C\left(|P_1(z_1)e^{w_1 z_1}| + \cdots + |P_m(z_1)e^{w_m z_1}|\right)\cdot e^{wR}\\ &\le mC|P_j(z_1)e^{w_j z_1}|(1+\exp(-\delta r))\cdot e^{wR}.\end{aligned} \tag{6.2.10}$$

Thus from Backlund's lemma together with (6.2.8) and (6.2.10), we obtain

$$\left|\mathrm{Re}\left\{\frac{1}{2\pi i}\int_{\Gamma_2}\frac{f'(z)}{f(z)}\,dz\right\}\right| \le \frac{\max_{|\zeta|\le R}\log\frac{|F(z_1+\zeta)|}{|F(z_1)|}}{2\log 2} + \frac{1}{2} \le \frac{\log\left(mC\cdot\frac{1+\exp(-\delta r)}{1-\exp(-\delta r)}\right) + wR}{2\log 2} + \frac{1}{2}.$$

We apply (6.2.9) once again to conclude that

$$\left|\mathrm{Re}\left\{\frac{1}{2\pi i}\int_{\Gamma_2}\frac{f'(z)}{f(z)}\,dz\right\}\right| \le wR + O(1) \le 4\varepsilon wr + O(1).$$

Next we use the argument principle, according to which the number $n(r,\Gamma,f)$ of zeros of f in a domain bounded by Γ is

$$n(r,\Gamma,f) = \frac{1}{2\pi i}\int_{\Gamma}\frac{f'(z)}{f(z)}\,dz.$$

Hence, putting everything together, it follows that

$$n(r,\Gamma,f) = \frac{b_j - b_k}{2\pi}r + \varepsilon(r). \tag{6.2.11}$$

Since counting functions are always nonnegative, it follows that $b_j - b_k = |b_j - b_k|$. The situation would be symmetric if we would have chosen the leading coefficients w_j and w_k the other way around in the reasoning above. Moreover, from (6.1.5) we get $\operatorname{Re}(w_j e^{i\theta^*}) = \operatorname{Re}(w_k e^{i\theta^*})$, which can be rewritten as

$$a_j \cos\theta^* - b_j \sin\theta^* = a_k \cos\theta^* - b_k \sin\theta^*.$$

Since we have chosen $\theta^* = 0$, it follows that $a_j = a_k$. Therefore, we have $|w_j - w_k| = |b_j - b_k|$, so that (6.2.11) now reads as

$$n(r,\Gamma,f) = \frac{|w_j - w_k|}{2\pi}r + \varepsilon(r).$$

Next, we replace Γ with Γ^s, where s is a constant defined by $\cos\varepsilon = \frac{r}{s}$; see Figure 6.2. Then

$$s = \frac{r}{\cos\varepsilon} \le \frac{r}{1 - \varepsilon^2/2} = r + \varepsilon(r).$$

Similarly as above, it follows that

$$n(s,\Gamma^s,f) = \frac{|w_j - w_k|}{2\pi}r + \varepsilon(r).$$

It is clear from Figure 6.2 that

$$n(r,\Gamma,f) \le n(r,S,f) \le n(s,\Gamma^s,f),$$

and so there exists an absolute constant $K > 0$ such that

$$\frac{|w_j - w_k|}{2\pi}r - \varepsilon Kr \le n(r,S,f) \le \frac{|w_j - w_k|}{2\pi}r + \varepsilon Kr \tag{6.2.12}$$

for all r large enough. To be exact, there could be at most countably many possible exceptional values of ε, r, s coming from the definition of the paths Γ and Γ^s, since they cannot contain any zeros of f. However, since $n(r,S,f)$ is a nondecreasing and nonnegative step function, these exceptional values can be ignored.

Note that the inequalities in (6.2.12) are valid for every $\varepsilon > 0$. Choosing a slightly smaller ε, the constant $\delta = \delta(\varepsilon)$ coming from the use of Lemma 6.2.3 also gets smaller, but we may always restrict to larger values of r such that $\exp(-\delta r) \leq 1/2$, for example. Thus the error terms $\pm\varepsilon K r$ in (6.2.12) can be replaced with $o(r)$ as $r \to \infty$. This completes the proof. □

Remark. Being able to choose an arbitrarily small $\varepsilon > 0$ in the previous proof suggests that there might be some leeway in replacing the critical sectors with smaller domains. Indeed, Schwengeler found in his PhD thesis [246] that the critical sectors can be replaced with *logarithmic strips*

$$\Lambda(c,\theta^*) = \left\{z = re^{i\theta} : r > 1,\ |\theta - \theta^*| < c\frac{\log r}{r}\right\},$$

where $c > 0$ is large enough. It is easy to see that $\Lambda(c,\theta^*)$ is essentially contained in $S(\varepsilon,\theta^*)$. However, such an improvement requires stronger versions of Lemmas 6.2.2 and 6.2.3 as well as a deeper proof. At the same time, the error term $o(r)$ can also be improved to $O(1)$ [58]. However, to keep things as simple as possible, we have intentionally avoided this approach. For more details, see [115, 117, 126, 127, 263].

Definition 6.2.5. For $\mathrm{co}(W)$, we define the quantity $C = C(\mathrm{co}(W))$ as follows. If $\mathrm{co}(W)$ is a polygon, then C is equal to the perimeter of $\mathrm{co}(W)$, whereas if $\mathrm{co}(W)$ is a line segment, then C is equal to two times the length of $\mathrm{co}(W)$.

The quantity C is the sum of the length of the sides of $\mathrm{co}(W)$ when $\mathrm{co}(W)$ is a polygon, and C is twice the length of $\mathrm{co}(W)$ when $\mathrm{co}(W)$ is a line segment. As an illustration of why the length of a line segment is counted twice, suppose that $\mathrm{co}(W)$ is a triangle, whose "top" vertex is barely above the base of the triangle. Then the perimeter of $\mathrm{co}(W)$ is only slightly more than two times the length of the base of the triangle. In the limiting case where the top vertex approaches the base of the triangle, C approaches twice the length of the base.

The following result is an immediate consequence of Theorem 6.2.1.

Corollary 6.2.6. *Let f be a transcendental exponential sum of the form* (6.2.1). *Then the number $n(r,1/f)$ of zeros of f in $\{|z| \leq r\}$ has the asymptotic growth*

$$n(r,1/f) = \frac{C}{2\pi}r + o(r), \quad r \to \infty.$$

Example 6.2.7. For the function

$$f(z) = e^{2z} + z^3 e^{iz} + e^{2iz} + (1+z)e^{-2z}$$

in Example 6.1.7, we have $C = 4(1+\sqrt{2})$, which gives

$$n(r,1/f) = (1+o(1))\frac{2(1+\sqrt{2})}{\pi}r, \quad r \to \infty.$$

6.3 Value distribution of exponential polynomials

An *exponential polynomial* f is an entire function of the form

$$f(z) = \sum_{j=1}^{m} P_j(z)e^{Q_j(z)}, \tag{6.3.1}$$

where $P_j(z)$, $Q_j(z)$ are polynomials for $1 \le j \le m$ such that $P_j(z) \not\equiv 0$ for all j and at most one of the exponents $Q_j(z)$ vanishes identically. We assume that the polynomials $Q_j(z)$ are pairwise different and normalized so that $Q_j(0) = 0$. In this case the terms $P_j(z)e^{Q_j(z)}$ in (6.3.1) are linearly independent over the field of rational functions, and the order of f is equal to $\max_{1\le j\le m}\{\deg Q_j\}$.

If f in (6.3.1) is transcendental, then it can be written in the *normalized form*

$$f(z) = F_0(z) + F_1(z)e^{w_1 z^q} + \cdots + F_n(z)e^{w_n z^q}, \tag{6.3.2}$$

where $q = \max\{\deg Q_j\} \ge 1$ is the order of f, the *frequencies* w_j of f are pairwise distinct nonzero constants, the *multipliers* $F_j(z)$ are exponential polynomials of order $\le q-1$ such that $F_j(z) \not\equiv 0$ for $1 \le j \le n$, and $n \le m$. The terms $P_j(z)e^{Q_j(z)}$ with $\deg Q_j \le q-1$, if any, are included in $F_0(z)$.

Remark. In Sections 6.1 and 6.2, we considered exponential sums of the form $f(z) = \sum_{j=1}^{m} P_j(z)e^{w_j z}$ having a set of conjugate frequencies $W = \{\overline{w}_1, \dots, \overline{w}_m\}$. At most one of the frequencies w_j was allowed to vanish, and if this was indeed the case, we assumed that $w_1 = 0$. In this section, we work with exponential polynomials of the normalized form (6.3.2). Due to the presence of the possible $F_0(z)$-term and to the variety of results to be discussed in this section, we will use a slightly different notation, which we introduce next.

Definition 6.3.1. Let f be given by (6.3.2), let $W = \{\overline{w}_1, \dots, \overline{w}_n\}$ denote the set of nonzero conjugate frequencies of f, and let $W_0 = W \cup \{\overline{w}_0 = 0\}$. If $F_0(z) \equiv 0$, then define $W_f = W$, and if $F_0(z) \not\equiv 0$, then set $W_f = W_0$. Let $\mathrm{co}(W)$ be the convex hull of W. Analogously to Definition 6.2.5, the quantity $C = C(\mathrm{co}(W))$ is the perimeter of a polygon $\mathrm{co}(W)$ and is twice the length of a line segment $\mathrm{co}(W)$. The quantities $C_0 = C(\mathrm{co}(W_0))$ and $C_f = C(\mathrm{co}(W_f))$ are defined similarly.

Recall from Lemma 3.9.4 that $\mathrm{co}(W)$ is the intersection of all the closed half-planes that contain W. In other words, $\mathrm{co}(W)$ is the smallest closed convex set containing W. The situation is analogous with $\mathrm{co}(W_0)$ and $\mathrm{co}(W_f)$.

The proof of the next result is similar to that of Lemma 6.1.2. The details are omitted.

Proposition 6.3.2. *If f is a transcendental exponential polynomial of the form* (6.3.2), *then the Phragmén–Lindelöf indicator function of f satisfies*

$$h_f(\theta) = \begin{cases} \max_{0\le j\le m}\{\mathrm{Re}(w_j e^{iq\theta})\}, & F_0 \not\equiv 0, \\ \max_{1\le j\le m}\{\mathrm{Re}(w_j e^{iq\theta})\}, & F_0 \equiv 0, \end{cases} \qquad \theta \in \mathbb{R}. \tag{6.3.3}$$

Suppose that f is an exponential polynomial of the form (6.3.2). Analogously as in Definition 6.1.3, we say that $\arg z = \theta^*$ is a *critical ray* for f if

$$h_f(\theta^*) = \operatorname{Re}(w_j e^{iq\theta^*}) = \operatorname{Re}(w_k e^{iq\theta^*}) \tag{6.3.4}$$

for a pair of indices $j \neq k$. The angle θ^* is called a *cusp*. Analogously as in Definition 6.1.4, we now define orthogonal rays. For reasons explained above, it is convenient to define them for the set W_f instead of the set W.

Definition 6.3.3. Let f be a transcendental exponential polynomial of the form (6.3.2), and let $\overline{w}_j$, $\overline{w}_k$ be two consecutive vertices of $\operatorname{co}(W_f)$. Then the ray $\arg z = \theta^{\perp}$ that is parallel to the outer normal vector of $\operatorname{co}(W_f)$ determined by the line through $\overline{w}_j$, $\overline{w}_k$ is called an *orthogonal ray* for f.

The number of critical rays for f is sq, where $q = \rho(f)$, and s is the number of vertices of $\operatorname{co}(W_f)$. Indeed, as in Corollary 6.1.6, there are s orthogonal rays $\arg z = \theta^{\perp}$, each of which induces q critical rays $\arg z = \theta_n^*$, where

$$\theta_n^* = \frac{\theta^{\perp} + 2n\pi}{q}, \qquad n = 0, \ldots, q-1. \tag{6.3.5}$$

This is easy to see because for each orthogonal ray $\arg z = \theta^{\perp}$, we have

$$\operatorname{Re}(w_j e^{i\theta^{\perp}}) = \operatorname{Re}(w_k e^{i\theta^{\perp}}) = \max_{\ell}\{\operatorname{Re}(w_\ell e^{i\theta^{\perp}})\},$$

and hence every $\theta^* = \theta_n^*$ satisfies (6.3.4). Thus, if $q \geq 2$ (i. e., when the exponential polynomial is not an exponential sum), then there are more critical rays than orthogonal rays, whereas if $q = 1$ (exponential sum), then the critical rays and orthogonal rays are the same by Lemma 6.1.5.

Around each critical ray $\arg z = \theta^*$, we define a *modified logarithmic strip*

$$\Lambda_p(c, \theta^*) = \left\{ z = re^{i\theta} : r > 1,\ |\theta - \theta^*| < c\frac{\log r}{r^p} \right\}, \tag{6.3.6}$$

where $c > 0$ and $p \in \mathbb{N}$. In general, it is obvious that $\Lambda_{p_2}(c, \theta^*) \subset \Lambda_{p_1}(c, \theta^*)$ for a pair of integers $p_1 \leq p_2$. Schwengeler [246] and Steinmetz [263] proved that the zeros of exponential sums and exponential polynomials lie in the domains $\Lambda_p(c, \theta^*)$ for $p = 1$, with at most $O(r^{q-1})$ exceptional zeros; see Theorems 6.3.4 and 6.3.9. The parameter p was discovered in [115], where it was used in the case where the multipliers $F_j(z)$ of f are exponential polynomials of order $\rho(F_j) \leq q - p$, where p is an integer satisfying $1 \leq p \leq q$. If $p = q$, then this assumption forces the multipliers $F_j(z)$ to be polynomials.

The zero distribution results in [115, 246, 263] rely on the fact that one exponential term at a time controls the growth of f outside the (modified) logarithmic strips. The situation is analogous to Lemma 6.2.3, where critical sectors have been used instead of critical logarithmic strips for simplicity. To "boost up" the dominance of those particular exponential terms, it needs to be assumed that the parameter $c > 0$ is large enough, which forces the variable z to be sufficiently far away from the critical rays. However, just by looking at the definition of the domains $\Lambda_p(c, \theta^*)$, we can see that the assumption "$c > 0$ is large enough" is rather mild.

In addition to assuming that $c > 0$ is large enough, the reasoning in [126] has another type of restriction for the parameter c. Indeed, it needs to be assumed that the boundary curves $\partial\Lambda_p(c, \theta^*)$ of the modified logarithmic strips avoid a certain collection of discs around the zeros of f. To achieve this, it is proved in [126] that the set $C \subset (0, \infty)$ of values c for which the boundary curves $\partial\Lambda_p(c, \theta^*)$ meet infinitely many of such discs has linear measure zero. This result is analogous to Lemma 1.4.1, according to which the set $\Theta \subset [0, 2\pi)$ of angles θ for which the ray $\arg z = \theta$ meets infinitely many discs of a given $\mathcal{E}$-set has linear measure zero.

The following theorem is an analogue of Theorem 6.2.1 for exponential polynomials, where the critical sectors can be replaced with domains of the form $\Lambda_p(c, \theta^*)$ around the critical rays. The result is true even if $q = 1$ (in which case $p = 1$). It is assumed that $c > 0$ is large enough and that c does not belong to an exceptional set C of linear measure zero. Judging by the statement of Theorem 6.3.4, we find both assumptions on c to be rather mild.

Theorem 6.3.4 ([126, Theorem 2.1]). *Let f be an exponential polynomial of the form* (6.3.2), *where the multipliers $F_j(z)$ are exponential polynomials of order $\rho(F_j) \le q - p$, where p is an integer satisfying $1 \le p \le q$. Then the number $n_{\lambda_p}(r)$ of zeros of f that lie outside the finitely many logarithmic strips $\Lambda_p(c, \theta^*)$ satisfies*

$$n_{\lambda_p}(r) = O(r^{q-p} + \log r). \tag{6.3.7}$$

Moreover, let $\overline{w}_j, \overline{w}_k$ be two consecutive vertices of $\mathrm{co}(W_f)$, *and let $\theta_0^\perp$ denote the argument of the ray parallel to the outer normal of* $\mathrm{co}(W_f)$ *corresponding to its side $[\overline{w}_j, \overline{w}_k]$. Let $\varepsilon > 0$, and let θ^* be a cusp of f associated with the angle $\theta_0^\perp$ by means of* (6.3.5). *Then the number of zeros $n(r, \Lambda_p)$ of f in $\Lambda_p(c, \theta^*) \cap \{|z| \le r\}$ satisfies*

$$n(r, \Lambda_p) = |w_j - w_k|\frac{r^q}{2\pi} + O(r^{q-p}\log^{3+\varepsilon} r). \tag{6.3.8}$$

The proof is similar to that of Theorem 6.2.1 but a lot more technical basically for two reasons: (1) The multipliers of f are now (transcendental) exponential polynomials themselves, so estimating their minimum modulus induces exceptional sets that need to be addressed. Meanwhile, the minimum modulus of a polynomial is obvious. (2) When constructing an analogue of the path of integration Γ appearing in the proof of Theorem 6.2.1, it is no longer enough to assume that there are no zeros of f on the path. Indeed, the path analogous to Γ needs to avoid certain discs around the zeros of f, which requires some precision. An interested reader can find more details in [115, 117, 126].

Remark. It seems plausible that $O(r^{q-p} + \log r)$ could be the sharp error term in (6.3.8), but this has not been confirmed. The log-term covers the case $p = q$.

Example 6.3.5 ([263]). An exponential polynomial may have infinitely many zeros outside the logarithmic strips $\Lambda(c, \theta^*)$. Indeed, let

$$f(z) = \left(e^{2\pi i z^2} - 1\right)\left(e^{(1+i)\pi z} - 1\right) =: F_0(z) + F_1(z)e^{2\pi i z^2},$$

where $F_0(z) = 1 - e^{(1+i)\pi z}$ and $F_1(z) = -F_0(z)$. Then the orthogonal rays for $\text{co}(W_f) = [0, -2\pi i]$ are $\arg z = 0$ and $\arg z = \pi$. The function f has infinitely many zeros $z_n = n + ni$, $n \in \mathbb{Z}$, lying outside the logarithmic strips around the critical rays $\arg z = 0, \pi/2, \pi, 3\pi/2$, with at most finitely many exceptions. Indeed, if $c > 0$ is large, then some of the zeros $z_n = n + ni$ of f may lie in the logarithmic strips near the origin, but there could of course be at most finitely many such zeros. This example illustrates the sharpness of (6.3.7) in the case where $q = 2$ and $p = 1$.

Corollary 6.2.6 extends to exponential polynomials as follows.

Corollary 6.3.6. *Under the assumptions of Theorem* 6.3.4, *the number* $n(r, 1/f)$ *of zeros of* f *has the asymptotic growth*

$$n(r, 1/f) = qC(\text{co}(W_f))\frac{r^q}{2\pi} + O(r^{q-p}\log^{3+\varepsilon} r). \tag{6.3.9}$$

Proof. Keeping in mind that the points $\overline{w}_j$, $\overline{w}_k$ in Theorem 6.3.4 are two consecutive vertices of $\text{co}(W_f)$, we obtain $\sum |w_j - w_k| = C(\text{co}(W_f))$, where the sum is taken over all the sides of $\text{co}(W_f)$. From (6.3.5) we see that the orthogonal ray associated with the pair $\overline{w}_j$, $\overline{w}_k$ induces q critical rays. Hence the statement follows directly from Theorem 6.3.4. □

Next, we focus on the Nevanlinna functions $T(r,f)$, $m(r, 1/f)$, $N(r, 1/f)$ of exponential polynomials f, but we will return back to considering the nonintegrated counting function $n(r, 1/f)$ at the end of this section.

Theorem 6.3.7 ([263]). *Let f be a transcendental exponential polynomial in the normalized form* (6.3.2). *Then*

$$T(r,f) = C_0\frac{r^q}{2\pi} + o(r^q). \tag{6.3.10}$$

The proof of Theorem 6.3.7 is rather deep and therefore is omitted. In addition to the original proof in [263], another proof involving an improved error term and a more precise analysis of the multipliers of f can be found in [126]. Note that [126, Lemma 4.3] is incorrect, but this issue is fixed by the reasoning in [117, Section 3.5]. A proof of any version of Theorem 6.3.7 requires Cauchy's formula for perimeters of convex hulls. This formula is discussed in detail in [117, Appendix B].

Example 6.3.8. If $f(z)$ is $\sin z$ or $\cos z$, then for a given $\varepsilon > 0$,

$$
\begin{aligned}
T(r,f) &= \frac{1}{2\pi}\left(\int_{-\pi+\varepsilon}^{-\varepsilon} + \int_{\varepsilon}^{\pi-\varepsilon}\right)\log^+|f(re^{i\theta})|\,d\theta + \frac{1}{2\pi}\left(\int_{|\theta-\pi|<\varepsilon} + \int_{|\theta|<\varepsilon}\right)\log^+|f(re^{i\theta})|\,d\theta \\
&= \frac{1}{2\pi}\int_{-\pi+\varepsilon}^{-\varepsilon}\log^+\left|e^{ire^{i\theta}}\right|d\theta + \frac{1}{2\pi}\int_{\varepsilon}^{\pi-\varepsilon}\log^+\left|e^{-ire^{i\theta}}\right|d\theta + O(\varepsilon r) \\
&= \frac{1}{2\pi}\int_{-\pi+\varepsilon}^{-\varepsilon}(-r\sin\theta)\,d\theta + \frac{1}{2\pi}\int_{\varepsilon}^{\pi-\varepsilon} r\sin\theta\,d\theta + O(\varepsilon r) = \frac{2r}{\pi} + O(\varepsilon r).
\end{aligned}
$$

Note that in this case, $W = \{-i, i\}$, $\mathrm{co}(W) = [-i, i]$, and $0 \in [-i, i]$. Hence $C = C_0 = 4$. Moreover, since $\varepsilon > 0$ was chosen arbitrarily, we have $O(\varepsilon r) = o(r)$. This calculation supports the conclusion of Theorem 6.3.7.

Theorem 6.3.9 ([263]). *Let f be a transcendental exponential polynomial of the forms* (6.3.1) *and* (6.3.2). *If $F_0(z) \not\equiv 0$, then*

$$m\left(r, \frac{1}{f}\right) = o(r^q) \tag{6.3.11}$$

and

$$N\left(r, \frac{1}{f}\right) = C_0\frac{r^q}{2\pi} + o(r^q), \tag{6.3.12}$$

whereas if $F_0(z) \equiv 0$, then

$$N\left(r, \frac{1}{f}\right) = C\frac{r^q}{2\pi} + o(r^q). \tag{6.3.13}$$

If one of the exponents $Q_1, \ldots, Q_m$ in (6.3.1) *vanishes identically, then* (6.3.11) *can be replaced with*

$$m\left(r, \frac{1}{f}\right) = O(\log r). \tag{6.3.14}$$

The proof of Theorem 6.3.9 requires a technical lemma, which involves Wronskian determinants. Recall that the Wronskian determinant of n given meromorphic functions $f_1, \ldots, f_n$ is defined as

$$W(f_1, \ldots, f_n) = \begin{vmatrix} f_1 & \cdots & f_n \\ f_1' & \cdots & f_n' \\ \vdots & \ddots & \vdots \\ f_1^{(n-1)} & \cdots & f_n^{(n-1)} \end{vmatrix}.$$

The next result is well known and useful.

Proposition 6.3.10 ([166, Proposition 1.4.2]). *Let $f_1,\dots,f_n$ be meromorphic functions. Then $W(f_1,\dots,f_n)\equiv 0$ on $\mathbb{C}$ if and only if $f_1,\dots,f_n$ are linearly dependent on $\mathbb{C}$.*

For $\nu = 0,\dots,n-1$, we denote by $W_\nu(f_1,\dots,f_n)$ the determinant that comes from $W(f_1,\dots,f_n)$ by replacing the row $(f_1^{(\nu)},\dots,f_n^{(\nu)})$ with $(f_1^{(n)},\dots,f_n^{(n)})$.

Lemma 6.3.11 ([166, Proposition 1.4.6]). *Let $f_1,\dots,f_n$ be linearly independent meromorphic functions, and define*

$$a_\nu = -\frac{W_\nu(f_1,\dots,f_n)}{W(f_1,\dots,f_n)}, \qquad \nu = 0,\dots,n-1. \tag{6.3.15}$$

Then each a_ν is a meromorphic function whose poles are among the poles of $f_1,\dots,f_n$ or the zeros of $W(f_1,\dots,f_n)$. Moreover, the functions $f_1,\dots,f_n$ are solutions of the linear differential equation

$$f^{(n)} + a_{n-1}(z)f^{(n-1)} + \cdots + a_1(z)f' + a_0(z)f = 0.$$

Proof. Observe that each a_ν in (6.3.15) is a well-defined meromorphic function because we have $W(f_1,\dots,f_n)\not\equiv 0$ on $\mathbb{C}$ by Proposition 6.3.10. The statement in the lemma about the poles of a_ν is easy to see from (6.3.15).

Since the last statement is clear for $n = 1$, we suppose that $n \geq 2$. Let h be any meromorphic function, and suppose first that n is even. Expanding the determinant $W(f_1,\dots,f_n,h)$ with respect to the last column, we obtain

$$\begin{aligned} W(f_1,\dots,f_n,h) &= h\begin{vmatrix} f_1' & \cdots & f_n' \\ f_1'' & \cdots & f_n'' \\ \vdots & \ddots & \vdots \\ f_1^{(n)} & \cdots & f_n^{(n)} \end{vmatrix} - h'\begin{vmatrix} f_1 & \cdots & f_n \\ f_1'' & \cdots & f_n'' \\ \vdots & \ddots & \vdots \\ f_1^{(n)} & \cdots & f_n^{(n)} \end{vmatrix} \\ &\quad + \cdots - h^{(n-1)}\begin{vmatrix} f_1 & \cdots & f_n \\ \vdots & \ddots & \vdots \\ f_1^{(n-2)} & \cdots & f_n^{(n-2)} \\ f_1^{(n)} & \cdots & f_n^{(n)} \end{vmatrix} + h^{(n)}W(f_1,\dots,f_n) \\ &=: hD_0 - h'D_1 + \cdots - h^{(n-1)}D_{n-1} + h^{(n)}W(f_1,\dots,f_n). \end{aligned}$$

Consider the determinant D_0. We swap the first and nth rows, then swap the second and nth rows, and so on, and finally we swap the $(n-1)$th and nth rows. Whenever two rows are swapped, the modified determinant will be multiplied by -1. Here we do an odd number of $n-1$ swaps, and hence $D_0 = -W_0(f_1,\dots,f_n)$. Consider next the determinant D_1. We swap the second and nth rows, then swap the third and nth rows, and so on, and finally we swap the $(n-1)$th and nth rows. Here we do an even number of $n-2$ swaps, and hence $D_1 = W_1(f_1,\dots,f_n)$. Continuing in this fashion, we obtain

$$D_\nu = (-1)^{\nu+1} W_\nu(f_1, \dots, f_n), \quad \nu = 0, \dots, n-1. \tag{6.3.16}$$

This leads to the equation

$$\frac{W(f_1, \dots, f_n, h)}{W(f_1, \dots, f_n)} = h^{(n)} - \sum_{\nu=0}^{n-1} \frac{W_\nu(f_1, \dots, f_n)}{W(f_1, \dots, f_n)} h^{(\nu)} = h^{(n)} + \sum_{\nu=0}^{n-1} a_\nu(z) h^{(\nu)}.$$

If h is any of the functions $f_1, \dots, f_n$, then $W(f_1, \dots, f_n, h)$ has two equal columns and hence vanishes identically. This proves the statement in the case where n is even. If n is odd, then the coefficient of $h^{(n)}$ is $-W(f_1, \dots, f_n)$, and (6.3.16) is to be replaced with

$$D_\nu = (-1)^\nu W_\nu(f_1, \dots, f_n), \quad \nu = 0, \dots, n-1.$$

The rest of the proof remains the same as above. □

Proof of Theorem 6.3.9. The proof is divided into three parts.

(1) Suppose that one of the exponents $Q_1, \dots, Q_m$ vanishes identically. Without loss of generality, we may suppose that $Q_1(z) \equiv 0$. This is a stronger assumption than $F_0(z) \not\equiv 0$ because F_0 could be transcendental. However, the conclusion (6.3.14) is also stronger than (6.3.11). Note that $m \geq 2$ in this case because f is transcendental.

As the functions $f_j(z) = P_{j+1}(z)e^{Q_{j+1}(z)}$, $j = 1, \dots, m-1$, are linearly independent, they solve the linear differential equation

$$L_n(f) := f^{(n)} + a_{n-1}(z) f^{(n-1)} + \cdots + a_0(z) f = 0, \quad n = m-1, \tag{6.3.17}$$

where the coefficients $a_j(z)$ are given by (6.3.15). Note that $f_j'(z) = R_{j+1}(z)e^{Q_{j+1}(z)}$, where

$$R_{j+1}(z) = P_{j+1}'(z) + Q_{j+1}'(z) P_{j+1}(z) \not\equiv 0.$$

For if $R_{j+1}(z) \equiv 0$, then $P_{j+1}'(z)/P_{j+1}(z) = -Q_{j+1}'(z)$, and integration on both sides reveals that $P_{j+1}(z) = C\exp(-Q_{j+1}(z))$ for some $C \in \mathbb{C} \setminus \{0\}$, which is a contradiction because $P_{j+1}(z), Q_{j+1}(z)$ are polynomials and $Q_{j+1}(z)$ is nonconstant. This means that the exponential term $e^{Q_{j+1}(z)}$ remains for derivatives of f_j of all orders. Therefore, from each column of each determinant

$$W(f_1, \dots, f_n),\ W_0(f_1, \dots, f_n),\ \dots,\ W_{n-1}(f_1, \dots, f_n),$$

we can pull out the corresponding exponential term as a common factor. Thus $\exp(Q_2(z) + \cdots + Q_m(z))$ is a common factor for each determinant above, which cancels out in the quotients (6.3.15). The entries of the remaining determinants are polynomials. Hence the coefficients $a_j(z)$ in (6.3.17) are rational functions.

If $P_1(z)$ were a solution of (6.3.17), it would have to be linearly dependent with the solutions $f_1, \dots, f_n$. Thus $P_1(z)$ would be transcendental, a contradiction. It follows that

$a_n(z) := L_n(P_1) \not\equiv 0$ is a rational function. We conclude that the exponential polynomial f is a solution of the nonhomogeneous equation

$$f^{(n)} + a_{n-1}(z)f^{(n-1)} + \cdots + a_0(z)f = a_n(z).$$

This can be rewritten as

$$\frac{1}{f} = \frac{1}{a_n}\left(\frac{f^{(n)}}{f} + a_{n-1}\frac{f^{(n-1)}}{f} + \cdots + a_0\right).$$

Using the first fundamental theorem and the lemma on the logarithmic derivative together with the fact that the coefficients are rational, we obtain

$$m\left(r, \frac{1}{f}\right) \le \sum_{j=0}^{n}\left(T(r, a_j) + m\left(r, \frac{f^{(j)}}{f}\right)\right) + O(1) = O(\log r),$$

which is (6.3.14).

(2) Suppose next that none of the exponents $Q_1, \ldots, Q_m$ vanishes identically and that $F_0(z) \not\equiv 0$. Then at least one of the exponents $Q_1, \ldots, Q_m$ must have degree $< q$. Without loss of generality, suppose that $\deg(Q_1) < q$. For convenience, set

$$g(z) = e^{-Q_1(z)}f(z) = P_1(z) + \sum_{k=2}^{m} P_k(z)e^{Q_k(z)-Q_1(z)}. \tag{6.3.18}$$

Now the discussion above can be applied to g, giving us $m(r, 1/g) = O(\log r)$, and by the first fundamental theorem,

$$N\left(r, \frac{1}{f}\right) = N\left(r, \frac{1}{g}\right) = T(r, g) + O(\log r). \tag{6.3.19}$$

Then applying Theorem 6.3.7 to f and using (6.3.18), we get

$$T(r,f) = C_0\frac{r^q}{2\pi} + o(r^q) = T(r,g) + o(r^q). \tag{6.3.20}$$

By the first fundamental theorem,

$$m\left(r, \frac{1}{f}\right) = T(r,f) - N\left(r, \frac{1}{f}\right) + O(1),$$

and we deduce estimate (6.3.11) by combining (6.3.19) and (6.3.20). Estimate (6.3.12) also follows from (6.3.19) and (6.3.20).

(3) Finally, suppose that none of the exponents $Q_1, \ldots, Q_m$ vanishes identically and that $F_0(z) \equiv 0$. In this case, for every j, we have $\deg(Q_j) = q$ and $w_j \neq 0$. Define g by (6.3.18), let

$$U = \{\overline{w}_2 - \overline{w}_1, \ldots, \overline{w}_m - \overline{w}_1\}$$

be the set of nonzero conjugate frequencies of g, and set $U_0 = U \cup \{0\}$. By simple vector calculus the convex hulls $\mathrm{co}(W)$ and $\mathrm{co}(U_0)$ have the same exact shape, with only a different location in the complex plane. The set $\mathrm{co}(U_0)$ is obtained by shifting $\mathrm{co}(W)$ by the vector $-\overline{w}_1$. Therefore, in particular,

$$C = C(\mathrm{co}(W)) = C_0(\mathrm{co}(U_0)).$$

Hence, applying Theorem 6.3.7 to g, we get

$$T(r,g) = C_0(\mathrm{co}(U_0))\frac{r^q}{2\pi} + o(r^q) = C\frac{r^q}{2\pi} + o(r^q). \tag{6.3.21}$$

The final statement (6.3.13) now follows from (6.3.21) and (6.3.19), where we observe that the assumption $F_0(z) \not\equiv 0$ in case (2) was not needed to prove (6.3.19). □

The error terms appearing in Theorem 6.3.9 have been improved in [126]. Here we have settled for larger error terms for two reasons:

(1) Maintaining simplicity in the reasoning.
(2) In many applications, the growth of the small error terms are irrelevant.

We now return back to the number $n(r, 1/f)$ of zeros of f. Corollary 6.3.6 is based on Theorem 6.3.4, which we did not prove. The following corollary is weaker than Corollary 6.3.6 because the error term is larger. However, we are able to give a rather straightforward proof for it based on Theorem 6.3.9, which we did prove.

Corollary 6.3.12 ([126]). *Let f be a transcendental exponential polynomial in the normalized form* (6.3.2). *Then*

$$n(r, 1/f) = \frac{qC}{2\pi}r^q + o(r^q).$$

Proof. By Corollary 6.2.6 we may suppose that $q \geq 2$. If $C = 0$, then the function f in (6.3.2) reduces to the form $f(z) = F(z)e^{wz^q}$, where $F(z)$ is an exponential polynomial of order $\leq q-1$ and $w \in \mathbb{C} \setminus \{0\}$. Since the statement is clear in this case, we may suppose that $C > 0$.

Write (6.3.13) in the form

$$C\frac{r^q}{2\pi} - \varepsilon(r)r^q \leq N\left(r, \frac{1}{f}\right) \leq C\frac{r^q}{2\pi} + \varepsilon(r)r^q, \tag{6.3.22}$$

where $\varepsilon : (0,\infty) \to (0,1)$ is a function such that $\varepsilon(r) \to 0$ as $r \to \infty$. Let $\delta : (0,\infty) \to (0,1)$ be any function such that $\delta(r) \to 0$ as $r \to \infty$. Using $\log x \geq \frac{x-1}{x}$ for $x \geq 1$, we have

$$
\begin{aligned}
n(r,1/f) &= \frac{n(r,1/f)}{\log(1+\delta(r))} \int\limits_r^{r(1+\delta(r))} \frac{dt}{t} \le \frac{1+\delta(r)}{\delta(r)} \int\limits_r^{r(1+\delta(r))} \frac{n(t,1/f)}{t}\,dt \\
&= \frac{1+\delta(r)}{\delta(r)} (N(r(1+\delta(r)),1/f) - N(r,1/f)).
\end{aligned}
$$

From the inequalities in (6.3.22) and the binomial formula, we get

$$
\begin{aligned}
n(r,1/f) &= \frac{C(1+\delta(r))}{2\pi\delta(r)} (r^q(1+\delta(r))^q - r^q + O(\varepsilon(r)r^q)) \\
&= \frac{C(1+\delta(r))}{2\pi\delta(r)} \left(r^q \sum_{j=1}^{q} \binom{q}{j} \delta(r)^j + O(\varepsilon(r)r^q) \right) \\
&= \frac{C(1+\delta(r))}{2\pi\delta(r)} \left(qr^q\delta(r) + r^q \sum_{j=2}^{q} \binom{q}{j} \delta(r)^j + O(\varepsilon(r)r^q) \right) \\
&= qC\frac{r^q}{2\pi} + O(\delta(r)r^q) + O(\varepsilon(r)r^q) + O\left(\frac{\varepsilon(r)r^q}{\delta(r)} \right).
\end{aligned}
$$

Similarly, using $\log x \le x - 1$ for $x \ge 1$, we get

$$
\begin{aligned}
n(r,1/f) &= \frac{n(r,1/f)}{-\log(1-\delta(r))} \int\limits_{r(1-\delta(r))}^{r} \frac{dt}{t} \ge \frac{1-\delta(r)}{\delta(r)} \int\limits_{r(1-\delta(r))}^{r} \frac{n(t,1/f)}{t}\,dt \\
&= qC\frac{r^q}{2\pi} + O(\delta(r)r^q) + O(\varepsilon(r)r^q) + O\left(\frac{\varepsilon(r)r^q}{\delta(r)} \right).
\end{aligned}
$$

Now choosing $\delta(r) = \sqrt{\varepsilon(r)}$, all the error terms above will have the form $o(r^q)$. □

We close the section with two examples to illustrate several concepts. The statements in these examples follow from the definitions, results, and discussion in this section. The first example has a lot of symmetry.

Example 6.3.13. Consider the exponential polynomial

$$f(z) = ze^z + e^{z^2} + (1+z^2)e^{iz^2} + e^{-z^2} + (z+e^z)e^{-iz^2}.$$

This function f has the form (6.3.2) with $q = 2$ and $F_0(z) = ze^z \not\equiv 0$. Then $W_f = W_0 = \{0, 1, -i, -1, i\}$, and the four orthogonal rays are

$$\arg z = \pm\pi/4, \pm 3\pi/4,$$

whereas the eight critical rays are

$$\arg z = \pm\pi/8, \pm 3\pi/8, \pm 5\pi/8, \pm 7\pi/8.$$

The majority of zeros of f lie in logarithmic strips around the critical rays.

For this function f, the frequencies and conjugate frequencies are the same. Since the vertices of $\mathrm{co}(W_f)$ are $1, -i, -1, i$, we have $C_0 = C = 4\sqrt{2}$. Then the Nevanlinna functions satisfy

$$T(r,f) = (1+o(1))\frac{2\sqrt{2}}{\pi}r^2, \quad m\left(r,\frac{1}{f}\right) = o(r^2),$$
$$N\left(r,\frac{1}{f}\right) = (1+o(1))\frac{2\sqrt{2}}{\pi}r^2, \quad n\left(r,\frac{1}{f}\right) = (1+o(1))\frac{4\sqrt{2}}{\pi}r^2.$$

Since $\rho(f) = 2$ and

$$\mathrm{Re}\left(e^{2i\theta}\right) = \cos(2\theta), \quad \mathrm{Re}\left(ie^{2i\theta}\right) = -\sin(2\theta),$$
$$\mathrm{Re}\left(-e^{2i\theta}\right) = -\cos(2\theta), \quad \mathrm{Re}\left(-ie^{2i\theta}\right) = \sin(2\theta),$$

the Phragmén–Lindelöf indicator function $h_f(\theta)$ is given by

$$h_f(\theta) = \begin{cases} \cos(2\theta), & \theta \in [-\pi/8, \pi/8] \cup [7\pi/8, 9\pi/8], \\ \sin(2\theta), & \theta \in [\pi/8, 3\pi/8] \cup [9\pi/8, 11\pi/8], \\ -\cos(2\theta), & \theta \in [3\pi/8, 5\pi/8] \cup [11\pi/8, 13\pi/8], \\ -\sin(2\theta), & \theta \in [5\pi/8, 7\pi/8] \cup [13\pi/8, 15\pi/8]. \end{cases} \tag{6.3.23}$$

From (6.3.23) we obtain

$$\max_{0\le\theta\le 2\pi}\{h_f(\theta)\} = 1,$$

which means that the type of f is $\tau(f) = 1$ by Corollary 4.2.7.

Example 6.3.14. Consider the exponential polynomial

$$f(z) = e^{-z^3} - \left(e^z + e^{z^2}\right)e^{-iz^3} + \left(z^4 + z + 1\right)e^{(1/\sqrt{3})z^3}.$$

This function f has the form (6.3.2) with $q = 3$ and $F_0(z) \equiv 0$. Then $W_f = W = \{-1, i, 1/\sqrt{3}\}$, and in this case the frequencies and conjugate frequencies are not the same set. The three orthogonal rays are

$$\arg z = \pi/6, 3\pi/4, 3\pi/2,$$

and the nine critical rays are

$$\arg z = \pi/18, 13\pi/18, 25\pi/18, \pi/4, 11\pi/12, 19\pi/12, \pi/2, 7\pi/6, 11\pi/6.$$

Since the vertices of $\mathrm{co}(W_f)$ are $-1, i, 1/\sqrt{3}$, we have $C = C_0 = \alpha$ where $\alpha = 1 + \sqrt{2} + \sqrt{3}$. Then the Nevanlinna functions satisfy

$$T(r,f) = (1+o(1))\frac{\alpha}{2\pi}r^3, \quad m\left(r,\frac{1}{f}\right) = o(r^3),$$
$$N\left(r,\frac{1}{f}\right) = (1+o(1))\frac{\alpha}{2\pi}r^3, \quad n\left(r,\frac{1}{f}\right) = (1+o(1))\frac{3\alpha}{2\pi}r^3.$$

Since $\rho(f) = 3$ and

$$\mathrm{Re}\left(-e^{3i\theta}\right) = -\cos(3\theta), \quad \mathrm{Re}\left(-ie^{3i\theta}\right) = \sin(3\theta), \quad \mathrm{Re}\left((1/\sqrt{3})e^{3i\theta}\right) = \frac{1}{\sqrt{3}}\cos(3\theta),$$

the Phragmén–Lindelöf indicator function $h_f(\theta)$ is given by

$$h_f(\theta) = \begin{cases} (1/\sqrt{3})\cos(3\theta), & \theta \in [-\pi/6, \pi/18] \cup [\pi/2, 13\pi/18] \cup [7\pi/6, 25\pi/18], \\ \sin(3\theta), & \theta \in [\pi/18, \pi/4] \cup [13\pi/18, 11\pi/12] \cup [25\pi/18, 19\pi/12], \\ -\cos(3\theta), & \theta \in [\pi/4, \pi/2] \cup [11\pi/12, 7\pi/6] \cup [19\pi/12, 11\pi/6]. \end{cases}$$

Therefore,

$$\max_{0\le\theta\le 2\pi} \{h_f(\theta)\} = 1,$$

which means that the type of f is $\tau(f) = 1$.

6.4 Exponential polynomials solve linear differential equations

Exponential polynomials are often presented as solutions of linear differential equations to both illustrate results and provide counterexamples. The general solution of a simple linear differential equation of the form

$$f^{(n)} + a_{n-1}f^{(n-1)} + \cdots + a_1 f' + a_0 f = 0 \tag{6.4.1}$$

with constants $a_0, \ldots, a_n$ is an exponential sum with polynomial coefficients whose frequencies are the distinct solutions of the characteristic equation

$$w^n + \sum_{j=0}^{n-1} a_j w^j = 0.$$

If at least one of the coefficients in (6.4.1) is a nonconstant polynomial, then (6.4.1) may not have exponential polynomial solutions. Airy's equation is an example.

The aim of this section is to show that in the converse direction, any nonconstant exponential polynomial f is a solution of a linear differential equation with polynomial coefficients, where the order of the differential equation can be estimated in terms of the form of f.

Some preparations are needed. For simplicity, let D denote the differential operator d/dz. For entire functions f, g, define

$$(D - g)f = f' - gf.$$

This operator can be applied repeatedly. For example,

$$(D - g_2)(D - g_1)f = (D - g_2)(f' - g_1 f) = f'' - (g_1 + g_2)f' + (g_1 g_2 - g_1')f.$$

We now state and prove the main result in this section.

Theorem 6.4.1 ([282]). *Let f be a transcendental exponential polynomial*

$$f(z)=\sum_{j=1}^{m}P_j(z)e^{Q_j(z)},\tag{6.4.2}$$

and set

$$p_j=1+\deg P_j \quad \text{and} \quad q=\max_{1\le j\le m}\{\deg Q_j\}.$$

Then f is a solution of a linear differential equation of the form

$$(D-Q_m')^{t_m}\cdots(D-Q_2')^{t_2}(D-Q_1')^{t_1}f=0\tag{6.4.3}$$

with polynomial coefficients, where $t_1,\dots,t_m$ are nonnegative integers with $t_1\ge p_1$. Moreover, the order n of equation (6.4.3) *satisfies*

$$n=\sum_{j=1}^{m}t_j\le\sum_{j=1}^{m}p_jq^{m-j}.\tag{6.4.4}$$

Proof. By the convention in (6.3.1) we have $P_j(z)\not\equiv 0$ for all $j=1,\dots,m$ in (6.4.2). The proof of the theorem is by induction on m and p_1.

Suppose that $m=1$. If $p_1=1$, then $f(z)=P_1e^{Q_1(z)}$, where $P_1\in\mathbb{C}\setminus\{0\}$, whereas if $p_1\ge 2$, then $f(z)=P_1(z)e^{Q_1(z)}$, where $P_1(z)$ is a polynomial of degree $p_1-1\ge 1$. In both cases,

$$\begin{aligned}(D-Q_1')f&=P_1'e^{Q_1}+P_1Q_1'e^{Q_1}-P_1Q_1'e^{Q_1}=P_1'e^{Q_1},\\ (D-Q_1')^2f&=(D-Q_1')P_1'e^{Q_1}=P_1''e^{Q_1},\\ &\vdots\\ (D-Q_1')^{p_1}f&=(D-Q_1')P_1^{(p_1-1)}e^{Q_1}=P_1^{(p_1)}e^{Q_1}\equiv 0.\end{aligned}$$

This proves statements (6.4.3) and (6.4.4) in the case where $m=1$ and $p_1\ge 1$ is any integer. Clearly, $t_1=p_1\ge 1$.

Assume then that the statements hold for some $m=k\ge 1$ and for any integer $p_1\ge 1$. We will now prove that the statements hold for $m=k+1$ and for any integer $p_1\ge 1$ by using induction on p_1.

(1) Suppose that $p_1=1$ and $m=k+1$. Set

$$F(z):=(D-Q_1')f(z)=\sum_{j=1}^{k+1}(D-Q_1')P_j(z)e^{Q_j(z)}=\sum_{j=1}^{k+1}S_j(z)e^{Q_j(z)},\tag{6.4.5}$$

where $S_j:=P_j'+(Q_j'-Q_1')P_j$ for $j=1,\dots,k+1$. Set

$$s_j := 1 + \deg S_j, \quad j = 1, \dots, k+1.$$

We have $s_1 = 1$ and $s_j \le p_j + q - 1$ for $j \ge 2$. Since $p_1 = 1$, we have $S_1(z) \equiv 0$, which means that $F(z)$ reduces to

$$F(z) = \sum_{j=1}^{k} S_{j+1}(z) e^{Q_{j+1}(z)} = \sum_{j=1}^{k} R_j(z) e^{T_j(z)},$$

where $R_j(z) = S_{j+1}(z)$ and $T_j(z) = Q_{j+1}(z)$. Set

$$r_j := 1 + \deg R_j, \quad j = 1, \dots, k.$$

Then

$$r_j = 1 + \deg S_{j+1} = s_{j+1} \le p_{j+1} + q - 1, \quad j = 1, \dots, k.$$

The induction hypothesis on $m = k$ now states that the function $F(z)$ solves a differential equation of the form

$$\begin{aligned} 0 &= (D - T_k')^{\tau_k} \cdots (D - T_2')^{\tau_2} (D - T_1')^{\tau_1} F \\ &= (D - Q_{k+1}')^{\tau_k} \cdots (D - Q_3')^{\tau_2} (D - Q_2')^{\tau_1} F, \end{aligned}$$

where $\tau_1, \dots, \tau_k$ are nonnegative integers with $\tau_1 \ge p_1$ such that the order $\tau_1 + \cdots + \tau_k$ of the equation satisfies

$$\begin{aligned} \sum_{j=1}^{k} \tau_j &\le \sum_{j=1}^{k} r_j q^{k-j} \le \sum_{j=1}^{k} (p_{j+1} + q - 1) q^{k-j} \\ &= \sum_{j=1}^{k} p_{j+1} q^{k-j} + (q-1) \sum_{j=1}^{k} q^{k-j} = \sum_{j=1}^{k} p_{j+1} q^{k-j} + q^k - 1 \\ &= \sum_{j=0}^{k} p_{j+1} q^{k-j} - 1 = \sum_{j=1}^{k+1} p_j q^{k+1-j} - 1, \end{aligned}$$

where we have used the fact that $p_1 = 1$. Since $F = (D - Q_1')f$, the function f satisfies a differential equation

$$(D - Q_{k+1}')^{\tau_k} \cdots (D - Q_3')^{\tau_2} (D - Q_2')^{\tau_1} (D - Q_1') f = 0 \tag{6.4.6}$$

of order

$$n = \sum_{j=1}^{k} \tau_j + 1 \le \sum_{j=1}^{k+1} p_j q^{k+1-j}.$$

Equation (6.4.6) corresponds to (6.4.3), where $m = k + 1$, $t_1 = 1$ and $t_j = \tau_{j-1}$ for $j = 2, \dots, k+1$. Thus (6.4.4) also holds for $m = k + 1$.

(2) Now assume that the statements hold for some $p_1 = \ell \geq 1$ and for $m = k + 1$.
(3) Suppose that $p_1 = \ell + 1$ and $m = k + 1$, and consider the function $F(z)$ in (6.4.5). Since $S_1 = P_1'$, we have $s_1 = p_1 - 1 = \ell$. Moreover, for $j \geq 2$, we have $s_j \leq p_j + q - 1$. The induction hypotheses in (2), namely that the statements hold for $p_1 = \ell$ and $m = k+1$, now state that the function $F(z)$ solves a differential equation of the form

$$(D - Q_{k+1}')^{\sigma_{k+1}} \cdots (D - Q_2')^{\sigma_2}(D - Q_1')^{\sigma_1}F = 0,$$

where $\sigma_1, \ldots, \sigma_{k+1}$ are nonnegative integers with $\sigma_1 \geq \ell$ such that the order $\sigma_1 + \cdots + \sigma_{k+1}$ of the equation satisfies

$$\sum_{j=1}^{k+1} \sigma_j \leq \sum_{j=1}^{k+1} s_j q^{k+1-j} \leq (p_1 - 1)q^k + \sum_{j=2}^{k+1} (p_j + q - 1)q^{k+1-j}$$
$$= \sum_{j=1}^{k+1} p_j q^{k+1-j} - q^k + (q-1)\sum_{j=2}^{k+1} q^{k+1-j} = \sum_{j=1}^{k+1} p_j q^{k+1-j} - 1.$$

Since $F = (D - Q_1')f$, the function f satisfies (6.4.3), where $t_1 = \sigma_1 + 1 \geq \ell + 1 = p_1$ and $t_j = \sigma_j$ for $j = 2, \ldots, k$. Thus (6.4.4) also holds for $p_1 = \ell + 1$.

We conclude that the statements hold for all integers $m \geq 1$ and $p_1 \geq 1$. □

Theorem 6.4.1 is about transcendental exponential polynomials. We close this section with a result about nonconstant polynomials (which are also exponential polynomials). It appears to be a new result.

Theorem 6.4.2. *Any nonconstant polynomial is a solution of a linear differential equation of the form*

$$f^{(n)} + P_{n-1}(z)f^{(n-1)} + \cdots + P_0(z)f = 0, \tag{6.4.7}$$

where each $P_k(z)$ is a polynomial with $P_0(z) \not\equiv 0$. The order n of equation (6.4.7) *satisfies $n = 1 + \deg f$.*

Proof. By [96, Example 6], if $n \geq 2$, then the equation

$$f^{(n)} - \frac{z^{n-1}}{(n-1)!}f^{(n-1)} + \frac{z^{n-2}}{(n-2)!}f^{(n-2)} - \cdots + (-1)^n f = 0, \tag{6.4.8}$$

possesses the $n-1$ linearly independent solutions

$$f_k(z) = z^k, \quad k = 1, \ldots, n-1.$$

Since a linear combination of solutions of (6.4.8) is also a solution of (6.4.8), we see that any nonconstant polynomial f of degree $n-1$ satisfying $f(0) = 0$ is a solution of equation (6.4.8).

In general, now suppose that $f(0)$ is any complex constant. Since f is a nonconstant polynomial, it follows by the fundamental theorem of algebra that there exists a constant $c \in \mathbb{C}$ such that $f(c) = 0$. Set $g(z) = f(z + c)$. Then $g(0) = 0$, and by (6.4.8), $g(z)$ is a polynomial solution of

$$g^{(n)} - \frac{(z+c)^{n-1}}{(n-1)!} g^{(n-1)} + \frac{(z+c)^{n-2}}{(n-2)!} g^{(n-2)} - \cdots + (-1)^n g = 0. \tag{6.4.9}$$

Then from (6.4.9) we can deduce that $f(z) = g(z - c)$ is a solution of an equation of the form (6.4.7) with $n = 1 + \deg f$. □

6.5 Dual exponential polynomials

We begin with examples, which will lead to defining dual exponential polynomials.

Example 6.5.1 ([116, 287]). *Second-order equations.* Let $P(z)$ be a polynomial. Then $f(z) = e^z + 1$ is a solution of

$$f'' - \left(1 + P(z)\left(1 + e^{-z}\right)\right)f' + P(z)f = 0.$$

Let m be a positive integer. For suitable constants $C_0, \ldots, C_m \in \mathbb{C} \setminus \{0\}$, the function $f(z) = C_0 + C_1 e^z + \cdots + C_m e^{mz}$ solves

$$f'' + e^{-z} f' - m^2 f = 0.$$

Example 6.5.2 ([287]). *Third-order equations.* The function $f(z) = 16 - 27e^{-2z} + 27e^{-3z}$ satisfies

$$f''' + (1/9)\left(9 + 9e^z + 4e^{2z}\right)f'' - 5f' + 3f = 0,$$

and the function $f(z) = 1 + 4e^z + 6e^{2z}$ satisfies

$$f''' + \left(e^{-z} + 1\right)f'' - 4f' - 4f = 0.$$

Example 6.5.3 ([287]). *Equations of arbitrary even order.* Let $n \geq 2$ be an even integer, and let μ be an integer satisfying $0 < \mu < n$. If $a_\mu(z) = e^{-z}$, $a_j = (-1)^j$ for $\mu < j < n$ and $a_j = (-1)^{j+1}$ for $0 \leq j < \mu$, then $f(z) = e^z + 1$ solves the equation

$$f^{(n)} + a_{n-1} f^{(n-1)} + \cdots + a_\mu(z) f^{(\mu)} + \cdots + a_1 f' + a_0 f = 0.$$

Example 6.5.4 ([287]). *Solutions of order of growth two.* The function $f(z) = \exp(z^2) - 1$ solves the following two equations:

$$f''' + \left(\exp(-z^2) - 2z - 1\right)f'' - 4f' + \left(4z^2 + 2\right)f = 0,$$
$$f''' - 2zf'' - \left(2 + 2e^{-z^2}\right)f' - 4zf = 0.$$

The function $f(z) = \exp(z^2/2 + z) + z + 1$ is a solution of

$$f''' + \left(\exp(-z^2/2 - z) - z - 1\right)f'' - f' - (z+1)f = 0.$$

All the above examples are characterized by a certain dual relationship between the solution f and the dominating coefficient, say $A(z)$. For instance, if f includes a term e^{cz} or e^{cz^2}, where $c > 0$ is a constant, then $A(z)$ includes, respectively, an opposite term of the form e^{-dz} or e^{-dz^2}, where $d > 0$ is a constant. In other words, if $|f(z)|$ is large in a certain sector, then $|A(z)|$ is small in the same sector, and vice versa. This leads to the following definition.

Definition 6.5.5 ([116, 287]). Let f be an exponential polynomial in the normalized form (6.3.2). If the nonzero frequencies $w_1, \dots, w_m$ of f all lie on a fixed ray $\arg w = \theta$, then f is called a *one-sided exponential polynomial*. If g is another one-sided exponential polynomial of the same order q as f such that the nonzero frequencies of g all lie on the opposite ray $\arg w = \theta + \pi$, then f and g are called *dual exponential polynomials.*

As observed in [117], by a careful examination of the proof of [287, Theorem 2], using Niino's theorem [299, Theorem 1.53] instead of Borel's theorem when proving [287, Lemma 2], we obtain the following result, which is illustrated by Examples 6.5.1–6.5.4.

Theorem 6.5.6 ([287]). *Suppose that f is an exponential polynomial solution of*

$$f^{(n)} + a_{n-1}(z)f^{(n-1)} + \cdots + a_\mu(z)f^{(\mu)} + \cdots + a_1(z)f' + a_0(z)f = 0, \tag{6.5.1}$$

where for precisely one index $\mu \in \{1, \dots, n-1\}$, $a_\mu(z)$ is a transcendental exponential polynomial, and $a_j(z)$ is an entire function satisfying $T(r, a_j) = S(r, a_\mu)$ for all $j \neq \mu$. Then one of the following statements hold:

(a) *f is a polynomial of degree $\leq \mu - 1$.*

(b) *f and $a_\mu(z)$ are dual exponential polynomials of order $q \in \mathbb{N}$. In this case, f has the normalized representation*

$$f(z) = S(z) + F_1(z)e^{w_1 z^q} + \cdots + F_m(z)e^{w_m z^q}, \tag{6.5.2}$$

where the nonzero constants w_j have the same argument, $m \in \mathbb{N}$, and $S(z)$ is a polynomial satisfying $S(z) \not\equiv 0$ and $\deg S \leq \mu - 1$.

In particular, if $\rho(a_\mu f^{(\mu)}) < q$ and $a_j(z)$ are polynomials for $j \neq \mu$, then we have case (b) *with the particular form*

$$f(z) = S(z) + Q(z)e^{P(z)} \quad \text{and} \quad a_\mu(z) = R(z)e^{-P(z)}, \tag{6.5.3}$$

where $P(z)$, $Q(z)$, $R(z)$ are polynomials with $\deg P = q$.

Note that $a_0(z)$ cannot be the dominating coefficient in Theorem 6.5.6 because in this case, all solutions $f \not\equiv 0$ of (6.5.1) would be of infinite order of growth. This is easy to see by writing

$$-a_0(z) = a_1(z)\frac{f'}{f} + \cdots + a_{n-1}(z)\frac{f^{(n-1)}}{f} + \frac{f^{(n)}}{f}$$

and then considering the lemma on the logarithmic derivative (Lemma 4.6.10) together with the assumption $T(r, a_j) = S(r, a_0)$ for $j \geq 1$ on the right-hand side, which would give a contradiction if f were of finite order.

For a deeper view on duality, we need to define the concepts of commensurable frequencies and strongly dual exponential polynomials.

Definition 6.5.7 ([170]). Let f in (6.3.2) be a one-sided exponential polynomial. If there exists a constant $w \in \mathbb{C} \setminus \{0\}$ such that w_j/w is a positive integer for every $j = 1, 2, \ldots, m$, then the (nonzero) frequencies of f are said to be *commensurable*, and w is called a *common factor*.

Example 6.5.8 ([116]). The functions $f(z) = e^{\pi z} + 3e^{2\pi z} + ze^{3\pi z}$ and $g(z) = e^{4iz} + e^{6iz}$ are one-sided exponential polynomials, where their frequencies are commensurable. Examples of common factors are π, $\pi/2$ for f and i, $2i$ for g. Thus common factors are not unique.

Note that if the nonzero frequencies of f are commensurable, then they are clearly linearly dependent over the rational numbers, but not the other way around. For example, the points $w_1 = 1$, $w_2 = \sqrt{2}$, $w_3 = \sqrt{2} - 1$ are linearly dependent over the rational numbers, but they are not commensurable.

Definition 6.5.9 ([116]). Suppose that f and g are dual exponential polynomials with commensurable frequencies $\{w_j\}$ ($j > 0$) and $\{\lambda_i\}$ ($i > 0$), respectively, sharing the same common factor w but with opposite signs. If for all i and j, the points $w_j + \lambda_i$ all lie on one ray whose initial point is the origin, then f and g are called *strongly dual exponential polynomials*.

Example 6.5.10 ([116]). The functions $f(z) = 1 + ze^z + 2e^{3z}$ and $g(z) = 1 - e^{-z}$ are strongly dual exponential polynomials, whereas $f(z)$ and $h(z) = g(z) + 2z^2e^{-2z}$ are not. Observe that f and h are dual exponential polynomials.

We proceed to state a stronger version of Theorem 6.5.6 in the second-order case, which is illustrated by Example 6.5.1 and by Example 6.5.3 with $n = 2$.

Theorem 6.5.11 ([116, 287]). *Suppose that f and $A(z)$ in the equation*

$$f'' + A(z)f' + B(z)f = 0$$

are transcendental exponential polynomials and that $B(z)$ is an entire function satisfying $T(r, B) = o(T(r, A))$. Then the following statements hold:

(a) *f and $A(z)$ are dual exponential polynomials of order $q \in \mathbb{N}$, f has the normalized representation*

$$f(z) = c + F_1(z)e^{w_1 z^q} + \cdots + F_m(z)e^{w_m z^q}, \quad m \in \mathbb{N},\ c \in \mathbb{C} \setminus \{0\},$$

and $B(z)$ is an exponential polynomial of order $\rho(B) \leq q - 1$.

(b) *The frequencies of f are commensurable if and only if the frequencies of $A(z)$ are commensurable. In both cases, f and $A(z)$ are strongly dual exponential polynomials.*

(c) *If $\rho(Af') < q$, then $q = 1$, and*

$$A(z) = ae^{-wz}, \quad B(z) = -w^2, \quad \text{and} \quad f(z) = c\left(1 + \frac{w}{a}e^{wz}\right), \tag{6.5.4}$$

where $w = w_1$ and $a \in \mathbb{C} \setminus \{0\}$.

6.6 Separation of zeros

By Corollary 6.3.12 the number of zeros (multiplicities included) of a transcendental exponential polynomial f of order q in the disc $|z| \leq r$ is asymptotically

$$n(r, 1/f) = \frac{qC}{2\pi}r^q + o(r^q),$$

where the constant $C > 0$ is defined in Section 6.3. For $z_0 \in \mathbb{C}$, let $n(z_0, r, f)$ denote the number of zeros (multiplicities included) of f in the disc $D(z_0, r) = \{z \in \mathbb{C} : |z - z_0| \leq r\}$. By applying the result above to the exponential polynomial $g(z) = f(z_0 + z)$, which has the same frequencies as f and therefore the same perimeter C as f, we find that

$$n(z_0, r, f) = \frac{qC}{2\pi}r^q + o(r^q).$$

This result does not say anything about the number of zeros of f in $D(z_0, r)$ for fixed values of r. The main goal of this section is to find such estimates. This will show that the zeros of f are uniformly separated in the complex plane.

Theorem 6.6.1 ([282]). *Let f be a transcendental exponential polynomial*

$$f(z) = \sum_{j=1}^{m} P_j(z)e^{Q_j(z)}, \tag{6.6.1}$$

and set

$$p_j = 1 + \deg P_j \quad \text{and} \quad q = \max_{1 \leq j \leq m}\{\deg Q_j\}.$$

Given $z_0 \in \mathbb{C}$, set

$$n = \sum_{j=1}^{m} p_j q^{m-j} \quad \text{and} \quad \psi(z) = \sum_{j=1}^{q} A_j z^j, \tag{6.6.2}$$

where

$$A_j = \max_{1\le p\le m}\left|\frac{Q_p^{(j)}(z_0)}{j!}\right|, \quad j = 1,\dots,q. \tag{6.6.3}$$

For all $\gamma > 1$ and $r > 0$, we have

$$\max_{|z|\le\gamma r}|f(z_0+z)| \le \gamma^{n-1}\left(\frac{\gamma^2}{\gamma^2-1}\right)^{1/2} \cdot \exp(\psi(r)+\psi(\gamma r))\max_{|z|\le r}|f(z_0+z)|. \tag{6.6.4}$$

Moreover, for all $r, t > 0$ and $s > 1$, we have

$$n(z_0, r, f) \le \frac{H}{\log s}, \tag{6.6.5}$$

where

$$H = (n-1)\log\left(s+1+\frac{s}{t}\right) + \psi(tr) + \psi((st+s+t)r) + \frac{t^2}{2(st+s)(st+s+2t)}.$$

Before proving Theorem 6.6.1, we discuss three corollaries, two of which will remarkably simplify the complicated estimate (6.6.5).

Corollary 6.6.2 ([282]). *Under the assumptions of Theorem* 6.6.1,

$$n(z_0, r, f) \le 3(n-1) + \psi(4r).$$

Proof. If $n = 1$, then f has no zeros, and the statement is trivial. Suppose that $n \ge 2$. We find that ψ is a superadditive function in the sense that $\psi(x) + \psi(y) \le \psi(x+y)$ for all $x, y > 0$. Further, choosing $s = 3$ and $t = 1/6$ in (6.6.5), we obtain

$$n(z_0, r, f) \le (n-1)\cdot\frac{\log 22}{\log 3} + \frac{1}{\log 3}\left(\psi\left(\frac{r}{6}\right) + \psi\left(\frac{11r}{3}\right) + \frac{1}{966}\right),$$

from which the statement follows. □

Corollary 6.6.3 ([282]). *Under the assumptions of Theorem* 6.6.1, *if $q \ge 2$, then*

$$n(z_0, r, f) \le (n-1)(1 + q\log(23q)) + 3q\cdot\psi(r).$$

Proof. Taking $s = e^{1/q}$ and $t = 1/(21q)$ in (6.6.5), it follows that

$$n(z_0, r, f) \le qG,$$

where

$$G = (n-1)\log(e^{1/q} + 1 + 21qe^{1/q}) + \psi\left(\frac{r}{21q}\right) + \psi\left(e^{1/q}\left(1+\frac{2}{21q}\right)r\right) + \frac{1}{882q^2}.$$

Here

$$\psi\left(\frac{r}{21q}\right)+\psi\left(e^{1/q}\left(1+\frac{2}{21q}\right)r\right)=\sum_{j=1}^{q}A_j\left(\left(\frac{1}{21q}\right)^j+e^{j/q}\left(1+\frac{2}{21q}\right)^j\right)r^j$$
$$\leq\max_{1\leq j\leq q}\left\{\left(\frac{1}{21q}\right)^j+e^{j/q}\left(1+\frac{2}{21q}\right)^j\right\}\psi(r).$$

Given $a,b>0$, the function $h(x)=a^x+b^x$ is twice differentiable and convex for $x>0$. Using this and the fact that $q\geq 2$, we infer

$$\max_{1\leq j\leq q}\left\{\left(\frac{1}{21q}\right)^j+e^{j/q}\left(1+\frac{2}{21q}\right)^j\right\}=\max\left\{\frac{1}{21q}+e^{1/q}\left(1+\frac{2}{21q}\right),\left(\frac{1}{21q}\right)^q+e\left(1+\frac{2}{21q}\right)^q\right\}.$$

Since $(1+1/x)^x<e$ for $x>0$, we obtain

$$\max_{1\leq j\leq q}\left\{\left(\frac{1}{21q}\right)^j+e^{j/q}\left(1+\frac{2}{21q}\right)^j\right\}\leq\max\left\{\frac{1}{42}+\frac{22}{21}e^{1/2},\left(\frac{1}{42}\right)^2+e^{23/21}\right\}<3.$$

Since we may assume that $n\geq 2$,

$$\begin{aligned}&q\left((n-1)\log(e^{1/q}+1+21qe^{1/q})+\frac{1}{882q^2}\right)\\&\leq q(n-1)\left(\log(e^{1/q}+1+21qe^{1/q})+\frac{1}{882q^2}\right)\\&=(n-1)\left(1+q\log\left((1+e^{-1/q}+21q)\exp\left(\frac{1}{882q^2}\right)\right)\right).\end{aligned}$$

Clearly, $1+e^{-1/q}+21q\leq(1/2+(1/2)e^{-1/q}+21)q\leq 22q$, and

$$\exp\left(\frac{1}{882q^2}\right)\leq\exp\left(\frac{1}{4\cdot 882}\right)<1.0003.$$

Therefore,

$$q\left((n-1)\log(e^{1/q}+1+21qe^{1/q})+\frac{1}{882q^2}\right)\leq(n-1)(1+q\log(23q)).$$

The statement follows by combining the estimates above. □

In the following corollary, we discuss the separation of zeros of exponential polynomials. This is a corrected form of [115, Lemma 5.2].

Corollary 6.6.4. *Let f be an exponential polynomial of the form* (6.6.1), *where $q\geq 1$ denotes the order of f, and let $\{\zeta_k\}$ be any sequence of nonzero complex points tending to infinity. Then for $p\geq q-1$, each disc of the form $|z-\zeta_k|\leq|\zeta_k|^{-p}$ contains at most $K=K(f,p)$ zeros of f, where K is independent of k.*

Proof. Choosing $z_0 = \zeta_k$ and using Corollary 6.6.2 or 6.6.3, we see that it suffices to show that $\psi(r)$ is uniformly bounded for $r = |\zeta_k|^{-p}$. Since the polynomials $Q_j(z)$ in (6.6.1) are of degree at most q, there exists a constant $C > 0$ such that the constants A_j in (6.6.3) satisfy

$$A_j \le C|\zeta_k|^{q-j}, \quad j = 1, \dots, q,$$

for all k. Therefore,

$$\psi(r) = \psi(|\zeta_k|^{-p}) \le C \sum_{j=1}^{q} |\zeta_k|^{q-j-jp} \le \begin{cases} qC|\zeta_k|^{q-1-p}, & |\zeta_k| \ge 1, \\ qC|\zeta_k|^{-qp}, & |\zeta_k| < 1. \end{cases}$$

Since $p \ge q-1$ and $|\zeta_k| \to \infty$ as $k \to \infty$, we have $\psi(r) \le K$, where $K = K(f,p)$ is independent of k. □

We now proceed to prove Theorem 6.6.1. We need three lemmas.

Lemma 6.6.5 ([282]). *Let g and h be analytic functions in a simply connected domain U, and suppose that f solves the equation*

$$(D - g(z))f(z) = h(z), \quad z \in U.$$

Then for all $z, z_0 \in U$,

$$f(z) = f(z_0)\exp\left(\int_{z_0}^{z} g(\zeta)\,d\zeta\right) + \int_{z_0}^{z} h(\zeta)\exp\left(\int_{\zeta}^{z} g(\xi)\,d\xi\right)d\zeta, \tag{6.6.6}$$

where the paths of integration lie in U.

Proof. Let $F(z)$ denote the right-hand side of (6.6.6). Then $F(z)$ is analytic in U and satisfies $F(z_0) = f(z_0)$ and $(D-g)F = h$ by Leibniz' integral rule. Thus $F \equiv f$ in U by Theorem 1.7.1. □

We use the symbol $\ll$ for term-by-term majoration of power series. For example,

$$\sum_{n=0}^{\infty} a_n z^n \ll \sum_{n=0}^{\infty} b_n z^n \tag{6.6.7}$$

means that $|a_n| \le b_n$ for every $n \ge 0$.

Lemma 6.6.6 ([282]). *Let $g_1, \dots, g_n$ be analytic functions in the disc $D(z_0, R_0)$, and let $\psi(z)$ be analytic in the disc $|z| < R_0$ such that $\psi(0) = 0$ and*

$$g_j(z_0 + \zeta) \ll \psi'(\zeta), \quad j = 1, \dots, n. \tag{6.6.8}$$

Suppose that f is a solution of the differential equation

$$(D - g_n) \cdots (D - g_1)f = 0 \tag{6.6.9}$$

in $D(z_0, R_0)$, and define the constants $a_j \in \mathbb{R}$ by equating the Taylor coefficients on both sides of the equation

$$\sum_{j=0}^{\infty} a_j z^j = \exp(\psi(z)) \sum_{j=0}^{\infty} \left| \frac{f^{(j)}(z_0)}{j!} \right| z^j. \tag{6.6.10}$$

Then for all $0 \le r < R_0$ and $0 \le \theta < 2\pi$,

$$|f(z_0 + re^{i\theta})| \le \sum_{j=0}^{n-1} a_j r^j \exp(\psi(r)).$$

Proof. We prove the result by induction on n. If $n = 1$, then solving the equation $f' - g_1 f = 0$, we obtain

$$f(z) = f(z_0) \exp\left(\int_{z_0}^{z} g_1(\zeta)\, d\zeta \right), \tag{6.6.11}$$

where $z \in D(z_0, R_0)$, and the integration path is the line segment $[z_0, z]$. We may write

$$g_1(z_0 + \zeta) = \sum_{n=0}^{\infty} c_n \zeta^n, \quad \psi(\zeta) = \sum_{n=0}^{\infty} d_n \zeta^n, \quad \psi'(\zeta) = \sum_{n=0}^{\infty} (n+1) d_{n+1} \zeta^n,$$

where $|\zeta| < R_0$, and $c_n \in \mathbb{C}$ and $d_n \ge 0$ satisfy $|c_n| \le (n+1) d_{n+1}$ for all $n \ge 0$ by assumption (6.6.8). Since $\psi(0) = 0$, we have $d_0 = 0$. Then by (6.6.11),

$$\begin{aligned} |f(z_0 + re^{i\theta})| &\le |f(z_0)| \exp\left(\int_0^r |g_1(z_0 + ue^{i\theta})|\, du \right) \\ &\le |f(z_0)| \exp\left(\int_0^r \sum_{n=0}^{\infty} |c_n| u^n\, du \right) \le |f(z_0)| \exp\left(\int_0^r \psi'(u)\, du \right), \end{aligned}$$

which gives

$$|f(z_0 + re^{i\theta})| \le a_0 \exp(\psi(r)).$$

This proves the statement for $n = 1$.

Suppose then that the statement is true for $n = k \ge 1$, and consider the case $n = k + 1 \ge 2$. Write

$$h(z) = f'(z) - g_1(z) f(z) = (D - g_1) f(z).$$

By Lemma 6.6.5, for $z = z_1 = z_0 + re^{i\theta}$,

$$f(z_1) = f(z_0)\exp\left(\int_{z_0}^{z_1} g_1(\zeta)\,d\zeta\right) + \int_{z_0}^{z_1} h(\zeta)\exp\left(\int_{\zeta}^{z_1} g_1(\xi)\,d\xi\right)d\zeta,$$

where the integration paths are the line segments $[z_0, z_1]$ and $[\zeta, z_1]$. Note that $\zeta \in [z_0, z_1]$. By elementary estimates and (6.6.8) when $j = 1$ we have

$$\begin{aligned}|f(z_1)| &\le |f(z_0)|\exp\left(\int_0^r |g_1(z_0 + ue^{i\theta})|\,du\right) + \int_0^r |h(z_0 + ue^{i\theta})|\exp\left(\int_u^r |g_1(z_0 + ve^{i\theta})|\,dv\right)du \\ &\le |f(z_0)|\exp(\psi(r)) + \int_0^r |h(z_0 + ue^{i\theta})|\exp(\psi(r) - \psi(u))\,du. \end{aligned} \tag{6.6.12}$$

By the definition of h and assumption (6.6.9), h is a solution of

$$(D - g_n)\cdots(D - g_2)h = 0$$

in $D(z_0, R_0)$. Therefore, if the constants $b_j \in \mathbb{R}$ are defined by the equation

$$\sum_{j=0}^{\infty} b_j z^j = \exp(\psi(z))\sum_{j=0}^{\infty}\left|\frac{h^{(j)}(z_0)}{j!}\right| z^j, \tag{6.6.13}$$

then by the induction hypothesis we have

$$|h(z_0 + ue^{i\theta})| \le \sum_{j=0}^{n-2} b_j u^j \exp(\psi(u)).$$

Substituting this into (6.6.12), we get

$$\begin{aligned}|f(z_1)| &\le |f(z_0)|\exp(\psi(r)) + \int_0^r \left(\sum_{j=0}^{n-2} b_j u^j\right)\exp(\psi(r))\,du \\ &= \left(|f(z_0)| + \sum_{j-0}^{n-2} \frac{b_j}{j+1} r^{j+1}\right)\exp(\psi(r)).\end{aligned}$$

It remains to prove that

$$|f(z_0)| + \sum_{j=0}^{n-2} \frac{b_j}{j+1} r^{j+1} \le \sum_{j=0}^{n-1} a_j r^j.$$

Since $a_0 = |f(z_0)|$, it suffices to prove that

$$\frac{b_j}{j+1} \le a_{j+1}, \quad j = 0, \ldots, n-2. \tag{6.6.14}$$

Writing $h(z_0+z) = f'(z_0+z) - g_1(z_0+z)f(z_0+z)$ in terms of Taylor series at the origin, we have

$$\sum_{j=0}^{\infty} \frac{h^{(j)}(z_0)}{j!} z^j = \sum_{j=0}^{\infty} \frac{f^{(j+1)}(z_0)}{j!} z^j - g_1(z_0+z) \sum_{j=0}^{\infty} \frac{f^{(j)}(z_0)}{j!} z^j,$$

and therefore, by (6.6.8),

$$\sum_{j=0}^{\infty} \left| \frac{h^{(j)}(z_0)}{j!} \right| z^j \ll \sum_{j=0}^{\infty} \left| \frac{f^{(j+1)}(z_0)}{j!} \right| z^j + \psi'(z) \cdot \sum_{j=0}^{\infty} \left| \frac{f^{(j)}(z_0)}{j!} \right| z^j.$$

This yields

$$\begin{aligned} \exp(\psi(z)) \sum_{j=0}^{\infty} \left| \frac{h^{(j)}(z_0)}{j!} \right| z^j &\ll \left(\sum_{j=0}^{\infty} \left| \frac{f^{(j+1)}(z_0)}{j!} \right| z^j + \psi'(z) \cdot \sum_{j=0}^{\infty} \left| \frac{f^{(j)}(z_0)}{j!} \right| z^j \right) \exp(\psi(z)) \\ &= \frac{d}{dz} \left(\left(\sum_{j=0}^{\infty} \left| \frac{f^{(j)}(z_0)}{j!} \right| z^j \right) \exp(\psi(z)) \right). \end{aligned}$$

By the definition of the constants a_j and b_j in (6.6.10) and (6.6.13), we get

$$\sum_{j=0}^{\infty} b_j z^j \ll \frac{d}{dz} \sum_{j=0}^{\infty} a_j z^j = \sum_{j=0}^{\infty} (j+1) a_{j+1} z^j.$$

This implies (6.6.14), and the proof is now complete. □

Lemma 6.6.7 ([272]). *Let $R, t > 0$ and $s > 1$, and let $f \not\equiv 0$ be an analytic function in a simply connected domain containing the closed disc $|z - z_0| \le (st + s + t)R$. Then*

$$n(z_0, R, f) \le \frac{1}{\log s} \cdot \log\left(\frac{\max_{|z| \le (st+s+t)R} |f(z_0+z)|}{\max_{|z| \le tR} |f(z_0+z)|} \right).$$

Proof. We prove the statement in the case $z_0 = 0$. Suppose that $\zeta \in \overline{D}(0, tR)$ is such that

$$|f(\zeta)| = \max_{|z| \le tR} |f(z)| = M(tR, f).$$

Recall Jensen's theorem for a function g analytic in the disc $D(0, \varrho)$ with $g(0) \neq 0$:

$$N(r, 1/g) = \frac{1}{2\pi} \int_0^{2\pi} \log|g(re^{i\theta})|\, d\theta - \log|g(0)|, \quad 0 \le r < \varrho.$$

Defining $g(z) = f(\zeta + z)$ and keeping in mind that $f(\zeta) \neq 0$, we have

$$\int_0^{sR} \frac{n(\zeta,t,f)}{t}\,dt = \frac{1}{2\pi}\int_0^{2\pi} \log|f(\zeta + sRe^{i\theta})|\,d\theta - \log|f(\zeta)|.$$

Since

$$\int_0^{sR} \frac{n(\zeta,t,f)}{t}\,dt \geq \int_R^{sR} \frac{n(\zeta,t,f)}{t}\,dt \geq n(\zeta,R,f)\log s,$$

we deduce that

$$n(\zeta,R,f) \leq \frac{1}{\log s}\cdot \max_{|z|\leq sR} \log \frac{|f(\zeta+z)|}{|f(\zeta)|}.$$

Clearly,

$$D(0,R) \subset D(\zeta,(1+t)R) \quad \text{and} \quad D(\zeta,(st+s)R) \subset D(0,(st+s+t)R),$$

and therefore,

$$\begin{aligned} n(0,R,f) \leq n(\zeta,(1+t)R,f) &\leq \frac{1}{\log s}\cdot \log \max_{|z|\leq sR(1+t)} \frac{|f(\zeta+z)|}{|f(\zeta)|} \\ &\leq \frac{1}{\log s}\cdot \log \frac{M((st+s+t)R,f)}{M(tR,f)}. \end{aligned}$$

This proves Lemma 6.6.7 in the case where $z_0 = 0$. The general case where $z_0 \neq 0$ follows by a change of variable. □

Formula (6.6.15) is known as the *Parseval identity*. Typically, the result is formulated for Fourier series, but we will need a version for analytic functions.

Theorem 6.6.8 ([204, p. 355]). *Let $f(z) = \sum_{n=0}^{\infty} a_n(z-z_0)^n$ be an analytic function in the disc $D(z_0,R)$, where $0 < R \leq \infty$ and $z_0 \in \mathbb{C}$. Then*

$$\sum_{n=0}^{\infty} |a_n|^2 r^{2n} = \frac{1}{2\pi}\int_0^{2\pi} |f(z_0 + re^{i\theta})|^2\,d\theta, \quad 0 \leq r < R. \tag{6.6.15}$$

Proof of Theorem 6.6.1. By Theorem 6.4.1, the function f solves the differential equation (6.4.3), which can be written in the form (6.6.9) with

$$\begin{aligned} g_1 \equiv \cdots \equiv g_{t_1} &\equiv Q_1', \\ g_{t_1+1} \equiv \cdots \equiv g_{t_1+t_2} &\equiv Q_2', \end{aligned}$$

and so on. The order n of (6.6.9) is given in (6.6.2). Developing $g(z) = Q_\ell(z_0 + z)$ at the origin, we find that

$$Q_\ell(z_0+z)=g(z)=\sum_{j=0}^{\deg Q_\ell}\frac{g^{(j)}(0)}{j!}z^j=\sum_{j=0}^{\deg Q_\ell}\frac{Q_\ell^{(j)}(z_0)}{j!}z^j,$$

and hence

$$Q'_\ell(z_0+z)=g'(z)=\sum_{j=1}^{\deg Q_\ell}\frac{Q_\ell^{(j)}(z_0)}{(j-1)!}z^{j-1}=\sum_{j=0}^{\deg Q_\ell-1}\frac{Q_\ell^{(j+1)}(z_0)}{j!}z^j.$$

If ψ is the function defined in (6.6.2), then $\psi(0)=0$, and

$$\psi'(z)=\sum_{j=0}^{q-1}(j+1)A_{j+1}z^j=\sum_{j=0}^{q-1}\max_{1\le p\le m}\left|\frac{Q_p^{(j+1)}(z_0)}{j!}\right|z^j.$$

Consequently,

$$Q'_\ell(z_0+\zeta)\ll\psi'(\zeta),\quad \ell=1,\dots,m.$$

By Lemma 6.6.6,

$$\max_{|z|\le\gamma r}|f(z_0+z)|\le\sum_{j=0}^{n-1}a_j(\gamma r)^j\exp(\psi(\gamma r)).\tag{6.6.16}$$

Applying the Cauchy–Schwarz inequality, for $0<r<\gamma r<R_0$, we have

$$\left(\sum_{j=0}^{n-1}a_j(\gamma r)^j\right)^2\le\sum_{j=0}^{n-1}(a_jr^j)^2\cdot\sum_{j=0}^{n-1}\gamma^{2j}\le\frac{\gamma^{2(n-1)}}{1-\gamma^{-2}}\cdot\sum_{j=0}^{n-1}(a_jr^j)^2.\tag{6.6.17}$$

Let g be the function defined by

$$g(z)=\sum_{j=0}^{\infty}a_jz^j,$$

where $a_j\ge 0$ are the constants discussed above and defined in (6.6.10), and where ψ is the polynomial defined in (6.6.2). As a product of two entire functions, g is entire. Applying Parseval's identity (6.6.15) to g, we obtain

$$\begin{aligned}\sum_{j=0}^{\infty}(a_jr^j)^2 &\overset{\text{Par.}}{=}\frac{1}{2\pi}\int_0^{2\pi}|g(re^{i\theta})|^2\,d\theta=\frac{1}{2\pi}\int_0^{2\pi}g(re^{i\theta})\overline{g(re^{i\theta})}\,d\theta\\ &=\frac{1}{2\pi}\int_0^{2\pi}\left(\sum_{j=0}^{\infty}a_j(re^{i\theta})^j\right)\left(\sum_{j=0}^{\infty}a_j(re^{-i\theta})^j\right)d\theta.\end{aligned}$$

Then, using (6.6.10), we have

$$\sum_{j=0}^{\infty}(a_j r^j)^2 = \frac{1}{2\pi}\int_0^{2\pi}\left|\sum_{j=0}^{\infty}\left|\frac{f^{(j)}(z_0)}{j!}\right|(re^{i\theta})^j\right|^2 \cdot |\exp(\psi(re^{i\theta}))|^2\,d\theta$$
$$\leq \exp(2\psi(r))\cdot\frac{1}{2\pi}\int_0^{2\pi}\left|\sum_{j=0}^{\infty}\left|\frac{f^{(j)}(z_0)}{j!}\right|(re^{i\theta})^j\right|^2 d\theta.$$

Let h be the entire function defined by

$$h(z) = \sum_{j=0}^{\infty} c_j z^j, \quad c_j = \left|\frac{f^{(j)}(z_0)}{j!}\right|.$$

Applying Parseval's identity (6.6.15) to h, we obtain

$$\sum_{j=0}^{\infty}\left|\frac{f^{(j)}(z_0)}{j!}\right|^2 r^{2j} = \sum_{j=0}^{\infty}|c_j|^2 r^{2j} \overset{\text{Par.}}{=} \frac{1}{2\pi}\int_0^{2\pi}|h(re^{i\theta})|^2\,d\theta$$
$$= \frac{1}{2\pi}\int_0^{2\pi}\left|\sum_{j=0}^{\infty}\left|\frac{f^{(j)}(z_0)}{j!}\right|(re^{i\theta})^j\right|^2 d\theta.$$

From this and from another application of Parseval's identity (6.6.15) applied to f, it follows that

$$\sum_{j=0}^{\infty}(a_j r^j)^2 \leq \exp(2\psi(r))\cdot\sum_{j=0}^{\infty}\left|\frac{f^{(j)}(z_0)}{j!}\right|^2 r^{2j}$$
$$\overset{\text{Par.}}{=} \exp(2\psi(r))\cdot\frac{1}{2\pi}\int_0^{2\pi}|f(z_0+re^{i\theta})|^2\,d\theta$$
$$\leq \exp(2\psi(r))\cdot\max_{|z|\leq r}|f(z_0+z)|^2.$$

Thus by (6.6.17),

$$\sum_{j=0}^{n-1} a_j(\gamma r)^j \leq \gamma^{n-1}\left(\frac{1}{1-\gamma^{-2}}\right)^{1/2}\cdot\exp(\psi(r))\cdot\max_{|z|\leq r}|f(z_0+z)|.$$

Combining this with (6.6.16) yields (6.6.4).

To prove (6.6.5), we use Lemma 6.6.7 with $\gamma = s+1+s/t$ and $r = tR$, giving us

$$n(z_0, r/t, f) \leq \frac{1}{\log s}\cdot\log\left(\frac{\max_{|z|\leq\gamma r}|f(z_0+z)|}{\max_{|z|\leq r}|f(z_0+z)|}\right).$$

Next, using (6.6.4), we have

$$n(z_0, r, f) \le \frac{1}{\log s} \cdot \log\left(y^{n-1} \left(\frac{1}{1-y^{-2}} \right)^{1/2} \exp(\psi(tr) + \psi(ytr)) \right)$$
$$= \frac{1}{\log s} \cdot \left((n-1)\log y + \frac{1}{2}\log\frac{1}{1-y^{-2}} + \psi(tr) + \psi(ytr) \right),$$

where

$$\log\frac{1}{1-y^{-2}} < \frac{y^{-2}}{1-y^{-2}} = \frac{1}{y^2-1}.$$

Statement (6.6.5) follows by substituting back $y = s + 1 + s/t$. □

We close this section with an estimate for the multiplicity of zeros.

Theorem 6.6.9 ([281, Example 1]). *Let f be a transcendental exponential polynomial of the form* (6.6.1), *where the P_j and Q_j are polynomials such that*

$$\max_{1\le j\le m}\{\deg P_j\} = p \quad \text{and} \quad \max_{1\le j\le m}\{\deg Q_j\} = q.$$

Then the multiplicity of any zero of f cannot exceed the number

$$m - 1 + mp + \frac{1}{2}m(m-1)(q-1).$$

7 Asymptotic and regular growth

We return back to considering the growth of entire functions f, here in more precise ways than in Chapter 2. The growth of the characteristic $T(r,f)$ is compared to the growth of the logarithmic maximum modulus $\log M(r,f)$, after which we show that an entire function f with Fabry or Fejér gaps satisfies $T(r,f) \sim \log M(r,f)$. We also discuss a result of Clunie, according to which, for any suitable function ϕ, we can find a transcendental entire function f such that $T(r,f) \sim \log M(r,f) \sim \phi(r)$.

The remaining part of this chapter is devoted to entire functions f of perfectly regular growth and of completely regular growth. The former means that the logarithmic maximum modulus $\log M(r,f)$ has regular growth, whereas the latter means that the modulus $|f(z)|$ has regular growth in a large portion of the complex plane. We then discuss some fundamental properties of these functions and prove that completely regular growth implies perfectly regular growth. Exponential polynomials turn out to be functions of completely regular growth.

7.1 Comparison of $T(r,f)$ and $\log M(r,f)$

We begin with a well-known comparison of $T(r,f)$ and $\log M(r,f)$.

Lemma 7.1.1 ([166, Proposition 2.2.2]). *Let f be a nonconstant entire function, and let $r_0 > 0$ be such that $M(r_0,f) > 1$. Then*

$$T(r,f) \le \log M(r,f) \le \frac{R+r}{R-r} T(R,f), \quad r_0 < r < R < \infty. \tag{7.1.1}$$

Proof. Since f is entire and $M(r_0,f) > 1$, we have

$$T(r,f) = m(r,f) = \frac{1}{2\pi} \int_0^{2\pi} \log^+ |f(re^{i\theta})|\, d\theta \le \log M(r,f), \quad r > r_0.$$

This proves the first inequality in (7.1.1).

To prove the second inequality, let $r > r_0$ and choose a point ζ that satisfies $\zeta = re^{i\theta}$ and $|f(\zeta)| = M(r,f)$. Let $\{a_j\}$ denote the zeros of f listed according to multiplicities and ordered with respect to increasing moduli. It is easy to see that

$$|R(z-a_j)|^2 = R^2(z-a_j)(\bar{z}-\bar{a}_j) < (R^2-\bar{a}_j z)(R^2 - a_j\bar{z}) = |R^2 - \bar{a}_j z|^2$$

for $a_j, z \in D(0,R)$. This gives

$$\frac{|R(z-a_j)|}{|R^2 - \bar{a}_j z|} < 1, \quad a_j, z \in D(0,R).$$

https://doi.org/10.1515/9783112249130-007

Therefore, by the Poisson–Jensen formula [300, Theorem 1.1], we have

$$\begin{aligned}\log M(r,f) = \log|f(\zeta)| &\leq \frac{1}{2\pi}\int_0^{2\pi} \frac{R^2 - r^2}{R^2 - 2Rr\cos(\theta-\varphi) + r^2}\log|f(Re^{i\varphi})|\,d\varphi \\ &= \frac{1}{2\pi}\int_0^{2\pi} \frac{(R+r)(R-r)}{(R-r)^2 + 2Rr(1-\cos(\theta-\varphi))}\log|f(Re^{i\varphi})|\,d\varphi \\ &\leq \frac{R+r}{R-r}m(R,f) = \frac{R+r}{R-r}T(R,f), \quad r < R.\end{aligned}$$

This completes the proof of (7.1.1). □

For infinite-order entire functions f, it is possible that

$$\lim_{r\to\infty}\frac{T(r,f)}{\log M(r,f)} = 0.$$

For example, if $f(z) = \exp(e^z)$, then clearly $\log M(r,f) = e^r$, whereas $T(r,f) \sim \frac{e^r}{\sqrt{2\pi^3 r}}$ by (4.6.1).

Regarding the next result, recall the definition of T-order of f from (4.6.5).

Lemma 7.1.2. *If f is an entire function, then*

$$\rho(f) = \rho_T(f). \tag{7.1.2}$$

Proof. If f is a constant, then $\rho(f) = \rho_T(f) = 0$ from the definitions of order.

Assume now that f is nonconstant. We obtain $\rho_T(f) \leq \rho(f)$ from the first inequality $T(r,f) \leq \log M(r,f)$ in (7.1.1). To prove the reverse inequality $\rho_T(f) \geq \rho(f)$, we choose $R = 2r$ in the second inequality in (7.1.1), which then reads as

$$\log M(r,f) \leq 3T(2r,f), \quad r > r_0.$$

Consequently,

$$\frac{\log\log M(r,f)}{\log r} \leq \frac{\log T(2r,f) + \log 3}{\log r} = \frac{\log T(2r,f)}{\log(2r) - \log 2} + \frac{\log 3}{\log r}.$$

Let $\varepsilon > 0$ be arbitrarily small, and choose $r_1 = r_1(\varepsilon) > r_0$ large enough such that

$$\frac{\log 2}{\log(2r)} < \varepsilon, \quad r > r_1.$$

Then

$$\log(2r) - \log 2 > (1-\varepsilon)\log(2r), \quad r > r_1,$$

and, consequently,

$$\rho(f) \leq \limsup_{r\to\infty} \frac{\log T(2r,f)}{(1-\varepsilon)\log(2r)} + \limsup_{r\to\infty} \frac{\log 3}{\log r} = \frac{\rho_T(f)}{1-\varepsilon}.$$

Letting $\varepsilon \to 0^+$, we conclude that $\rho(f) \leq \rho_T(f)$. □

We next give the definition of type of a meromorphic function.

Definition 7.1.3. If f is a meromorphic function satisfying $\rho = \rho_T(f) \in (0,\infty)$, then the *T-type of f* is the quantity

$$\tau_T(f) = \limsup_{r\to\infty} \frac{T(r,f)}{r^\rho} \in [0,\infty].$$

Identity (7.1.2) allows us to compare the T-type $\tau_T(f)$ with the usual type $\tau(f)$ of an entire function f. These two quantities are not always equal; for example, $\tau(e^z) = 1$ and $\tau_T(e^z) = 1/\pi$. In Section 7.3, we will discuss entire functions f such that $T(r,f) \sim \log M(r,f)$, in which case the two types are equal. However, we do have the following general property.

Lemma 7.1.4 ([106, p. 19]). *Let f be an entire function of order $\rho \in (0,\infty)$. Then $\tau_T(f) < \infty$ if and only if $\tau(f) < \infty$.*

Proof. First suppose that $\tau(f) < \infty$. From the first inequality in (7.1.1) we immediately get that $\tau_T(f) \leq \tau(f)$. Hence $\tau_T(f) < \infty$.

Conversely, suppose that $\tau_T(f) < \infty$. Let $\varepsilon > 0$. Then choosing $R = 2r$ in the second inequality in (7.1.1) and using the definition of τ_T-type, we obtain

$$\begin{aligned}\log M(r,f) &\leq 3T(2r,f) \leq 3(\tau_T + \varepsilon)(2r)^\rho \\ &= 3 \cdot 2^\rho (\tau_T + \varepsilon) r^\rho, \quad r > r_0(\varepsilon).\end{aligned}$$

This clearly implies $\tau(f) < \infty$. □

The next theorem was originally proved by Shimizu in 1929.

Theorem 7.1.5 ([106, Theorem 1.8]). *Let f be a nonconstant entire function, and let $K > 1$. Then*

$$\liminf_{r\to\infty} \frac{\log M(r,f)}{T(r,f)(\log T(r,f))^K} = 0.$$

As Hayman pointed out [106, p. 20], the function $\exp(e^z)$ shows that Theorem 7.1.5 does not hold when $K = 1$; see (4.6.1). The proof of Theorem 7.1.5 requires one technical lemma on real-valued functions.

Lemma 7.1.6 ([106, Lemma 1.2]). *Let $a \in \mathbb{R}$, and let $\phi : [a, \infty) \to (1, \infty)$ be an unbounded function that is bounded on every interval $[a, b]$, $a < b < \infty$. Then for each $K > 1$, there exists a sequence $\{x_n\}$ such that $x_n \to \infty$ and*

$$\phi(x) < K\phi(x_n), \quad x_n < x < x_n + \frac{1}{(\log \phi(x_n))^K} + \frac{1}{\phi(x_n)}.$$

Proof. Suppose that our statement is false. Then for every sufficiently large x, there exists ξ such that

$$x < \xi < x + \frac{1}{(\log \phi(x))^K} + \frac{1}{\phi(x)} \quad \text{and} \quad \phi(\xi) \geq K\phi(x).$$

Suppose that this holds for $x \geq x_1$ for some $x_1 \geq a$. Then if x_n is defined, we take x_{n+1} such that

$$x_n < x_{n+1} < x_n + \frac{1}{(\log \phi(x_n))^K} + \frac{1}{\phi(x_n)} \quad \text{and} \quad \phi(x_{n+1}) \geq K\phi(x_n).$$

The existence of such x_{n+1} is guaranteed by our assumption that the statement is false. We may assume that $\phi(x_1) \geq K$ because $\phi(x)$ is unbounded on $[a, \infty)$. Inductively,

$$\phi(x_{n+1}) \geq K\phi(x_n) \geq K^n\phi(x_1) \geq K^{n+1}.$$

Thus for all n, we have $\log \phi(x_n) \geq n \log K$, and, consequently,

$$x_{n+1} - x_n \leq \frac{1}{(\log \phi(x_n))^K} + \frac{1}{\phi(x_n)} \leq \frac{1}{(n \log K)^K} + \frac{1}{K^n}.$$

Hence the telescoping series $\sum_n (x_{n+1} - x_n)$ converges, so that the sequence $\{x_n\}$ is bounded from above. Meanwhile, $\phi(x_n) \to \infty$ as $n \to \infty$, which contradicts the assumption that ϕ is bounded on finite intervals. This completes the proof. □

Proof of Theorem 7.1.5. Since f is nonconstant, $\log M(r, f)$ is an unbounded function by Liouville's theorem. Then so is $T(r, f)$ by Lemma 7.1.1. Define $\phi(x) = T(e^x, f)$. Then for some $a > 0$, $\phi(x)$ satisfies the hypothesis of Lemma 7.1.6, which means that there exists a sequence $\{x_n\}$ such that $x_n \to \infty$ and

$$\phi(x) < K\phi(x_n), \quad x_n < x < x_n + \frac{1}{(\log \phi(x_n))^K} + \frac{1}{\phi(x_n)}.$$

Obviously, $\phi(x_n) \to \infty$ as $n \to \infty$. Choose $\log r = x_n$ and $\log R = x_n + (\log \phi(x_n))^{-K}$ in Lemma 7.1.1. This gives

$$\log M(r, f) \leq \frac{R + r}{R - r} T(R, f) = \frac{\exp((\log \phi(x_n))^{-K}) + 1}{\exp((\log \phi(x_n))^{-K}) - 1} T(R, f)$$

$$\begin{aligned}
&< K\frac{\exp((\log\phi(x_n))^{-K})+1}{\exp((\log\phi(x_n))^{-K})-1}T(r,f)\\
&= K\frac{\exp((\log T(r,f))^{-K})+1}{\exp((\log T(r,f))^{-K})-1}T(r,f)\\
&\sim \frac{2KT(r,f)}{\exp((\log T(r,f))^{-K})-1}\\
&\sim 2KT(r,f)(\log T(r,f))^{K}, \quad r = e^{x_n},\ n\to\infty.
\end{aligned}$$

Thus

$$\liminf_{r\to\infty}\frac{\log M(r,f)}{T(r,f)(\log T(r,f))^K} < \infty. \tag{7.1.3}$$

Finally, suppose that L is a constant satisfying $1 < L < K$. Since $L > 1$, from (7.1.3) we obtain

$$\liminf_{r\to\infty}\frac{\log M(r,f)}{T(r,f)(\log T(r,f))^L} < \infty. \tag{7.1.4}$$

Then from (7.1.3) and (7.1.4) we have

$$\liminf_{r\to\infty}\frac{\log M(r,f)}{T(r,f)(\log T(r,f))^K} = \liminf_{r\to\infty}\frac{\log M(r,f)}{T(r,f)(\log T(r,f))^L}\cdot(\log T(r,f))^{L-K} = 0.$$

The statement follows. □

Hayman and Stewart [113] have shown that if, in addition to the hypothesis of Theorem 7.1.5, f has no zeros, then the conclusion of Theorem 7.1.5 holds for $K > 1/2$. The example $f(z) = \exp(e^z)$ shows that the result is no longer true for $K = 1/2$; see (4.6.1).

Up to this point, we have compared $T(r,f)$ and $\log M(r,f)$ in terms of growth only. Next, we focus on other kinds of properties. To this end, we recall the well-known Cartan identity [300, p. 11]

$$T(r,f) = \frac{1}{2\pi}\int_0^{2\pi} N\left(r,\frac{1}{f-e^{i\theta}}\right)d\theta + \log^+|f(0)|. \tag{7.1.5}$$

Observe that [277] $T(r,f)$ is differentiable for $r > 0$. Then differentiating (7.1.5), we obtain

$$\frac{dT(r,f)}{d\log r} = \frac{1}{2\pi}\int_0^{2\pi} n\left(r,\frac{1}{f-e^{i\theta}}\right)d\theta. \tag{7.1.6}$$

Since $n(r,\frac{1}{f-e^{i\theta}})$ is a nondecreasing function of r for each θ, it follows from (7.1.6) and Lemma 1.3.7(b) that $T(r,f)$ is a convex function in $\log r$.

According to Proposition 2.1.2, $M(r,f)$ is a strictly increasing and continuous function of $r > 0$. The rest of this section is devoted to convexity and differentiability properties of $M(r,f)$.

Lemma 7.1.7. *If f is a nonconstant entire function, then $\log M(r,f)$ is a convex function in $\log r$.*

Proof. By Hadamard's three-circles theorem [273, pp. 172–173], for all $1 < t_1 < t < t_2 < \infty$, we have

$$\log M(t) \le \frac{\log t_2 - \log t}{\log t_2 - \log t_1} \log M(t_1) + \frac{\log t - \log t_1}{\log t_2 - \log t_1} \log M(t_2),$$

where $M(s) = M(s,f)$. Set $t_1 = e^{r_1}$, $t_2 = e^{r_2}$, and $t = e^r$. Then $0 < r_1 < r < r_2 < \infty$, and

$$\log M(e^r) \le \frac{r_2 - r}{r_2 - r_1} \log M(e^{r_1}) + \frac{r - r_1}{r_2 - r_1} \log M(e^{r_2}).$$

Noting that $r = \alpha r_1 + (1 - \alpha) r_2$ for some $\alpha \in (0,1)$, we get

$$\log M(e^r) \le \alpha \log M(e^{r_1}) + (1 - \alpha) \log M(e^{r_2}).$$

This shows that $\log M(e^r)$ is convex, which in turn implies the statement. □

By Lebesgue's theorem [236, p. 112], every nondecreasing function defined on an open interval is differentiable almost everywhere. This applies to $M(r,f)$ as well, but even more can be said about the differentiability of $M(r,f)$.

Theorem 7.1.8 ([279, p. 27]). *Let f be an entire function. Then the maximum modulus $M(r,f)$ has one-sided derivatives for every $r > 1$, and they are equal except perhaps for at most a countable number of points.*

Proof. We may suppose that f is nonconstant, since otherwise the statement is trivial. Then by Lemma 7.1.7 we have that $g(r) := \log M(e^r,f)$ is a convex function. By Lemma 1.3.9, g has one-sided derivatives for every $r > 0$, and they are equal except perhaps for at most a countable number of points. This implies the statement. □

Finally, note that the logarithmic maximum modulus of a nonconstant entire function f has the integral representation

$$\log M(r,f) = \log M(r_0,f) + \int_{r_0}^{r} \frac{u(t)}{t}\, dt, \quad r > r_0 \ge 1,$$

where $u : (0,\infty) \to \mathbb{R}$ is a nondecreasing and left-continuous function. This identity is stated in [279, p. 27], but it also follows from (1.3.3) with $\psi(s) = \log s$.

Historical notes

Walter Kurt Hayman (1926–2020) had numerous important papers and books that greatly contributed to the theory of functions. As stated in [41], "Walter Hayman was, for decades, the dominant figure in the study of entire and meromorphic functions and associated fields." Always happy to discuss mathematics with mathematicians of all levels, his friendly, approachable personality made him a *magnet* at conferences and caused a steady stream of people to visit him at Imperial College in London and participate in his Monday Seminar.

7.2 Asymptotic relation $T(r,f) \sim \log M(r,f)$

We consider entire functions f given in terms of Taylor series having Fejér or Fabry gaps (defined below). We will prove that such functions satisfy

$$T(r,f) \sim \log M(r,f), \tag{7.2.1}$$

as $r \to \infty$ outside a possible exceptional set.

Definition 7.2.1. Let S be a strictly increasing sequence $\{\lambda_n\}_{n=1}^{\infty}$ of positive integers. We say that S is a *Fejér gap sequence* if

$$\sum_{n=1}^{\infty} \lambda_n^{-1} < \infty \tag{7.2.2}$$

and that S is a *Fabry gap sequence* if $\lambda_n/n \to \infty$ as $n \to \infty$.

Suppose that $S = \{\lambda_n\}_{n=1}^{\infty}$ is a Fejér gap sequence. Since the terms of the convergent series (7.2.2) are positive and strictly decreasing, the Abel–Pringsheim theorem [102, p. 350] implies that

$$\lim_{n\to\infty} n\lambda_n^{-1} = 0.$$

Hence S is a Fabry gap sequence. The converse is obviously not true in general. Simple examples: (a) $\{n^2\}$ is a Fejér gap sequence, and (b) $\{n\lfloor\log(n+2)\rfloor\}$ is a Fabry gap sequence but not a Fejér gap sequence, where $\lfloor x \rfloor$ is the integer part of x.

Definition 7.2.2. Let f be an entire function given in terms of a Taylor series $f(z) = \sum_{n=0}^{\infty} a_n z^{\lambda_n}$, where $a_n \in \mathbb{C}$ and $\lambda_0 = 0$. We say that f has *Fejér gaps* (resp., *Fabry gaps*) if $\{\lambda_n\}_{n=1}^{\infty}$ is a Fejér (resp., Fabry) gap sequence.

Even though it is not needed in this book, for completeness, we mention that a Taylor series $f(z) = \sum_{n=0}^{\infty} a_n z^{\lambda_n}$ is said to have *Hadamard gaps* if for some absolute constant q satisfying $q > 1$, we have

$$\frac{\lambda_{n+1}}{\lambda_n} \geq q > 1, \quad n \geq 1. \tag{7.2.3}$$

Applying (7.2.3) inductively, we obtain

$$0 \leq \frac{n}{\lambda_n} \leq \frac{n}{q\lambda_{n-1}} \leq \cdots \leq \frac{n}{q^{n-1}\lambda_1} \to 0, \quad n \to \infty.$$

Hence a Hadamard gap series is a Fabry gap series.

In general, a Taylor series $f(z) = \sum_{n=0}^{\infty} a_n z^{\lambda_n}$ is called a *gap series* if the sequence $\{\lambda_n\}_{n=1}^{\infty}$ omits infinitely many positive integers. Formulas for the order and the lower order of such a series can be found in [155].

So far we have discussed gap series developed at $z = 0$, but in an analogous way, a Taylor series developed at any point $z = a$ can be a gap series. This leads us to the following conjecture of Pólya:

> If f is a transcendental entire function and if the Taylor series of f at $z = a$ has Fabry gaps, then the Taylor series of f at $z = b$ cannot have Fabry gaps whenever $b \neq a$.

Pólya's conjecture was confirmed by Rényi in 1956 [231].

Murai [215] has proved that *entire functions with Fejér gaps have no finite deficient values*. This is a very profound result, but, unfortunately, its proof is beyond the scope of this book. We will focus on (7.2.1) instead.

Let $L(r,f) = \min_{|z|=r} |f(z)|$ denote the *minimum modulus* of an entire function f. For $r > 0$, it is clear that

$$\log^+ L(r,f) \leq \frac{1}{2\pi} \int_0^{2\pi} \log^+ |f(re^{i\theta})| \, d\theta = T(r,f) \leq \log^+ M(r,f). \tag{7.2.4}$$

We now state two main results of this section, which were originally proved by Fuchs and Murai. Both results imply (7.2.1) through appropriate values of r by (7.2.4).

Theorem 7.2.3 ([72]). *Let f be an entire function of finite order given by a Fabry gap series. Then for $\varepsilon > 0$,*

$$\log L(r,f) > (1-\varepsilon) \log M(r,f)$$

for all $r \notin E \cup [0,1]$, where $E \subset (1,\infty)$ satisfies $\overline{\log \operatorname{dens}}(E) = 0$.

Theorem 7.2.4 ([215]). *Let f be an entire function given by a Fejér gap series. Then for $\varepsilon > 0$,*

$$T(r,f) > (1-\varepsilon) \log M(r,f)$$

for all $r \notin E \cup [0,1]$, where $E \subset (1,\infty)$ is a set of finite logarithmic measure.

To prove Theorem 7.2.3, we need two lemmas, the first of which is an almost forgotten result by Kövari, even though it improves earlier results by Pólya, Turán, and others. The second lemma is known as *Fuchs' small arcs lemma*, and it is rather well known to people working in Nevanlinna theory and its applications. Exact references to the original papers can be found in [72].

Lemma 7.2.5 ([72]). *Let f be an entire function of finite order given by a Fabry gap series, and denote*

$$M(r,\gamma,\delta,f) = \max_{\gamma \le \theta \le \gamma+\delta} |f(re^{i\theta})|,$$

where $\delta > 0$. For each fixed $\eta > 0$, we have

$$\log M(r,\gamma,\delta,f) > (1-\eta)\log M(r,f) \tag{7.2.5}$$

uniformly in γ for all $r \notin E_1 \cup [0,1]$, where $E_1 \subset (1,\infty)$ satisfies $\overline{\log\text{dens}}(E_1) = 0$.

Lemma 7.2.6 ([72]). *Let f be a meromorphic function of finite order ρ. For all $\zeta > 0$ and $\delta \in (0,1/2)$, there exist a constant $K(\rho,\zeta) > 0$ and a set $E_2 \subset (1,\infty)$ with $\underline{\log\text{dens}}(E_2) > 1-\zeta$ such that*

$$r\int_J \left|\frac{f'(re^{i\theta})}{f(re^{i\theta})}\right| d\theta < K(\rho,\zeta)\left(\delta\log\frac{1}{\delta}\right)T(r,f), \quad r \in E_2, \tag{7.2.6}$$

for every interval $J \subset \mathbb{R}$ of length δ.

Proof of Theorem 7.2.3. Choose $\zeta > 0$ and $\eta \in (0,\varepsilon/2)$, and set $\rho = \rho(f)$. By L'Hospital's rule we have $\delta\log\frac{1}{\delta} \to 0$ as $\delta \to 0^+$. Hence, with the notation in Lemma 7.2.6, there is $\delta \in (0,1/2)$ such that

$$K(\rho,\zeta)\delta\log(1/\delta) < \varepsilon/2. \tag{7.2.7}$$

Let E_1 and E_2 be as in Lemmas 7.2.5 and 7.2.6, and denote $F = E_2 \setminus E_1$. Then

$$\underline{\log\text{dens}}(F) \ge \underline{\log\text{dens}}(E_2) - \overline{\log\text{dens}}(E_1) > 1-\zeta.$$

Let $r \in F$. Then the inequalities in (7.2.5) and (7.2.6) are simultaneously true. In particular, for all $\phi \in \mathbb{R}$, there exists $\psi \in \mathbb{R}$ such that $\phi - \delta < \psi < \phi$ and

$$\log|f(re^{i\psi})| > (1-\eta)\log M(r,f) > (1-\varepsilon/2)\log M(r,f).$$

Using

$$\left|\frac{d}{d\theta}\log|f(re^{i\theta})|\right| = \left|\text{Re}\left(\frac{d}{d\theta}\log f(re^{i\theta})\right)\right| \le \left|\frac{d}{d\theta}\log f(re^{i\theta})\right| = r\left|\frac{f'(re^{i\theta})}{f(re^{i\theta})}\right|,$$

we obtain

$$\begin{aligned}\log|f(re^{i\phi})| &= \log|f(re^{i\psi})| + \int_{\psi}^{\phi} \frac{d}{d\theta}\log|f(re^{i\theta})|\,d\theta \\ &\geq \log|f(re^{i\psi})| - \int_{\psi}^{\phi} \left|\frac{d}{d\theta}\log|f(re^{i\theta})|\right| d\theta \\ &> (1-\varepsilon/2)\log M(r,f) - r\int_{\psi}^{\phi}\left|\frac{f'(re^{i\theta})}{f(re^{i\theta})}\right| d\theta. \end{aligned} \tag{7.2.8}$$

Recall from (7.1.1) that $T(r,f) \leq \log M(r,f)$ for every r such that $M(r,f) > 1$. Since f is transcendental as an infinite series, we may assume without loss of generality that this inequality holds for every $r \in F$. Hence, putting (7.2.6), (7.2.7), and (7.2.8) together, we deduce

$$\log|f(re^{i\phi})| > (1-\varepsilon)\log M(r,f), \quad r \in F.$$

Since $\phi \in \mathbb{R}$ was chosen arbitrarily, and the lower bound does not depend on ϕ, we obtain

$$\log L(r,f) > (1-\varepsilon)\log M(r,f), \quad r \in F.$$

The exceptional set E in the statement is the set $E = (1,\infty) \setminus F$, and it clearly satisfies $\overline{\log \operatorname{dens}}(E) < \zeta$ by (1.2.3). Finally, since $\zeta > 0$ can be chosen arbitrarily small, the statement follows. □

The proof of Theorem 7.2.4 is more complicated than that of Theorem 7.2.3, and hence it requires more preparation.

Lemma 7.2.7 ([215]). *Let $A : (0,\infty) \to [1,\infty)$ be a nondecreasing continuous function, and let $\rho : [0,\infty) \to (0,\infty)$ be a continuous function that satisfies*

$$\int_0^\infty \rho(r)\,dr < \infty \quad \text{and} \quad \rho(r+h) \leq e^h \rho(r) \tag{7.2.9}$$

for all $r > 0$ and $h \geq 0$. Then for all $q > 1$, the set

$$E = \{r \geq 1 : A(r + r\rho(\log A(r))) \geq qA(r)\}$$

has finite logarithmic measure.

The second condition in (7.2.9) is slightly stronger and possibly more clear than the analogous condition assumed by Murai in [215, Lemma 4]. However, when Lemma 7.2.7 is actually applied later in this section, we will see that this stronger form can be assumed.

The conclusion in Lemma 7.2.7 is reminiscent of the Borel-type lemmas used in Nevanlinna theory. Typical to all Borel-type lemmas is an assumption on monotonicity, which in this case would be the monotonicity of the functions $A(r)$ and $\rho(r)$. In particular, $\rho(r)$ should be nonincreasing or at least eventually nonincreasing. However, the conditions in (7.2.9) do not seem to imply this. Interestingly, Murai does not prove Lemma 7.2.7 in his paper [215] but refers to the proof of a similar result in Kövari's paper [165] instead, where monotonicity is assumed. Luckily, the assumption on the monotonicity of $\rho(r)$ is not needed here.

Proof of Lemma 7.2.7. We begin by ruling out some trivial cases. If $A(r)$ is a bounded function, then since $A(r)$ is also nondecreasing and positive, the limit

$$L = \lim_{r\to\infty} A(r)$$

exists as a finite positive number. Since the function $A(r + r\rho(\log A(r)))$ approaches the same limit L as $r \to \infty$, and since $q > 1$, we find that E is a bounded set, and the statement follows. Hence we may suppose that $A(r)$ is an unbounded function. Also, if $E = \emptyset$ or if $\sup(E) < \infty$, then we are done, and therefore we may suppose that $\sup(E) = \infty$.

Let $r_1 = \inf(E)$. It is clear that $r_1 \geq 1$. Suppose that r_n is defined for some integer $n \geq 1$, and then define

$$r_n' = r_n + r_n\rho(\log A(r_n)) \quad \text{and} \quad r_{n+1} = \inf(E \setminus [1, r_n')).$$

Obviously, $r_n < r_n' \leq r_{n+1}$ for every n. By the definition of the points r_n and the continuity of $A(r)$ and $\rho(r)$, we have $r_n \in E$ for every $n \in \mathbb{N}$. Thus, since $A(r)$ is also nondecreasing, we obtain

$$A(r_{n+1}) \geq A(r_n') \geq qA(r_n) \geq qA(r_{n-1}') \geq \cdots \geq q^n A(r_1) \geq q^n,$$

from which it is clear that $r_n \to \infty$. Moreover, by the construction there are no points in E that lie in the intervals (r_n', r_{n+1}), some of which could be empty. Therefore, $E \subset \bigcup_{n=1}^{\infty}[r_n, r_n']$, and, consequently,

$$\int_E \frac{dr}{r} \leq \sum_{n=1}^{\infty} \int_{r_n}^{r_n'} \frac{dr}{r} \leq \sum_{n=1}^{\infty} \frac{r_n' - r_n}{r_n} = \sum_{n=1}^{\infty} \rho(\log(A(r_n))).$$

It remains to prove that the series in the upper bound converges.

For every $n \in \mathbb{N}$, there exists a unique integer $k_n \geq 0$ such that

$$k_n \log q \leq \log A(r_n) < (k_n + 1) \log q.$$

As observed above, we have $A(r_{n+1}) \geq qA(r_n)$, and hence

$$\log A(r_{n+1}) \geq \log q + \log A(r_n) \geq (k_n + 1) \log q.$$

We see that $k_n + 1$ is a positive integer satisfying $\log A(r_{n+1}) \geq (k_n + 1) \log q$, whereas, by definition, k_{n+1} is the largest nonnegative integer satisfying $\log A(r_{n+1}) \geq k_{n+1} \log q$. Hence it follows that

$$k_{n+1} \geq k_n + 1. \tag{7.2.10}$$

Using (7.2.10) inductively, we find that

$$k_{n+1} \geq k_1 + n \geq n, \quad n \in \mathbb{N}.$$

This means, in particular, that $k_n \geq 1$ for all $n \geq 2$ and $k_n \to \infty$ as $n \to \infty$.

Fix $n \in \mathbb{N} \setminus \{1\}$. For each r satisfying $(k_n - 1) \log q \leq r \leq k_n \log q$, we have

$$0 \leq \log A(r_n) - k_n \log q \leq \log A(r_n) - r \leq (k_n + 1) \log q - (k_n - 1) \log q = 2 \log q,$$

so that, by the second condition in (7.2.9),

$$\rho(\log A(r_n)) = \rho(r + (\log A(r_n) - r)) \leq e^{\log A(r_n) - r} \rho(r) \leq q^2 \rho(r).$$

Integrating both sides with respect to r, we get

$$\rho(\log A(r_n)) \log q \leq q^2 \int_{(k_n-1)\log q}^{k_n \log q} \rho(r)\, dr, \quad n \geq 2.$$

From (7.2.10) we see that the open intervals $((k_n - 1) \log q, k_n \log q)$ are disjoint with respect to n. Hence

$$\begin{aligned}\int_E \frac{dr}{r} &\leq O(1) + \sum_{n=2}^{\infty} \rho(\log(A(r_n))) \leq O(1) + \sum_{n=2}^{\infty} \frac{q^2}{\log q} \int_{(k_n-1)\log q}^{k_n \log q} \rho(r)\, dr \\ &\leq O(1) + \frac{q^2}{\log q} \int_0^{\infty} \rho(r)\, dr < \infty.\end{aligned}$$

This completes the proof. □

Lemma 7.2.7 is needed to prove the following lemma, which in turn is needed to prove Lemma 7.2.14.

Lemma 7.2.8 ([215]). *Let $A : (0,\infty) \to [1,\infty)$ be a nondecreasing continuous function, and let $\gamma : [1,\infty) \to [1,\infty)$ be a nondecreasing function satisfying*

$$\int_1^\infty \frac{\gamma(t)}{t^2}\,dt < \infty.$$

Set

$$\Gamma(r) = \int_1^r \frac{\gamma(t)}{t}\,dt, \quad r \geq 1,$$

and for each $r > 0$, define v_r as the number satisfying $\Gamma(v_r) = A(r)$. Then v_r is well defined and unique, and for any $q > 1$, the set

$$F = \{r \geq 1 : A(r + r\Gamma(v_r)/v_r) \geq qA(r)\}$$

has finite logarithmic measure.

Proof. Since $\gamma(r)$ is nondecreasing and $\gamma(r) \geq 1$, we see that $\Gamma(r) \geq \log r$ for $r \geq 1$. Thus $\Gamma(r) \to \infty$ as $r \to \infty$. Since $\Gamma'(r) = \frac{\gamma(r)}{r} > 0$ for $r > 1$, the function $\Gamma(r)$ is differentiable on $(1,\infty)$ and strictly increasing on $[1,\infty)$. From $\Gamma(1) = 0$ and the continuity of Γ, it follows by the intermediate value theorem that the function $\Gamma(r)$ attains every value of $[0,\infty)$ for $r \geq 1$. Since $A(r) \in [1,\infty)$ for all $r \in (0,\infty)$, the values v_r are well defined. The values v_r are also unique because Γ is strictly increasing.

Note that

$$\frac{\gamma(r)}{r} = \gamma(r)\int_r^\infty \frac{dt}{t^2} \leq \int_r^\infty \frac{\gamma(t)}{t^2}\,dt = o(1), \quad r \to \infty. \tag{7.2.11}$$

Thus $\gamma(r) = o(r)$, and, consequently,

$$\Gamma(r) = o(r), \quad r \to \infty.$$

Since $\Gamma(r)$ is strictly increasing on $[1,\infty)$ and differentiable on $(1,\infty)$, it has an inverse function $\alpha : [0,\infty) \to [1,\infty)$, which is strictly increasing on $[0,\infty)$, differentiable on $(0,\infty)$, and satisfies $\alpha(0) = 1$.

Now set

$$\rho(r) = \frac{e^r}{\alpha(e^r)}, \quad r \geq 0.$$

Differentiating $\Gamma(\alpha(t)) = t$, we obtain $1 = \Gamma'(\alpha(t))\alpha'(t) = \frac{\gamma(\alpha(t))}{\alpha(t)}\alpha'(t)$. Using this identity and changing the variable twice, it follows that

$$\int_0^\infty \rho(r)\,dr = \int_0^\infty \frac{e^r}{\alpha(e^r)}\,dr \overset{t=e^r}{=} \int_1^\infty \frac{dt}{\alpha(t)} = \int_1^\infty \frac{dt}{\alpha'(t)\gamma(\alpha(t))}$$
$$= \int_1^\infty \frac{\gamma(\alpha(t))}{\alpha(t)^2}\alpha'(t)\,dt \overset{s=\alpha(t)}{=} \int_{\alpha(1)}^\infty \frac{\gamma(s)}{s^2}\,ds < \infty,$$

where the last integral converges because of the given hypothesis. Moreover, for all $r > 0$ and $h \geq 0$,

$$\rho(r+h) = \frac{e^h e^r}{\alpha(e^h e^r)} \leq e^h \rho(r),$$

where we have used the monotonicity of $\alpha(r)$. Therefore, $\rho(r)$ satisfies the assumptions in (7.2.9). Finally, since

$$\rho(\log A(r)) = \frac{A(r)}{\alpha(A(r))} = \frac{\Gamma(\nu_r)}{\nu_r},$$

$\int_F \frac{dr}{r} < \infty$ by Lemma 7.2.7. □

We proceed to discuss *Valiron's theorem*. The version in [150, Satz 4.6] is restricted to functions of finite order of growth, whereas the version below does not have this restriction. Lemma 7.2.7 will be used in the proof. Note that the maximal term $\mu(r,f)$ of a polynomial f is defined in the same way as for a transcendental entire f.

Theorem 7.2.9. *Let f be a nonconstant entire function, and let $\varepsilon \in (0,1)$. Then there exists a set $E \subset (1,\infty)$ of finite logarithmic measure such that*

$$\log \mu(r,f) \geq (1-\varepsilon)\log M(r,f), \quad r \notin E \cup [0,1].$$

Proof. If f is a polynomial or a transcendental entire function of finite order, then the statement follows by [150, Satz 4.6]. Hence we may suppose that f has infinite order.

Our starting point is another result of Valiron [150, Satz 4.4], according to which

$$M(r,f) \leq \mu(r,f)\left(\nu(R,f) + \frac{R}{R-r}\right), \quad 0 < r < R < \infty, \tag{7.2.12}$$

where $\nu(R,f)$ is the central index of f. We may apply Lemma 7.2.7 with $q = 2$ and

$$\rho(r) = \frac{1}{(r+1)\log^2(r+e)}.$$

For this choice of $\rho(r)$, it is clear that (7.2.9) holds. This allows us to choose

$$R = r + \frac{r}{(\log \nu(r,f) + 1)\log^2(\log \nu(r,f) + e)}, \qquad r > r_0,$$

where $r_0 > 0$ is chosen so that $\nu(r_0,f) > 1$. Recall that the maximal term and central index are both nondecreasing and unbounded functions of r. For the above choice of R, estimate (7.2.12) gives

$$\begin{aligned} M(r,f) &\le \mu(r,f)\left(\nu(R,f) + \frac{2r}{R-r}\right) \\ &= \mu(r,f)(\nu(R,f) + 2(\log \nu(r,f) + 1)\log^2(\nu(r,f) + e)). \end{aligned} \tag{7.2.13}$$

Using Lemma 7.2.7 in (7.2.13) with $A(r) = \nu(r,f)$ and $q = 2$, we conclude that

$$M(r,f) \le \mu(r,f)(2\nu(r,f) + 2(\log \nu(r,f) + 1)\log^2(\nu(r,f) + e))$$

for $r \notin E \cup [0,1]$, where $E \subset (1,\infty)$ has finite logarithmic measure. By trivial estimates, we have

$$M(r,f) \le 5\mu(r,f)\nu(r,f)\log^2 \nu(r,f), \quad r \notin E \cup [0,1],$$

where, in addition, we may have to assume that r is large enough, which can be done by assuming a slightly larger set E. Now set $\varepsilon_0 = \frac{\varepsilon}{1-\varepsilon}$. Then by Lemma 4.4.7, we obtain

$$M(r,f) \le \mu(r,f)^{1+\varepsilon_0}, \quad r \notin E \cup [0,1],$$

where $E \subset (1,\infty)$ has finite logarithmic measure but might be slightly larger than above. Taking logarithms on both sides, we obtain the result. □

Lemma 7.2.10 ([215]). *For any trigonometric polynomial $P(t) = \sum_k C_k e^{ikt}$, $t \in \mathbb{R}$, $k \in \mathbb{Z}$, with n nonzero coefficients $C_k \in \mathbb{C}$, there exists a constant $C > 0$ such that*

$$m_0(P) := \frac{1}{2\pi}\int_0^{2\pi} \log^+|P(t)|\,dt \ge \log^+ \max_k |C_k| - nC. \tag{7.2.14}$$

The lower bound in (7.2.14) can in some cases be negative, but when this estimate will be applied in proving Theorem 7.2.4, the lower bound turns out to be positive.

Partial proof of Lemma 7.2.10. The statement can be proved by induction. The case $n = 1$ follows immediately, and so we will prove the case $n = 2$, which is similar to the proof of the general case for all n. Since $m_0(e^{ilt}P) = m_0(P)$ for all $l \in \mathbb{Z}$, we may suppose that P is of the form

$$P(t) = C_0 + C_k e^{ikt}, \quad C_0, C_k \in \mathbb{C} \setminus \{0\},\ k \in \mathbb{N}.$$

Moreover, by considering $Q(t) = e^{ikt}P(-t) = C_k + C_0e^{ikt}$ in place of $P(t)$, if necessary, we may suppose that $|C_k| \geq |C_0|$. Then

$$\log^+ |C_k| = m_0(P'/k) = m_0\left(\frac{PP'}{Pk}\right) \leq m_0(P) + m_0\left(\frac{P'}{Pk}\right), \tag{7.2.15}$$

and

$$2\pi m_0\left(\frac{P'}{Pk}\right) = \int_0^{2\pi} \log^+\left|\frac{C_ke^{ikt}}{C_0 + C_ke^{ikt}}\right| dt = \int_0^{2\pi} \log^+\left|\frac{1}{D + e^{ikt}}\right| dt,$$

where $D = C_0/C_k$ satisfies $0 < |D| \leq 1$. If $0 < |D| < 1$, then it is clear that there exists a constant $C > 0$ such that $m_0(P'/Pk) \leq C$, and, consequently, by (7.2.15),

$$m_0(P) \geq \log^+ |C_k| - C. \tag{7.2.16}$$

Hence we may suppose that $|D| = 1$. Further, by appealing to a change of variable we may as well assume that $D = 1$. Then

$$\begin{aligned} m_0\left(\frac{P'}{Pk}\right) &= \frac{1}{4\pi}\int_0^{2\pi} \log^+\left|\frac{1}{1+e^{ikt}}\right|^2 dt = \frac{1}{4\pi}\int_0^{2\pi} \log^+\left(\frac{1}{2+2\cos(kt)}\right) dt \\ &\overset{s=kt}{=} \frac{1}{4k\pi}\int_0^{2k\pi} \log^+\left(\frac{1}{2+2\cos s}\right) ds \\ &= \frac{1}{4\pi}\int_0^{2\pi} \log^+\left(\frac{1}{2+2\cos s}\right) ds \\ &= \frac{1}{4\pi}\int_{2\pi/3}^{4\pi/3} \log\left(\frac{1}{2+2\cos s}\right) ds, \end{aligned}$$

because $2 + 2\cos s \geq 1$ for $0 \leq s \leq \frac{2\pi}{3}$ and $\frac{4\pi}{3} \leq s \leq 2\pi$. Using

$$\cos x \geq -1 + \frac{1}{2}(x-\pi)^2 - \frac{1}{24}(x-\pi)^4, \quad \frac{2\pi}{3} \leq x \leq \frac{4\pi}{3},$$

we obtain

$$\begin{aligned} \int_{2\pi/3}^{4\pi/3} \log\left(\frac{1}{2+2\cos s}\right) ds &\leq \int_{2\pi/3}^{4\pi/3} \log\left(\frac{1}{(s-\pi)^2 - \frac{1}{12}(s-\pi)^4}\right) ds \\ &= \int_{2\pi/3}^{4\pi/3} \log\left(\frac{1}{(s-\pi)^2(1-\frac{1}{12}(s-\pi)^2)}\right) ds \end{aligned}$$

$$\leq \int_{2\pi/3}^{4\pi/3} \log\left(\frac{10}{9(s-\pi)^2}\right) ds \leq 5.$$

Hence $m_0(\frac{P'}{Pk}) \leq C$ for $C = \frac{5}{4\pi}$, and again (7.2.16) holds by (7.2.15). This proves the case $n = 2$. The proof of the general case is a laborious computation [215], although it follows the same lines of reasoning as above. The details are omitted. □

Definition 7.2.11 ([215]). For a sequence $S = \{\lambda_n\}_{n=1}^{\infty}$ of positive integers, set

$$\omega(r, S) = \text{“the number of integers } n \geq 1 \text{ such that } \lambda_n < r\text{”}, \ r \geq 0,$$

$$\Omega(r, S) = \int_0^r \frac{\omega(t, S)}{t}\, dt, \ r \geq 0.$$

Note that $\omega(r, S) = 0$ for $0 \leq r < 1$, and hence $\Omega(r, S)$ is well defined.

Lemma 7.2.12. *Let $S = \{\lambda_n\}_{n=1}^{\infty}$ be a strictly increasing sequence of positive integers. Then the following three statements are equivalent:*

(a) *S is a Fejér gap sequence.*

(b) *The integral $\int_0^{\infty} \frac{d\omega(t,S)}{t}$ converges.*

(c) *The integral $\int_0^{\infty} \frac{\omega(t,S)}{t^2}\, dt$ converges.*

Moreover, we have $\omega(r, S) = o(r)$ as $r \to \infty$ if S is a Fejér gap sequence.

Proof. A simple Riemann–Stieltjes integration and integration by parts give

$$\sum_{\lambda_n < r} \frac{1}{\lambda_n} = \int_0^r \frac{d\omega(t, S)}{t} = \frac{\omega(r, S)}{r} + \int_0^r \frac{\omega(t, S)}{t^2}\, dt.$$

Since clearly $\omega(r, S) \leq r$, we find that (a), (b), (c) are all equivalent. Then, in fact, by a calculation similar to that in (7.2.11) we deduce that $\omega(r, S) = o(r)$. □

Lemma 7.2.13 ([215]). *For any Fejér gap sequence T there exists a Fejér gap sequence S containing T as a subsequence such that for some constant $C > 0$,*

$$\sqrt{r} \leq \omega(r, S) \leq C\Omega(r, S), \quad r \geq 2. \tag{7.2.17}$$

Proof. For $x \geq 0$, let $\lceil x \rceil$ denote the smallest integer k satisfying $k \geq x$, and let $T_0 = \{\gamma_n\}_{n=1}^{\infty}$ denote the strictly increasing sequence consisting of $T = \{\lambda_n\}_{n=1}^{\infty}$ and all the integers in the set

$$V = \bigcup_{j=1}^{\infty} [2^j, 2^j + \lceil 2^{(j+1)/2} \rceil)$$

that do not belong to T. Note that $2^j + \lceil 2^{(j+1)/2} \rceil \le 2^{j+1}$ for all $j \ge 1$, so that the semiopen intervals $[2^j, 2^j + \lceil 2^{(j+1)/2} \rceil)$ in V are disjoint for all j. Then

$$\sum_{n=1}^{\infty} \gamma_n^{-1} = \sum_{n=1}^{\infty} \lambda_n^{-1} + \sum_{\gamma_n \in V} \gamma_n^{-1} \le O(1) + \sum_{n=1}^{\infty} \frac{2^{(n+1)/2} + 1}{2^n} < \infty,$$

so that T_0 is a Fejér gap sequence. If $r \ge 8$, then there exists an integer $j \ge 3$ such that $2^j \le r \le 2^{j+1}$, and hence $2^{j/2} \le \sqrt{r} \le 2^{(j+1)/2}$. Observing that each of the semiopen intervals in the set V contains exactly $\lceil 2^{(j+1)}/2 \rceil$ integers (which all belong to T_0), we obtain

$$\begin{aligned} \omega(r, T_0) \ge \omega(2^j, T_0) &\ge \sum_{k=1}^{j-1} 2^{(k+1)/2} = \sum_{k=1}^{j-1} (\sqrt{2})^{k+1} \\ &= 2 \cdot \frac{(\sqrt{2})^{j-1} - 1}{\sqrt{2} - 1} \ge 2(\sqrt{2})^{j-1} = 2^{(j+1)/2} \ge \sqrt{r}. \end{aligned}$$

In addition, we may suppose that the integers $1, \dots, 8$ are included in T_0, so that also $\omega(r, T_0) \ge \sqrt{r}$ for $2 \le r \le 8$. Thus, to begin with, we will assume without loss of generality that

$$\omega(r, T) \ge \sqrt{r}, \quad r \ge 2. \tag{7.2.18}$$

It remains to construct a Fejér gap sequence $S \supset T$ satisfying the second inequality in (7.2.17) for some $C > 0$ because such S would automatically satisfy $\omega(r, S) \ge \sqrt{r}$ for $r \ge 2$.

For $j \ge 1$, let σ_j denote the number of integers in the sequence T that lie in the interval $[2^{j-1}, 2^j)$. Then, in fact,

$$\sigma_j = \omega(2^j, T) - \omega(2^{j-1}, T).$$

From the discussion in the previous paragraph we may suppose that each interval $[2^{j-1}, 2^j)$ contains at least $\lceil 2^{j/2} \rceil$ points of T. Hence

$$\sigma_j \ge \lceil 2^{j/2} \rceil \ge 2^{j/2} \ge 2^5 = 32 > 0, \quad j \ge 10.$$

With $\beta = 2/5$, set

$$\tau_j = \min\left\{ \sigma_j + \sum_{\ell=1}^{\infty} \beta^{\ell} \sigma_{j+\ell}, \ 2^{j-1} \right\}, \quad j \ge 1.$$

It is clear that $\tau_j > 0$ for $j \ge 10$.

We proceed to prove two properties satisfied by the constants τ_j.

(1) We prove that

$$\sum_{j=10}^{\infty} \tau_j 2^{-j} < \infty. \tag{7.2.19}$$

From the definition of the constants τ_j, by grouping terms diagonally, we obtain

$$\begin{aligned}\sum_{j=10}^{\infty} \tau_j 2^{-j} &\le \sum_{j=10}^{\infty} 2^{-j} \sum_{\ell=0}^{\infty} \beta^{\ell} \sigma_{j+\ell} \\ &= 2^{-10}(\sigma_{10} + \beta\sigma_{11} + \beta^2\sigma_{12} + \beta^3\sigma_{13} + \cdots) \\ &\quad + 2^{-11}(\sigma_{11} + \beta\sigma_{12} + \beta^2\sigma_{13} + \beta^3\sigma_{14} + \cdots) \\ &\quad + 2^{-12}(\sigma_{12} + \beta\sigma_{13} + \beta^2\sigma_{14} + \beta^3\sigma_{15} + \cdots) + \cdots \\ &= 2^{-10}\sigma_{10} + 2^{-11}\sigma_{11}(2\beta + 1) + 2^{-12}\sigma_{12}((2\beta)^2 + 2\beta + 1) + \cdots \\ &= \sum_{j=10}^{\infty} \sigma_j 2^{-j} \sum_{k=0}^{j-10} (2\beta)^k \le \sum_{j=10}^{\infty} \sigma_j 2^{-j} \sum_{k=0}^{\infty} (2\beta)^k = 5 \sum_{j=10}^{\infty} \sigma_j 2^{-j}.\end{aligned}$$

By Riemann–Stieltjes integration, recalling that $\omega(r,T) = 0$ for $0 \le r < 1$, it follows that

$$\begin{aligned}\sum_{j=10}^{\infty} \tau_j 2^{-j} &\le 5 \sum_{j=10}^{\infty} 2^{-j}(\omega(2^j,T) - \omega(2^{j-1},T)) \\ &= 5 \sum_{j=10}^{\infty} \frac{1}{2^j} \int_{2^{j-1}}^{2^j} d\omega(r,T) \le 5 \int_0^{\infty} \frac{d\omega(r,T)}{r}.\end{aligned}$$

Since T is a Fejér gap sequence, it follows from Lemma 7.2.12 that this last integral converges, which in turn proves (7.2.19).

(2) We prove that

$$\tau_{j+1} \le 3\tau_j, \quad j \ge 10. \tag{7.2.20}$$

If $\tau_j = 2^{j-1}$, then $\tau_{j+1} \le 2^j \le 3\tau_j$, and if $\tau_j = \sigma_j + \sum_{\ell=1}^{\infty} \beta^{\ell} \sigma_{j+\ell}$, then

$$\begin{aligned}\tau_{j+1} &\le \sigma_{j+1} + \sum_{\ell=1}^{\infty} \beta^{\ell} \sigma_{j+\ell+1} = \sigma_{j+1} + \beta^{-1} \sum_{\ell=2}^{\infty} \beta^{\ell} \sigma_{j+\ell} \\ &= \beta^{-1} \sum_{\ell=1}^{\infty} \beta^{\ell} \sigma_{j+\ell} \le 3 \sum_{\ell=1}^{\infty} \beta^{\ell} \sigma_{j+\ell} \le 3\tau_j.\end{aligned}$$

This proves (7.2.20).

Let T^c be the sequence of all the positive integers that do not appear in the sequence T. We now construct the sequence S in the statement as follows: We add all the points in T^c that lie in the interval $[1, 2^9)$ to the sequence T. Next, for every $j \ge 10$, we add arbitrarily chosen $\tau_j - \sigma_j$ points in T^c that lie in the interval $[2^{j-1}, 2^j)$ to T and call the resulting sequence S. Recall that σ_j is the number of points in the sequence T that lie in $[2^{j-1}, 2^j)$ and that $\tau_j \le 2^{j-1} = 2^j - 2^{j-1}$. The elements in S are organized so that it is a strictly increasing sequence of positive integers. Denote $S = \{\mu_n\}_{n=1}^{\infty}$. At first glance, it may seem that S contains all positive integers. However, from (7.2.19) and the fact that T is a Fejér gap sequence, we obtain

$$\begin{aligned}\sum_{n=1}^{\infty} \mu_n^{-1} &= \sum_{\mu_n \in T} \mu_n^{-1} + \sum_{\mu_n \in T^c} \mu_n^{-1} \le O(1) + \sum_{n=10}^{\infty} \frac{\tau_n - \sigma_n}{2^{n-1}} \\ &\le O(1) + 2\sum_{n=10}^{\infty} \tau_n 2^{-n} < \infty.\end{aligned}$$

This proves that S also is a Fejér gap sequence.

Finally, if $2^{j-1} \le r < 2^j$ and $j \ge 13$, then by (7.2.20)

$$\begin{aligned}\omega(r,S) &\le \omega(2^j,S) \le \omega(2^{j-3},S) + \tau_{j-2} + \tau_{j-1} + \tau_j \le \omega(2^{j-3},S) + 39\tau_{j-3} \\ &\le 40\omega(2^{j-3},S) \le 40\omega(r/e,S) \le 40\int_{r/e}^{r} \frac{\omega(t,S)}{t}\,dt \le 40\Omega(r,S).\end{aligned}$$

On the other hand, if $2 \le r < 2^{12}$, then by using $\Omega(2,S) \ge \int_1^2 \frac{dt}{t} = \log 2$ we have

$$\omega(r,S) \le 2^{12} \le 2^{13}\Omega(2,S) \le 2^{13}\Omega(r,S).$$

Hence we may choose $C = 2^{13}$ in (7.2.17). □

If an entire function f has Fejér gaps, then we may express it as

$$f(z) = \sum_{n=0}^{\infty} a_n z^{\lambda_n}, \quad a_n \in \mathbb{C},\ \lambda_0 = 0, \tag{7.2.21}$$

where $S = \{\lambda_n\}_{n=1}^{\infty}$ is a Fejér gap sequence. To simplify the notation, for the rest of this section, we agree that

$$\begin{gathered} m(r) = m(r,f), \quad M(r) = M(r,f), \quad \mu(r) = \mu(r,f), \\ \omega(r) = \omega(r,S), \quad \Omega(r) = \Omega(r,S). \end{gathered}$$

Lemma 7.2.14 ([215]). *Let f be an entire function of the form* (7.2.21), *where $S = \{\lambda_n\}_{n=1}^{\infty}$ satisfies* (7.2.17), *and let $\omega^* : (0,\infty) \to [0,\infty)$ be a nondecreasing function such that $\omega^*(r) = 0$ for $0 < r < 1$, $4\omega(r) \le \omega^*(r)$ for $r > 0$,*

$$\lim_{r\to\infty} \frac{\omega^*(r)}{\omega(r)} = \infty, \quad \textit{and} \quad \int_0^{\infty} \frac{\omega^*(t)}{t^2}\,dt < \infty.$$

Define

$$\Omega^*(r) = \int_0^r \frac{\omega^*(t)}{t}\,dt, \quad r > 0,$$

and for large values of $r > 0$, define u_r to be the number that satisfies $\Omega^*(u_r) = 5\log\mu(r)$. Then u_r is well defined and unique. Set

$$\Lambda(r) = \sum^* |a_n| r^{\lambda_n},$$

where $\sum^*$ denotes the summation over all indices n such that $\lambda_n > u_r$. Then there exists a set $E \subset (1,\infty)$ of finite logarithmic measure such that

$$\Lambda(r) \le 1, \quad r \notin E \cup [0,1]. \tag{7.2.22}$$

Remark. Functions ω^* satisfying the assumptions in Lemma 7.2.14 do exist. For example, keeping in mind that $\omega(r) = 0$ for $0 \le r < 1$, define $\omega^*(r) = \omega(r)/\sqrt{I(r)}$, where

$$I(r) = \int_r^\infty \frac{\omega(t)}{t^2}\,dt \to 0, \quad r \to \infty.$$

It is clear that $\omega^*(r)$ is nonnegative and satisfies $\omega^*(r)/\omega(r) \to \infty$ as $r \to \infty$. As a ratio of a nondecreasing function $\omega(r)$ and a nonincreasing function $\sqrt{I(r)}$, $\omega^*(r)$ is nondecreasing. Further, $I'(r) = -\frac{\omega(r)}{r^2}$ by the Leibniz integral rule. Thus

$$\int_0^r \frac{\omega^*(t)}{t^2}\,dt = -\int_0^r \frac{I'(t)}{\sqrt{I(t)}}\,dt = 2\sqrt{I(0)} - 2\sqrt{I(r)} \le 2\sqrt{I(0)}.$$

Finally, it is just a matter of normalization to obtain $4\omega(r) \le \omega^*(r)$ for $r > 0$.

Proof of Lemma 7.2.14. By the continuity, unboundedness, and the intermediate value theorem, the function $\Omega^*(r)$ attains every value in $(0,\infty)$ for $r \in (0,\infty)$. Since the maximal term $\mu(r)$ is nondecreasing (Lemma 4.4.3) and unbounded, we may suppose that $\mu(r) > 1$ for $r > r_0 \ge 1$. Since $(\Omega^*)'(r) = \frac{\omega^*(r)}{r}$, we can assume that $\Omega^*(r)$ is strictly increasing for $r > r_0$. Hence the points u_r are well defined and unique for $r > r_0$.

Since $\omega^*(r) = 0$ for $0 < r < 1$, we make use of Lemma 7.2.8 with $A(r) = 5\log\mu(r)$, $\gamma(r) = \omega^*(r)$, $\Gamma(r) = \Omega^*(r)$, and $q = 2$. It follows that the set

$$F = \{r > r_0 : \mu(r + r\sigma_r^*) \ge \mu(r)^2\}$$

with $\sigma_r^* = \Omega^*(u_r)/u_r$ has finite logarithmic measure. Then by a geometric series argument and the estimate $\log(1+x) \ge \frac{x}{1+x}$ for $x > 0$, it follows that for all $r \notin F \cup [0, r_0]$,

$$\begin{aligned}
\Lambda(r) &= \sum^* |a_n| (r + r\sigma_r^*)^{\lambda_n} (1+\sigma_r^*)^{-\lambda_n} \le \mu(r)^2 \sum^* (1+\sigma_r^*)^{-\lambda_n} \\
&= \mu(r)^2 \sum_{\lambda_n > u_r} (1+\sigma_r^*)^{-\lambda_n} \le \mu(r)^2 \sum_{n > u_r} (1+\sigma_r^*)^{-n} \\
&\le \mu(r)^2 (1+\sigma_r^*)^{-u_r} \sum_{n=0}^\infty (1+\sigma_r^*)^{-n} = \mu(r)^2 \frac{1+\sigma_r^*}{\sigma_r^*} \exp(-u_r \log(1+\sigma_r^*))
\end{aligned}$$

$$\leq \mu(r)^2 \frac{1+\sigma_r^*}{\sigma_r^*} \exp\left(-\frac{u_r \sigma_r^*}{1+\sigma_r^*}\right) = \mu(r)^2 \frac{1+\sigma_r^*}{\sigma_r^*} \exp\left(-\frac{\Omega^*(u_r) u_r}{\Omega^*(u_r) + u_r}\right).$$

A similar calculation as in (7.2.11) shows that $\omega^*(r) = o(r)$, and hence $\Omega^*(r) = o(r)$ and $\sigma_r^* = o(1)$. Recall that $\Omega^*(r)$ is an unbounded function and strictly increasing for $r > r_0$. By choosing a larger r_0, if necessary, we may suppose that

$$1 \leq \Omega^*(u_r) \leq u_r/2 \quad \text{and} \quad \sigma_r^* \leq 1, \quad r \notin F \cup [0, r_0].$$

Now by Lemma 7.2.9 we have

$$\mu(r) \geq M(r)^{3/4}, \quad r \notin G \cup [0, 1],$$

where $G \subset (1, \infty)$ is a set of finite logarithmic measure. Setting $F_1 = F \cup G$ and using the assumption $\Omega^*(u_r) = 5 \log \mu(r)$, we obtain that for $r \notin F_1 \cup [0, r_0]$,

$$\begin{aligned}\Lambda(r) &\leq \frac{2\mu(r)^2}{\sigma_r^*} \exp\left(-\frac{2}{3}\Omega^*(u_r)\right) = \frac{2u_r\mu(r)^2}{\Omega^*(u_r)} \exp\left(-\frac{10}{3}\log \mu(r)\right) \\ &\leq 2u_r\mu(r)^{-4/3} \leq 2u_r M(r)^{-1}.\end{aligned}$$

Clearly, F_1 has finite logarithmic measure.

Finally, by L'Hospital's rule and the assumption $\omega^*(r)/\omega(r) \to \infty$ as $r \to \infty$,

$$\lim_{r\to\infty} \frac{\Omega(r)}{\Omega^*(r)} = \lim_{r\to\infty} \frac{\omega(r)}{\omega^*(r)} = 0. \tag{7.2.23}$$

Hence by (7.2.17) and the fact that $\mu(r,f) \leq M(r,f)$ we have

$$u_r \leq \omega(u_r)^2 \leq C^2\Omega(u_r)^2 = o(\Omega^*(u_r)^2) = o(\log^2 \mu(r)) = o(M(r)).$$

We conclude that $\Lambda(r) \leq 1$ for $r \notin F_1 \cup [0, r_0]$, where $F_1 \subset (1, \infty)$ is of finite logarithmic measure, and $r_0 > 1$ is possibly slightly larger than above. Hence we may choose $E = F_1 \cup (1, r_0]$ for the set E in the statement. □

Proof of Theorem 7.2.4. Without loss of generality, we may suppose that f is of the form (7.2.21), where $a_0 = 1$, and the Fejér gap sequence $S = \{\lambda_n\}_{n=1}^{\infty}$ satisfies (7.2.17). The reason why we may do this is because of (7.2.22) in Lemma 7.2.14, according to which adding finitely many new terms to a Fejér gap series f has an insignificant contribution to the growth of the functions $T(r,f)$ and $\log M(r,f)$ in the statement of Theorem 7.2.4.

Let $\omega^*(r)$, $\Omega^*(r)$, u_r, and $\Lambda(r)$ be as in Lemma 7.2.14. By the proof of Lemma 7.2.14 we have $\Lambda(r) \leq 1$ for $r \notin F_1 \cup [0, r_0]$, where $F_1 \subset (1, \infty)$ has finite logarithmic measure. Hence

$$\max\{|a_n|r^{\lambda_n} : \lambda_n \leq u_r\} = \mu(r), \quad r \notin F_1 \cup [0, r_0].$$

Recall that r_0 is large enough, so that u_r is well defined and $\mu(r) > 1$ for all $r > r_0$. For $r > r_0$, define

$$f_*(z) = \sum_* a_n z^{\lambda_n}, \quad f^*(z) = \sum^* a_n z^{\lambda_n}, \quad \text{and} \quad P_r(t) = f_*(re^{it}),$$

where $\sum_*$ denotes the summation over all indices n such that $\lambda_n \le u_r$, and $\sum^*$ denotes the summation over all indices n such that $\lambda_n > u_r$.

By (7.2.17) and (7.2.23), there exists a constant $C > 0$ such that

$$\omega(u_r) \le C\Omega(u_r) = o(\Omega^*(u_r)) = o(\log \mu(r)).$$

Then by Theorem 7.2.9 and Lemma 7.2.10, there exists a constant $K > 0$ such that

$$\begin{aligned} m_0(P_r) &\ge \log^+ \max\{|a_n| r^{\lambda_n} : \lambda_n \le u_r\} - K\omega(u_r) \\ &= (1 - o(1)) \log \mu(r) = (1 - o(1)) \log M(r) \end{aligned} \tag{7.2.24}$$

as $r \to \infty$, $r \notin F_2 \cup [0, r_0]$, where $F_2 = F_1 \cup E$. Obviously, F_2 has finite logarithmic measure.

Let $\varepsilon \in (0, 1/2)$, and define the auxiliary function

$$h(x) = (x+1)^{1-\varepsilon} - x^{1-2\varepsilon}, \quad x \ge 0.$$

Since $1 - \varepsilon > 1 - 2\varepsilon$, it is clear that $h(x)$ is unbounded and is eventually nondecreasing. Hence there exists $R > 0$ such that $h(x) \ge 1$ for all $x \ge R$. Now by Lemma 7.2.14 and (7.2.24), there exists a set $G \subset (1, \infty)$ of finite logarithmic measure such that

$$m_0(P_r) \ge (1 - \varepsilon^2) \log(M(r) + 1), \ \Lambda(r) \le 1, \quad \text{and} \quad (M(r)+1)^{1-\varepsilon} \ge M(r)^{1-2\varepsilon} + 1$$

for all $r \notin G \cup [0, 1]$. Set

$$H_r = \{t \in [0, 2\pi) : \log^+ |P_r(t)| \ge (1 - \varepsilon) \log(M(r) + 1)\}, \quad r > 0.$$

Then for all $r \notin G \cup [0, 1]$,

$$\begin{aligned} 2\pi(1 - \varepsilon^2) \log(M(r) + 1) \le 2\pi m_0(P_r) &= \left(\int_{H_r} + \int_{H_r^c} \right) \log^+ |P_r(t)| \, dt \\ &\le \operatorname{meas}(H_r) \log(M(r) + 1) + \operatorname{meas}(H_r^c)(1 - \varepsilon) \log(M(r) + 1), \end{aligned}$$

where $H_r^c = [0, 2\pi) \setminus H_r$ is the complement of H_r. Next, we cancel out $\log(M(r) + 1)$ and use $\operatorname{meas}(H_r^c) = 2\pi - \operatorname{meas}(H_r)$ to obtain $\operatorname{meas}(H_r) \ge 2\pi(1 - \varepsilon)$. Now, if $t \in H_r$ and $r \notin G \cup [0, 1]$, then

$$\log^+ |f(re^{it})| = \log^+ |f_*(re^{it}) + f^*(re^{it})| \ge \log^+(|f_*(re^{it})| - |f^*(re^{it})|)$$

$$\geq \log^+(|P_r(t)| - \Lambda(r)) \geq \log((M(r)+1)^{1-\varepsilon} - 1)$$
$$\geq (1-2\varepsilon)\log M(r).$$

Consequently, by choosing $\varepsilon' = (3-2\varepsilon)\varepsilon > 0$ we have

$$m(r) \geq \frac{1}{2\pi}\int_{H_r} \log^+|f(re^{it})|\,dt \geq (1-2\varepsilon)(1-\varepsilon)\log M(r) = (1-\varepsilon')\log M(r)$$

for all $r \notin G \cup [0,1]$. This completes the proof. □

7.3 Prescribed asymptotic growth

Keeping in mind the asymptotic equation (7.2.1), valid for Taylor series with Fejér/Fabry gaps, we arrive at the following general problem: For a function $\phi : [r_0, \infty) \to (0, \infty)$, is it possible to find a transcendental entire function f such that

$$T(r,f) \sim \log M(r,f) \sim \phi(r)?$$

For this problem to make sense, we need to assume something about the function ϕ. To begin with, we find by the maximum modulus principle and Liouville's theorem that ϕ must be increasing and unbounded. In fact, since f is transcendental, we must have $\phi(r) \neq O(\log r)$ by Lemma 2.1.5. Moreover, since $T(r,f)$ and $\log M(r,f)$ are both convex in $\log r$ (see Section 7.1), so should ϕ.

We now state the main result of this section, originally due to Clunie, and give a partial proof for it.

Theorem 7.3.1 ([55]). *For $r_0 \geq 0$, let $\phi : (r_0, \infty) \to (0, \infty)$ be a strictly increasing function and convex in $\log r$ such that $\phi(r) \neq O(\log r)$. Then there exists a transcendental entire function f such that*

$$T(r,f) \sim \log M(r,f) \sim \phi(r). \tag{7.3.1}$$

Clunie's paper [55] is based on the ideas of Edrei and Fuchs [63], which in turn is based on the work of Valiron [279, p. 130]. Improvements and extensions of Theorem 7.3.1 can be found in [56].

Partial proof of Theorem 7.3.1. We prove that there exists an entire function f for which $\log M(r,f) \sim \phi(r)$. This part is relatively easy. The proof of the complete statement is rather lengthy, and we refer to the original work [55] for details.

By (1.3.3) and the fact that ϕ is convex in $\log r$, we may assume that

$$\phi(r) = \int_1^r \varphi(t)\,d(\log t) = \int_1^r \frac{\varphi(t)}{t}\,dt,$$

where $\varphi : [1, \infty) \to [0, \infty)$ is a nondecreasing, unbounded, and integrable function with $\varphi(1) = 0$. The function φ is unbounded because if it were bounded, then we would obtain that $\phi(r) = O(\log r)$, which would contradict the hypothesis.

Let $\{r_n\}$ be any strictly increasing unbounded sequence of positive real numbers satisfying $\varphi(r_n) = n$. As φ is not necessarily strictly increasing, there might be more than one such sequence. Next, define

$$F(z) = \sum_{n=1}^{\infty} a_n z^n, \quad a_n = \frac{1}{r_1 \cdots r_n}.$$

Since

$$\lim_{n\to\infty} \left| \frac{a_{n+1} z^{n+1}}{a_n z^n} \right| = \lim_{n\to\infty} \left| \frac{z}{r_{n+1}} \right| = 0 < 1$$

for any fixed $z \in \mathbb{C}$, the series F is an entire function. Let $|z| = r$. Since F is entire, $a_n r^n \to 0$ as $n \to \infty$. Therefore, we are able to define the maximal term

$$\mu(r, F) = \max_n \{a_n r^n\}.$$

Suppose that $r_n \le r \le r_{n+1}$. If $m < n$, then

$$a_m r^m \le a_n r^n \iff r_{m+1} \cdots r_n \le r^{n-m},$$

where the latter inequality is true because $r \ge r_n$. If $m > n$, then

$$a_m r^m \le a_n r^n \iff r^{m-n} \le r_{n+1} \cdots r_m,$$

where the latter inequality is true because $r \le r_{n+1}$. Hence we have proved that

$$\mu(r, F) = a_n r^n, \quad r_n \le r \le r_{n+1}. \tag{7.3.2}$$

During this proof of Theorem 7.3.1, we will state and prove three lemmas, starting with the next lemma.

Lemma 7.3.2. $\log \mu(r, F) \sim \phi(r)$.

Proof. For every $r \ge r_1$ there exists a positive integer N such that $r_N \le r \le r_{N+1}$. Hence by (7.3.2),

$$\log \mu(r, F) = N \log r - \sum_{n=1}^{N} \log r_n. \tag{7.3.3}$$

Next, we find that

$$\phi(r) = \sum_{n=1}^{N-1} \int_{r_n}^{r_{n+1}} \frac{\varphi(t)}{t}\, dt + \int_{r_N}^{r} \frac{\varphi(t)}{t}\, dt + O(1).$$

Since $\varphi(r_n) = n$ and since φ is nondecreasing, we find the following upper and lower estimates for $\phi(r)$:

$$\begin{aligned}\phi(r) &\le \sum_{n=1}^{N-1} (n+1) \log \frac{r_{n+1}}{r_n} + (N+1) \log \frac{r}{r_N} + O(1)\\ &= (N+1)\log r - \sum_{n=1}^{N} \log r_n + O(1),\\ \phi(r) &\ge \sum_{n=1}^{N-1} n \log \frac{r_{n+1}}{r_n} + N \log \frac{r}{r_N} + O(1)\\ &= N \log r - \sum_{n=1}^{N} \log r_n + O(1).\end{aligned}$$

Thus by (7.3.3),

$$O(1) \le \phi(r) - \log \mu(r, F) \le \log r + O(1).$$

Since $\phi(r) \ne O(\log r)$, it follows that $\log \mu(r, F) \sim \phi(r)$. □

Next, we define a strictly increasing sequence of positive integers $\{\lambda_n\}$ as follows. Choose $\lambda_1 = 1$, and suppose that $\lambda_1, \ldots, \lambda_n$ have been specified. If

$$a_{\lambda_n+1} r_{\lambda_n+1}^{\lambda_n+1} > 2a_{\lambda_n} r_{\lambda_n}^{\lambda_n},$$

then choose $\lambda_{n+1} = \lambda_n + 1$. Otherwise, choose λ_{n+1} to be the largest integer m for which

$$a_m r_m^m \le 2a_{\lambda_n} r_{\lambda_n}^{\lambda_n}.$$

Since $\{a_n r_n^n\}$ is a strictly increasing sequence and diverges to ∞ as $n \to \infty$, the sequence $\{\lambda_n\}$ is well defined.

Define

$$f(z) = \sum_{n=1}^{\infty} \frac{a_{\lambda_n} z^{\lambda_n}}{n^2}.$$

Since the series $F(z)$ converges absolutely uniformly on compact subsets of $\mathbb{C}$, so does the associated gap series $\sum_{n=1}^{\infty} a_{\lambda_n} z^{\lambda_n}$, and therefore $f(z)$ converges by the comparison test. Hence f is entire. It will be proved that $\log M(r,f) \sim \phi(r)$, which will complete the first part of the proof.

Lemma 7.3.3. $\log n = o(\log a_{\lambda_n} r_{\lambda_n}^{\lambda_n})$.

Proof. From the construction of the sequence $\{\lambda_n\}$ we see that

$$a_{\lambda_{n+2}} r_{\lambda_{n+2}}^{\lambda_{n+2}} > 2 a_{\lambda_n} r_{\lambda_n}^{\lambda_n}, \quad n \geq 1.$$

Using these inequalities for even and odd natural numbers n, we find that

$$\begin{aligned} \sum_{n=1}^{\infty} (a_{\lambda_n} r_{\lambda_n}^{\lambda_n})^{-\delta} &= \sum_{n=1}^{\infty} (a_{\lambda_{2n}} r_{\lambda_{2n}}^{\lambda_{2n}})^{-\delta} + \sum_{n=1}^{\infty} (a_{\lambda_{2n-1}} r_{\lambda_{2n-1}}^{\lambda_{2n-1}})^{-\delta} \\ &< \sum_{n=1}^{\infty} (2^{n-1} a_{\lambda_2} r_{\lambda_2}^{\lambda_2})^{-\delta} + \sum_{n=1}^{\infty} (2^{n-1} a_{\lambda_1} r_{\lambda_1}^{\lambda_1})^{-\delta} < \infty \end{aligned}$$

for all $\delta > 0$. Since the terms of this convergent series are positive and strictly decreasing, the Abel–Pringsheim theorem [102, p. 350] yields

$$\lim_{n\to\infty} n (a_{\lambda_n} r_{\lambda_n}^{\lambda_n})^{-\delta} = 0.$$

In particular, there exists $N \geq 1$ such that $n \leq (a_{\lambda_n} r_{\lambda_n}^{\lambda_n})^{\delta}$ for all $n \geq N$. Since $\delta > 0$ can be arbitrarily small, this implies the statement. □

Lemma 7.3.4. $\log M(r,f) \sim \log \mu(r,F)$.

Proof. Consider the interval $r_\nu \leq r \leq r_{\nu+1}$ in two cases.

(1) Suppose that for some n, we have $\nu = \lambda_n$. Then, similarly as in (7.3.2),

$$a_{\lambda_n} r^{\lambda_n} = \mu(r,F), \quad r_\nu \leq r \leq r_{\nu+1}.$$

Clearly,

$$a_{\lambda_m} r^{\lambda_m} \leq \mu(r,F), \quad m \geq 1,\ r \geq 0,$$

and since the Taylor series of f has positive coefficients,

$$\frac{\mu(r,F)}{n^2} \leq M(r,f) = \sum_{n=1}^{\infty} \frac{a_{\lambda_n} r^{\lambda_n}}{n^2} \leq \mu(r,F) \sum_{n=1}^{\infty} \frac{1}{n^2} = \frac{\pi^2}{6} \mu(r,F). \tag{7.3.4}$$

From Lemma 7.3.3 we have

$$\log n = o(\log \mu(r,F)), \quad r_\nu \leq r \leq r_{\nu+1},\ r \to \infty. \tag{7.3.5}$$

This, together with (7.3.4), proves that $\log M(r,f) \sim \log \mu(r,F)$ as $r \to \infty$ through the values under consideration.

(2) Suppose then that there does not exist a natural number k such that $\nu = \lambda_k$. In this case, $\nu \geq 2$ because $\lambda_1 = 1$. Let λ_n be the largest λ_k satisfying $\lambda_k < \nu$. Then by the construction,

$$a_{\nu+1} r_{\nu+1}^{\nu+1} \le 2a_{\lambda_n} r_{\lambda_n}^{\lambda_n}.$$

Now

$$a_{\lambda_n} r_{\lambda_n}^{\lambda_n} \le \mu(r,F) \le 2a_{\lambda_n} r_{\lambda_n}^{\lambda_n}, \quad r_\nu \le r \le r_{\nu+1}, \tag{7.3.6}$$

and, consequently,

$$\frac{\mu(r,F)}{2n^2} \le M(r,f) \le \frac{\pi^2}{6}\mu(r,F), \quad r_\nu \le r \le r_{\nu+1}. \tag{7.3.7}$$

From Lemma 7.3.3 and (7.3.6) we again obtain (7.3.5). This, together with (7.3.7), proves that $\log M(r,f) \sim \log \mu(r,F)$ as $r \to \infty$ through the values under consideration. □

The statement $\log M(r,f) \sim \phi(r)$ follows from Lemmas 7.3.2 and 7.3.4. This completes the first part of the proof of Theorem 7.3.1, that is, it proves part of (7.3.1). See [55] for the complete proof. □

7.4 Perfectly regular growth

In this section, we consider entire functions f for which the logarithmic maximum modulus $\log M(r,f)$ has regular growth.

Definition 7.4.1 ([279, pp. 41–46], [249]). We say that an entire function f of order $\rho \in (0,\infty)$ is of *regular growth* (r. g.) if

$$\lim_{r\to\infty} \frac{\log\log M(r,f)}{\log r} = \rho$$

and is of *perfectly regular growth* (p. r. g.) if

$$\lim_{r\to\infty} \frac{\log M(r,f)}{r^\rho} \in (0,\infty).$$

For functions of r. g., the order and lower order are equal, whereas for functions of p. r. g., the type and lower type are equal.

It is clear that p. r. g. implies r. g., but r. g. does not imply p. r. g. in general. The following example proves the latter claim.

Example 7.4.2. Let $\phi(r) = r^\rho \exp(\sqrt{\log r})$, where $\rho \in (0,\infty)$ and $r > e$. Since

$$\begin{aligned}\frac{d}{dr}\left(\frac{d\phi(r)}{d(\log r)}\right) &= \frac{d}{dr}\left(\phi(r)\left(\rho + \frac{1}{2\sqrt{\log r}}\right)\right)\\ &\ge \left(\frac{1}{\log r} - \frac{1}{(\log r)^{3/2}}\right)\frac{\phi(r)}{4r} \ge 0,\end{aligned}$$

we see that $\frac{d\phi(r)}{d(\log r)}$ is nondecreasing for $r > e$. Hence the function ϕ is convex in $\log r$ by Lemma 1.3.7. Thus by Theorem 7.3.1, there exists a transcendental entire function f such that

$$\log M(r,f) = (1 + o(1))\phi(r) = (1 + o(1))r^{\rho} \exp(\sqrt{\log r}).$$

Then

$$\log\log M(r,f) = \rho \log r + \sqrt{\log r} + o(1).$$

Clearly, f is of order ρ and of r. g., but f is not of p. r. g.

Proposition 7.4.3. *Transcendental exponential polynomials are of p. r. g.*

Proof. Suppose that f is an exponential polynomial of order $q \geq 1$. The Phragmén–Lindelöf indicator $h_f(\theta)$ of f is given by (6.3.3). Since $h_f(\theta)$ is bounded for $\theta \in [0, 2\pi]$, Corollary 4.2.7 implies that the type τ of f satisfies

$$\tau = \max_{0 \leq \theta \leq 2\pi} h_f(\theta) \in (0, \infty).$$

In particular, there exists $\theta_0 \in [0, 2\pi]$ such that

$$\tau = h_f(\theta_0).$$

We proceed to prove that

$$\log M(r,f) \sim \tau r^q, \quad r \to \infty. \tag{7.4.1}$$

Let $\varepsilon > 0$. By the definition of type, there exists a constant $R_1(\varepsilon) > 0$ such that

$$\log M(r,f) \leq (\tau + \varepsilon) r^q, \quad r \geq R_1(\varepsilon).$$

Thus to prove (7.4.1), it suffices to prove that there exists a constant $R_2(\varepsilon) \geq R_1(\varepsilon)$ such that

$$(\tau - \varepsilon) r^q \leq \log M(r,f), \quad r \geq R_2(\varepsilon).$$

Suppose on the contrary that there exists a strictly increasing sequence $\{r_n\}$ of positive real numbers tending to infinity such that

$$\log M(r_n,f) < (\tau - \varepsilon) r_n^q, \quad n \in \mathbb{N}.$$

By Lemma 6.2.3, we find that the indicator function of a transcendental exponential sum g can be expressed as the usual limit

$$h_g(\theta) = \lim_{r \to \infty} r^{-1} \log|g(re^{i\theta})|$$

for all $\theta \in [0, 2\pi)$ with finitely many exceptions. The exceptional values of θ correspond to the critical rays of g. An analogous result to Lemma 6.2.3 for exponential polynomials of order q can be found in [126, Lemma 7.2]. In this analogue the representation for $f(z)$ holds outside the critical rays and outside an $\mathcal{E}$-set. By Lemma 1.4.1 this means that

$$h_f(\theta) = \lim_{r\to\infty} r^{-q} \log|f(re^{i\theta})| \tag{7.4.2}$$

for almost every $\theta \in [0, 2\pi)$.

By (7.4.2) and the fact that $h_f(\theta)$ is continuous (Theorem 4.2.6), there exists a value θ_1 near θ_0 such that

$$h_f(\theta_1) = \lim_{r\to\infty} r^{-q} \log|f(re^{i\theta_1})| \geq \tau - \varepsilon/3.$$

Hence there exists a constant $R_3(\varepsilon) \geq R_2(\varepsilon)$ such that

$$\log|f(re^{i\theta_1})| \geq (\tau - \varepsilon/2)r^q, \quad r \geq R_3(\varepsilon).$$

Let N be an integer such that $r_N \geq R_3(\varepsilon)$. Now we can summarize everything:

$$(\tau - \varepsilon/2)r_N^q \leq \log|f(r_N e^{i\theta_1})| \leq \log M(r_N, f) \leq (\tau - \varepsilon)r_N^q,$$

which is obviously a contradiction. This proves (7.4.1). □

The next two results show that entire functions f satisfying $T(r,f) \sim \log M(r,f)$ do not need to have regular growth. As far as we know, these results are new.

Theorem 7.4.4. *There exist entire functions f that do not have regular growth and satisfy $T(r,f) \sim \log M(r,f)$ as $r \to \infty$.*

Proof. We will construct an auxiliary function ϕ to be an infinite number of line segments that are attached together, where each line segment has a positive slope, such that the slopes of two successive line segments satisfy that the slope of the first line segment is less than the slope of the second. This will make ϕ a continuous, unbounded, and convex function. Then $\phi(e^x)$ is a convex function as a composition of two convex functions, which by definition means that ϕ is convex in $\log x$. The line segments will bounce back and forth between the curves $y = x^2$ and $y = x^3$, which will cause ϕ to not have regular growth.

Let the first line segment L_1 be the line segment between $(2, 4)$ on $y = x^2$ and $(3, 27)$ on $y = x^3$. The slope m_1 of L_1 is $m_1 = 23$. Let the second line segment L_2 be the line segment between $(3, 27)$ on $y = x^3$ and $(30, 900)$ on $y = x^2$. The slope of L_2 is $m_2 = 32.\overline{3}$. Now let the third line segment L_3 be the line segment between $(30, 900)$ on $y = x^2$ and $(40, 64\,000)$ on $y = x^3$. The slope of L_3 is $m_3 = 6\,310$. Observe that

$$0 < m_1 < m_2 < m_3.$$

We now use mathematical induction to finish the construction, that is, to obtain all the desired line segments. There are two cases to consider.

Case 1. First, suppose that L_n is a line segment between (a, a^2) on $y = x^2$ and (b, b^3) on $y = x^3$, where $b > a \geq 2$. Then the slope of L_n is

$$m_n = \frac{b^3 - a^2}{b - a} > 0.$$

We construct the next line segment L_{n+1} by joining the point (b, b^3) on $y = x^3$ to some point (c, c^2) on $y = x^2$, where $c > b$ and c is chosen large enough so that the slope m_{n+1} of L_{n+1} satisfies $m_{n+1} > m_n$. This will be accomplished when

$$m_{n+1} = \frac{c^2 - b^3}{c - b} > m_n = \frac{b^3 - a^2}{b - a}. \tag{7.4.3}$$

Since a, b, m_n are three fixed constants, it is clear that c can be chosen large enough so that (7.4.3) is satisfied, since the left side behaves like c as $c \to \infty$. This proves the induction step for case 1.

Case 2. Second, suppose that L_n is a line segment between (a, a^3) on $y = x^3$ and (b, b^2) on $y = x^2$, where $b > a \geq 2$, and where the slope of L_n satisfies

$$m_n = \frac{b^2 - a^3}{b - a} > 0.$$

We construct the next line segment L_{n+1} by joining the point (b, b^2) on $y = x^2$ to some point (c, c^3) on $y = x^3$, where $c > b$, and c is chosen large enough so that the slope of L_{n+1} satisfies

$$m_{n+1} = \frac{c^3 - b^2}{c - b} > m_n = \frac{b^2 - a^3}{b - a}. \tag{7.4.4}$$

Since a, b, m_n are fixed constants, we see that c can be chosen large enough so that (7.4.4) is satisfied, since the left side behaves like c^2 as $c \to \infty$. This proves the induction step for case 2.

Cases 1 and 2 complete the construction of the function ϕ. Since the slopes of the line segments are positive and strictly increasing, ϕ is clearly convex and therefore convex in $\log x$. The function ϕ is obviously continuous and unbounded. From the construction of ϕ we notice that there exist two positive sequences $\{u_n\}$ and $\{v_n\}$ such that $\phi(u_n) = u_n^2$ and $\phi(v_n) = v_n^3$ for all n, where $u_n \to \infty$ and $v_n \to \infty$. By Theorem 7.3.1 there exists an entire function f such that $T(r,f) \sim \log M(r,f) \sim \phi(r)$ as $r \to \infty$. Hence

$$\log M(u_n,f) \sim u_n^2 \quad \text{and} \quad \log M(v_n,f) \sim v_n^3$$

as $n \to \infty$. It follows that $\underline{\rho}(f) \le 2$ and $\rho(f) \ge 3$, which shows that f is not of regular growth. □

Theorem 7.4.5. *There exist entire functions f that have regular growth but do not have perfectly regular growth, where $T(r,f) \sim \log M(r,f)$.*

Proof. We will construct an auxiliary function ϕ by relying on the reasoning used in constructing [151, Example 1]. Let $x_n = 2^{2^n}$ for $n \in \mathbb{N}$, so that $x_{n+1} = x_n^2$. We define ϕ to be the piecewise linear function inscribed in x^2, that is, $\phi(x_n) = x_n^2$, $\phi(x_{n+1}) = x_{n+1}^2$, and ϕ is linear on the intervals $[x_n, x_{n+1}]$ with

$$\begin{aligned}\phi(x) &= \phi(x_n) + \frac{\phi(x_{n+1}) - \phi(x_n)}{x_{n+1} - x_n}(x - x_n)\\ &= x_n^2 + \frac{x_{n+1}^2 - x_n^2}{x_{n+1} - x_n}(x - x_n)\\ &= x_n^2 + (x_{n+1} + x_n)(x - x_n), \quad x \in [x_n, x_{n+1}].\end{aligned}$$

It is obvious that ϕ is increasing, continuous, and convex. As a convex function, ϕ is also convex in $\log x$.

Let $\alpha \in [0,1]$, and set $y_n = \alpha x_n + (1-\alpha)x_{n+1}$. Then

$$\phi(y_n) = x_n^2 + (1-\alpha)(x_{n+1}^2 - x_n^2) = (1-\alpha)x_{n+1}^2 + \alpha x_n^2.$$

We have

$$\frac{\log \phi(x_n)}{\log x_n} = 2, \quad n \in \mathbb{N},$$

and

$$\frac{\log \phi(y_n)}{\log y_n} \sim \frac{\log((1-\alpha)x_{n+1}^2)}{\log((1-\alpha)x_{n+1})} \sim 2, \quad n \in \mathbb{N},\ \alpha \in (0,1).$$

For every real $x \ge x_1$ there exists $N \in \mathbb{N}$ such that $x_N \le x \le x_{N+1}$. In other words, $x = y_N$ for some $\alpha \in [0,1]$. Thus by the continuity of ϕ,

$$\lim_{x\to\infty} \frac{\log \phi(x)}{\log x} = 2. \tag{7.4.5}$$

Moreover,

$$\frac{\phi(x_n)}{x_n^2} = 1, \quad n \in \mathbb{N},$$

and

$$\frac{\phi(y_n)}{y_n^2} \sim \frac{(1-\alpha)x_{n+1}^2}{(1-\alpha)^2 x_{n+1}^2} = \frac{1}{1-\alpha}, \qquad n \in \mathbb{N},\ \alpha \in (0,1).$$

The latter limit will be larger than any given positive number when α is close enough to 1. This shows that

$$\liminf_{x\to\infty} \frac{\phi(x)}{x^2} = 1 \quad \text{and} \quad \limsup_{x\to\infty} \frac{\phi(x)}{x^2} = \infty. \tag{7.4.6}$$

By Theorem 7.3.1 there exists an entire function f such that $T(r,f) \sim \log M(r,f) \sim \phi(r)$ as $r \to \infty$. From (7.4.5) and (7.4.6) we have

$$\lim_{r\to\infty} \frac{\log\log M(r,f)}{\log r} = 2, \quad \liminf_{r\to\infty} \frac{\log M(r,f)}{r^2} = 1, \quad \limsup_{r\to\infty} \frac{\log M(r,f)}{r^2} = \infty,$$

which means that

$$\rho(f) = \underline{\rho}(f) = 2, \quad \tau(f) = \infty, \quad \underline{\tau}(f) = 1.$$

Thus f is of regular growth but not of perfectly regular growth. □

Theorem 7.3.1 allows us to construct examples of entire functions of p. r. g. of any finite positive order of growth. The next result gives more information on slowly growing entire functions of p. r. g. with negative real zeros.

Theorem 7.4.6 ([37, Theorem 4.1.1]). *Let f be an entire function of order $\rho \in (0,1)$ such that $f(0) = 1$. Suppose that all zeros of f are real and negative, and that $n(r) = n(r, 1/f) \sim ar^\rho$ for some $a > 0$. Then*

$$\log M(r,f) \sim \frac{a\pi r^\rho}{\sin(\rho\pi)}, \tag{7.4.7}$$

and

$$h_f(\theta) = \frac{a\pi\cos(\rho\theta)}{\sin(\rho\pi)}, \qquad -\pi < \theta < \pi. \tag{7.4.8}$$

Proof. Let $\{z_n\}$ denote the sequence of (negative) zeros of f, listed according to multiplicities and organized with respect to increasing moduli. Thus if $r_n = |z_n|$, then $0 < r_1 \le r_2 \le r_3 \le \cdots$. Then by the Hadamard factorization,

$$f(z) = \prod_{n=1}^{\infty}\left(1 + \frac{z}{r_n}\right).$$

Define the expressions

$$G(z) = z\int_0^\infty \frac{n(t)}{t(z+t)}\,dt \quad \text{and} \quad H(z) = z\int_0^\infty \frac{at^\rho}{t(z+t)}\,dt, \qquad z \in \mathbb{C}\setminus(-\infty, 0].$$

It will be proved that $G(z)$ and $H(z)$ are analytic in the slit plane $\mathbb{C} \setminus (-\infty, 0]$.

Consider first the case where z is real and positive. Using a telescoping argument and the fact that $n(t)$ is a step function, we may write

$$\begin{aligned}\log f(r) &= \sum_{n=1}^{\infty} \log\left(1+\frac{r}{r_n}\right) = \sum_{n=1}^{\infty} n\cdot\left(\log\left(1+\frac{r}{r_n}\right) - \log\left(1+\frac{r}{r_{n+1}}\right)\right)\\ &= \sum_{n=1}^{\infty} n\cdot\left(\int_{r_n}^{r_{n+1}} \frac{dt}{t} - \int_{r_n}^{r_{n+1}} \frac{dt}{r+t}\right) = \sum_{n=1}^{\infty} n\cdot\left(\int_{r_n}^{r_{n+1}} \frac{r\,dt}{t(r+t)}\right)\\ &= r\sum_{n=1}^{\infty}\int_{r_n}^{r_{n+1}} \frac{n\,dt}{t(r+t)} = r\sum_{n=1}^{\infty}\int_{r_n}^{r_{n+1}} \frac{n(t)}{t(r+t)}\,dt\\ &= r\int_0^{\infty} \frac{n(t)}{t(r+t)}\,dt = G(r),\end{aligned} \tag{7.4.9}$$

where $n(t) = 0$ for $0 \le t < r_1$, and where the branches of the logarithms are principal branches, so they are real on the positive real axis. With this choice of branches, we are dealing with real logarithms just like in real analysis. An alternative proof for the representation $\log f(r) = G(r)$ is by means of Riemann–Stieltjes integration; see [37, p. 55]. Since the zeros of f are real and negative, we obtain

$$\log M(r,f) = \log f(r) = G(r), \quad r > 0.$$

It is known and can be verified (by using a suitable computer software) that

$$r\int_0^{\infty} \frac{t^{\rho}}{t(r+t)}\,dt = \frac{\pi r^{\rho}}{\sin(\rho\pi)}. \tag{7.4.10}$$

Therefore,

$$H(r) = r\int_0^{\infty} \frac{a t^{\rho}}{t(r+t)}\,dt = \frac{a\pi r^{\rho}}{\sin(\rho\pi)}.$$

Let $\varepsilon > 0$. Since $n(r) \sim a r^{\rho}$ by assumption, there exists $R_0 = R_0(\varepsilon) > r_1$ such that

$$(a-\varepsilon)r^{\rho} \le n(r) \le (a+\varepsilon)r^{\rho}, \quad r \ge R_0.$$

Hence, if we replace $n(t)$ in $G(r)$ with $a t^{\rho}$, then the difference of the two integrals will satisfy

$$|G(r) - H(r)| = \left| r\int_0^{\infty} \frac{n(t) - a t^{\rho}}{t(r+t)}\,dt \right| \le r\int_0^{\infty} \frac{|n(t) - a t^{\rho}|}{t(r+t)}\,dt$$

$$\leq r\int_0^{R_0}\frac{n(t)+at^\rho}{t(r+t)}\,dt+\varepsilon r\int_{R_0}^{\infty}\frac{t^\rho}{t(r+t)}\,dt$$
$$\leq r\int_{r_1}^{R_0}\frac{n(t)}{t(r+t)}\,dt+ar\int_0^{R_0}\frac{t^\rho}{t(r+t)}\,dt+\varepsilon r\int_0^{\infty}\frac{t^\rho}{t(r+t)}\,dt$$
$$\leq \int_{r_1}^{R_0}\frac{n(t)}{t}\,dt+a\int_0^{R_0}t^{\rho-1}\,dt+\varepsilon r\int_0^{\infty}\frac{t^\rho}{t(r+t)}\,dt=O(\varepsilon r^\rho),$$

where (7.4.10) was used. This yields (7.4.7).

To prove (7.4.8), we consider the expressions $G(z)$ and $H(z)$ for a complex variable $z=re^{i\theta}$, where $-\pi<\theta<\pi$. Using (7.4.10) and the change of variable $t=ue^{i\theta}$, in which case the path of integration is along the ray $\arg z=\theta$, we obtain

$$H(z)=z\int_0^{\infty}\frac{at^\rho}{t(z+t)}\,dt=re^{i\theta}\int_0^{\infty}\frac{at^\rho}{t(re^{i\theta}+t)}\,dt$$
$$=are^{i\rho\theta}\int_0^{\infty}\frac{u^\rho}{u(r+u)}\,du=\frac{a\pi e^{i\rho\theta}r^\rho}{\sin(\rho\pi)}=\frac{a\pi z^\rho}{\sin(\rho\pi)},$$

where $z^\rho=\exp(\rho\log z)$ is defined for the principal branch of the logarithm, that is, the logarithm is real on the real axis, and $-\pi<\arg z<\pi$. This representation of $H(z)$ shows that the function $H(z)$ is analytic in the slit plane $\mathbb{C}\setminus(-\infty,0]$. Moreover,

$$|G(z)-H(z)|\leq|z|\int_0^{\infty}\frac{|n(t)-at^\rho|}{t|z+t|}\,dt,\quad z\in\mathbb{C}\setminus(-\infty,0].$$

Using the estimate

$$|z+t|^2=t^2+2tr\cos\theta+r^2=(t+r)^2+2tr(\cos\theta-1)$$
$$=(t+r)^2\Big(1+\frac{2tr}{(t+r)^2}(\cos\theta-1)\Big)\geq(t+r)^2\Big(1+\frac{1}{2}(\cos\theta-1)\Big)$$
$$=\frac{1}{2}(t+r)^2(\cos\theta+1)>0,\quad -\pi<\theta<\pi,$$

we obtain $|G(z)-H(z)|=O(\varepsilon r^\rho)$ for a complex variable $z=re^{i\theta}\in\mathbb{C}\setminus(-\infty,0]$, similarly as in the real case above. This line of reasoning also proves that the expression $G(z)$ is uniformly convergent for z on compact subsets of $\mathbb{C}\setminus(-\infty,0]$. Consequently, $G(z)$ is analytic in $\mathbb{C}\setminus(-\infty,0]$; see [169, p. 308] or [185, p. 369].

Since $\log f(z)$ and $G(z)$ are both analytic in $\mathbb{C}\setminus(-\infty,0]$, and since they coincide on the positive real axis by (7.4.9), they coincide in $\mathbb{C}\setminus(-\infty,0]$ by the uniqueness theorem. It follows that

$$\log f(re^{i\theta}) \sim \frac{a\pi e^{i\rho\theta} r^\rho}{\sin(\rho\pi)}, \qquad -\pi < \theta < \pi. \tag{7.4.11}$$

Taking the real parts on both sides of (7.4.11), we obtain (7.4.8). □

The condition $n(r) \sim ar^\rho$ in Theorem 7.4.6 can be relaxed to the condition $n(r) \sim L(r)r^\rho$, where $L(r)$ is a slowly varying function [261]; see Definition 11.5.2 for the definition of a slowly varying function.

Example 7.4.7. Let $\rho \in (0,1)$, and let $\{z_n\}$ be the sequence defined by $z_n = -n^{1/\rho}$. The exponent of convergence of $\{z_n\}$ is clearly $\lambda = \rho$. Further, let f be the Hadamard product

$$f(z) = \prod_{n=1}^{\infty}(1 - z/z_n). \tag{7.4.12}$$

Then f is entire of order $\rho(f) = \lambda = \rho$. Fix $r \geq 2$. Then there exists $N \in \mathbb{N}$ such that $N \leq r \leq N+1$. Thus the number $n(r)$ of zeros z_n of f in $|z| \leq r$ has the upper bound

$$n(r) \leq \lceil (N+1)^\rho \rceil \leq (N+1)^\rho + 1 \leq N^\rho + 2 \leq r^\rho + 2,$$

where Lemma 9.4.6 has been deployed. Using

$$r^\rho - (r-1)^\rho = \rho \int_{r-1}^{r} x^{\rho-1}\,dx \leq \rho(r-1)^{\rho-1},$$

we also obtain the lower bound

$$n(r) \geq \lfloor N^\rho \rfloor \geq N^\rho - 1 \geq (r-1)^\rho - 1 \geq r^\rho - \rho(r-1)^{\rho-1} - 1 \geq r^\rho - 2.$$

Thus $r^\rho - 2 \leq n(r) \leq r^\rho + 2$ for all $r \geq 2$, which gives us $n(r) \sim r^\rho$. Consequently,

$$\log M(r,f) \sim \frac{\pi r^\rho}{\sin(\pi\rho)} \quad \text{and} \quad h_f(\theta) = \frac{\pi\cos(\rho\theta)}{\sin(\rho\pi)}$$

by Theorem 7.4.6. This conclusion is particularly true for Lindelöf functions of order $\rho \in (0,1)$ discussed in Example 5.5.1.

Theorems 2.5.5 and 2.5.6 allow us to express the orders and types of an entire function f in terms of its Taylor coefficients. These conditions can also be used to express the regularity of growth of f.

Corollary 7.4.8 ([153, 250]). *Let $f(z) = \sum_{n=0}^{\infty} a_n z^n$ be an entire function. If the limit*

$$K = \lim_{n\to\infty} \frac{n\log n}{-\log|a_n|}$$

exists on $[0,\infty)$, then f is of r. g. and of order K. If the limit

$$M = \lim_{n\to\infty} n\left|\frac{a_{n+1}}{a_n}\right|^\rho$$

exists on $[0,\infty)$ for some $\rho \in (0,\infty)$, then f is of p. r. g., order ρ, and type M/ρ.

The exponential polynomial $f(z) = e^{z^2} + e^z$ discussed in Example 2.5.4 shows that an entire function can be of p. r. g. even though the limits K and M in Corollary 7.4.8 fail to exist.

7.5 Completely regular growth

In this section, we consider entire functions f for which the modulus $|f(z)|$ has regular growth in a large portion of the plane $\mathbb{C}$.

Definition 7.5.1 ([25], [183, pp. 139–140], [232, pp. 6–8]). We say that an entire function f of order $\rho \in (0, \infty)$ and of mean type is of *completely regular growth* (c. r. g.) if there exists a sequence of Euclidean discs $D(z_k, r_k)$ satisfying

$$\sum_{|z_k| \le r} r_k = o(r) \tag{7.5.1}$$

such that

$$\log|f(re^{i\theta})| = (h_f(\theta) + o(1))r^\rho, \quad re^{i\theta} \notin \bigcup_k D(z_k, r_k), \tag{7.5.2}$$

as $r \to \infty$ uniformly in θ.

A subset of $\mathbb{C}$ that can be covered by a sequence of discs $D(z_k, r_k)$ satisfying (7.5.1) is known as a C_0-set. Let E be the circular projection of the discs $D(z_k, r_k)$ onto the non-negative real axis. Then E is covered by the intervals $(|z_k| - r_k, |z_k| + r_k)$ of length $2r_k$. Consequently,

$$\overline{\operatorname{dens}}(E) \le \limsup_{r \to \infty} r^{-1} \sum_{|z_k| \le r} 2r_k = 0.$$

Subsets of $\mathbb{R}_+$ of zero upper linear density are called E_0-sets [183, p. 139].

It is natural to ask how p. r. g. and c. r. g. are related to each other. The following result is probably known to experts in this field, but neither the statement nor the proof seems to appear in the literature.

Theorem 7.5.2. *c. r. g. implies p. r. g.*

Proof. Let f be entire and of c. r. g. with respect to its order $\rho \in (0, \infty)$. Then by (7.5.2),

$$\log|f(re^{i\theta})| = (h_f(\theta) + o(1))r^\rho$$

outside a possible C_0-set whose circular projection E is an E_0-set. Since f is of mean type $\tau \in (0, \infty)$, its indicator function $h_f(\theta)$ is bounded, and $\tau = \max_{0 \le \theta \le 2\pi} h_f(\theta)$ by Corollary 4.2.7. Thus

$$\log M(r,f) = (\tau + o(1))r^\rho, \quad r \notin E.$$

Let $\varepsilon \in (0,\tau)$. Then

$$(\tau - \varepsilon)r^\rho \le \log M(r,f) \le (\tau + \varepsilon)r^\rho \tag{7.5.3}$$

for all $r \notin E$ large enough. By applying Lemma 1.2.7 with $\psi(t) = t$ in (7.5.3), it follows that for every $\alpha > 1$, there is a constant $r_0 = r_0(\alpha) > 0$ such that

$$(\tau - \varepsilon)\left(\frac{r}{\alpha}\right)^\rho \le \log M(r,f) \le (\tau + \varepsilon)(\alpha r)^\rho, \quad r > r_0.$$

Since α can be arbitrarily close to 1 and ε can be arbitrarily close to 0, it follows that $r^{-\rho} \log M(r,f) \to \tau$ as $r \to \infty$. This yields the statement. □

Example 7.5.3. Entire functions of p. r. g. need not be of c. r. g. Indeed, if f and g are of c. r. g., then they are both of p. r. g. However, $f - ag$ for $a \in \mathbb{C}$ is not necessarily of c. r. g. [80]. On the other hand, if either (a) $\rho(f) \neq \rho(g)$ or if (b) $\rho(f) = \rho(g)$ but $\tau(f) \neq \tau(g)$, then the linear combination $f - ag$ is of p. r. g. for every $a \in \mathbb{C}$.

Next, we define c. r. g. on rays.

Definition 7.5.4 ([183, p. 139], [81, p. 32]). An entire function f of order $\rho \in (0,\infty)$ is said to be of *completely regular growth on a ray* $\arg z = \theta$ if its indicator function satisfies

$$h_f(\theta) = \overset{*}{\lim_{r\to\infty}} \frac{\log|f(re^{i\theta})|}{r^\rho},$$

where the notation with * refers to the limit $r \to \infty$ taken outside a possible E_0-set.

It can be shown that if f is of c. r. g. on a dense set of rays, then f is of c. r. g. in the whole plane. The difficulty in proving this profound result lies in proving that the E_0-set is uniform for all rays, that is, the same E_0-set can be used for every ray; see [183, pp. 139–142].

For an exponential polynomial f of order $q \ge 1$, we have seen that the limit in (7.4.2) holds for almost every $\theta \in [0, 2\pi)$. Thus f is of completely regular growth on almost every ray, and, consequently, f is of c. r. g.

The essence of functions of c. r. g. lies in the fact that their zero sequences are regularly distributed. We now define what this means.

Definition 7.5.5 ([232, pp. 6–7]). Let $Z = \{w_k\}$ be an infinite sequence of complex numbers ordered by nondecreasing modulus such that $|w_k| \to \infty$ as $k \to \infty$. Suppose that the counting function $n(r)$ of the points $\{w_k\}$ has order $\rho \in (0,\infty)$ and is of finite type with respect to ρ. Let $n_Z(r,\theta_1,\theta_2)$ denote the number of points $\{w_k\}$ in the region

$$\theta_1 < \arg z < \theta_2, \quad |z| \le r,$$

each point being counted according to its multiplicity. We say that $\{w_k\}$ has an *angular density* with respect to ρ if for all but a countable set of values θ_1, θ_2, the limit

$$\lim_{r\to\infty} \frac{n_Z(r,\theta_1,\theta_2)}{r^\rho} \tag{7.5.4}$$

exists as a finite number.

If $\rho \notin \mathbb{N}$, then the sequence $\{w_k\}$ is called *regularly distributed* if it has an angular density. For $\rho \in \mathbb{N}$, the sequence $\{w_k\}$ is called *regularly distributed* if $\{w_k\}$ has an angular density and, moreover, the limit

$$\lim_{r\to\infty} \sum_{0<|w_k|\le r} w_k^{-\rho} \tag{7.5.5}$$

exists as a finite number.

Note that the order of the counting function $n(r)$ in Definition 7.5.5 equals the exponent of convergence of $\{w_k\}$ by Theorem 3.1.6.

Example 7.5.6. For the zero sequence $\{w_k\}$ of $\sin z$, the series $\sum_{w_k\neq 0} w_k^{-1}$ is not absolutely convergent, but the limit in (7.5.5) for $\rho = 1$ clearly exists. Thus $\{w_k\}$ is regularly distributed. Similarly, the zeros of $\cos z$ are regularly distributed.

For entire functions f, regular distribution of zeros is equivalent to f being of c. r. g.

Theorem 7.5.7 ([232, Theorem 1.2.1]). *Let f be an entire function of finite order $\rho > 0$ and of mean type, and suppose that f has an infinite number of zeros. Then f is of c. r. g. if and only if the zero sequence of f is regularly distributed.*

The proof of Theorem 7.5.7 in [232] is long and uses significant Nevanlinna theory.

Example 7.5.8. We give an elementary proof that entire functions of r. g. need not be of c. r. g. Define the sequence $\{w_k\}$ by $w_k = -k$, $k \in \mathbb{N}$. Then $\{w_k\}$ has angular density with $\rho = 1$, but $\{w_k\}$ is not regularly distributed because the limit in (7.5.5) does not exist as a finite number. Hence by Theorem 7.5.7, the Hadamard product f constructed from the zeros $\{w_k\}$ is not of c. r. g. Since $n(r) = r + O(1)$ by the reasoning in Example 7.4.7, it follows by Jensen's theorem and Lemma 3.5.12 that

$$\begin{aligned}
r + O(\log r) = N(r) \le \log M(r,f) &= O(r^2)\int_0^\infty \frac{n(t)}{t^2(r+t)}\,dt \\
&= O(r^2)\left(\int_1^\infty \frac{dt}{t(t+r)} + \int_1^\infty \frac{dt}{t^2(r+t)}\right) \\
&= O(r)\left(\left[\log\frac{t}{t+r}\right]_1^\infty + \int_1^\infty \frac{dt}{t^2}\right) = O(r\log r).
\end{aligned}$$

Thus the function f is of order one and of r. g.

The following theorem is known to experts, being only a special case of more general results that rely on intense methods; see [183, Chapter II]. An elementary proof is given because the result will be needed in Section 9.4.

Theorem 7.5.9. *For every $\rho \in (0,\infty)$, there exists an entire function f of c. r. g. such that $\rho(f) = \rho$, where $h_f(\theta) > 0$ for all $\theta \in \mathbb{R}$.*

Proof. Suppose first that $\rho \in (0,1/2)$. Let f be the function in (7.4.12), having negative real zeros $\{z_n\}$ given by $z_n = -n^{1/\rho}$. Then by Theorem 7.4.6,

$$h_f(\theta) = \frac{\pi\cos(\rho\theta)}{\sin(\rho\pi)}, \quad -\pi < \theta < \pi, \tag{7.5.6}$$

where (7.5.6) holds for "lim" in place of "lim sup" by the proof of Theorem 7.4.6. Note that representation (7.5.6) extends to the case $\theta = \pi$ by the continuity of the indicator (see Theorem 4.2.6), but the identity might no longer hold for "lim" in place of "lim sup" in this case.

Since f is of c. r. g. on all rays, except possibly on the negative real axis, it is of c. r. g. in $\mathbb{C}$. Alternatively, f is of c. r. g. by Theorem 7.5.7. In addition, it follows from (7.5.6) by the 2π-periodicity and the continuity of the indicator that $h_f(\theta) > 0$ for all $\theta \in \mathbb{R}$. In particular, the indicator $h_f(\theta)$ is a 2π-periodic function even though $\cos(\rho\theta)$ is not. Thus the statement follows in the case $\rho \in (0,1/2)$.

Suppose then that $\rho \in [1/2,\infty)$, and let n be the smallest integer such that $2\rho < n$. Set $\rho_0 = \rho/n$, and let f_0 be the entire function of order $\rho_0 \in (0,1/2)$ in (7.4.12) discussed above. Then the function $f(z) = f_0(z^n)$ is of order ρ and of c. r. g. such that its indicator satisfies

$$h_f(\theta) = \frac{\pi\cos(n\rho_0\theta)}{\sin(\rho_0\pi)}, \quad -\frac{\pi}{n} < \theta < \frac{\pi}{n}.$$

Now the period of $h_f(\theta)$ is $\frac{2\pi}{n}$, so that $h_f(\theta) > 0$ for all $\theta \in \mathbb{R}$. □

Exponential polynomials are of c. r. g., but this is not easy to prove in detail. Steinmetz [263] proved that if f is an exponential polynomial of order $q \geq 1$, then the limit

$$\lim_{r\to\infty} \frac{\log|f(re^{i\theta})|}{r^q}$$

exists as a finite number for every $\theta \in (-\pi,\pi]$ with at most finitely many exceptions. As such, f is of c. r. g. on almost every ray, and hence it is of c. r. g. in $\mathbb{C}$.

The reasoning in the previous paragraph leaves a lot to the imagination. With Theorem 7.5.7 in mind, it is therefore of interest to prove that the zero sequences of exponential polynomials are regularly distributed.

Theorem 7.5.10 ([117, Lemma 3.9]). *If an exponential polynomial f has an infinite zero-sequence Z, then Z is regularly distributed.*

Proof. Denote $q = \rho(f) \geq 1$. If the normalized form of f in (6.3.2) has only one term, then f is of the form $f(z) = F(z)e^{wz^q}$, where the zeros of f are precisely the zeros of $F(z)$, which is either a polynomial or an exponential polynomial of order $\leq q-1$. Thus we may suppose that the normalized form of f has at least two terms, in which case $\lambda(f) = \rho(f) = q \geq 1$ by Theorem 6.3.9. Let $\{z_n\}$ denote the zero-sequence Z of f, where the points z_n are listed according to multiplicities and organized with respect to increasing modulus. Without loss of generality, we may suppose that $z_n \neq 0$ for all n, that is, $f(0) \neq 0$.

Let $\theta_1 < \theta_2$. Since countably many exceptions for θ_1, θ_2 are allowed, we may suppose that $\theta_1, \theta_2 \neq \theta^*$, where $\arg z = \theta^*$ is one of the critical rays. There are two cases to consider.

(1) Suppose that the sector $S(\theta_1, \theta_2) = \{z : \theta_1 < \arg z < \theta_2\}$ does not contain any of the critical rays $\arg z = \theta^*$. Let $n_\Lambda(r)$ denote the nonintegrated counting function of the zeros of f in $|z| \leq r$ lying outside the finite union of modified logarithmic strips $\Lambda(\theta^*, c)$ defined in (6.3.6). Then

$$n_Z(r, \theta_1, \theta_2) \leq n_\Lambda(r) \leq \frac{1}{\log 2} \int_r^{2r} \frac{n_\Lambda(t)}{t}\, dt \leq \frac{1}{\log 2} N_\Lambda(2r) = O(r^{q-1} + \log r),$$

(see [115, Theorem 2.4] or [263, Satz 4]), so that the limit in (7.5.4) exists and is equal to zero. Here the $O(\log r)$ error term is needed only if $q = 1$.

(2) Suppose that $S(\theta_1, \theta_2)$ contains at least one of the critical rays $\arg z = \theta^*$. It is clear that $S(\theta_1, \theta_2)$ essentially contains the logarithmic strip $\Lambda(c, \theta^*) = \Lambda_1(c, \theta^*)$ for each $\theta^* \in (\theta_1, \theta_2)$. Hence the limit in (7.5.4) exists by Theorem 6.3.4 with $p = 1$. Note that $p = 1$ is not an additional restriction here due to the fact that $\Lambda_1(c, \theta^*)$ contains all domains of the form $\Lambda_p(c, \theta^*)$ for $p \in \mathbb{N}$.

We have now proved that the sequence Z has angular density. It remains to prove that the limit

$$\lim_{r\to\infty} \sum_{0<|z_n|\leq r}{}' z_n^{-\rho} \tag{7.5.7}$$

exists as a finite complex number. To prove this, we use Theorem 6.3.4 with $p = 1$. Without loss of generality, we may from now on assume that $|z_n| > 1$.

Recall that f is given in the normalized form (6.3.2). If $F_0(z) \equiv 0$, then define $W_f = \{\overline{w}_1, \ldots, \overline{w}_n\}$, and if $F_0(z) \not\equiv 0$, then set $W_f = \{\overline{w}_0 = 0, \overline{w}_1, \ldots, \overline{w}_n\}$. As in (6.3.5), the set $\text{co}(W_f)$ has s orthogonal rays (parallel to the outer normals of $\text{co}(W_f)$), say,

$$\arg z = \varphi_j, \quad j = 0, \ldots, s-1,$$

where $s \geq 2$ is the number of vertices of $\mathrm{co}(W_f)$. (If $\mathrm{co}(W_f)$ is a line segment, then $s = 2$.) By appealing to a rotation, if necessary, we may suppose that $\varphi_0 = 0$. Further, we may suppose that the angles φ_j are organized in an increasing order such that

$$0 = \varphi_0 < \varphi_1 < \cdots < \varphi_{s-1} < 2\pi. \tag{7.5.8}$$

Each orthogonal ray induces q critical rays $\arg z = \theta_{n,j}$ such that

$$\theta_{n,j} = \frac{\varphi_j + 2n\pi}{q}, \quad j = 0, \dots, s-1;\ n = 0, \dots, q-1. \tag{7.5.9}$$

From Theorem 6.3.4 (with $p = 1$), we know that the majority of the points z_n lie in finitely many logarithmic strips

$$\Lambda(c, \theta^*) = \left\{ re^{i\theta} : r > 1,\ |\theta - \theta^*| < c\frac{\log r}{r} \right\},$$

where $c > 0$ is a large enough fixed constant, and where $\arg z = \theta^*$ denotes a critical ray for f. Indeed, if $\{w_n\}$ denotes the subsequence of $\{z_n\}$ that does not lie in these strips, then by Theorem 6.3.4 (with $p = 1$), the exponent of convergence of $\{w_n\}$ is $\leq q-1$. Clearly,

$$\sum_{n=1}^{\infty} |w_n|^{-q} < \infty,$$

and, consequently, $\sum_{n=1}^{\infty} w_n^{-q}$ converges by Lemma 3.3.7. Thus to prove that the limit in (7.5.7) exists as a finite number, we may suppose without loss of generality that all the points z_n lie in the aforementioned logarithmic strips $\Lambda(\theta^*, c)$ about the critical rays $\arg z = \theta^*$, where θ^* represents one of the angles in (7.5.9).

Denote $z_n = r_n e^{i\phi_n}$. Then

$$\sum_{|z_n| \leq r} z_n^{-q} = \sum_{r_n \leq r} r_n^{-q} e^{-i\phi_n q} = \sum_{r_n \leq r} r_n^{-q} \cos(\phi_n q) - i \sum_{r_n \leq r} r_n^{-q} \sin(\phi_n q). \tag{7.5.10}$$

Obviously, it suffices to prove that the real and imaginary parts both converge to finite limits as $r \to \infty$. We will divide the proof into four cases.

Case 1. Suppose that $s = 2$ and $q = 1$. In this case, $\mathrm{co}(W_f)$ is a line segment parallel to the imaginary axis by the global assumption $\varphi_0 = 0$. As such, $\mathrm{co}(W_f)$ has precisely two vertices, say $\overline{w}_1$ and $\overline{w}_2$. Denote

$$\alpha = |w_2 - w_1| = |\overline{w}_2 - \overline{w}_1| = \frac{C(\mathrm{co}(W_f))}{2}.$$

Since $q = 1$, the critical rays and orthogonal rays coincide by Lemma 6.1.5, both being either the positive or the negative real axis. All the zeros z_n of f are now located in the following two logarithmic strips:

$$\Lambda_0 := \Lambda(c, 0) = \left\{ re^{i\theta} : r > 1,\ |\theta| < c\frac{\log r}{r} \right\},$$

$$\Lambda_1 := \Lambda(c, \pi) = \left\{ re^{i\theta} : r > 1,\ |\theta - \pi| < c\frac{\log r}{r} \right\}.$$

By (6.3.8) (with $q = p = 1$), we have

$$n(r, \Lambda_j) = \frac{ar}{2\pi} + O(\log^{3+\varepsilon} r), \quad j = 0, 1,$$

where $\varepsilon > 0$ is a small fixed constant. This means that there are constants $C, R_0 > 1$ such that

$$\frac{ar}{2\pi} - C\log^{3+\varepsilon} r \le n(r, \Lambda_j) \le \frac{ar}{2\pi} + C\log^{3+\varepsilon} r, \quad r \ge R_0,\ j = 0, 1.$$

Denote $\Lambda_j(s, r) := \Lambda_j \cap \{z : s < |z| \le r\}$ for $j = 0, 1$, where $r > s \ge 1$.

We begin with the real part in (7.5.10). If $z_n = r_n e^{i\phi_n} \in \Lambda_0$, then

$$1 - \frac{c^2 \log^2 r_n}{2r_n^2} \le 1 - \frac{\phi_n^2}{2} \le \cos\phi_n \le 1.$$

Let $R = R(\varepsilon) \ge R_0$ be large enough so that

$$\left| \frac{n(t, \Lambda_j)}{t} - \frac{a}{2\pi} \right| < \frac{\varepsilon}{4}, \quad t \ge R,\ j = 0, 1,$$

$$\int_R^\infty \frac{C\log^{3+\varepsilon} t}{t^2}\,dt < \frac{\varepsilon}{4}, \quad \text{and} \quad \sum_{r_n > R} \frac{c^2 \log^2 r_n}{2r_n^3} < \frac{\varepsilon}{4}.$$

Then for $r > R$, we have

$$\left| \left[\frac{n(t, \Lambda_0)}{t} \right]_R^r \right| \le \left| \frac{n(r, \Lambda_0)}{r} - \frac{a}{2\pi} \right| + \left| \frac{a}{2\pi} - \frac{n(R, \Lambda_0)}{R} \right| < \frac{\varepsilon}{4} + \frac{\varepsilon}{4} = \frac{\varepsilon}{2},$$

and, consequently,

$$\begin{aligned}
\sum_{z_n \in \Lambda_0(R,r)} \frac{\cos\phi_n}{r_n} &\le \sum_{z_n \in \Lambda_0(R,r)} \frac{1}{r_n} = \int_R^r \frac{dn(t, \Lambda_0)}{t} = \left[\frac{n(t, \Lambda_0)}{t} \right]_R^r + \int_R^r \frac{n(t, \Lambda_0)}{t^2}\,dt \\
&< \int_R^r \frac{n(t, \Lambda_0)}{t^2}\,dt + \frac{\varepsilon}{2} \le \frac{a}{2\pi} \int_R^r \frac{dt}{t} + \int_R^\infty \frac{C\log^{3+\varepsilon} t}{t^2}\,dt + \frac{\varepsilon}{2} \\
&< \frac{a}{2\pi} \log\frac{r}{R} + \frac{3\varepsilon}{4},
\end{aligned}$$

$$\sum_{z_n \in \Lambda_0(R,r)} \frac{\cos\phi_n}{r_n} \ge \sum_{z_n \in \Lambda_0(R,r)} \left(\frac{1}{r_n} - \frac{c^2 \log^2 r_n}{2r_n^3} \right) \ge \int_R^r \frac{dn(t, \Lambda_0)}{t}\,dt - \sum_{r_n > R} \frac{c^2 \log^2 r_n}{2r_n^3}$$

$$> \int_R^r \frac{dn(t,\Lambda_0)}{t}\,dt - \frac{\varepsilon}{4} > \int_R^r \frac{n(t,\Lambda_0)}{t^2}\,dt - \frac{3\varepsilon}{4}$$

$$\geq \frac{\alpha}{2\pi}\int_R^r \frac{dt}{t} - \int_R^\infty \frac{C\log^{3+\varepsilon} t}{t^2}\,dt - \frac{3\varepsilon}{4} > \frac{\alpha}{2\pi}\log\frac{r}{R} - \varepsilon.$$

Combining these two estimates, we obtain

$$\sum_{z_n\in\Lambda_0(R,r)} \frac{\cos\phi_n}{r_n} = \frac{\alpha}{2\pi}\log\frac{r}{R} + \gamma_0(R), \tag{7.5.11}$$

where $\gamma_0(R)$ is a real quantity with $|\gamma_0(R)| < \varepsilon$. Next, if $z_n = r_n e^{i\phi_n} \in \Lambda_1$, then

$$-1 \leq \cos\phi_n \leq -1 + \frac{(\phi_n - \pi)^2}{2} \leq -1 + \frac{c^2\log^2 r_n}{2r_n^2},$$

and similarly as above, we obtain

$$\sum_{z_n\in\Lambda_1(R,r)} \frac{\cos\phi_n}{r_n} = -\frac{\alpha}{2\pi}\log\frac{r}{R} + \gamma_1(R), \tag{7.5.12}$$

where $\gamma_1(R)$ is a real quantity with $|\gamma_1(R)| < \varepsilon$. From (7.5.11) and (7.5.12) we see that

$$\begin{aligned}\sum_{r_n\leq r} \frac{\cos\phi_n}{r_n} &= \sum_{1<r_n\leq R} \frac{\cos\phi_n}{r_n} + \sum_{z_n\in\Lambda_0(R,r)} \frac{\cos\phi_n}{r_n} + \sum_{z_n\in\Lambda_1(R,r)} \frac{\cos\phi_n}{r_n}\\ &= \sum_{1<r_n\leq R} \frac{\cos\phi_n}{r_n} + \gamma_0(R) + \gamma_1(R), \quad r > R.\end{aligned}$$

This means, in particular, that

$$\left|\sum_{R<r_n\leq r} \frac{\cos\phi_n}{r_n}\right| = \left|\sum_{r_n\leq r} \frac{\cos\phi_n}{r_n} - \sum_{1<r_n\leq R} \frac{\cos\phi_n}{r_n}\right| < 2\varepsilon, \quad r > R. \tag{7.5.13}$$

Since the sequence $\{r_n\}$ is nondecreasing and tends to infinity, there exists $N = N(\varepsilon) \in \mathbb{N}$ such that $r_N > R$. Let m and l be arbitrary integers with $m > l > N$. We proceed to apply (7.5.13) in proving that the expression

$$\left|\sum_{n=l}^{m} \frac{\cos\phi_n}{r_n}\right|$$

is less than a positive constant times ε. This will then match the assumption in Cauchy's convergence test, which will be needed. To begin with, keeping in mind that $\{r_n\}$ is non-decreasing, we may write

$$\sum_{n=l}^{m} \frac{\cos\phi_n}{r_n} = \sum_{r_l \le r_n \le r_m} \frac{\cos\phi_n}{r_n} - \sum_{r_n=r_l,\, n<l} \frac{\cos\phi_n}{r_n} - \sum_{r_n=r_m,\, n>m} \frac{\cos\phi_n}{r_n}. \tag{7.5.14}$$

Indeed, f may have multiple zeros on the circles $|z| = r_l$ and $|z| = r_m$ that are included in the sum $\sum_{r_l \le r_n \le r_m} \frac{\cos\phi_n}{r_n}$ but do not appear in the sum $\sum_{n=l}^{m} \frac{\cos\phi_n}{r_n}$. However, it is also possible that one (or both) of the sums $\sum_{r_n=r_l,\, n<l} \frac{\cos\phi_n}{r_n}$ and $\sum_{r_n=r_m,\, n>m} \frac{\cos\phi_n}{r_n}$ is empty. From Corollary 6.2.6 we can deduce that the number of zeros of f on any given circle $|z| = r$ is at most $o(r)$ whenever $r > 0$ is large enough. (Note that $q = 1$ is assumed in case 1, so that f is an exponential sum.) Hence, in addition to the previous restrictions for the constant $R > 0$, we may suppose that $R > 0$ is chosen large enough so that the number of zeros of f on $|z| = r$ is $\le \varepsilon r$ whenever $r > R$. This gives us

$$\left| \sum_{r_n=r_l,\, n<l} \frac{\cos\phi_n}{r_n} \right| + \left| \sum_{r_n=r_m,\, n>l} \frac{\cos\phi_n}{r_n} \right| \le \frac{\varepsilon r_l}{r_l} + \frac{\varepsilon r_m}{r_m} = 2\varepsilon,$$

because $r_m \ge r_l \ge r_N > R$. Therefore, by (7.5.13) and (7.5.14),

$$\begin{aligned} \left| \sum_{n=l}^{m} \frac{\cos\phi_n}{r_n} \right| &< \left| \sum_{R<r_n\le r_m} \frac{\cos\phi_n}{r_n} - \sum_{R<r_n<r_l} \frac{\cos\phi_n}{r_n} \right| + 2\varepsilon \\ &\le \left| \sum_{R<r_n\le r_m} \frac{\cos\phi_n}{r_n} \right| + \left| \sum_{R<r_n<r_l} \frac{\cos\phi_n}{r_n} \right| + 2\varepsilon < 6\varepsilon. \end{aligned}$$

Hence, using Cauchy's convergence test and keeping in mind that $q = 1$ is assumed, the real part in (7.5.10) converges to a finite limit as $r \to \infty$.

As for the imaginary part in (7.5.10), we have

$$\begin{aligned} \sum_{r_n \le r} \frac{|\sin\phi_n|}{r_n} &= \sum_{z_n\in\Lambda_0(r)} \frac{|\sin\phi_n|}{r_n} + \sum_{z_n\in\Lambda_1(r)} \frac{|\sin\phi_n|}{r_n} \\ &\le \sum_{z_n\in\Lambda_0(r)} \frac{|\phi_n|}{r_n} + \sum_{z_n\in\Lambda_1(r)} \frac{|\phi_n - \pi|}{r_n} \\ &\le C \sum_{r_n\le r} \frac{\log r_n}{r_n^2} \le C \sum_{n=1}^{\infty} \frac{\log r_n}{r_n^2} < \infty, \end{aligned}$$

which shows that $\sum_{n=1}^{\infty} \frac{\sin\phi_n}{r_n}$ converges by Lemma 3.3.7. It follows from (7.5.10) that the limit in (7.5.7) exists when $s = 2$ and $q = 1$.

Case 2. Suppose that $s \ge 3$ and $q = 1$. In this case, $\mathrm{co}(W_f)$ is a convex polygon with a nonempty interior. The critical rays and orthogonal rays coincide by Lemma 6.1.5, both being parallel to the outer normals of $\mathrm{co}(W_f)$, that is, the critical rays are $\arg z = \varphi_j$, where the angles φ_j are organized as in (7.5.8). For $j \in \{0, \ldots, s-1\}$, let a_j be the length of the side of $\mathrm{co}(W_f)$ corresponding to the ray $\arg z = \varphi_j$. Then

$$a_0 + a_1 + \cdots + a_{s-1} = C(\mathrm{co}(W_f)).$$

All the zeros z_n of f are located in the logarithmic strips

$$\Lambda_j := \Lambda(c, \varphi_j) = \left\{re^{i\theta} : r > 1,\ |\theta - \varphi_j| < c\frac{\log r}{r}\right\}, \quad j = 0, \dots, s-1.$$

From (6.3.8) (with $q = p = 1$) we have

$$n(r, \Lambda_j) = \frac{a_j r}{2\pi} + O(\log^{3+\varepsilon} r), \quad j = 0, \dots, s-1,$$

where $\varepsilon > 0$ is a small fixed constant. Denote $\Lambda_j(R, r) := \Lambda_j \cap \{z : R < |z| \le r\}$ for $j = 0, \dots, s-1$ and $r > R > 1$. Set $z_n = r_n e^{i\phi_n}$. If $z_n \in \Lambda_j$, then $\phi_n = \arg z_n$ satisfies

$$\begin{aligned}
\cos\phi_n &= \cos(\phi_n - \varphi_j + \varphi_j) = \cos(\phi_n - \varphi_j)\cos\varphi_j - \sin(\phi_n - \varphi_j)\sin\varphi_j \\
&= (1 + \varepsilon(r_n))\cos\varphi_j + \varepsilon(r_n)\sin\varphi_j, \\
\sin\phi_n &= \sin(\phi_n - \varphi_j + \varphi_j) = \sin(\phi_n - \varphi_j)\cos\varphi_j + \cos(\phi_n - \varphi_j)\sin\varphi_j \\
&= \varepsilon(r_n)\cos\varphi_j + (1 + \varepsilon(r_n))\sin\varphi_j,
\end{aligned}$$

where $\varepsilon(r_n)$ is a real quantity that satisfies

$$|\varepsilon(r_n)| = O\left(\frac{\log r_n}{r_n}\right)$$

but is not necessarily the same quantity each time it appears. Analogously as in case 1, for $R = R(\varepsilon) > 1$ large enough, let $\gamma(R)$ denote a real quantity with $|\gamma(R)| < \varepsilon$, not necessarily the same quantity at each occurrence. We obtain

$$\begin{aligned}
\sum_{R<r_n\le r} \frac{\cos\phi_n}{r_n} &= \sum_{j=0}^{s-1} \sum_{z_n \in \Lambda_j(R,r)} \frac{\cos\phi_n}{r_n} = \sum_{j=0}^{s-1} \sum_{R<r_n\le r} \frac{\cos\varphi_j}{r_n} + \gamma(R) \\
&= \sum_{j=0}^{s-1} \cos\varphi_j \int_R^r \frac{dn(t, \Lambda_j)}{t} + \gamma(R) \\
&= \frac{1}{2\pi} \log\frac{r}{R} \sum_{j=0}^{s-1} a_j \cos\varphi_j + \gamma(R),
\end{aligned} \tag{7.5.15}$$

$$\begin{aligned}
\sum_{R<r_n\le r} \frac{\sin\phi_n}{r_n} &= \sum_{j=0}^{s-1} \sum_{z_n \in \Lambda_j(R,r)} \frac{\sin\phi_n}{r_n} = \sum_{j=0}^{s-1} \sum_{R<r_n\le r} \frac{\sin\varphi_j}{r_n} + \gamma(R) \\
&= \sum_{j=0}^{s-1} \sin\varphi_j \int_R^r \frac{dn(t, \Lambda_j)}{t} + \gamma(R) \\
&= \frac{1}{2\pi} \log\frac{r}{R} \sum_{j=0}^{s-1} a_j \sin\varphi_j + \gamma(R).
\end{aligned} \tag{7.5.16}$$

The unit outer normal vector corresponding to the jth side of $\mathrm{co}(W_f)$ (in the direction of the orthogonal ray $\arg z = \varphi_j$) is

$$\mathbf{n}_j = (\cos\varphi_j, \sin\varphi_j), \quad j = 0, \ldots, s-1.$$

The length of the jth side of $\mathrm{co}(W_j)$ is a_j, so the outer normal vector of length a_j corresponding to the jth side of $\mathrm{co}(W_f)$ is

$$\boldsymbol{N}_j = (a_j \cos\varphi_j, a_j \sin\varphi_j), \quad j = 0, \ldots, s-1.$$

Let $\zeta_0, \ldots, \zeta_{s-1}$ denote the vertices of $\mathrm{co}(W_f)$, and let $\zeta_s := \zeta_0$. We may suppose that the vertices are listed in counterclockwise order so that the vector $\boldsymbol{v}_j = \zeta_{j+1} - \zeta_j$ from ζ_j to ζ_{j+1} is perpendicular to $\boldsymbol{N}_j$ for $j = 0, \ldots, s-1$. As a telescoping sum, the sum of the vectors $\boldsymbol{v}_j$ must be equal to the zero vector. Due to this setup and the fact that the vectors $\boldsymbol{v}_j$ and $\boldsymbol{N}_j$ have the same length a_j, it follows that $\boldsymbol{N}_j = -i\boldsymbol{v}_j$ for $j = 0, \ldots, s-1$. This implies that the sum of the vectors $\boldsymbol{N}_j$ must also be equal to the zero vector, which gives us

$$\sum_{j=0}^{s-1} a_j \cos\varphi_j = 0 \quad \text{and} \quad \sum_{j=0}^{s-1} a_j \sin\varphi_j = 0.$$

Hence from (7.5.15) and (7.5.16), it follows that

$$\sum_{R<r_n\le r} \frac{\cos\phi_n}{r_n} = \gamma(R) \quad \text{and} \quad \sum_{R<r_n\le r} \frac{\sin\phi_n}{r_n} = \gamma(R).$$

As in case 1, using Cauchy's convergence test and keeping in mind that $q = 1$, we find that the real and imaginary parts of (7.5.10) both converge as $r \to \infty$. This means that the limit in (7.5.7) exists as a finite number when $s \ge 3$ and $q = 1$.

Case 3. Suppose that $s = 2$ and $q \ge 2$. Again, $\mathrm{co}(W_f)$ is a line segment parallel to the imaginary axis by the global assumption $\varphi_0 = 0$. Let $\overline{w}_1, \overline{w}_2$, and a be as in case 1. From (7.5.9) (with $\varphi_0 = 0$ and $\varphi_1 = \pi$), we find that there are now $2q$ critical rays $\arg z = \theta_k$, where

$$\theta_k = \frac{k\pi}{q}, \quad k = 0, \ldots, 2q-1.$$

All the zeros z_n of f are located in the $2q$ logarithmic strips

$$\Lambda_k := \Lambda(c, \theta_k) = \left\{re^{i\theta} : r > 1,\ |\theta - \theta_k| < c\frac{\log r}{r}\right\}, \quad k = 0, \ldots, 2q-1.$$

From (6.3.8) (with $q \ge 2$ and $p = 1$) we have

$$n(r, \Lambda_k) = \frac{ar^q}{2\pi} + O(r^{q-1}\log^{3+\varepsilon} r), \quad k = 0, \ldots, 2q-1,$$

where $\varepsilon > 0$ is a small fixed constant. Denote $\Lambda_k(R, r) := \Lambda_k \cap \{z : R < |z| \le r\}$ for $k = 0, \ldots, 2q-1$ and $r > R > 1$. Set $z_n = r_n e^{i\phi_n}$. If $z_n \in \Lambda_k$, then $\phi_n q = \arg(z_n^q)$ satisfies

$$\begin{aligned}\cos(\phi_n q) &= \cos(\phi_n q - \theta_k q + \theta_k q)\\ &= \cos(\phi_n q - \theta_k q)\cos(\theta_k q) - \sin(\phi_n q - \theta_k q)\sin(\theta_k q)\\ &= \cos(\phi_n q - \theta_k q)\cos(k\pi) = (-1)^k + \varepsilon(r_n),\\ \sin(\phi_n q) &= \sin(\phi_n q - \theta_k q + \theta_k q)\\ &= \sin(\phi_n q - \theta_k q)\cos(\theta_k q) + \cos(\phi_n q - \theta_k q)\sin(\theta_k q)\\ &= \sin(\phi_n q - \theta_k q)\cos(k\pi) = \varepsilon(r_n),\end{aligned}$$

where $\varepsilon(r_n)$ is a real quantity as in case 2. The remaining details in case 3 form a q-analogue of case 1 in the following sense:

- Since $\sin(\phi_n q) = \varepsilon(r_n)$, the imaginary part of (7.5.10) converges absolutely as $r \to \infty$. Thus by Lemma 3.3.7, the imaginary part of (7.5.10) converges in the usual sense as $r \to \infty$.
- The real part of (7.5.10) divides into $2q$ sums according to k. The sums related to the two consecutive logarithmic strips Λ_k and Λ_{k+1} cancel out leaving an error term $\gamma(R)$ satisfying $|\gamma(R)| < \varepsilon$, provided that $R = R(\varepsilon) > 1$ is large enough. This is due to the fact that

$$\begin{aligned}\sum_{z_n \in \Lambda_k(R,r)} \frac{\cos(\phi_n q)}{r_n^q} &= (-1)^k \int_R^r \frac{dn(t, \Lambda_k)}{t^q} + \sum_{z_n \in \Lambda_k(R,r)} \frac{\varepsilon(r_n)}{r_n^q}\\ &= (-1)^k \frac{\alpha q}{2\pi} \log \frac{r}{R} + \gamma(R), \quad r > R.\end{aligned}$$

  We conclude that the real part of (7.5.10) converges as $r \to \infty$.

It follows from (7.5.10) that the limit in (7.5.7) exists when $s = 2$ and $q \ge 2$.

Case 4. Suppose that $s \ge 3$ and $q \ge 2$. We now have two sets of critical rays as described in (7.5.9), one set for the parameter s and another set for the parameter q. This leads to a combination of the techniques discussed in cases 2 and 3. The details are omitted. □

8 Linear differential equations with polynomial coefficients

In this chapter, we consider solutions of linear differential equations of the form

$$f^{(n)} + P_{n-1}(z)f^{(n-1)} + \cdots + P_0(z)f = 0, \quad n \geq 2, \tag{8.0.1}$$

where $P_0(z), P_1(z), \ldots, P_{n-1}(z)$ are polynomials with $P_0(z) \not\equiv 0$. Recall from Corollary 1.7.2 that all solutions f of (8.0.1) are entire functions.

We have the restriction $n \geq 2$ in equation (8.0.1) for two reasons. (1) When $n = 1$ in (8.0.1), we can integrate the equation $f' + P_0(z)f = 0$ directly and obtain that any nontrivial solution has the form $f(z) = e^{Q(z)}$, where $Q(z)$ is a polynomial satisfying $Q'(z) = -P_0(z)$. Since this completely describes the situation regarding nontrivial solutions of (8.0.1) when $n = 1$, there is no loss to assume that $n \geq 2$. (2) When we use a standard normalizing transformation of (8.0.1) in Section 8.2, it will be necessary to have $n \geq 2$.

In Section 8.1, we discuss the possible orders of growth of solutions of (8.0.1), originating from the classical work of Wittich. The oscillation of solutions of (8.0.1) is considered in Section 8.2 in terms of the exponent of convergence. In Section 8.3, we discuss the asymptotic integration of the particular second-order case $f'' + P(z)f = 0$, originating from the work of Hille. The method of asymptotic integration gives information on the asymptotic growth and the frequency of zeros of solutions in sectors as well as in modified sectors. Generalizations of the Airy integral are given in Section 8.4. Both Sections 8.4 and 8.5 contain contour integral solutions of equations of the form (8.0.1), which often have interesting properties, as we saw in Section 5.3 with the Airy integral. Finally, a straightforward proof, based on asymptotic integration, is given in Section 8.6 for the known fact that the zeros of solutions of $f'' + P(z)f = 0$ are regularly distributed. This in turn implies that these solutions are of completely regular growth, which was proved independently by Petrenko and Steinmetz for the general linear differential equation (8.0.1) by using intense methods.

8.1 Growth of solutions

Wittich proved the following classical result.

Theorem 8.1.1 ([166, Theorem 4.1]). *Suppose that the functions $A_0(z), \ldots, A_{n-1}(z)$ are entire. Then all the solutions of*

$$f^{(n)} + A_{n-1}(z)f^{(n-1)} + \cdots + A_0(z)f = 0 \tag{8.1.1}$$

have finite order if and only if all the coefficients $A_0(z), \ldots, A_{n-1}(z)$ are polynomials.

https://doi.org/10.1515/9783112249130-008

The rest of the section is focused on equation (8.0.1). It is known [166, p. 127] that every solution f of (8.0.1) satisfies the sharp estimate

$$\rho(f) \leq 1 + \max_{0 \leq j \leq n-1} \frac{\deg P_j}{n-j}. \tag{8.1.2}$$

The proof of (8.1.2) given in [166] is based on Wiman–Valiron theory, and another proof can easily be constructed from the following general growth estimate.

Theorem 8.1.2 ([118, Theorem 5.1]). *Let f be a solution of* (8.1.1), *where the coefficients $A_j(z)$ are analytic in $|z| < R$ for some $0 < R \leq \infty$. Let $m \in \{1, \ldots, n\}$ be the number of coefficients $A_j(z)$ satisfying $A_j(z) \not\equiv 0$, and let $\theta \in [0, 2\pi)$ and $\varepsilon > 0$. If $z_\theta = \nu e^{i\theta}$ with $0 \leq \nu < R$ is such that $A_j(z_\theta) \neq 0$ for some $j = 0, \ldots, n-1$, then for all $\nu < r < R$,*

$$|f(re^{i\theta})| \leq C \exp\left(m \int_\nu^r \max_{j=0,\ldots,n-1} |A_j(te^{i\theta})|^{\frac{1}{n-j}}\, dt \right), \tag{8.1.3}$$

where C is a constant that satisfies

$$C \leq (1+\varepsilon) \max_{j=0,\ldots,n-1} \left(\frac{|f^{(j)}(z_\theta)|}{m^j \max_{k=0,\ldots,n-1} |A_k(z_\theta)|^{\frac{j}{n-k}}} \right). \tag{8.1.4}$$

We proceed to state a theorem of Valiron.

Theorem 8.1.3 ([279, p. 108]). *Every transcendental solution f of* (8.0.1) *satisfies*

$$\log M(r,f) = (1 + o(1))Cr^q$$

as $r \to \infty$, where $C > 0$ is a constant, and $q > 0$ is a rational number.

From Theorem 8.1.3 we see that f has finite rational order and finite positive type, and that f is of p. r. g.

Wittich proved the following result.

Theorem 8.1.4 ([292, 293]). *For an equation of the form* (8.0.1), *there exists a finite set of positive rational numbers such that the order of any transcendental entire function f must be one of these rational numbers.*

In his proof of Theorem 8.1.4, Wittich used the method of Frobenius, Wiman–Valiron theory, the theory of algebraic functions, and the Newton–Puiseux diagram, where the rational numbers are obtained from the Newton–Puiseux diagram. Other proofs were given by Helmrath and Nikolaus [137] and Jank and Volkmann [150].

Using simple arithmetic with the degrees of the polynomials $P_j(z)$ in (8.0.1), Wiman–Valiron theory, estimates for logarithmic derivatives, and the method of order reduction, Gundersen, Steinbart, and Wang [96] proved Theorem 8.1.5, which includes Theorem 8.1.4. This proof did not use the Newton–Puiseux diagram or theory of algebraic functions. To state Theorem 8.1.5 and related results, we first need the following definitions and notation.

In equation (8.0.1), set $d_j = \deg P_j$ if $P_j \not\equiv 0$ and $d_j = -\infty$ if $P_j \equiv 0$, $0 \le j \le n-1$. Define the strictly decreasing finite sequence of nonnegative integers

$$s_1 > s_2 > \cdots > s_p \ge 0 \tag{8.1.5}$$

as follows. Choose s_1 to be the unique integer satisfying

$$\frac{d_{s_1}}{n-s_1} = \max_{0\le k\le n-1} \frac{d_k}{n-k} \quad \text{and} \quad \frac{d_{s_1}}{n-s_1} > \frac{d_k}{n-k}, \quad 0 \le k < s_1. \tag{8.1.6}$$

Then given $s_j, j \ge 1$, define s_{j+1} to be the unique integer satisfying

$$\frac{d_{s_{j+1}} - d_{s_j}}{s_j - s_{j+1}} = \max_{0\le k<s_j} \frac{d_k - d_{s_j}}{s_j - k} > -1 \quad \text{and} \quad \frac{d_{s_{j+1}} - d_{s_j}}{s_j - s_{j+1}} > \frac{d_k - d_{s_j}}{s_j - k}, \quad 0 \le k < s_{j+1}. \tag{8.1.7}$$

For a certain p, the integer s_p will exist, but the integer s_{p+1} will not exist, and then the sequence $s_1, s_2, \ldots, s_p$ terminates with s_p. Observe that $p \le n$ and that (8.1.5) holds.

Now define the rational numbers $\alpha_1, \alpha_2, \ldots, \alpha_p$ by

$$\alpha_j = 1 + \frac{d_{s_j} - d_{s_{j-1}}}{s_{j-1} - s_j}, \tag{8.1.8}$$

where

$$s_0 = n \quad \text{and} \quad d_{s_0} = d_n = 0. \tag{8.1.9}$$

From (8.1.7) and (8.1.8) we see that $\alpha_j > 0$ for all $1 \le j \le p$.

Observe that the integers $s_1, s_2, \ldots, s_p$ can be expressed as follows:

$$s_1 = \min\left\{j : \frac{d_j}{n-j} = \max_{0\le k\le n-1} \frac{d_k}{n-k}\right\},$$

and given $s_j, j \ge 1$, we get

$$s_{j+1} = \min\left\{i : \frac{d_i - d_{s_j}}{s_j - i} = \max_{0\le k<s_j} \frac{d_k - d_{s_j}}{s_j - k} > -1\right\}.$$

Theorem 8.1.5 ([96]). *For equation* (8.0.1), *we have the following:*
(a) *If f is a transcendental solution of* (8.0.1), *then $\rho(f) = \alpha_j$ for some j satisfying $1 \le j \le p$.*
(b) *If $s_1 \ge 1$ and $p \ge 2$, then we have the following inequalities:*

$$\alpha_1 > \alpha_2 > \cdots > \alpha_p \ge \frac{1}{s_{p-1} - s_p} \ge \frac{1}{s_1 - s_p} \ge \frac{1}{s_1}.$$

(c) *If* $s_1 = 0$, *then any nontrivial solution* f *of* (8.0.1) *satisfies*

$$\rho(f) = 1 + d_0/n.$$

Examples below show that (1) there exist equations of the form (8.0.1) that possess solutions with orders that equal all the possible orders α_j in (a) and that (2) there exist other equations of the form (8.0.1) where not every α_j in (a) is an order of a solution of the equation.

Example 8.1.6 ([96, Example 5]). The equation

$$f^{(4)} + (8z^3 - 13)f'' - (16z^4 + 16z^3 + 12z^2 + 4z + 2)f' + (16z^4 + 8z^3 + 12z^2 + 4z + 14)f = 0$$

possesses the solutions $f_1 = e^z$ and $f_2 = e^{z^2}$, and by Corollary 8.1.12, there exists a solution f_3 satisfying $\rho(f_3) = 5/2$. For this equation, we have $p = 3$, $s_1 = 2$, $s_2 = 1$, $s_3 = 0$, $\alpha_1 = 5/2$, $\alpha_2 = 2$, and $\alpha_3 = 1$. Thus in this example, all the possible rational numbers in (8.1.8) are orders of solutions of the equation.

Example 8.1.7 ([96, Example 8]). This example shows that not every value α_j in (8.1.8) is necessarily attained as the order of a solution of (8.0.1). Consider the equation

$$f'' + z^3 f' - (z^4 + z^2)f = 0, \tag{8.1.10}$$

where $p = 2$, $\alpha_1 = 4$, and $\alpha_2 = 2$. Thus, 4 and 2 are the only possible orders for transcendental solutions of this equation. Obviously, (8.1.10) does not possess a nontrivial polynomial solution. If f is a transcendental solution of (8.1.10), and if we set

$$v(z) = f(z)e^{-\frac{1}{2}z^2}, \tag{8.1.11}$$

then v is a solution of

$$v'' + (z^3 + 2z)v' + v = 0. \tag{8.1.12}$$

Equation (8.1.12) does not have a nontrivial polynomial solution. From Theorem 8.1.5 we obtain that all transcendental solutions v to (8.1.12) have order 4. Hence from (8.1.11) we have $\rho(f) = 4$, which shows that all nontrivial solutions f of the original equation (8.1.10) have order 4. Thus $\alpha_2 = 2$ is not reached.

Remark. It is shown in [96] that for any positive integers $q_1, q_2, \ldots, q_n$ satisfying

$$0 < q_1 < q_2 < \cdots < q_n,$$

there exists an equation of the form (8.0.1) that possesses n distinct solutions $f_1, f_2, \ldots, f_n$ such that $\rho(f_k) = q_k$ for $1 \le k \le n$.

We state two corollaries of Theorem 8.1.5.

Corollary 8.1.8 ([96]). *There can exist at most* $s_1 + 1$ *distinct orders of transcendental solutions of* (8.0.1).

Example 8.1.6 shows that Corollary 8.1.8 is sharp. For more comments and examples concerning the sharpness of Corollary 8.1.8, see [96, p. 1227].

Corollary 8.1.9 ([96, 292, 293]). *Every transcendental solution f of* (8.0.1) *satisfies*

$$\rho(f) \geq \frac{1}{n-1}. \tag{8.1.13}$$

Example 8.1.10 ([96, Example 1]). The contour integral

$$G(z) = \int_{|w|=1} \exp\left\{\frac{z}{w} - \frac{1}{2w^2} - w\right\} dw \tag{8.1.14}$$

defines an entire function $G(z)$ with $\rho(G) = 1/2$, and $f = G(z)$ is a solution of the equation

$$f''' - zf'' - f = 0. \tag{8.1.15}$$

This example gives an equality in (8.1.13).

Wittich [292], [293, pp. 65–68] indicated that for each $n \geq 2$, there exists a solution ψ of an equation of the form (8.0.1) such that $\rho(\psi) = 1/(n-1)$, which shows that (8.1.13) is sharp for every n.

Theorem 8.1.11 ([96]). *For any $j = 1, 2, \ldots, p$, there can exist at most s_j linearly independent solutions f of* (8.0.1) *satisfying $\rho(f) < \alpha_j$.*

Example 8.1.6 and examples on pp. 1242–1244 of [96] show that Theorem 8.1.11 is sharp. The next two results are corollaries of Theorem 8.1.11.

Corollary 8.1.12 ([96]). *For each equation of the form* (8.0.1), *there exists a solution f of* (8.0.1) *that satisfies $\rho(f) = \beta$, where*

$$\beta = 1 + \max_{0 \leq k \leq n-1} \frac{\deg P_k}{n-k}.$$

Corollary 8.1.13 ([96]). *There can exist at most s_p linearly independent polynomial solutions of* (8.0.1).

It is worth mentioning that Corollary 8.1.12 shows that an equation of the form (8.0.1) cannot have all polynomial solutions. Corollary 8.1.13 is sharp because the following example shows that it is possible for an equation of the form (8.0.1) to possess exactly s_p linearly independent polynomial solutions for any value that the integer s_p could have.

Example 8.1.14 ([96, Example 6]). Let q and n be integers satisfying $2 \leq q \leq n$. Then the equation

$$f^{(n)} - \frac{1}{(q-1)!} z^{q-1} f^{(q-1)} + \frac{1}{(q-2)!} z^{q-2} f^{(q-2)} - \cdots + (-1)^{q-1} zf' + (-1)^q f = 0$$

possesses the $q-1$ linearly independent polynomial solutions

$$f_k(z) = z^k, \quad k = 1, 2, \ldots, q-1.$$

For this equation, we have $p = 1$, $s_p = s_1 = q - 1$, and $\alpha_1 = n/(n-q+1)$. Therefore, this gives a sharp example for Corollary 8.1.13 and Theorem 8.1.11.

It was shown in [96] that the following result is a consequence of Theorem 8.1.11 and Corollary 8.1.13.

Corollary 8.1.15 ([96]). *If* $\{f_1, f_2, \ldots, f_n\}$ *is a fundamental set of solutions of equation* (8.0.1), *then*

$$\sum_{k=1}^{n} \rho(f_k) \geq n + d_0. \tag{8.1.16}$$

When $q = n \geq 2$, Example 8.1.14 gives an equation of the form (8.0.1) that possesses a fundamental set of solutions consisting of $n-1$ polynomials and one transcendental solution of order n, by using Corollary 8.1.12. This gives the equality $n = n$ in (8.1.16). For more discussion and examples concerning the sharpness of Corollary 8.1.15, see [96, p. 1229].

Example 8.1.16 ([96, Example 3]). The equation

$$f''' - f'' + zf' - zf = 0$$

possesses the solution $f_1(z) = e^z$, and by Corollary 8.1.12, another solution f_2 satisfying $\rho(f_2) = 3/2$. By Corollary 8.1.15 there must exist a third solution f_3 satisfying $\rho(f_3) \geq 3/2$ such that $\{f_1, f_2, f_3\}$ is a fundamental set of solutions. It follows from Corollary 8.1.12 that $\rho(f_3) = 3/2$. Therefore, inequality (8.1.16) becomes the equality $4 = 4$ for this fundamental set $\{f_1, f_2, f_3\}$. This example also illustrates the sharpness of Corollary 8.1.8 and Theorem 8.1.11 (see [96, p. 1243]).

The next theorem will be needed in Section 8.2. The proof uses many results discussed in this section.

Theorem 8.1.17. *Let* $\rho \geq 1$ *be any fixed constant, and suppose that the coefficients* $A_0(z), \ldots, A_{n-1}(z)$ *of* (8.1.1) *are entire. Then* $A_0(z), \ldots, A_{n-1}(z)$ *are polynomials satisfying*

$$\deg A_j \leq (n-j)(\rho - 1), \quad j = 0, \ldots, n-1, \tag{8.1.17}$$

if and only if all solutions f *of* (8.1.1) *satisfy*

$$\rho(f) \leq \rho. \tag{8.1.18}$$

Proof. From (8.1.2) it is clear that (8.1.17) implies (8.1.18). Conversely, suppose that (8.1.18) holds. Then from Theorem 8.1.1 we conclude that all the coefficients $A_0(z), \ldots, A_{n-1}(z)$ of

(8.1.1) are polynomials. Suppose on the contrary to (8.1.17) that one of the coefficients $A_{k_0}(z)$ satisfies

$$\deg A_{k_0} > (n-k_0)(\rho-1).$$

Then

$$1 + \max_{0\le k\le n-1} \frac{\deg A_k}{n-k} \ge 1 + \frac{\deg A_{k_0}}{n-k_0} > \rho \ge \rho(f),$$

which contradicts the conclusion of Corollary 8.1.12. This proves (8.1.17). □

For more results and examples that have a similar nature to those in this section, see [96].

8.2 Oscillation of solutions

When conducting an investigation of the zeros of solutions of (8.0.1), it can usually be assumed that $P_{n-1}(z) \equiv 0$, by the following reasoning. If $\Phi(z) = \int^z P_{n-1}(\zeta)\,d\zeta$ denotes a primitive function of $P_{n-1}(z)$, then the standard transformation

$$f(z) = g(z)\exp\left(-\frac{1}{n}\Phi(z)\right), \tag{8.2.1}$$

which does not change the zeros, transforms (8.0.1) into an equation of the form

$$g^{(n)} + Q_{n-2}(z)g^{(n-2)} + \cdots + Q_0(z)g = 0, \quad n \ge 2, \tag{8.2.2}$$

where $Q_0(z), \dots, Q_{n-2}(z)$ are polynomials, which we now discuss.

The reason why the term $Q_{n-1}(z)g^{(n-1)}$ is missing in (8.2.2) is as follows. If f is a solution of (8.0.1) and if we set $P_n(z) \equiv 1$, then

$$\begin{aligned} 0 &= \sum_{k=0}^{n} P_k(z) f^{(k)}(z) = \sum_{k=0}^{n} P_k(z)\left(g(z)\exp\left(-\frac{1}{n}\Phi(z)\right)\right)^{(k)} \\ &= \sum_{k=0}^{n} P_k(z) \sum_{j=0}^{k} \binom{k}{j} g^{(j)}(z)\left(\exp\left(-\frac{1}{n}\Phi(z)\right)\right)^{(k-j)} \\ &= P_n(z)\sum_{j=0}^{n}\binom{n}{j} g^{(j)}(z)\left(\exp\left(-\frac{1}{n}\Phi(z)\right)\right)^{(n-j)} \\ &\quad + P_{n-1}(z)\sum_{j=0}^{n-1}\binom{n-1}{j} g^{(j)}(z)\left(\exp\left(-\frac{1}{n}\Phi(z)\right)\right)^{(n-1-j)} \\ &\quad + \sum_{k=0}^{n-2} P_k(z)\sum_{j=0}^{k}\binom{k}{j} g^{(j)}(z)\left(\exp\left(-\frac{1}{n}\Phi(z)\right)\right)^{(k-j)}. \end{aligned} \tag{8.2.3}$$

There are precisely two terms in (8.2.3) that include the derivative $g^{(n-1)}(z)$. The contribution of these two terms in (8.2.3) is

$$ng^{(n-1)}(z)\left(\exp\left(-\frac{1}{n}\Phi(z)\right)\right)' + P_{n-1}(z)g^{(n-1)}(z)\exp\left(-\frac{1}{n}\Phi(z)\right) = 0$$

by the definition of the function $\Phi(z)$. In other words, the terms involving $g^{(n-1)}(z)$ cancel out from (8.2.3).

The coefficients $Q_0(z), \ldots, Q_{n-2}(z)$ can be expressed explicitly. To see this, first note that the coefficient of $g^{(n)}$ in (8.2.3) is

$$P_n(z)\exp\left(-\frac{1}{n}\Phi(z)\right) = \exp\left(-\frac{1}{n}\Phi(z)\right).$$

Thus equation (8.2.3) needs to be divided by this term to obtain (8.2.2). From (8.2.3) we can deduce that

$$Q_j(z) = \frac{\sum_{k=0}^{n} P_k(z)\binom{k}{j}(\exp(-\frac{1}{n}\Phi(z)))^{(k-j)}}{\exp(-\frac{1}{n}\Phi(z))}$$

for $j = 0, \ldots, n-2$. This explicit form shows that the coefficients $Q_0(z), \ldots, Q_{n-2}(z)$ are polynomials.

Before we state the results in this section, we mention that the transformation of equation (8.0.1) into equation (8.2.2) by formula (8.2.1) normally results in $Q_0(z) \not\equiv 0$, but there are some rare exceptions when it turns out that $Q_0(z) \equiv 0$. We give two illustrations.

Example 8.2.1. (a) For the equation

$$f''' + 3zf'' + 4z^2 f' + 2z^3 f = 0,$$

the transformation

$$f(z) = g(z)\exp\left(-\frac{1}{2}z^2\right)$$

gives the following equation:

$$g''' + (z^2 - 3)g' = 0.$$

(b) Let $P(z) \not\equiv 0$ be a polynomial and consider the equation

$$f''' + P(z)f'' + \left(P'(z) + \frac{1}{3}P(z)^2\right)f' + \left(\frac{1}{3}P''(z) + \frac{1}{3}P(z)P'(z) + \frac{1}{27}P(z)^3\right)f = 0.$$

The transformation

$$f(z) = g(z)\exp\left(-\frac{1}{3}R(z)\right),$$

where $R(z)$ is any polynomial satisfying $R'(z) = P(z)$, gives the trivial equation

$$g''' = 0,$$

where all the solutions are polynomials.

The reason we mention this rare possibility is because the proofs of most results in this section require the condition $Q_0(z) \not\equiv 0$ as these proofs use results about equation (8.0.1) which has the condition $P_0(z) \not\equiv 0$.

The next result is an analogue of Theorem 8.1.17 for the exponent of convergence.

Theorem 8.2.2 ([123, Theorem 1.1]). *Let $\lambda \geq 1$ be any fixed constant, and suppose that the coefficients $Q_0(z), \ldots, Q_{n-2}(z)$ in* (8.2.2) *are entire functions with $Q_0(z) \not\equiv 0$. Then $Q_0(z), \ldots, Q_{n-2}(z)$ are polynomials satisfying*

$$\deg Q_j \leq (n-j)(\lambda - 1), \quad j = 0, \ldots, n-2, \tag{8.2.4}$$

if and only if all solutions g of (8.2.2) *satisfy*

$$\lambda(g) \leq \lambda. \tag{8.2.5}$$

If (8.2.4) is satisfied, then (8.1.2) implies that $\lambda(g) \leq \rho(g) \leq \lambda$, and hence (8.2.5) is satisfied. A nontrivial proof that (8.2.5) implies (8.2.4) was obtained in [123]. We proceed to state an analogue of Corollary 8.1.12 for the exponent of convergence.

Corollary 8.2.3. *For each equation of the form* (8.2.2) *with polynomial coefficients $Q_0(z), \ldots, Q_{n-2}(z)$, $Q_0(z) \not\equiv 0$, there exists a solution g of* (8.2.2) *that satisfies $\lambda(g) = \gamma$, where*

$$\gamma = 1 + \max_{0 \leq k \leq n-2} \frac{\deg Q_k}{n-k}.$$

Proof. All solutions g of (8.2.2) satisfy $\lambda(g) \leq \rho(g) \leq \gamma$ by (8.1.2). Suppose that $\lambda(g) < \alpha < \gamma$ for all solutions g. Then the proof of Theorem 8.2.2 in [123], which is several pages long, can be used to show that

$$\deg Q_k \leq (n-k)(\alpha - 1), \quad k = 0, \ldots, n-2,$$

which contradicts the definition of the constant γ. □

Corollary 8.2.4 ([123, Corollary 1.2]). *Let $\lambda \geq 1$ be any fixed constant, and suppose that the coefficients $Q_0(z), \ldots, Q_{n-2}(z)$ of* (8.2.2) *are entire and $Q_0(z) \not\equiv 0$. Then the following three statements are equivalent:*

(a) *The coefficients $Q_0(z), \ldots, Q_{n-2}(z)$ are polynomials such that the condition* (8.2.4) *is satisfied, and there exists at least one j for which* $\deg Q_j = (n-j)(\lambda-1)$.

(b) *All solutions g satisfy $\lambda(g) \leq \lambda$, and there exists a solution g for which $\lambda(g) = \lambda$.*

(c) *All solutions g satisfy $\rho(g) \leq \lambda$, and there exists a solution g for which $\rho(g) = \lambda$.*

Proof. Statements (a) and (b) are equivalent by Theorem 8.2.2 and Corollary 8.2.3, and the equivalence of (a) and (c) follows from Corollary 8.1.12 and Theorem 8.1.17. □

Corollary 8.2.4 addresses equation (8.2.2) when (8.2.2) has polynomial coefficients. As discussed in Section 8.1, the possible orders of growth of transcendental solutions of such equations are contained within a certain finite set of rational numbers. Hence, if the constant λ in Corollary 8.2.4 is not one of these allowed values, then all the statements (a)–(c) are false. Conversely, if any one of the statements (a)–(c) is valid, then the constant λ must be the largest one of these allowed values.

Equation (8.2.2) with polynomial coefficients and $Q_0(z) \not\equiv 0$ may possess a zero-free solution base $\{g_1, \ldots, g_n\}$. However, according to Corollaries 8.1.12 and 8.2.3, there exists a linear combination g of the functions $g_1, \ldots, g_n$ that has the maximal amount of zeros in the sense of $\lambda(g) = \rho(g)$.

Example 8.2.5 ([284]). Let $c_1, \ldots, c_n$ be the n distinct roots of $c^n = 1$. Then the functions

$$g_j(z) = \exp(c_j z), \quad j = 1, \ldots, n,$$

form a zero-free solution base for $g^{(n)} - g = 0$. As an extra observation, note that the product $g_1(z) \cdots g_n(z) \equiv 1$.

Exponential polynomials are examples of entire functions whose zeros are distributed near finitely many rays, as described in Theorem 6.3.4. Moreover, every exponential polynomial solves a linear differential equation with polynomial coefficients by Theorem 6.4.1. We use the following definition in the next two results.

Definition 8.2.6 ([266]). Let g be an entire function with finite positive order ρ, and denote the counting function of the zeros of g in the intersection of a domain $S \subset \mathbb{C}$ and the disc $|z| \leq r$ by $n(r, S, 1/g)$. We say that the zeros of g are *distributed near the rays* $\arg z = \varphi_j, j = 1, \ldots, k$, where

$$0 \leq \varphi_1 < \varphi_2 < \cdots < \varphi_k < \varphi_{k+1} = \varphi_1 + 2\pi, \tag{8.2.6}$$

if for every sufficiently small $\delta > 0$,

$$n(r, \varphi_j + \delta < \arg z < \varphi_{j+1} - \delta, 1/g) = o(r^\rho)$$

as $r \to \infty$, for $j = 1, \ldots, k$.

The above discussion brings us to the following result.

Theorem 8.2.7 ([266, Theorem 2]). *Let $\{g_1, \ldots, g_n\}$ be a fundamental solution base of* (8.2.2)*, where the coefficients $Q_0(z), \ldots, Q_{n-2}(z)$ are polynomials with $Q_0(z) \not\equiv 0$. Suppose that the zeros of $g_1, \ldots, g_n$ are distributed near finitely many rays* $\arg z = \varphi_j, j = 1, \ldots, k$, *satisfying* (8.2.6)*. Then either*

$$\rho(g_j) \leq \frac{\pi}{\min_{1\leq m\leq k}\{\varphi_{m+1} - \varphi_m\}}, \quad j = 1, \ldots, n, \tag{8.2.7}$$

or the coefficients of (8.2.2) *are constants.*

We have an immediate consequence of Theorem 8.2.7.

Corollary 8.2.8 ([266, Corollary 3]). *Let* $\{g_1, \ldots, g_n\}$ *be a fundamental solution base of* (8.2.2), *where the coefficients* $Q_0(z), \ldots, Q_{n-2}(z)$ *are polynomials with* $Q_0(z) \not\equiv 0$. *Suppose that the zeros of* $g_1, \ldots, g_n$ *are distributed near the real axis. Then the coefficients of* (8.2.2) *are constants.*

Proof. Since the zeros of $g_1, \ldots, g_n$ are distributed near the real axis, inequality (8.2.7) now reads as

$$\rho(g_j) \leq 1, \quad j = 1, \ldots, n.$$

However, by Corollary 8.1.12, this is not possible unless the coefficients of (8.2.2) are constants. □

8.3 Asymptotic integration

The classical Airy equation and the parabolic cylinder equations are examples of equations of the form

$$f'' + P(z)f = 0, \tag{8.3.1}$$

where $P(z) = p_n z^n + p_{n-1}z^{n-1} + \cdots + p_0$ is a nonconstant polynomial of degree $n \geq 1$. The next result is well known.

Proposition 8.3.1 ([166, Proposition 5.1]). *If f is a nontrivial solution of equation* (8.3.1), *then*

$$\rho(f) = \frac{n+2}{2}.$$

Proof. It is obvious that a nonconstant polynomial cannot be a solution of (8.3.1). Hence f is transcendental. The result now follows by dividing (8.3.1) by f and applying Wiman–Valiron theory. □

For a nontrivial solution f of (8.3.1), we have by Proposition 8.3.1 that $\lambda(f) \leq \rho(f) = (n+2)/2$. In addition, we will see (among other things) that if f has infinitely many zeros, then $n(r, 1/f)$ and $N(r, 1/f)$ are both asymptotically comparable to $r^{(n+2)/2}$. This happens, in particular, when n is odd, because then $\rho(f)$ is not an integer, which means that f has an infinite number of zeros by Corollary 3.5.9.

If n is even, then (8.3.1) can have zero-free solutions. Indeed, if $Q(z)$ is a nonconstant polynomial, then $f = e^Q$ solves (8.3.1) with $P(z) = -Q''(z) - Q'(z)^2$. Here $\deg P = 2(\deg Q - 1)$ is an even number (or zero).

Hille's theory of asymptotic integration reveals the location of the zeros of nontrivial solutions of (8.3.1). The original reasoning in [138, Section 7.4] is rather difficult to follow, which was the motivation for writing a simplified version of the theory in [92]. The principal results will be highlighted in this section. We will also state and prove a new lemma, which will be used in our proof of Theorem 8.6.3. At the end of this section, we discuss the key ideas behind the theory of asymptotic integration. Complete proofs with details can be found in [92]. Although the focus of this section is on the work in [138, Section 7.4] and [92], see also [221, 259, 285].

We begin by defining $n + 2$ *critical sectors* by

$$W_j(\varepsilon) = \{z : |\arg z - \theta_j| < \varepsilon\}, \quad \theta_j = \frac{2\pi j - \arg p_n}{n+2}, \tag{8.3.2}$$

where $-\pi < \arg p_n \leq \pi$, $\varepsilon > 0$ is arbitrarily small, and $j = 0, \dots, n+1$. The rays $\arg z = \theta_j$ are called *critical rays*. Hille [138, Section 7.4] used his theory of asymptotic integration to prove that all but at most finitely many zeros of any nontrivial solution of (8.3.1) lie in the union of the sectors $W_j(\varepsilon)$. This result has recently been improved in [92] together with updated proofs of some known asymptotic properties of solutions of (8.3.1). We proceed to review these results.

Set

$$c = \frac{p_{n-1}}{np_n}, \quad q = \frac{n+2}{2}, \quad d = \frac{(p_n)^{1/2}}{q}, \tag{8.3.3}$$

where $(p_n)^{1/2} = \sqrt{|p_n|}\exp(i\frac{\arg p_n}{2})$ with $-\pi < \arg p_n \leq \pi$, and denote

$$e_n(r) = \begin{cases} r^{-2}, & n > 2, \\ r^{-2}\log r, & n = 2, \\ r^{-3/2}, & n = 1. \end{cases} \tag{8.3.4}$$

We say that an entire function f of order $\rho \in (0, \infty)$ *blows up exponentially* on the ray $\arg z = \theta$ or *decays to zero exponentially* on the ray $\arg z = \theta$, respectively, if

$$\liminf_{r\to\infty} \frac{\log|f(re^{i\theta})|}{r^\rho} > 0 \quad \text{or} \quad \liminf_{r\to\infty} \frac{-\log|f(re^{i\theta})|}{r^\rho} > 0.$$

Theorem 8.3.2 ([92]). *Let f be a nontrivial solution of* (8.3.1). *Then the following statements hold:*

(a) *In any given open sector S between consecutive critical rays, f either* (i) *blows up exponentially on all the rays $\arg z = \theta$ in S or* (ii) *decays exponentially to zero on all* *the rays* $\arg z = \theta$ *in* S. *Specifically, on all the rays* $\arg z = \theta$ *in* S, f *is asymptotically comparable to either*

$$E_1(z) = \exp(idz^q(1 + o(1))) \quad \text{or} \quad E_2(z) = \exp(-idz^q(1 + o(1))),$$

where d *is defined in* (8.3.3). *Moreover, in each such sector* S, *there exist solutions of* (8.3.1) *of both types* (i) *and* (ii).

(b) *In any two adjacent sectors* $S(\theta_{j-1}, \theta_j)$ *and* $S(\theta_j, \theta_{j+1})$ *that border one common critical ray* $\arg z = \theta_j$, *there cannot exist a ray in* $S(\theta_{j-1}, \theta_j)$ *and another ray in* $S(\theta_j, \theta_{j+1})$ *such that* f *decays exponentially to zero on both rays. Here* $\theta_{-1} = \theta_{n+1}$.

Any branch cut outside the sector S in (a) and any branch for the square roots in the expressions $E_1(z)$ and $E_2(z)$ can be chosen as long as they are the same for both $E_1(z)$ and $E_2(z)$; see [92] and the discussion at the end of this section.

The ε-sectors in (8.3.2) can be replaced with the domains

$$\Lambda_j = \{z = re^{i\theta} : r > R,\ |\theta - \theta_j| < Ce_n(r)\}$$

and their translates $\Lambda_{j,c} = \{z : z + c \in \Lambda_j\}$. Here $R > 0$ is large enough, $C = C(n, R) > 0$, and c is defined in (8.3.3). See Figure 8.1.

Theorem 8.3.3 ([92]). *Let* f *be a nontrivial solution of* (8.3.1). *Then all but at most finitely many zeros of* f *lie in the union*

$$\bigcup_{j=0}^{n+1} \Lambda_{j,c}.$$

If f *has infinitely many zeros in a particular* $\Lambda_{j,c}$, *then*

$$n(r, \Lambda_{j,c}, 1/f) = \frac{\sqrt{|p_n|}}{q\pi} r^q(1 + o(1)), \quad r \to \infty,$$
$$N(r, \Lambda_{j,c}, 1/f) = \frac{\sqrt{|p_n|}}{q^2\pi} r^q(1 + o(1)), \quad r \to \infty,$$

where the counting function $n(r, \Lambda_{j,c}, 1/f)$ *refers to only those zeros in* $\Lambda_{j,c}$ *that satisfy* $|z| \le r$, *and* $N(r, \Lambda_{j,c}, 1/f)$ *is the corresponding integrated counting function. Here* q *is defined in* (8.3.3).

Remark. Bank [10] made the following observation. If a nontrivial solution f of (8.3.1) possesses an infinite number of zeros in some critical sector of the form (8.3.2), where $\arg z = \theta_j$ is a critical ray, then these zeros must approach the translate $\arg(z + c) = \theta_j$ of the critical ray, where c is defined in (8.3.3). We can see that this property follows from Theorem 8.3.3. For estimates on how rapidly these zeros of f must approach this translate of the critical ray, see [11, 132].

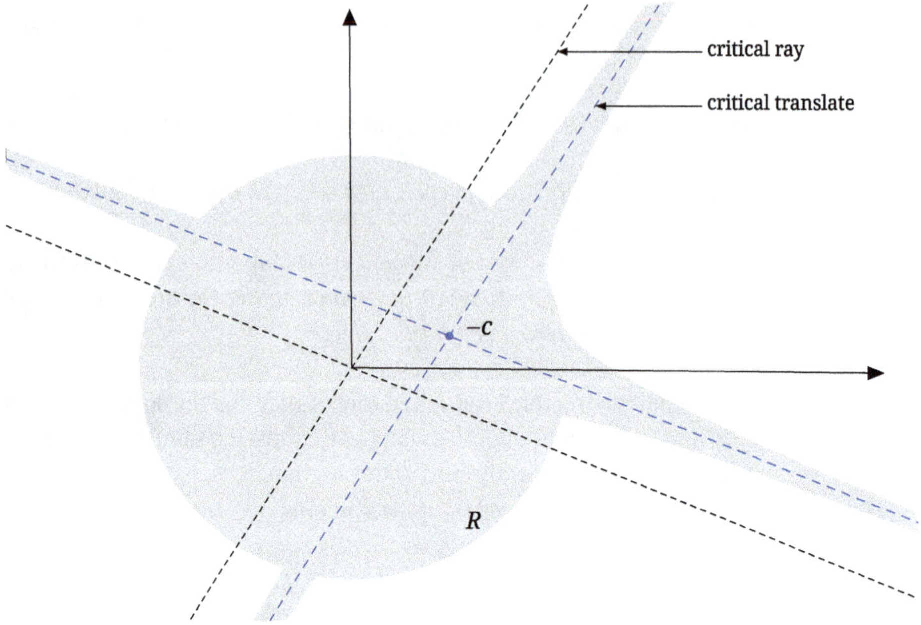

Figure 8.1: Domains $\Lambda_{j,c}$ around the critical translates in the case $n = 2$. All the zeros of f lie in the shaded area when $R > 0$ is large enough. This figure is reproduced from [92, Fig. 1], published under CC BY 4.0 license.

Definition 8.3.4 ([85]). If a solution f of (8.3.1) has finitely many zeros in $W_j(\varepsilon)$, then the critical ray $\arg z = \theta_j$ is called a *shortage ray* of f; otherwise, it is a *nonshortage ray* of f.

Theorem 8.3.5 ([92]). *Let f be a nontrivial solution of* (8.3.1). *Then the following statements hold:*

(a) *If f blows up exponentially on each ray in two adjacent sectors $S(\theta_{j-1}, \theta_j)$ and $S(\theta_j, \theta_{j+1})$ that border a common critical ray* $\arg z = \theta_j$, *then the critical ray* $\arg z = \theta_j$ *is a nonshortage ray.*

(b) *If f decays to zero exponentially on all the rays in a sector $S(\theta_j, \theta_{j+1})$, then both critical rays* $\arg z = \theta_j, \theta_{j+1}$ *are shortage rays.*

We proceed with more observations regarding shortage rays, some of which might be new. As in [85], we say that f is a *shortage solution* of (8.3.1) if f is a nontrivial solution of (8.3.1) that has at least one shortage ray. It is known [85, Theorem 5] that any nontrivial solution of (8.3.1) has an even number of shortage rays and that “most” solutions of (8.3.1) are not shortage solutions.

We use the following notation. If $\arg z = \theta_j$ and $\arg z = \theta_{j+1}$ are adjacent critical rays ($\theta_j < \theta_{j+1}$), then we say that the sector

$$S(\theta_j, \theta_{j+1}) = \{z : \theta_j < \arg z < \theta_{j+1}\}$$

is a + *sector* if f blows up exponentially on each ray $\arg z = \phi$ in $S(\theta_j, \theta_{j+1})$, and we say that $S(\theta_j, \theta_{j+1})$ is a – *sector* if f decays exponentially to zero on each ray $\arg z = \phi$ in $S(\theta_j, \theta_{j+1})$. We can see that the next result follows from inspection of Theorems 8.3.2 and 8.3.5.

Corollary 8.3.6. *Let f be a nontrivial solution of* (8.3.1), *and suppose that* $\arg z = \theta_{j-1}$, $\arg z = \theta_j$, *and* $\arg z = \theta_{j+1}$ *are three consecutive critical rays ($\theta_{j-1} < \theta_j < \theta_{j+1}$). Then the following properties hold:*

(a) *If $S(\theta_{j-1}, \theta_j)$ and $S(\theta_j, \theta_{j+1})$ are both + sectors, then* $\arg z = \theta_j$ *is a nonshortage ray.*
(b) *If $S(\theta_{j-1}, \theta_j)$ is a – sector and $S(\theta_j, \theta_{j+1})$ is a + sector, then* $\arg z = \theta_j$ *is a shortage ray.*
(c) *If $S(\theta_{j-1}, \theta_j)$ is a + sector and $S(\theta_j, \theta_{j+1})$ is a – sector, then* $\arg z = \theta_j$ *is a shortage ray.*
(d) *It is not possible that both $S(\theta_{j-1}, \theta_j)$ and $S(\theta_j, \theta_{j+1})$ are – sectors.*

We use Corollary 8.3.6 in the proof of the following result, which possibly is new, because we have not seen it in the literature.

Theorem 8.3.7. *Let f be a nontrivial solution of* (8.3.1), *where n is even and* $p_n > 0$. *Then the following properties hold:*

(a) *There are an even number of shortage rays of f, which occur in adjacent pairs* $\{\theta_j, \theta_{j+1}\}$ *that satisfy*

$$\theta_{j+1} - \theta_j = \frac{\pi}{q}, \tag{8.3.5}$$

where $q = (n + 2)/2$.

(b) *There are an even number of nonshortage rays of f, which occur in pairs* $\{\theta_j, \theta_k\}$ *that satisfy*

$$q\theta_j = 2\nu\pi \quad \text{and} \quad q\theta_k = (2\mu + 1)\pi, \tag{8.3.6}$$

where ν and μ are integers.

Proof. Since $p_n > 0$, it follows that the critical ray $\arg z = \theta_0$ is the positive real axis. Since n is even, f has an even number $(= n + 2)$ of critical rays, and $q = (n + 2)/2$ is a positive integer.

Suppose that f is not a shortage solution. Then f has no shortage rays, and f has an infinite number of zeros in each of the ε-sectors $W_j(\varepsilon)$ enclosing all $n+2$ critical rays of f. Since the critical rays are equally spaced and $q = (n+2)/2$, we can deduce that there are exactly q critical rays $\arg z = \alpha$ and exactly q critical rays $\arg z = \beta$ satisfying $q\alpha = 2\nu\pi$ and $q\beta = (2\mu + 1)\pi$, where ν, μ are integers. This gives (8.3.6) and completes the proof when f is not a shortage solution.

For the rest of the proof, we assume that f is a shortage solution. Observe that if all the critical rays of f are shortage rays, then parts (a) and (b) are obviously true. Therefore, we assume that f has at least one shortage ray and at least one nonshortage ray. There is no loss in generality to assume that $\arg z = \psi$ is a shortage ray and $\arg z = \psi - \pi/q$

is a nonshortage ray. Then from Corollary 8.3.6 and Theorem 8.3.5, we can deduce that $S(\psi - \pi/q, \psi)$ must be a + *sector*, $S(\psi, \psi + \pi/q)$ must be a − *sector*, and $\arg z = \psi + \pi/q$ must be a shortage ray. Hence

$$\{\arg z = \psi,\ \arg z = \psi + \pi/q\} \tag{8.3.7}$$

is a pair of adjacent shortage rays.

Next, consider the critical ray $\arg z = \psi + 2\pi/q$, which can be either a shortage ray or a nonshortage ray. If it is a nonshortage ray, then this initial process would stop here. Now suppose that $\arg z = 2\pi/q$ is a shortage ray. Since $S(\psi, \psi+\pi/q)$ is a − *sector*, we obtain from Corollary 8.3.6 that $S(\psi+\pi/q, \psi+2\pi/q)$ must be a + *sector* and $S(\psi+2\pi/q, \psi+3\pi/q)$ must be a − *sector*. Therefore, $\{\arg z = \psi + 2\pi/q, \arg z = \psi + 3\pi/q\}$ is another pair of adjacent shortage rays. Then from (8.3.7) we obtain the following two pairs of adjacent shortage rays:

$$\{\arg z = \psi,\ \arg z = \psi + \pi/q\}, \quad \{\arg z = \psi + 2\pi/q,\ \arg z = \psi + 3\pi/q\}. \tag{8.3.8}$$

After this, we would consider the next critical ray $\arg z = \psi + 4\pi/q$ and proceed in the same manner. Specifically, if $\arg z = \psi + 4\pi/q$ is a nonshortage ray, then this initial process would stop here, and if $\arg z = \psi+4\pi/q$ is a shortage ray, then the same reasoning will show that $\arg z = \psi + 5\pi/q$ is a shortage ray. Note that $\arg z = \psi + (n+1)\pi/q$ is a nonshortage ray, since it represents the same ray as $\arg z = \psi - \pi/q$, which we assumed to be a nonshortage ray. Therefore, it follows that this initial process will stop after a finite number of steps. No matter where this initial process stops, we can deduce that there exist a positive integer d such that we have the following collection of one or more pairs of adjacent shortage rays:

$$\{\arg z = \psi,\ \arg z = \psi + \pi/q\}, \ldots, \{\arg z = \psi + (d-1)\pi/q,\ \arg z = \psi + d\pi/q\}, \tag{8.3.9}$$

where $\arg z = \psi + (d+1)\pi/q$ is a nonshortage ray, and $d+1 \le n+1$. By comparing (8.3.7), (8.3.8), and (8.3.9) we see that d is an odd integer. Thus $d \le n$, where d is odd, and n is even. Hence $d \le n-1$. We have two cases.

Case 1. $d = n-1$. Then the list of pairs in (8.3.9) contains n shortage rays, and there are only two other critical rays,

$$\arg z = \psi + (d+1)\pi/q \quad \text{and} \quad \arg z = \psi + (d+2)\pi/q. \tag{8.3.10}$$

From the definition of d we have that $\arg z = \psi + (d+1)\pi/q$ is a nonshortage ray. Also, since $d = n-1$, we have $\arg z = \psi + (d+2)\pi/q = \psi + (n+1)\pi/q$, which means that $\arg z = \psi + (d+2)\pi/q$ is a nonshortage ray. Therefore, these two remaining rays (8.3.10) are both nonshortage rays, and we see that parts (a) and (b) both hold. In this case, only one initial process is used.

Case 2. $d < n - 1$. Then we have at least four critical rays left over,

$$\arg z = \psi + (d+1)\pi/q, \quad \arg z = \psi + (d+2)\pi/q, \ldots, \arg z = \psi + (n+1)\pi/q. \tag{8.3.11}$$

If all these remaining rays are nonshortage rays, then parts (a) and (b) will follow again like before. Now suppose that at least one of these remaining rays in (8.3.11) is a shortage ray. Then we can use the same process as before, with this smaller list rather than with all $n+2$ critical rays, to get a second list of adjacent pairs of shortage rays, which can then be added to the first list in (8.3.9). It is clear that if we use this process a finite number of times, then we will get a complete list of adjacent pairs of shortage rays, which proves part (a).

Finally, since there are an even number ($= n + 2$) of critical rays, and the shortage rays occur in adjacent pairs from part (a), it follows that there must be an even number of nonshortage rays. Since all the shortage rays occur in adjacent pairs that satisfy (8.3.5), we can deduce that the nonshortage rays can be arranged in a list of pairs that satisfy (8.3.6). This proves part (b). In case 2, we use two or more of the above processes instead of just the initial process. □

The following two examples give illustrations of both parts (a) and (b) in Theorem 8.3.7. The first example shows that the pairs $\{\theta_j, \theta_k\}$ in part (b) cannot always be listed in adjacent pairs.

Example 8.3.8. From Example 5.4.1, there exists an infinite sequence of positive constants β_k satisfying $0 < \beta_0 < \beta_1 < \beta_2 < \cdots < \beta_k < \cdots$, where $\beta_k \to +\infty$ as $k \to +\infty$, such that for each k, the equation

$$f'' + (z^4 - \beta_k)f = 0 \tag{8.3.12}$$

possesses a solution $f = f_k(z)$ that has an infinite number of real positive zeros, an infinite number of real negative zeros, and at most finitely many nonreal zeros. Since the six critical rays of equation (8.3.12) are $\arg z = 0, \pm\pi/3, \pm 2\pi/3, \pi$, it follows that $\arg z = 0, \pi$ are two nonshortage rays of f_k and $\arg z = \pm\pi/3, \pm 2\pi/3$ are four shortage rays of f_k. Observe that the two nonshortage rays are not adjacent critical rays and that the four shortage rays can be listed as adjacent pairs as follows:

$$\{\arg z = \pi/3,\ \arg z = 2\pi/3\} \quad \text{and} \quad \{\arg z = -2\pi/3,\ \arg z = -\pi/3\}.$$

This gives illustrations of both parts (a) and (b) in Theorem 8.3.7.

The next example shows that the pairs $\{\theta_j, \theta_k\}$ in part (b) can be adjacent pairs.

Example 8.3.9. The justification, including references, of some statements in this example can be found in Section 4 of [95]. It is known that the equation

$$f'' - \frac{1}{4}z^2 f = 0 \tag{8.3.13}$$

possesses the solution $f = D_{-1/2}(z)$, where $D_{-1/2}(z)$ is the classical parabolic cylinder function, where it is known that

$$D_{-1/2}(x) \to 0 \quad \text{as} \quad x \to +\infty \tag{8.3.14}$$

along the positive real axis. The critical rays of (8.3.13) are $\arg z = \pm\pi/4, \pm 3\pi/4$. Hence from (8.3.14) it follows by Theorem 8.3.2 that $S(-\pi/4, \pi/4)$ must be a − *sector*. Then by Theorem 8.3.5, $\arg z = \pm\pi/4$ are two shortage rays of $D_{-1/2}(z)$. A proof that the remaining two critical rays $\arg z = \pm 3\pi/4$ are nonshortage rays can be found in [95, p. 744]. Clearly, these two nonshortage rays are adjacent critical rays, and the shortage rays $\arg z = \pm\pi/4$ are also adjacent critical rays.

We now give some discussion about the key steps in the proofs of Theorems 8.3.2–8.3.5. First, we may assume that the polynomial coefficient $P(z)$ in (8.3.1) is normalized, that is, we may consider the equation

$$g'' + Q(z)g = 0, \tag{8.3.15}$$

where

$$Q(z) = \begin{cases} z, & n = 1, \\ z^n + a_{n-2}z^{n-2} + \cdots + a_0, & n \geq 2. \end{cases} \tag{8.3.16}$$

There is no loss of generality in doing this because if f is a solution of (8.3.1), then from $c = \frac{p_{n-1}}{np_n}$ in (8.3.3), it is easy to see that $g(z) = f(\mu z - c)$ is a solution of (8.3.15), where μ is a constant satisfying $\mu^{n+2} = p_n^{-1}$. With this normalization, the critical rays are now of the form $\arg z = \psi_j$, where

$$\psi_j = \frac{2\pi j}{n+2}, \quad j = 0, \ldots, n+1.$$

Second, following Hille ([138, p. 340], [140, p. 179]), a transformation

$$\begin{cases} \zeta = L(z) = \int_{z_0}^{z} Q(\xi)^{1/2}\, d\xi, \\ w(\zeta) = Q(\zeta)^{1/4} g(\zeta), \end{cases} \tag{8.3.17}$$

introduced by Liouville in 1837 is needed. This transformation was originally used on the positive real line, where there are no questions about branches. However, for the complex variables in (8.3.17), questions about branches have to be addressed.

To deal with the branching issue, we rewrite $Q(z)$ in (8.3.16) as

$$Q(z) = z^n(1 + \ell(z)), \quad n \geq 1, \tag{8.3.18}$$

where $\ell(z) \equiv 0$ if $n = 1$ and

$$\ell(z) = \frac{a_{n-2}}{z^2} + \cdots + \frac{a_0}{z^n}, \quad n \geq 2.$$

It is easy to see that $\arg(1+\ell(z)) \in (-\pi/2, \pi/2)$ for all $|z|$ large enough, say $|z| > R$. Next, we define

$$G_j(R) = \{z \in \mathbb{C} : |z| > R,\ \psi_{j-1} < \arg z < \psi_{j+1}\}, \quad j = 0, \ldots, n+1,$$

with the convention that $\psi_{-1} = \psi_{n+1} - 2\pi$ and $\psi_{n+2} = \psi_0 + 2\pi$. For a nontrivial solution g of (8.3.15), denote the restriction of g to $G_j(R)$ by $g_j(z)$. Then Liouville's transformation takes the form

$$\begin{cases} \zeta = L_j(z) = \int_{z_0}^{z} \xi^{n/2}(1+\ell(\xi))^{1/2} d\xi, \\ w_j(\zeta) = z^{n/4}(1+\ell(z))^{1/4} g_j(z), \end{cases} \tag{8.3.19}$$

where $z_0 = 2Re^{i\psi_j}$ and $z \in G_j(R)$. The path of integration between $z_0, z \in G_j(R)$ is a polygonal path, which consists of at most two line segments lying in $G_j(R)$. The branches of the square roots and the fourth roots in (8.3.19) are fixed as follows:

(a) For $(1+\ell(z))^{1/2}$ and $(1+\ell(z))^{1/4}$, we always use the principal branch

$$-\pi < \arg(1+\ell(z)) \leq \pi. \tag{8.3.20}$$

(b) The square root $w^{n/2}$ is defined by $w^{n/2} = (w^{1/2})^n$, where

$$w^{1/2} = |w|^{1/2} \exp\left(i\frac{\arg w}{2}\right), \quad \psi_j - \pi < \arg w \leq \psi_j + \pi. \tag{8.3.21}$$

(c) The fourth root $w^{n/4}$ is defined by $w^{n/4} = (w^{1/4})^n$, where

$$w^{1/4} = |w|^{1/4} \exp\left(i\frac{\arg w}{4}\right), \quad \psi_j - \pi < \arg w \leq \psi_j + \pi. \tag{8.3.22}$$

Note that the domain $G_j(R)$ is trivially included in $\psi_j - \pi < \arg w \leq \psi_j + \pi$.

The following lemma gives an asymptotic representation for the function $L_j(z)$ in (8.3.19) when z is confined to $G_j(R)$. This representation is used in [92] to prove that the function $w_j(\zeta)$ in (8.3.19) is well-defined, among other things.

Lemma 8.3.10 ([92, Lemma 2.1]). *The function $\zeta = L_j(z)$ in (8.3.19) satisfies*

$$\zeta = L_j(z) = \frac{2}{n+2} z^{(n+2)/2}(1+K(z)), \quad z \in G_j(R),$$

where $|K(z)| = O(e_n(|z|))$ as $z \to \infty$ in $G_j(R)$, where $e_n(r)$ is in (8.3.4). Moreover, we have $|\zeta| \sim \frac{2}{n+2}|z|^{(n+2)/2}$ as $|z| \to \infty$ and

$$|\arg(1+K(z))| = O(e_n(|z|)), \quad |z| \to \infty.$$

It turns out that the normalization (8.3.16), together with Liouville's transformation (8.3.19), transforms equation (8.3.15) into a perturbed sine equation of the form

$$w'' + (1 - T(\zeta))w = 0, \tag{8.3.23}$$

where $T(\zeta) = O(\zeta^{-2})$ as $\zeta \to \infty$ [92]. Recall that the sine equation

$$w'' + w = 0$$

has three types of solutions, namely $e^{i\zeta}$, $e^{-i\zeta}$, and $\sin(\zeta - \zeta_0)$, any two of which are linearly independent. Moreover, by a process called asymptotic integration, it turns out that the solutions of (8.3.23) are asymptotic to the solutions of the sine equation [92]. This is the crucial idea behind the proofs of Theorems 8.3.2–8.3.5.

Next, we focus on radial oscillation of solutions of (8.3.1).

Definition 8.3.11 ([296, p. 294]). Let f be an entire function, and let $\arg z = \theta \in \mathbb{R}$ be a ray. For $\varepsilon > 0$, the exponent of convergence of the zeros of f in the sector $S(\theta - \varepsilon, \theta + \varepsilon)$ is denoted by $\lambda_{\theta,\varepsilon}(f)$. The *radial exponent of convergence* of the zeros of f is given by

$$\lambda_\theta(f) = \lim_{\varepsilon \to 0^+} \lambda_{\theta,\varepsilon}.$$

Example 8.3.12. Let $f(z) = e^z - 1$. Then the zeros of f are at the points $z_n = 2\pi n i$, $n \in \mathbb{Z}$, on the imaginary axis. We have $\lambda_{\pi/2}(f) = \lambda_{-\pi/2}(f) = 1$ and $\lambda_\theta(f) = 0$ for every $\theta \in [-\pi, \pi) \setminus \{\pi/2, -\pi/2\}$.

Theorem 8.3.13 ([296, Theorem 1]). *Let $P(z)$ be a polynomial of degree $n \geq 1$, and let f_1 and f_2 be two linearly independent solutions of* (8.3.1). *If for some $\theta_0 \in \mathbb{R}$,*

$$\limsup_{r \to \infty} \frac{\log\log|E(re^{i\theta_0})|}{\log r} = \frac{n+2}{2},$$

where $E = f_1 f_2$, then there exist $\theta_1, \theta_2 \in \mathbb{R}$ with $\theta_1 \leq \theta_0 \leq \theta_2$ such that $\theta_2 - \theta_1 = \frac{2\pi}{n+2}$ and $\lambda_{\theta_1}(f) = \lambda_{\theta_2}(f) = \frac{n+2}{2}$.

8.4 Generalizations of the Airy integral

In Chapter 5, we encountered entire functions that can be represented by contour integrals, such as the Mittag-Leffler function, the reciprocal of the gamma function, and the Airy integral. In particular, from Section 5.3, the Airy integral

$$\mathrm{Ai}(z) = \frac{1}{2\pi i} \int_C \exp\left(\frac{1}{3}w^3 - zw\right) dw, \tag{8.4.1}$$

where C is the contour that goes from ∞ to 0 along $\arg w = -\pi/3$ and then from 0 to ∞ along $\arg w = \pi/3$, is a solution of Airy's differential equation

$$f'' - zf = 0.$$

In this and the next section, we discuss several examples of entire functions that are represented as contour integrals which are solutions of linear differential equations with polynomial coefficients, which we call *contour integral solutions*. All the contour integrals in this and the next section can be shown to be entire functions, whose derivatives of all orders can be found by differentiating under the integral sign. The proof of this can be accomplished by using the reasoning in the proof of Theorem 5.3.2, which showed that these properties hold for the Airy integral $\mathrm{Ai}(z)$.

The Airy integral is a shortage solution, which has an infinite number of real negative zeros and no other zeros. Since Airy's equation has the form (8.3.1), it follows from the asymptotic integration results in Section 8.3 that there exists a sector of maximal opening where the Airy integral decays exponentially to zero and that, on the other hand, "most" of the solutions of Airy's equation do not have this property. The Airy integral is one of many examples which illustrate that contour integral solutions often have interesting special properties. The Airy integral is well known in mathematical physics and Airy's equation has applications in the theory of diffraction, the dispersion of water waves, and the turning-point problem; see [152, 268].

In this section, we discuss two classes of contour integral solutions, which are generalizations of the Airy integral.

8.4.1 Change z to z^n in Airy's equation

Consider the differential equation

$$f'' - z^n f = 0, \tag{8.4.2}$$

where $n = 1, 2, \ldots$. This is Airy's equation when $n = 1$.

Theorem 8.4.1 ([95]). *Let $n \geq 1$ be a fixed integer. Let m_1 and m_2 be integers such that $m_1 - m_2$ is not an integer multiple of $n + 2$. Let $G(z)$ denote the contour integral function defined by*

$$G(z) = \frac{1}{2\pi i} \int_C e^{P(z,w)}\, dw \tag{8.4.3}$$

with

$$P(z, w) = \sum_{k=0}^{\lfloor (n+2)/2 \rfloor} \frac{(-1)^k}{n+2-k} \binom{n+2-k}{k} z^k w^{n+2-2k}, \tag{8.4.4}$$

where $\lfloor (n+2)/2 \rfloor$ is the greatest integer $\leq (n+2)/2$, and $C = C(m_1, m_2)$ is the contour that runs from ∞ to 0 along $\arg w = (2m_1 + 1)\pi/(n+2)$ and then from 0 to ∞ along $\arg w = (2m_2 + 1)\pi/(n+2)$.

Then $G(z)$ is a shortage solution of equation (8.4.2) *with exactly two shortage rays,*

$$\arg z = \frac{2m_1 + 2m_2 + 1}{n+2}\pi \quad \text{and} \quad \arg z = \frac{2m_1 + 2m_2 + 3}{n+2}\pi. \tag{8.4.5}$$

If B is a collection of pairwise linearly independent shortage solutions of (8.4.2) such that any shortage solution of (8.4.2) is a constant multiple of a function in B, then B is called a *basis of shortage solutions* of (8.4.2). From Theorem 8.4.1 it follows that there are different representations of a basis of shortage solutions of (8.4.2) such that each solution in each basis is a contour integral function of the form (8.4.3). The following result gives such a representation.

Theorem 8.4.2 ([95]). *For j an integer, let $G_j(z)$ denote the contour integral function*

$$G_j(z) = \frac{1}{2\pi i}\int_{C_j} e^{P(z,w)}\,dw, \tag{8.4.6}$$

where C_j is the contour that runs from ∞ to 0 along the ray $\arg w = (2\lfloor j/2\rfloor - 1)\pi/(n+2)$ *and then from 0 to ∞ along the ray* $\arg w = (2\lfloor (j+1)/2\rfloor + 1)\pi/(n+2)$, *and $P(z,w)$ is defined in* (8.4.4).

Then the following statements hold:

(a) *For each j, $G_j(z)$ is a shortage solution of* (8.4.2) *with exactly two shortage rays,*

$$\arg z = \frac{2j-1}{n+2}\pi \quad \textit{and} \quad \arg z = \frac{2j+1}{n+2}\pi.$$

(b) *The functions $G_j(z)$, $j = 1, 2, \ldots, n+1$, are a basis of shortage solutions of* (8.4.2).

The proofs of these two theorems involve a lot of calculations with combinatorial sums of the type found in [65], where the Cauchy integral formula is frequently used.

Since $G_0(z)$ is a solution of equation (8.4.2), and it is known that the initial values of $G_0(z)$ are [95, Section 8]

$$G_0(0) = \frac{1}{(n+2)^{\frac{n+1}{n+2}}\Gamma\left(\frac{n+1}{n+2}\right)} \quad \text{and} \quad G_0'(0) = \frac{1}{(n+2)^{\frac{1}{n+2}}\Gamma\left(\frac{1}{n+2}\right)},$$

it follows that $G_0(z)$ is real on the real axis. This agrees with the Airy integral, where we observe that when $n = 1$, we have $G_0(z) \equiv \mathrm{Ai}(z)$.

In the case where $n = 2$ in (8.4.2), it is well known that the shortage solutions of the equation $f'' - z^2 f = 0$ can all be expressed in terms of the classical parabolic cylinder function D_ν for $\nu = -1/2$. For the details regarding the relationships of these well-known formulas and the above theorems, see [95, Section 4].

Theorem 8.4.3 ([95, 268]). *For each $n \geq 1$, let $G_0(z)$ be defined as in* (8.4.6).

(a) *The nonreal zeros of $G_0(z)$ are all contained in the open set S given by $S = \bigcup_k S_k$, where S_k is the double sector*

$$S_k = \left\{z : \frac{2k}{n+1}\pi < |\arg z| < \frac{2k+1}{n+2}\pi\right\}$$

with k running over all the integers satisfying $0 < 2k < n+1$.

(b) *If n is even, then $G_0(z)$ has no real zeros.*
(c) *If n is odd, then $G_0(z)$ has an infinite number of negative real zeros and no nonnegative real zeros.*

For each integer j, there exists a constant K_j such that

$$G_j(z) \equiv K_j G_0(a_j z) \tag{8.4.7}$$

with

$$a_j = \exp\left(-j\frac{2\pi i}{n+2}\right),$$

where K_j can be calculated; see [95, p. 740].

Using Theorems 8.4.2(b) and 8.4.3 together with (8.4.7), we can make the following conclusions. First, when n is odd, it follows that there is one special critical ray

$$L_j : \arg z = \frac{n+2+2j}{n+2}\pi, \tag{8.4.8}$$

which has the following properties: (i) $G_j(z)$ has an infinite number of zeros that lie on L_j, (ii) all the zeros that lie "near" L_j must actually lie on L_j, and (iii) $G_j(z)$ cannot have any zeros on any other critical ray besides L_j. Second, when n is even, any shortage solution does not possess a zero on any critical ray.

Theorem 8.4.4 ([95]). *Let a and b be distinct zeros of $G_j(z)$, where $ab \neq 0$.*
(i) *If n is even, then* $\arg a \neq \arg b$.
(ii) *If n is odd, then either* $\arg a \neq \arg b$, *or a and b both lie on L_j.*

Proof. We use Lommell's method. Suppose that $G_0(a) = G_0(b)$, where $\arg a = \arg b$ and $0 < |a| < |b|$. Since $G_0(z)$ is a solution of (8.4.2), we can see that

$$\frac{d}{dz}(\overline{a}G_0(az)G_0'(\overline{a}z) - aG_0(\overline{a}z)G_0'(az)) = (\overline{a}^{n+2} - a^{n+2})z^n G_0(az)G_0(\overline{a}z).$$

Set $\mu = b/a$. Then μ is real with $\mu > 1$, and

$$(\overline{a}^{n+2} - a^{n+2})\int_1^{\mu} x^n G_0(ax)G_0(\overline{a}x)\,dx = [\overline{a}G_0(ax)G_0'(\overline{a}x) - aG_0(\overline{a}x)G_0'(ax)]_1^{\mu}.$$

Since $G_0(z)$ is real on the real axis and $G_0(a) = 0$, it follows from the reflection principle that $G_0(\overline{a}) = 0$. Also, since

$$\overline{G_0(\overline{a}\mu)} = G_0(a\mu) = G_0(b) = 0,$$

we have $G_0(\overline{a}\mu) = 0$. Therefore, we get

$$(\overline{a}^{n+2} - a^{n+2}) \int_1^{\mu} x^n |G_0(ax)|^2 \, dx = 0.$$

Hence we must have $\overline{a}^{n+2} = a^{n+2}$, which means that a^{n+2} is a real number. From Theorem 8.4.3 we can deduce that a must be a real number, a contradiction. Thus the statements hold for $G_0(z)$. Then from (8.4.7) and (8.4.8), we see that the statements hold for $G_j(z)$ for any integer j. □

8.4.2 A class of equations with all orders ≥ 2

Consider the class of linear differential equations of the form

$$f^{(n)} + (-1)^{n+1} b f^{(k)} + (-1)^{n+1} z f = 0, \quad 0 < k < n, \quad b \in \mathbb{C}, \tag{8.4.9}$$

and the associated contour integral function

$$\varphi(z) = \frac{1}{2\pi i} \int_C \exp(-wz + b\alpha w^{k+1} + \beta w^{n+1}) \, dw \tag{8.4.10}$$

with

$$\alpha = \frac{(-1)^{k+1}}{k+1} \quad \text{and} \quad \beta = \frac{1}{n+1},$$

where the contour C runs from ∞ to 0 along $\arg w = -\pi/(n+1)$ and then from 0 to ∞ along $\arg w = \pi/(n+1)$.

In the particular case where $n = 2$ and $b = 0$, the function φ in (8.4.10) is the Airy integral $\mathrm{Ai}(z)$, and (8.4.9) is Airy's equation.

Theorem 8.4.5 ([91]). *The function $f = \varphi(z)$ in* (8.4.10) *is a nontrivial entire solution of* (8.4.9).

Proof. Since $0 < k < n$, from (8.4.10) we obtain

$$\begin{aligned}
&2\pi i(\varphi^{(n)}(z) + (-1)^{n+1} b \varphi^{(k)}(z) + (-1)^{n+1} z\varphi(z)) \\
&\quad = \int_C ((-w)^n + (-1)^{n+1} b(-w)^k + (-1)^{n+1} z) \exp(-wz + b\alpha w^{k+1} + \beta w^{n+1}) \, dw \\
&\quad = (-1)^n \int_C \frac{\partial}{\partial w} \exp(-wz + b\alpha w^{k+1} + \beta w^{n+1}) \, dw \\
&\quad = (-1)^n [\exp(-wz + b\alpha w^{k+1} + \beta w^{n+1})]_C = 0,
\end{aligned}$$

where the last equality holds by the following reasoning. Define

$$F(z,w) = \exp(-wz + baw^{k+1} + \beta w^{n+1}).$$

Since C is a continuous curve consisting of two rays that meet at the point $w = 0$, and since for any fixed z, $F(z,w) \to 0$ as $w \to \infty$ along either $\arg w = \pi/(n+2)$ or $\arg w = -\pi/(n+2)$, we obtain

$$[\exp(-wz + baw^{k+1} + \beta w^{n+1})]_C = 0 - F(z,0) + F(z,0) - 0 = 0.$$

It follows that $f = \varphi(z)$ is a solution of (8.4.9).

To complete the proof of Theorem 8.4.5, it remains to show that $\varphi(z) \not\equiv 0$. A lengthy proof of this is in [91]. □

Next, we discuss certain properties satisfied by the contour integral φ in (8.4.10). It is clear that equation (8.4.9) cannot possess nontrivial polynomial solutions, and hence φ must be transcendental. It then follows from Theorems 8.4.5 and 8.1.5 that

$$\rho(\varphi) = 1 + 1/n. \tag{8.4.11}$$

It is well known that the Airy integral $\mathrm{Ai}(z)$ is real on the real axis, and this property also holds, more generally, for φ in (8.4.10) whenever b is real [91, Theorem 5].

Since the Airy integral $\mathrm{Ai}(z)$ satisfies $\rho(\mathrm{Ai}) = 3/2$ by Theorem 8.3.1 and has only negative real zeros by Theorem 5.3.4, where the three critical rays of Airy's equation $f'' - zf = 0$ are $\arg z = \pi, \pm\pi/3$, it follows from asymptotic theory of integration that $\arg z = \pm\pi/3$ are the shortage rays of $\mathrm{Ai}(z)$ and that $|\mathrm{Ai}(z)|$ has exponential decay growth of order $\exp(-|z|^{3/2})$ on any ray in the sector $-\pi/3 < \arg z < \pi/3$. Furthermore, the only solutions of Airy's equation that have an exponential decay growth in $-\pi/3 < \arg z < \pi/3$ are constant multiples of $\mathrm{Ai}(z)$, and all the other nontrivial solutions blow up exponentially in this sector. The special exponential decay property of $|\mathrm{Ai}(z)|$ in $-\pi/3 < \arg z < \pi/3$ is a particular case of the following general result.

Theorem 8.4.6 ([91]). *Let $n \geq 2$. For each θ satisfying*

$$-\frac{n\pi}{2n+2} < \theta < \frac{n\pi}{2n+2}, \tag{8.4.12}$$

there exist positive constants $K = K(\theta)$ and $M = M(\theta)$ that depend only on θ, such that φ in (8.4.10) *satisfies*

$$|\varphi(z)| \leq M\exp(-K|z|^{1+1/n}), \quad z = |z|e^{i\theta}. \tag{8.4.13}$$

Remark. Since $\rho(\varphi) = 1+1/n$ by (8.4.11), it follows from the Phragmén–Lindelöf theory that the exponent $1 + 1/n$ in (8.4.13) cannot be replaced with a larger exponent and the length $n\pi/(n+1)$ of the interval in (8.4.12) cannot be larger. For more comments about Theorem 8.4.6, see Section 8.5, where a paper of Steinmetz [267] is discussed.

Proof of Theorem 8.4.6. We will give the proof and diagrams in [91]. This proof entails using Cauchy's theorem twice to show that for any fixed θ satisfying (8.4.12), the contour C in (8.4.10) can be replaced by two other contours without changing the value of $\varphi(z)$ when $\arg z = \theta$.

Let θ be a fixed constant satisfying (8.4.12), and then let $\mu = \mu(\theta)$ and $\tau = \tau(\theta)$ be the constants

$$\mu = -\frac{\pi}{n+1} - \frac{\theta}{n} \quad \text{and} \quad \tau = \frac{\pi}{n+1} - \frac{\theta}{n}. \tag{8.4.14}$$

From (8.4.12) and (8.4.14), we obtain the following four double inequalities:

$$-\frac{3\pi}{2} < (n+1)\mu < -\frac{\pi}{2} \quad \text{and} \quad -\frac{\pi}{2} < \mu + \theta < \frac{\pi}{2}, \tag{8.4.15}$$

$$\frac{\pi}{2} < (n+1)\tau < \frac{3\pi}{2} \quad \text{and} \quad -\frac{\pi}{2} < \tau + \theta < \frac{\pi}{2}. \tag{8.4.16}$$

Now let z be a point satisfying $\arg z = \theta$, i. e., $z = |z|e^{i\theta}$. Then let $C(\mu)$ be the arc of the circle $|w| = |z|$ from $w = |z|e^{-i\pi/(n+1)}$ to $w = |z|e^{i\mu}$. From (8.4.12), (8.4.14), and (8.4.15) we can deduce that

$$\cos(n+1)\xi \le \cos(n+1)\mu < 0$$

for any point $w = |z|e^{i\xi}$ lying on $C(\mu)$. Hence with α and β in (8.4.10),

$$\begin{aligned}
&\left|\frac{1}{2\pi i}\int\limits_{C(\mu)} \exp(-wz + b\alpha w^{k+1} + \beta w^{n+1})\,dw\right| \\
&\quad \le \frac{|z|}{2\pi}\int\limits_{C(\mu)} \exp(|z|^2 + |b\alpha||z|^{k+1} + \beta|z|^{n+1}\cos(n+1)\xi)\,|d\xi| \\
&\quad \le |z|\exp(|z|^2 + |b\alpha||z|^{k+1} + \beta|z|^{n+1}\cos(n+1)\mu).
\end{aligned}$$

Since $2 \le k+1 < n+1$, $\beta > 0$, and $\cos(n+1)\mu < 0$, we obtain

$$\frac{1}{2\pi i}\int\limits_{C(\mu)} \exp(-wz + b\alpha w^{k+1} + \beta w^{n+1})\,dw \to 0 \quad \text{as} \quad |z| \to \infty. \tag{8.4.17}$$

Similarly, using (8.4.12), (8.4.14), and (8.4.16), we get

$$\frac{1}{2\pi i}\int\limits_{C(\tau)} \exp(-wz + b\alpha w^{k+1} + \beta w^{n+1})\,dw \to 0 \quad \text{as} \quad |z| \to \infty, \tag{8.4.18}$$

where $C(\tau)$ is the arc of the circle $|w| = |z|$ from $w = |z|e^{i\pi/(n+1)}$ to $w = |z|e^{i\tau}$.

Now observe from Cauchy's theorem that

$$\frac{1}{2\pi i}\int_{E_j} \exp(-wz + baw^{k+1} + \beta w^{n+1})\,dw = 0, \quad j = 1, 2,$$

where E_1 is the closed curve consisting of the line segment from 0 to $|z|e^{-i\pi/(n+1)}$, the arc $C(\mu)$, and the line segment from $|z|e^{i\mu}$ to 0, and E_2 is the closed curve consisting of the line segment from 0 to $|z|e^{i\pi/(n+1)}$, the arc $C(\tau)$, and the line segment from $|z|e^{i\tau}$ to 0. See Figure 8.2. By letting $|z| \to \infty$, we can deduce from (8.4.10), (8.4.17), and (8.4.18) that

$$\varphi(z) = \frac{1}{2\pi i}\int_{L} \exp(-wz + baw^{k+1} + \beta w^{n+1})\,dw, \tag{8.4.19}$$

where the contour $L = L(\theta)$ runs from ∞ to 0 along $\arg z = \mu$ and then from 0 to ∞ along $\arg z = \tau$. Note that $L = C$ when $\theta = 0$. See Figure 8.2.

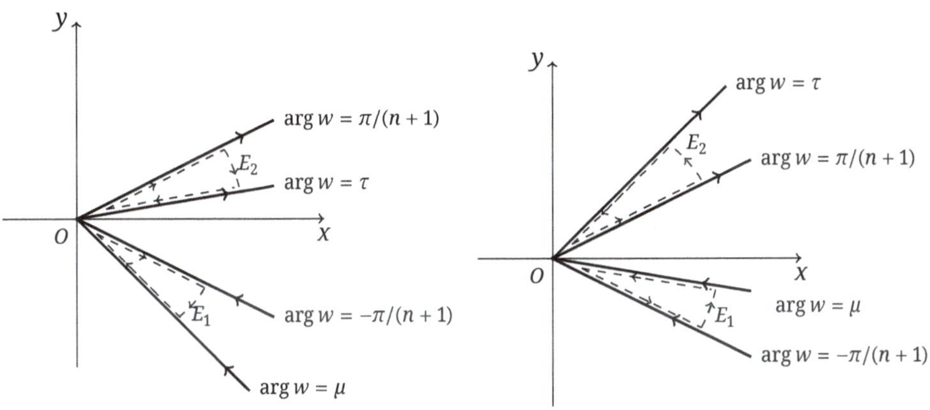

(a) The case $\theta > 0$. **(b)** The case $\theta < 0$.

Figure 8.2: Contours L, C and closed curves E_1, E_2. This figure is similar to [91, Figure 1].

We now apply Cauchy's theorem again, this time to (8.4.19), and obtain further that for any constant $R > 0$, we have

$$\varphi(z) = \frac{1}{2\pi i}\int_{J} \exp(-wz + baw^{k+1} + \beta w^{n+1})\,dw, \tag{8.4.20}$$

where $J = J_1 + J_2 + J_3$ is the contour in Figure 8.3 with

$$\begin{aligned} J_1 &: w = re^{i\mu}, && r \text{ goes from } \infty \text{ to } R,\\ J_2 &: w = Re^{i\xi}, && \xi \text{ goes from } \mu \text{ to } \tau,\\ J_3 &: w = re^{i\tau}, && r \text{ goes from } R \text{ to } \infty. \end{aligned}$$

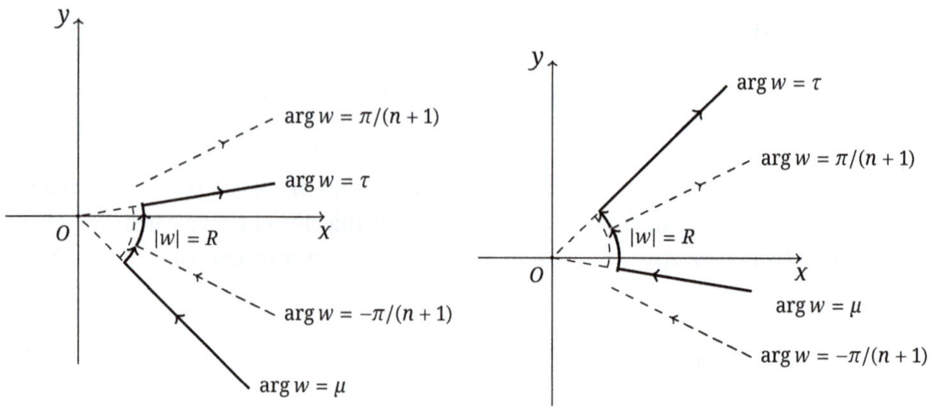

(a) The case $\theta > 0$. **(b)** The case $\theta < 0$.

Figure 8.3: Contour J. This figure is similar to [91, Figure 2].

From (8.4.20) we have

$$\begin{aligned}\varphi(z) &= \frac{1}{2\pi i}\int_{\infty}^{R} \exp(-r|z|e^{i(\mu+\theta)} + bar^{k+1}e^{i(k+1)\mu} + \beta r^{n+1}e^{i(n+1)\mu})e^{i\mu}\,dr \\ &+ \frac{1}{2\pi}\int_{\mu}^{\tau} \exp(-R|z|e^{i(\xi+\theta)} + b\alpha R^{k+1}e^{i(k+1)\xi} + \beta R^{n+1}e^{i(n+1)\xi})Re^{i\xi}\,d\xi \\ &+ \frac{1}{2\pi i}\int_{R}^{\infty} \exp(-r|z|e^{i(\tau+\theta)} + bar^{k+1}e^{i(k+1)\tau} + \beta r^{n+1}e^{i(n+1)\tau})e^{i\tau}\,dr \\ &=: I_1 + I_2 + I_3.\end{aligned}$$

Now let $A = A(\theta)$ be the constant

$$A = \min\{\cos(\mu+\theta), \cos(\tau+\theta)\} > 0, \tag{8.4.21}$$

where $A > 0$ from (8.4.15) and (8.4.16). Then let $B = B(\theta)$ be any fixed constant that satisfies

$$0 < B < ((n+1)A)^{1/n}. \tag{8.4.22}$$

Choose $R = B|z|^{1/n}$. Since $\cos(\mu+\theta) > 0$ and $|\alpha| < 1 < 2\pi$,

$$|I_1| < \int_{B|z|^{1/n}}^{\infty} \exp(-r|z|\cos(\mu+\theta) + |b|r^{k+1} + \beta r^{n+1}\cos(n+1)\mu)\,dr$$

$$\leq \exp(-B|z|^{1+1/n}\cos(\mu+\theta)) \int_{B|z|^{1/n}}^{\infty} \exp(|b|r^{k+1} + \beta r^{n+1}\cos(n+1)\mu)\,dr.$$

We have $\cos(n+1)\mu < 0$ from (8.4.15), and since $\beta > 0$ and $k < n$,

$$|I_1| < P\exp(-B|z|^{1+1/n}\cos(\mu+\theta)), \tag{8.4.23}$$

where $P = P(\theta)$ is the positive constant

$$P = \int_0^{\infty} \exp(|b|r^{k+1} + \beta r^{n+1}\cos(n+1)\mu)\,dr.$$

Similarly, since $\cos(\tau+\theta) > 0$ and $\cos(n+1)\tau < 0$, from (8.4.16) we will get

$$|I_3| < Q\exp(-B|z|^{1+1/n}\cos(\tau+\theta)), \tag{8.4.24}$$

where $Q = Q(\theta)$ is the positive constant

$$Q = \int_0^{\infty} \exp(|b|r^{k+1} + \beta r^{n+1}\cos(n+1)\tau)\,dr.$$

Finally,

$$|I_2| \leq \frac{B|z|^{1/n}}{2\pi} \int_{\mu}^{\tau} \exp(B(\beta B^n - \cos(\xi+\theta))|z|^{1+1/n} + |b|B^{k+1}|z|^{(k+1)/n})\,d\xi.$$

From (8.4.15) and (8.4.16) we have

$$-\frac{\pi}{2} < \mu+\theta \leq \xi+\theta \leq \tau+\theta < \frac{\pi}{2}, \quad \mu \leq \xi \leq \tau.$$

Hence from (8.4.21), it follows that

$$\cos(\xi+\theta) \geq A, \quad \mu \leq \xi \leq \tau.$$

Thus

$$|I_2| \leq B|z|^{1/n}\exp(B(\beta B^n - A)|z|^{1+1/n} + |b|B^{k+1}|z|^{(k+1)/n}), \tag{8.4.25}$$

where $B(\beta B^n - A) < 0$ from $\beta = 1/(n+1)$ and (8.4.22). Since $\varphi(z) = I_1 + I_2 + I_3$ when $\arg z = \theta$, we now deduce that (8.4.13) holds from an observation of (8.4.23), (8.4.24), and (8.4.25). This proves Theorem 8.4.6. □

In the remainder of this section, we discuss other generalizations of the Airy integral. Let $u(z)$ be the function $\varphi(z)$ in (8.4.10) when $b = 0$, that is,

$$u(z) = \frac{1}{2\pi i}\int_C \exp\left(-wz + \frac{1}{n+1}w^{n+1}\right)dw, \tag{8.4.26}$$

where the contour C runs from ∞ to 0 along $\arg w = -\pi/(n+1)$ and then from 0 to ∞ along $\arg w = \pi/(n+1)$. By using Theorems 8.4.5 and 8.1.5, we see that $u(z)$ is a transcendental solution of the equation

$$f^{(n)} + (-1)^{n+1}zf = 0. \tag{8.4.27}$$

When $n = 2$, $u(z)$ is the Airy integral $\mathrm{Ai}(z)$ in (8.4.1), and (8.4.27) is Airy's equation.

If $\beta_1, \beta_2, \ldots, \beta_{n+1}$ are the $n+1$ distinct roots of unity and if

$$f_j(z) = u(\beta_j z), \quad j = 1, 2, \ldots, n+1, \tag{8.4.28}$$

then $f_1, f_2, \ldots, f_{n+1}$ are solutions of equation (8.4.27). The following result is a generalization of the known property of the Airy integral $\mathrm{Ai}(z)$ in (8.4.1).

Theorem 8.4.7 ([91]). *The functions $f_1, f_2, \ldots, f_{n+1}$ in* (8.4.28) *are solutions of* (8.4.27) *of order $1 + 1/n$, and any n of these $n+1$ functions are linearly independent.*

Regarding the proof of Theorem 8.4.7, since $u(z)$ in (8.4.26) is the function $\varphi(z)$ in (8.4.10) when $b = 0$, it follows from Theorem 8.4.5, Theorem 8.1.5 (or Wiman–Valiron theory), and (8.4.28) that each f_j is a solution of (8.4.27) of order $1+1/n$. For the proof that any n of the $n+1$ functions $f_1, f_2, \ldots, f_{n+1}$ are linearly independent, see [91, p. 336].

Theorem 8.4.7 shows that every solution of equation (8.4.27) is a contour integral function. By varying the contour C in (8.4.26), the functions $f_1, f_2, \ldots, f_{n+1}$ can be expressed in a different form; see formula (4.19) in [91].

As noted earlier, the Airy integral $\mathrm{Ai}(z)$ is real on the real axis. More generally, this property holds for $\varphi(z)$ whenever b is real [91, Theorem 5].

8.5 More contour integral solutions

It may be difficult to determine information about the history of contour integral solutions of linear differential equations. As Steinmetz [267, p. 567] noted, although this is part of "mathematical folklore", it is a topic that deserves to be better known. Regarding its history, Steinmetz said that there are a few remarks in Wasow [285, p. 123 ff.], and there are presentations in Ince [144, Ch. XVIII] and Hille [140, p. 216 ff.], but with a constraint that removes a very important class of equations.

In this section, we briefly discuss more contour integral solutions of linear differential equations. More examples can be found in [18, 19].

8.5.1 A class of equations with large orders

The justification of the statements in this subsection can be found in [91]. Consider a linear differential equation of the form

$$f^{(n)} - zf^{(k)} - f = 0, \quad 1 < k < n, \tag{8.5.1}$$

and let $\psi = \psi(n,k)$ denote the contour integral function

$$\psi(z) = \frac{1}{2\pi i} \int\limits_{|w|=1} w^{k-2} \exp\left(\frac{z}{w} - \frac{1}{(n-k+1)w^{n-k+1}} - \frac{w^{k-1}}{k-1}\right) dw. \tag{8.5.2}$$

Theorem 8.5.1 ([91]). *The function $f = \psi(z)$ is a transcendental solution of* (8.5.1).

Proof. As in the proof of Theorem 8.4.5, we obtain

$$\begin{aligned}
&(2\pi i)(\psi^{(n)}(z) - z\psi^{(k)}(z) - \psi(z)) \\
&= \int\limits_{|w|=1} \left(\frac{1}{w^{n-k+2}} - \frac{z}{w^2} - w^{k-2}\right) \exp\left(\frac{z}{w} - \frac{1}{(n-k+1)w^{n-k+1}} - \frac{w^{k-1}}{k-1}\right) dw \\
&= \left[\exp\left(\frac{z}{w} - \frac{1}{(n-k+1)w^{n-k+1}} - \frac{w^{k-1}}{k-1}\right)\right]_{|w|=1} = 0,
\end{aligned}$$

since $|w| = 1$ is a closed curve. Thus $\psi(z)$ satisfies (8.5.1).

It is not difficult to show that $\psi \not\equiv 0$; see [91]. Last, ψ is transcendental because (8.5.1) clearly cannot possess a nontrivial polynomial solution. □

Example 8.1.10 is a particular case of Theorem 8.5.1 because when $n = 3$ and $k = 2$ in (8.5.2), we obtain $2\pi i\psi(z) \equiv G(z)$, where $G(z)$ is the function in (8.1.14).

In [91], it is shown that the function ψ in (8.5.2) has the following properties: (a) $\rho(\psi) = 1 - 1/k$. (b) If n is even and k is odd, then ψ is an even function that is unbounded on every ray from the origin. (c) ψ is real on the real axis.

Section 6 in [91] discusses contour integral solutions of the equation

$$f^{(4)} - zf''' - f = 0, \tag{8.5.3}$$

which is a particular case of (8.5.1). One such solution is when $n = 4$ and $k = 3$ in (8.5.2), which we denote by $U(z)$. By Theorem 8.5.1,

$$U(z) = \frac{1}{2\pi i} \int\limits_{|w|=1} w \exp\left(\frac{z}{w} - \frac{1}{2w^2} - \frac{w^2}{2}\right) dw \tag{8.5.4}$$

satisfies (8.5.3). Another contour integral function that satisfies (8.5.3) is

$$H(z) = \frac{1}{2\pi i} \int_A w \exp\left(\frac{z}{w} - \frac{1}{2w^2} - \frac{w^2}{2} \right) dw, \tag{8.5.5}$$

where $A = A_1 + A_2 + A_3$ is the contour with

$$\begin{aligned} A_1 &: w = r, \quad r \text{ goes from } +\infty \text{ to } 1, \\ A_2 &: w = e^{i\theta}, \quad \theta \text{ goes from } 0 \text{ to } \pi, \\ A_3 &: w = -r, \quad r \text{ goes from } 1 \text{ to } +\infty. \end{aligned}$$

Theorem 8.5.2 ([91]). *The three contour integral functions $U(z), H(z), H(-z)$ are pairwise linearly independent solutions of* (8.5.3) *satisfying*

$$U(z) \equiv H(z) + H(-z). \tag{8.5.6}$$

Moreover,

$$\rho(U(z)) = \rho(H(z)) = \rho(H(-z)) = 2/3, \quad \textit{and} \quad \tau(U(z)) = 3/2. \tag{8.5.7}$$

It is also known [91] that $U(z)$ is an even function that is unbounded on every ray from the origin and that $U(z)$ is real on both the real and imaginary axes. Further, $U(z)$ has its largest growth on the imaginary axis, where it satisfies

$$\log|U(ir)| = \log|U(-ir)| = \log M(r, U) = \left(\frac{3}{2} + o(1) \right) r^{2/3}, \quad r \to \infty. \tag{8.5.8}$$

Equation (8.5.8) comes from (6.3) in [91], which did not contain the three log signs (this was a typo in [91]).

8.5.2 Laplace contour integral solutions

The Laplace transform from a beginning course in real differential equations is often the first time that students see a solution of a differential equation defined by an integral. Laplace transforms can also be put in the setting of the complex plane [207].

Steinmetz proved a general result [267, 2.1 Theorem], which shows that when certain conditions are satisfied, a linear differential equation with polynomial coefficients possesses a Laplace contour integral solution (defined below). He points out that this general result is the most useful when studying equations of the form

$$f^{(n)} + \sum_{j=0}^{n-1} (a_j z + b_j) f^{(j)} = 0, \quad a_0 z + b_0 \not\equiv 0, \tag{8.5.9}$$

where each a_j and b_j is a constant. In particular, if certain conditions are satisfied, then equation (8.5.9) possesses a *Laplace contour integral solution* of the form

$$f(z) = \frac{1}{2\pi i} \int_C \Phi(w) e^{-zw} \, dw \tag{8.5.10}$$

for some function $\Phi(w)$ and an appropriate infinite contour C; see [267]. The calculations in [267] exhibit properties of these solutions (8.5.10), such as the order of growth, the Phragmén–Lindelöf indicator function, the zero distribution, asymptotic expansions, subnormal and polynomial solutions, etc.

Steinmetz [267, p. 570] showed that Airy's equation, equation (8.4.9), and equation (8.5.1), which are satisfied, respectively, by the Airy integral, the contour integral (8.4.10), and the contour integral (8.5.2), are all examples of his general result about Laplace contour integral solutions of equation (8.5.9). See [267] for further examples.

From Section 8.3 we recall that solutions of equations of the form $f''+P(z)f = 0$ (with a polynomial $P(z) \not\equiv 0$) which have an exponential decay growth in a sector $S(a,b)$, are shortage solutions, which are unusual solutions because in a very understandable sense, "most" solutions are not shortage solutions. Therefore, we think of such solutions as special solutions, and it is natural to expect that a solution of the more general equation (8.0.1) that has exponential decay growth in a sector $S(a,b)$ will also be a special solution.

As discussed in Section 8.4, the Airy integral $\mathrm{Ai}(z)$ has order 3/2 and has only negative real zeros, where the three critical rays of Airy's equation $f'' - zf = 0$ are $\arg z = \pi, \pm\pi/3$, and $|\mathrm{Ai}(z)|$ has exponential decay growth of order $\exp(-|z|^{3/2})$ on any ray in the sector $-\pi/3 < \arg z < \pi/3$. Furthermore, any nonconstant solution of Airy's equation that is not a constant multiple of $\mathrm{Ai}(z)$ blows up exponentially in this sector. This special decay property of $|\mathrm{Ai}(z)|$ is a particular case of Theorem 8.4.6. Since we make several comments concerning Theorem 8.4.6 below, we repeat Theorem 8.4.6 here for the convenience of the reader.

Theorem 8.5.3 ([91]). *Let $n \geq 2$. For each θ satisfying*

$$-\frac{n\pi}{2n+2} < \theta < \frac{n\pi}{2n+2}, \tag{8.5.11}$$

there exist positive constants $K = K(\theta)$ and $M = M(\theta)$, depending only on θ, such that φ in (8.4.10) *satisfies*

$$|\varphi(z)| \leq M \exp(-K|z|^{1+1/n}), \quad z = |z|e^{i\theta}. \tag{8.5.12}$$

Steinmetz observed that the proof of Theorem 8.5.3 for $n \geq 2$ in [91] does not work for $n = 1$, and he then used a different proof to show that Theorem 8.5.3 also holds for $n = 1$; see [267, p. 583]. Furthermore, as part of a more general result [267, 5.1 Theorem], he showed that for all $n \geq 1$, the Phragmén–Lindelöf indicator function of the solution φ in Theorem 8.5.3 has the form

$$h_\varphi(\theta) = -\frac{1}{\rho}\cos(\rho\theta), \quad |\theta| < \pi,$$

where $\rho = \rho(\varphi) = 1 + 1/n$. This result also shows that φ in (8.4.10) has exponential decay in the sector (8.5.11). In addition, [267, 5.1 Theorem] exhibits other properties, including that φ has at most finitely many zeros in any sector of the form

$$|\arg z - \pi| < \varepsilon,$$

where $\varepsilon > 0$ is any positive number. Similar information about other contour integral solutions of differential equations can be found in [267].

As mentioned above, Steinmetz presented several other contour integral solutions of differential equations.

Example 8.5.4 ([267]). The equation

$$f^{(4)} + (z-1)f''' - 8f'' - zf' + 2f = 0 \tag{8.5.13}$$

possesses the contour integral solution

$$f_1(z) = \frac{1}{2\pi i} \int\limits_{C_{2,\pi/2}} \frac{\exp(\frac{w^2}{2} + w - zw)}{w^3(w-1)^3(w+1)^4}\, dw,$$

where the contour $C_{2,\pi/2}$ goes from ∞ to $2i$ along $\arg w = \pi/2$, then from $2i$ to $-2i$ along the arc of the circle $|z| = 2$ in the right half-plane, and then from $-2i$ to ∞ along $\arg w = -\pi/2$. In addition, equation (8.5.13) has a fundamental system of solutions f_1, f_2, f_3, f_4, where

$$f_2(z) = (2z^3 - 27z^2 + 150z - 324)e^z, \quad f_3(z) = (z^2 + 6z + 14)e^{-z}, \quad f_4(z) = z^2 + 8.$$

For more examples of Laplace contour integral solutions, see [267]. The Airy integral has often been used both to illustrate differential equation results and to show that some differential equations theorems are sharp. The papers [91, 267] provide more examples of contour integral solutions of differential equations that have special properties, which may have the potential to be illustrative examples for future differential equations results.

Historical notes

Pierre-Simon Laplace (1749–1827) was a French mathematician and astronomer. His five-volume work *Mécanique céleste* showed how celestial mechanics could be transformed from a geometrical investigation to a calculus investigation. Simmons [260, p. 413] said this about Laplace: "His main interests were celestial mechanics, the theory of probability, and personal advancement."

8.6 Completely regular solutions

It is natural to guess that the transcendental solutions of equation (8.0.1) have regular growth. In this section, we shed light on this, beginning with a result of Valiron.

Theorem 8.6.1 ([279, p. 108]). *Every transcendental solution of* (8.0.1) *is of p. r. g.*

The next result was proved independently by Petrenko and Steinmetz.

Theorem 8.6.2 ([225, 265]). *Every transcendental solution of* (8.0.1) *is of c. r. g.*

From Lemma 7.5.2 we know that c. r. g. implies p. r. g. This shows that Theorem 8.6.1 is a particular case of Theorem 8.6.2. Both proofs of Theorem 8.6.2 in [225, 265] are very intense. The main goal of this section is to show how the asymptotic integration theory can be used to give a new proof of Theorem 8.6.2 for the particular case where $n = 2$. For convenience, we write a precise statement of the result.

Theorem 8.6.3. *If $P_0(z)$ and $P_1(z)$ are polynomials such that $P_0(z) \not\equiv 0$, then every nontrivial solution of*

$$f'' + P_1(z)f' + P_0(z)f = 0 \tag{8.6.1}$$

is of c. r. g.

Proof. Similarly as in Section 8.2, we see that

$$f(z) = h(z)\exp\left(-\frac{1}{2}\int^{z} P_1(\zeta)\,d\zeta\right)$$

transforms (8.6.1) into

$$h'' + P(z)h = 0, \tag{8.6.2}$$

where $P(z)$ is a polynomial that can be explicitly written as

$$P(z) = P_0(z) - \frac{1}{4}P_1(z)^2 - \frac{1}{2}P_1'(z).$$

If $P(z) \equiv 0$, then we obtain $f(z) = (az + b)e^{R(z)}$, where $R(z)$ is a polynomial, and $a, b \in \mathbb{C}$. But then f is of c. r. g. by the definition of c. r. g. If $P(z) \not\equiv 0$ is a constant function, then (8.6.2) reduces to a solvable sine equation, whose nontrivial solutions are clearly of c. r. g. Consequently, f is of c. r. g. Hence we may suppose that $\deg P = n \geq 1$.

For the rest of the proof, we use the asymptotic integration theory in Section 8.3, where we refer back to the formulas in that section.

As observed in (8.3.15) and (8.3.16), we may assume that the polynomial

$$P(z) = p_n z^n + p_{n-1}z^{n\ 1} + \cdots + p_0$$

in (8.6.2) is normalized, that is, instead of (8.6.2), we may consider the equation

$$g'' + Q(z)g = 0, \tag{8.6.3}$$

where

$$Q(z) = \begin{cases} z, & n = 1, \\ z^n + a_{n-2}z^{n-2} + \cdots + a_0, & n \geq 2. \end{cases}$$

Indeed, if h is a solution of (8.6.2), then $g(z) = h(\mu z - c)$ is a solution of (8.6.3), where $c = \frac{p_{n-1}}{np_n}$, and μ is a constant satisfying $\mu^{n+2} = p_n^{-1}$. Moreover, if g is of c. r. g., then h is of c. r. g., and, consequently, f is of c. r. g.

If a solution g of (8.6.3) has at most finitely many zeros, then since $\rho(g)$ is finite by Proposition 8.3.1, we see by Hadamard's theorem that $g = Se^T$ for some polynomials S, T, and hence g is of c. r. g. by the definition of c. r. g. Therefore, we may suppose that g has an infinite zero sequence $\{z_k\}$. By Theorem 8.3.3 all the zeros (with at most finitely many exceptions) of any nontrivial solution g of (8.6.3) lie in finitely many ε-sectors about the critical rays. Then by Definition 7.5.5, the zero sequence $\{z_k\}$ of g has angular density.

Suppose first that n is odd. Then $q = \frac{n+2}{2} \notin \mathbb{N}$, so that by Definition 7.5.5, the sequence $\{z_k\}$ is regularly distributed. By Theorem 8.1.3, g has finite positive order and finite type, and thus by Theorem 7.5.7, g is of c. r. g.

For the rest of the proof, we suppose that n is even, so that $q \in \mathbb{N}$. From Definition 7.5.5 we see that for $\{z_k\}$ to be regularly distributed, and hence for g to be of c. r. g., it suffices to prove that the limit

$$\lim_{r\to\infty} \sum_{0<|z_k|\le r} z_k^{-q} \tag{8.6.4}$$

exists as a finite complex number. The convergence of (8.6.4) will depend only on the nonshortage rays of g, and by Theorem 8.3.7(b), these nonshortage rays can be listed in pairs that satisfy (8.3.6). Now let $\{\arg z = \theta,\ \arg z = \phi\}$ be any of these pairs of nonshortage rays of g such that $q\theta = 2m\pi$ and $q\phi = (2l + 1)\pi$, where m, l are integers.

Let $\{a_k\}$ denote the zeros of g that approach the nonshortage ray $\arg z = \theta$ satisfying $q\theta = 2m\pi$. Since $\arg z = \theta$ is one of the critical rays, we have $\theta = \psi_j$ for some j, where

$$\psi_j = \frac{2\pi j}{n+2}, \quad j = 0, \dots, n+1.$$

As in Section 8.3, $g_j(z)$ denotes a restriction of g to the domain

$$G_j(R) = \{z \in \mathbb{C} : |z| > R,\ \psi_{j-1} < \arg z < \psi_{j+1}\}, \quad j = 0, \dots, n+1.$$

We will use the modified Liouville transformation (8.3.19), where the index j corresponds to the index j satisfying $\theta = \psi_j$. Let $\alpha_k = L_j(a_k)$, where $L_j(z)$ is defined in (8.3.19). Since $\psi_j = 2m\pi/q$, the points α_k approach the positive real axis by Lemma 8.3.10. From (8.3.19) we have $w_j(\alpha_k) = 0$. Then the first part of [92, Lemma 3.2] shows that $w_j(z)$ has precisely one zero in each square

$$Q_{k,\gamma} = \{z = x + iy : |x - x_0 - k\pi| < \gamma,\ |y - y_0| < \gamma\},$$

where $x_0, y_0 \in \mathbb{R}$ are fixed, $k \in \mathbb{N}$, and $\gamma > 0$ is a small constant. By reorganizing the points α_k, if necessary, we may suppose that $\alpha_k \in Q_{k,\gamma}$ for every $k \in \mathbb{N}$. This gives

$$\alpha_k = k\pi + O(1), \quad k \in \mathbb{N}.$$

Thus by Lemma 8.3.10,

$$\alpha_k = \frac{1}{q} a_k^q + O(a_k^\mu), \quad k \to \infty,$$

where $0 \leq \mu \leq q - 1 < q$. Hence

$$k\pi = \frac{1}{q} a_k^q + O(a_k^\mu), \quad k \to \infty,$$

which gives

$$a_k^q = qk\pi + O(k^\tau), \quad k \to \infty, \tag{8.6.5}$$

where $0 \leq \tau \leq \mu/q < 1$.

Let $\{b_k\}$ denote the zeros of g that approach the nonshortage ray $\arg z = \phi$ satisfying $q\theta = (2l+1)\pi$. Since $\arg z = \phi$ is one of the critical rays, we have $\phi = \psi_i$ for some integer i. Set $\beta_k = L_i(b_k)$. Since $\psi_i = (2l+1)\pi/q$, the points β_k approach the negative real axis by Lemma 8.3.10. From (8.3.19) we have $w_i(\beta_k) = 0$, and similarly to the above, the first part of [92, Lemma 3.2] shows that

$$\beta_k = -k\pi + O(1), \quad k \in \mathbb{N}.$$

By using the same argument as for the zeros $\{a_k\}$, we can deduce that

$$b_k^q = -qk\pi + O(k^\tau), \quad k \to \infty, \tag{8.6.6}$$

where $0 \leq \tau \leq \mu/q < 1$.

From (8.6.5) and (8.6.6) we have

$$\sum_{k=1}^{\infty} \left(\frac{1}{a_k^q} + \frac{1}{b_k^q} \right) = \sum_{k=1}^{\infty} O\left(\frac{1}{k^\nu} \right),$$

where $\nu = 2 \quad \tau > 1$. Hence the series

$$\sum_{k=1}^{\infty} \left(\frac{1}{a_k^q} + \frac{1}{b_k^q} \right) \tag{8.6.7}$$

converges. Since we showed that (8.6.7) holds for this arbitrary pair of nonshortage rays that satisfy (8.3.6), and since all the nonshortage rays can be listed in pairs that satisfy (8.3.6), it follows that the limit (8.6.4) exists as a finite number, which completes the proof. □

9 Linear differential equations with entire coefficients

In this chapter, we consider solutions of linear differential equations of the form

$$f^{(n)} + A_{n-1}(z)f^{(n-1)} + \cdots + A_0(z)f = 0, \tag{9.0.1}$$

where n is a positive integer, and $A_0(z), A_1(z), \ldots, A_{n-1}(z)$ are entire functions with $A_0(z) \not\equiv 0$. Recall from Corollary 1.7.2 that all the solutions of (9.0.1) are entire functions.

In Section 9.1, we give a general description of the multiplicities of the zeros of solutions of (9.0.1), where a proof of an almost forgotten result of Gol'berg from 1994 is given. In Section 9.2, we discuss results and examples that address the following question: How many finite-order solutions of (9.0.1) are possible when at least one of the coefficients is transcendental? This includes Frei's classical theorem and more recent results. In Section 9.3, we discuss the important work of Bank and Laine on the solutions of the equation $f'' + A(z)f = 0$, where $A(z) \not\equiv 0$ is entire, which includes their landmark paper (1982) and the useful Bank–Laine equation. Their work in this section leads naturally to an interesting class of entire functions, called the Bank–Laine functions, which are discussed later in Section 12.2. In Section 9.4, we give details of Bergweiler's recent construction (2023) for the fact that if $A(z)$ and $B(z)$ are of completely regular growth, then the transcendental finite-order solutions of $f'' + A(z)f' + B(z)f = 0$ are not always of completely regular growth. This answers a question of Gol'dberg and Ostrovskii from 1994.

9.1 Multiplicities of zeros of solutions

Solutions of (9.0.1) cannot have zeros of arbitrarily large multiplicities.

Lemma 9.1.1. *Let f be a nontrivial solution of equation* (9.0.1). *If f has a zero of multiplicity m at z_0, then $m \leq n - 1$.*

Proof. Suppose on the contrary that $m \geq n$ and write (9.0.1) in the form

$$\frac{f^{(n)}}{f} + A_{n-1}(z)\frac{f^{(n-1)}}{f} + \cdots + A_0(z) = 0. \tag{9.1.1}$$

Since $m \geq n$, the quotient $f^{(j)}/f$ has a pole of multiplicity j at z_0 for $j = 1, \ldots, n$. In particular, $f^{(n)}/f$ has a pole of multiplicity n at z_0, whereas the remaining terms in (9.1.1) have a pole of multiplicity $\leq n-1$ at z_0. This contradiction shows that $m \leq n-1$, as stated. □

Conversely to the situation above, if f is an entire function with zeros of multiplicity $\leq n - 1$, then there exist entire functions $A_0(z), \ldots, A_{n-1}(z)$ such that f solves equation (9.0.1), where $A_0(z) \not\equiv 0$. This result is attributed to Gol'dberg [103, p. 300], where a sketch of the proof can also be found. The proof below is based on this sketch and applies the interpolation theory.

https://doi.org/10.1515/9783112249130-009

Theorem 9.1.2 ([103, p. 300]). *Let f be a nonconstant entire function, and let n be a positive integer. If the multiplicity of every zero of f is at most $n-1$, then f satisfies an equation of the form* (9.0.1).

Proof. Suppose first that f has no zeros. Then $f = e^g$ for some nonconstant entire function g. Consequently, f solves the equation $f' + A_0(z)f = 0$, where $A_0(z) = -g'(z) \not\equiv 0$. This gives the statement.

We may now suppose that f has at least one zero. Let p_k denote the multiplicity of the zero z_k of f. By the hypothesis, $p_k \leq n-1$ for every k.

For the convenience of the reader, we prove the case $n = 2$ separately. In this case the function f has only simple zeros, which means that $p_k = 1$ for every k. We will construct entire functions $A_1(z)$ and $A_0(z) \not\equiv 0$ such that f solves

$$f'' + A_1(z)f' + A_0(z)f = 0.$$

For $A_1(z)$ to be entire, we need $A_1(z)$ to solve the following interpolation problem at the zeros z_k of f:

$$A_1(z_k) = -\frac{f''(z_k)}{f'(z_k)} = \sigma_k, \tag{9.1.2}$$

where $\sigma_k \in \mathbb{C}$. Such a function $A_1(z)$ exists: If $\{z_k\}$ is a finite sequence, then $A_1(z)$ can be chosen as the Lagrange interpolation polynomial, whereas if $\{z_k\}$ is an infinite sequence, then $A_1(z)$ can be constructed by means of Theorem 4.7.2. Fix some constant $\zeta \in \mathbb{C}$ that satisfies $f(\zeta)f'(\zeta) \neq 0$. Along with (9.1.2), we may additionally require that

$$A_1(\zeta) = 1 - \frac{f''(\zeta)}{f'(\zeta)}. \tag{9.1.3}$$

Note that the constant 1 is irrelevant in (9.1.3) as any constant $c \neq 0$ could be chosen in place of 1 to guarantee that

$$f''(z) + A_1(z)f'(z) \not\equiv 0. \tag{9.1.4}$$

Having found an entire function $A_1(z)$ satisfying (9.1.2) and (9.1.3), we define

$$A_0(z) = -\frac{f''(z) + A_1(z)f'(z)}{f(z)},$$

which is entire and $\not\equiv 0$ by (9.1.2), (9.1.3), and (9.1.4). This completes the proof for the case $n = 2$.

We now prove the remaining cases $n \geq 3$. To this end, we define complex numbers $\zeta_{j,k}, j = 1, \ldots, p_k$, as follows: Set

$$\zeta_{p_k,k} = -\frac{f^{(n)}(z_k)}{f^{(p_k)}(z_k)}.$$

Since $f^{(p_k)}(z_k) \neq 0$, we have $\zeta_{p_k,k} \in \mathbb{C}$. If $p_k \geq 2$, then we define, recursively for every $j \in \{1, \dots, p_k - 1\}$, the points

$$\zeta_{p_k-j,k} = -\frac{f^{(n+j)}(z_k) + \zeta_{p_k,k} f^{(p_k+j)}(z_k) + \cdots + \zeta_{p_k-j+1,k} f^{(p_k+1)}(z_k)}{f^{(p_k)}(z_k)}.$$

The coefficients $A_0(z), \dots, A_{n-1}(z)$ are then defined in four steps.

(1) In the case where all the zeros z_k of f satisfy $p_k \leq n-2$, we may apply the Weierstrass factorization theorem to construct entire functions $A_{p_k+1}(z), \dots, A_{n-1}(z)$ such that each of them has a zero of multiplicity p_k at each point z_k. If $p_k = n - 1$ for some k, then this step does not apply.

(2) By Theorem 4.7.4, we may choose entire functions $A_1(z), \dots, A_{p_k}(z)$ such that $A_1(z_k) = \zeta_{1,k}, \dots, A_{p_k}(z_k) = \zeta_{p_k,k}$, where their derivatives $A_1'(z), \dots, A_{p_k}'(z)$ have a zero of multiplicity $p_k - 1$ at each point z_k. The latter condition implies that for any fixed k, there exist entire functions $B_1(z), \dots, B_{p_k}(z)$ such that

$$A_j'(z) = (z - z_k)^{p_k-1} B_j(z) \quad \text{and} \quad B_j(z_k) \neq 0, \quad j = 1, \dots, p_k.$$

These conditions imply, by the Leibniz integral rule, that

$$A_j(z) = \zeta_{j,k} + \int_{z_k}^{z} (w - z_k)^{p_k-1} B_j(w)\, dw, \quad j = 1, \dots, p_k, \tag{9.1.5}$$

where the integration path is along the line segment $[z_k, z]$. Note that the integral in (9.1.5) is an entire function with a zero of multiplicity p_k at z_k.

(3) In addition to the requirements for $A_1(z)$ in step (2), we require that for a fixed $\zeta \in \mathbb{C}$ satisfying $f(\zeta)f'(\zeta) \neq 0$,

$$A_1(\zeta) = 1 - \frac{f^{(n)}(\zeta) + A_{n-1}(\zeta) f^{(n-1)}(\zeta) + \cdots + A_2(z) f''(\zeta)}{f'(\zeta)}.$$

Similarly as in the proof of the case $n = 2$, any constant $c \neq 0$ could be chosen in place of 1. This additional requirement for $A_1(z)$ guarantees that

$$f^{(n)}(z) + A_{n-1}(z) f^{(n-1)}(z) + \cdots + A_1(z) f'(z) \not\equiv 0. \tag{9.1.6}$$

(4) Having now defined the entire coefficients $A_1(z), \dots, A_{n-1}(z)$, we define

$$A_0(z) = -\frac{f^{(n)}(z) + A_{n-1}(z) f^{(n-1)}(z) + \cdots + A_1(z) f'(z)}{f(z)}.$$

Then $A_0(z) \not\equiv 0$ by (9.1.6).

Now that all the coefficients are defined, it remains to prove that $A_0(z)$ is entire. Clearly, $A_0(z)$ is meromorphic, with all of its possible poles at the zeros z_k of f. Thus it suffices to prove that

$$F(z) := f^{(n)}(z) + A_{n-1}(z)f^{(n-1)}(z) + \cdots + A_1(z)f'(z)$$

has a zero of multiplicity $\geq p_k$ at every z_k. For $m \in \{0, \dots, p_k - 1\}$, we have

$$\begin{aligned} F^{(m)}(z) &= f^{(n+m)}(z) + \sum_{j=1}^{n-1} \left(A_j(z)f^{(j)}(z)\right)^{(m)} \\ &= f^{(n+m)}(z) + \sum_{j=1}^{n-1} \sum_{\ell=0}^{m} \binom{m}{l} A_j^{(m-\ell)}(z) f^{(j+\ell)}(z). \end{aligned} \tag{9.1.7}$$

From step (2) we have $A_j^{(m-\ell)}(z_k) = 0$ in (9.1.7) whenever $\ell \neq m$. Moreover, since $f^{(j)}(z_k) = 0$ for every $j \in \{0, \dots, p_k - 1\}$ or, in other words, $f^{(j+m)}(z_k) = 0$ in (9.1.7) whenever $j + m \leq p_k - 1$, it follows that

$$\begin{aligned} F^{(m)}(z_k) &= f^{(n+m)}(z_k) + \sum_{j=1}^{n-1} A_j(z_k) f^{(j+m)}(z_k) \\ &= f^{(n+m)}(z_k) + \sum_{j=p_k-m}^{n-1} A_j(z_k) f^{(j+m)}(z_k), \end{aligned} \tag{9.1.8}$$

where $m \in \{0, \dots, p_k - 1\}$. We consider the following two cases.

(a) Suppose that $p_k = n - 1$, and recall that $n \geq 3$. Then by using step (2), (9.1.8) becomes

$$F^{(m)}(z_k) = f^{(n+m)}(z_k) + \sum_{j=n-1-m}^{n-1} \zeta_{j,k} f^{(j+m)}(z_k),$$

where $m \in \{0, \dots, n - 2\}$. We have

$$\begin{aligned} F(z_k) &= f^{(n)}(z_k) + \zeta_{n-1,k} f^{(n-1)}(z_k) \\ &= f^{(n)}(z_k) - \frac{f^{(n)}(z_k)}{f^{(n-1)}(z_k)} \cdot f^{(n-1)}(z_k) = 0, \end{aligned}$$

and if $m \in \{1, \dots, n-2\}$, then the definition of $\zeta_{n-1-m,k}$, together with the assumption $p_k = n - 1$, gives us

$$F^{(m)}(z_k) = f^{(n+m)}(z_k) + \sum_{j=n-m}^{n-1} \zeta_{j,k} f^{(j+m)}(z_k) + \zeta_{n-1-m,k} f^{(n-1)}(z_k) = 0.$$

This proves that $F(z)$ has a zero of multiplicity $\geq p_k = n - 1$ at z_k.

(b) Suppose that $p_k \leq n-2$. Using (9.1.8), step (1), and step (2), we have

$$F^{(m)}(z_k) = f^{(n+m)}(z_k) + \sum_{j=p_k-m}^{p_k} \zeta_{j,k} f^{(j+m)}(z_k),$$

where $m \in \{0,\ldots,p_k-1\}$. A similar argument as in case (a) shows that $F(z)$ has a zero of multiplicity $\geq p_k$ at z_k.

In both cases the function $F(z)$ has a zero of multiplicity $\geq p_k$ at z_k. □

9.2 At least one transcendental coefficient

Consider a linear differential equation of the form

$$f^{(n)} + A_{n-1}(z)f^{(n-1)} + \cdots + A_0(z)f = 0, \tag{9.2.1}$$

where n is a positive integer, and $A_0(z), A_1(z), \ldots, A_{n-1}(z)$ are entire functions with $A_0(z) \not\equiv 0$ such that at least one $A_j(z)$ is transcendental. By Theorem 8.1.1, there exists at least one infinite-order solution of (9.2.1). The next result of Frei gives an upper bound for the maximum number of linearly independent finite-order solutions of equations of the form (9.2.1).

Theorem 9.2.1 ([69], [70], [166, p. 60]). *If in equation* (9.2.1), *$A_j(z)$ is transcendental and $A_k(z)$ is a polynomial whenever $k > j$, then* (9.2.1) *possesses at most j linearly independent finite-order solutions.*

Example 9.2.2. The function $f(z) = e^{-z}$ satisfies

$$f'' + e^z f' + (e^z - 1)f = 0. \tag{9.2.2}$$

By Theorem 9.2.1 or Theorem 8.1.1, any solution of (9.2.2) that is linearly independent with e^{-z} must have infinite order.

There are numerous examples for $n = 2$ that illustrate Theorem 9.2.1. The next two examples illustrate Theorem 9.2.1 for $n = 3$ and $n = 4$.

Example 9.2.3 ([287]). The functions $f_1(z) = e^z + z$, $f_2(z) = e^z - 1$, and $f_3(z) = z + 1$ are finite-order solutions of the third-order equation

$$f''' + (z - 1 + e^{-z})f'' - (z+1)f' + f = 0,$$

and any two of f_1, f_2, f_3 are linearly independent.

Example 9.2.4. The functions $f_1(z) = z, f_2(z) = z^2$, and $f_3(z) = e^z + 1$ are linearly independent finite-order solutions of the fourth-order equation

$$f^{(4)} + \left(e^{-z} + \frac{1}{2}z^2 - z\right)f''' - \frac{1}{2}z^2 f'' + zf' - f = 0.$$

Example 9.2.4 is a particular case of the following general example.

Example 9.2.5. For any integer $n \geq 4$, the functions

$$f_1(z) = z,\ f_2(z) = z^2, \ldots, f_{n-2}(z) = z^{n-2},\ f_{n-1}(z) = e^z + 1$$

are $n-1$ linearly independent finite-order solutions of the nth-order equation

$$f^{(n)} + A(z)f^{(n-1)} + \frac{(-1)^{n-1}}{(n-2)!}z^{n-2}f^{(n-2)} + \frac{(-1)^{n-2}}{(n-3)!}z^{n-3}f^{(n-3)} + \cdots + zf' - f = 0,$$

where

$$A(z) = e^{-z} + \frac{(-1)^{n-2}}{(n-2)!}z^{n-2} + \frac{(-1)^{n-3}}{(n-3)!}z^{n-3} + \cdots - z.$$

Here we have $j = n - 1$ in Theorem 9.2.1.

The above examples show that for any $n \geq 1$, it is possible for an equation of the form (9.2.1) to have $n-1$ linearly independent finite-order solutions, which is the maximum allowed in Theorem 9.2.1. The most general question regarding examples of Theorem 9.2.1 would be the following:

Question 9.2.6. For any $n \geq 2$ and j satisfying $0 < j < n$, does there exist an equation of the form (9.2.1) where $A_j(z)$ is transcendental and $A_k(z)$ is a polynomial for $k > j$, that possesses exactly j linearly independent finite-order solutions?

The answer to Question 9.2.6 is yes, because the above examples show that the answer is yes when $j = n - 1$, and the next two examples show that the answer is yes when $0 < j < n - 1$. Therefore the number j in Theorem 9.2.1 is sharp.

Example 9.2.7. For any $n \geq 2$, $f(z) = e^z + 1$ satisfies the equation

$$f^{(n)} + e^{-z}f' - f = 0.$$

Here we have $j = 1$ and $n \geq 2$.

Example 9.2.8. Let $n \geq 4$, and let j satisfy $2 \leq j \leq n - 2$. Then the functions

$$f_1(z) = z,\ f_2(z) = z^2, \ldots, f_{j-1}(z) = z^{j-1},\ f_j(z) = e^z + 1$$

are j linearly independent finite-order solutions of the equation

$$f^{(n)} + A_j(z)f^{(j)} + \frac{(-1)^j}{(j-1)!}z^{j-1}f^{(j-1)} + \frac{(-1)^{j-1}}{(j-2)!}z^{j-2}f^{(j-2)} + \cdots + zf' - f = 0,$$

where

$$A_j(z) = e^{-z} + \frac{(-1)^{j-1}}{(j-1)!}z^{j-1} + \frac{(-1)^{j-2}}{(j-2)!}z^{j-2} + \cdots - z.$$

The conditions on the coefficients in Theorem 9.2.1 can be refined. For instance, by expressing the dominance of $A_j(z)$ in different ways and allowing the coefficients $A_{j+1}(z), \ldots, A_{n-1}(z)$ to be transcendental, it is shown [128] that the conclusion of Frei's theorem still holds under an additional estimation on the asymptotic lower bound for the growth of the solutions. Theorem 9.2.1 is a particular case of the following two theorems.

Theorem 9.2.9 ([128]). *Let j be the smallest index in* (9.2.1) *such that*

$$\limsup_{r\to\infty} \sum_{k=j+1}^{n-1} \frac{T(r,A_k)}{T(r,A_j)} < 1.$$

(If $j = n-1$, then the above sum is considered to be equal to 0.*) Then $A_j(z)$ is transcendental, and every solution base of* (9.2.1) *has at least $n-j$ solutions f for which*

$$T(r,A_j) \lesssim \log T(r,f) \lesssim \frac{R+r}{R-r}T(R,A_j), \quad r \notin E,$$

where $E \subset [0,\infty)$ has finite linear measure, and $r < R < \infty$. For these solutions, the value 0 *is the only possible finite deficient value.*

Theorem 9.2.10 ([128]). *Let j be the smallest index in* (9.2.1) *such that*

$$\limsup_{r\to\infty} \sum_{k=j+1}^{n-1} \frac{\log^+ M(r,A_k)}{\log^+ M(r,A_j)} < 1.$$

(If $j = n-1$, then the above sum is considered to be equal to 0.*) Then $A_j(z)$ is transcendental, and every solution base of* (9.2.1) *has at least $n-j$ solutions f for which*

$$\log T(r,f) \asymp \log M(r,A_j), \quad r \notin E,$$

where $E \subset [0,\infty)$ has finite linear measure. For these solutions, the value 0 *is the only possible finite deficient value.*

The value 0 being the only possible finite deficient value in Theorems 9.2.9 and 9.2.10 originates from a 1968 theorem of Wittich, which is one of the cornerstones in oscillation theory. Wittich's result was originally formulated for rational coefficients in [293, p. 54] and then generalized to small meromorphic coefficients in [166, p. 62]. The following version is for small entire coefficients.

Theorem 9.2.11. *Suppose that a solution f of* (9.2.1) *is admissible in the sense that*

$$T(r,A_j) = o(T(r,f)), \quad r \notin E,\ j = 0,\dots,n-1, \tag{9.2.3}$$

where $E \subset [0,\infty)$ is a set of finite linear measure. Then the value 0 is the only possible finite deficient value for f.

Analogues of Theorem 9.2.11, where the Nevanlinna deficient value is replaced with a Petrenko deviated value or a Eremenko deviated value, can be found in [122].

Sometimes Theorems 9.2.9 and 9.2.10 give a larger number of linearly independent solutions of infinite order than can be ascertained from Theorem 9.2.1. The next example illustrates this.

Example 9.2.12 ([128]). The equation

$$f'' - (2e^z + 1)f' + e^{2z}f = 0 \tag{9.2.4}$$

is satisfied by the infinite-order linearly independent functions

$$f_1(z) = \exp(e^z) \quad \text{and} \quad f_2(z) = \exp(z + e^z),$$

which means that all nontrivial solutions of (9.2.4) have infinite order. This agrees with Theorems 9.2.9 and 9.2.10, which both conclude that equation (9.2.4) has at least two linearly independent solutions of infinite order. Indeed, the coefficient functions $A_0(z) = e^{2z}$ and $A_1(z) = -(2e^z + 1)$ in (9.2.4) satisfy

$$\limsup_{r\to\infty} \frac{T(r,A_1)}{T(r,A_0)} = \limsup_{r\to\infty} \frac{\log M(r,A_1)}{\log M(r,A_0)} = \frac{1}{2} < 1.$$

Also in this example, all nontrivial solutions f are admissible in the sense of (9.2.3), but without an exceptional set.

Numerous authors have investigated the following two general problems.

Question 9.2.13. What conditions on the coefficients in (9.2.1) will guarantee that every nontrivial solution of equation (9.2.1) has infinite order?

Question 9.2.14. If f is a nontrivial finite-order solution of (9.2.1), then how do the properties of the coefficients of (9.2.1) affect the properties of f?

Because there is so much literature on these two questions, due to space limitations, we will discuss a few results and give references to other results. We discuss some results concerning the two questions for second-order equations

$$f'' + A(z)f' + B(z)f = 0, \tag{9.2.5}$$

where $A(z)$ and $B(z) \not\equiv 0$ are entire functions such that at least one of $A(z)$, $B(z)$ is transcendental, and we also give some generalizations to equations of the form (9.2.1). It is easy to prove the next result.

Theorem 9.2.15 ([87]). *If in equation* (9.2.5), *either* (a) $\rho(A) < \rho(B)$ *or* (b) $A(z)$ *is a polynomial and* $B(z)$ *is transcendental, then every nontrivial solution of* (9.2.5) *has infinite order.*

Proof. Suppose that f is a nontrivial finite-order solution of (9.2.5). Rewriting (9.2.5) in the form

$$B(z) = -A(z)\frac{f'}{f} - \frac{f''}{f} \tag{9.2.6}$$

and using Nevanlinna's estimate of the logarithmic derivative (Lemma 4.6.10), we obtain

$$T(r,B) \le T(r,A) + O(\log r).$$

Thus, if (a) or (b) holds, then we get a contradiction. This means that $\rho(f) = \infty$. □

Frei proved the following result.

Theorem 9.2.16 ([71, 287]). *If C is a nonzero constant, then the equation*

$$f'' + e^{-z}f' + Cf = 0$$

possesses a nontrivial finite-order solution f if and only if $C = -k^2$, where k is a positive integer. Such a finite-order solution f is an exponential sum of the form

$$f(z) = C_0 + C_1e^z + \cdots + C_ke^{kz},$$

where all the constants C_j are nonzero.

Langley proved the next result, which completed partial results of Ozawa, Amemiya and Ozawa, and Gundersen.

Theorem 9.2.17 ([171]). *If $Q(z)$ is a nonconstant polynomial and $A \neq 0$ is a constant, then every nontrivial solution of*

$$f'' + Ae^{-z}f' + Q(z)f = 0 \tag{9.2.7}$$

has infinite order.

Tao and Long [269] showed that if the coefficient Ae^{-z} in (9.2.7) is replaced by the more general exponential sum $\sum_{j=1}^{m} A_je^{jz}$, where each $A_j \in \mathbb{C}$ ($A_m \neq 0$), then Theorem 9.2.17 still holds. Moreover, Zhang [303] proved that Ae^{-z} in (9.2.7) can be replaced with $e^{P(z)}$ under some restrictions on the degrees of the polynomials $P(z)$ and $Q(z)$. Theorem 9.2.17 gives an answer to the next question for the particular equation (9.2.7).

Question 9.2.18 ([89, Question 5.1]). If $A(z)$ is an entire function satisfying $\lambda(A) < \rho(A) < \infty$ and $Q(z)$ is a nonconstant polynomial, then does every nontrivial solution of

$$f'' + A(z)f' + Q(z)f = 0$$

have infinite order?

The earlier results of Ozawa, Amemiya and Ozawa, and Gundersen gave partial results toward the goal of proving Theorem 9.2.17 by showing that Theorem 9.2.17 holds for various conditions on the polynomial $Q(z)$ in (9.2.7). The next result of Long, Shi, Wu, and Zhang reminds us of this process, because it shows that certain conditions on polynomials give partial results on Question 9.2.18.

Theorem 9.2.19 ([197]). *Let $A(z) = v(z)e^{P(z)}$, where $v(z) \not\equiv 0$ is an entire function, and $P(z) = a_n z^n + a_{n-1}z^{n-1} + \cdots + a_0$ is a nonconstant polynomial of degree $n \geq 1$ such that $\rho(v) < n$. Let $Q(z) = b_m z^m + b_{m-1}z^{m-1} + \cdots + b_0$ be a polynomial of degree $m \geq 1$. Then all nontrivial solutions of the equation*

$$f'' + A(z)f' + Q(z)f = 0 \tag{9.2.8}$$

have infinite order, provided that one of the following conditions holds:

(a) $m + 2 < 2n$.

(b) $m + 2 > 2n$ *and* $m + 2 \neq 2kn$ *for all integers* k.

(c) $m + 2 = 2n$ *and* a_n^2/b_m *is not a negative real number.*

From Theorem 9.2.19 we see that the answer to Question 9.2.18 is yes when $m+2 < 2n$, and we get partial results on Question 9.2.18 for the cases where $m + 2 \geq 2n$. For related results, see [162, 197].

The next result shows, among other things, that a nonconstant finite-order solution of (9.2.5) has an asymptotic value (and so does its derivatives) when $A(z)$ dominates the growth of $B(z)$ in a sector. Asymptotic values of entire functions are discussed in general in Section 12.1.

Theorem 9.2.20 ([87]). *Let $A(z)$ and $B(z) \not\equiv 0$ be entire functions such that for real constants $\alpha, \beta, \theta_1, \theta_2$ satisfying $\alpha > 0$, $\beta > 0$, and $\theta_1 < \theta_2$, we have*

$$|A(z)| \geq \exp\left((1 + o(1))\alpha|z|^{\beta}\right)$$

and

$$|B(z)| \leq \exp\left(o(1)|z|^{\beta}\right)$$

as $z \to \infty$ in $\theta_1 \leq \arg z \leq \theta_2$. Let $\varepsilon > 0$ be a small constant, and let $S(\varepsilon)$ denote the sector $\theta_1 + \varepsilon \leq \arg z \leq \theta_2 - \varepsilon$.

If f is a nontrivial solution of (9.2.5) *of finite order, then the following conditions hold:*

(a) *There exists a constant $b \neq 0$ such that $f(z) \to b$ as $z \to \infty$ in $S(\varepsilon)$. Furthermore,*

$$|f(z) - b| \leq \exp(-(1 + o(1))\alpha|z|^{\beta}) \tag{9.2.9}$$

as $z \to \infty$ in $S(\varepsilon)$.

(b) *For each integer* $k \geq 1$,

$$|f^{(k)}(z)| \leq \exp(-(1+o(1))\alpha|z|^{\beta}) \tag{9.2.10}$$

as $z \to \infty$ *in* $S(\varepsilon)$.

Theorem 9.2.20 generalized and improved results of Ozawa [223]. Inequalities (9.2.9) and (9.2.10) are sharp.

Example 9.2.21 ([87]). For any constant $b \neq 0, f(z) = e^z + b$ satisfies the equation

$$f'' + \left(e^z - be^{-z} + b - 2\right)f' + \left(1 - e^z\right)f = 0.$$

This example shows that (9.2.9) and (9.2.10) are sharp in the sense that neither of the two numbers α, β in (9.2.9) or (9.2.10) can be replaced by a larger number.

For more examples that illustrate that (9.2.9) and (9.2.10) are sharp in the above sense, see [87, Example 2].

Theorem 9.2.20 was generalized to equation (9.2.1) in the following manner. Belaïdi and Hamouda [22] showed that the result holds when $A_0(z)$ dominates the growth of the other coefficients $A_j(z), j = 1, \ldots, n-1$. Then Laine and Yang showed that the next result holds for any dominating coefficient $A_s(z)$ in (9.2.1) such that $1 \leq s \leq n-1$.

Theorem 9.2.22 ([168]). *Let* $n \geq 2$ *be an integer, and let* $\delta > 0$ *be a real number satisfying* $n\delta < 1$. *Let* $\theta_1, \theta_2, \alpha, \beta$ *be real constants such that* $\theta_1 < \theta_2$, $\alpha > 0$, *and* $\beta > 0$. *Suppose that in equation* (9.2.1), *there exists a particular coefficient* $A_s(z)$ *for some* $s \in \{1, \ldots, n-1\}$ *such that*

$$|A_s(z)| \geq \exp\left((1+\delta)\alpha|z|^{\beta}\right)$$

and

$$|A_j(z)| \leq \exp\left(\delta\alpha|z|^{\beta}\right) \quad \text{for } j \neq s$$

whenever $|z| \geq R(\delta)$ *in the sector* $\theta_1 \leq \arg z \leq \theta_2$.

If $\varepsilon > 0$ *is small enough and if* f *is a transcendental finite-order solution of* (9.2.1), *then the following conditions hold:*

(a) *There exist an integer* $j \in \{0, \ldots, s-1\}$ *and a constant* $b_j \neq 0$ *such that* $f^{(j)}(z) \to b_j$ *as* $z \to \infty$ *in* $S(\varepsilon)$. *More precisely,*

$$|f^{(j)}(z) - b_j| \leq \exp(-(1-n\delta)\alpha|z|^{\beta}) \tag{9.2.11}$$

in $S(\varepsilon)$, *provided that* $|z|$ *is large enough. Here* $S(\varepsilon) = \{z : \theta_1 + \varepsilon \leq \arg z \leq \theta_2 - \varepsilon\}$.

(b) *For each integer* $m \geq j+1$,

$$|f^{(m)}(z)| \leq \exp(-(1-n\delta)\alpha|z|^{\beta}) \tag{9.2.12}$$

in $S(3\varepsilon)$ *for all* $|z|$ *large enough.*

Example 9.2.23 ([168]). The function $f(z) = e^z + 1$ is a solution of

$$f''' + 2e^{-z}f'' - e^z f' + (e^z - 2)f = 0.$$

Choose $\alpha = \frac{3}{4} \cdot (1 - \frac{\pi^2}{18}) > 0$, and let $\delta \in (0, 1/3)$. Then the coefficient $A_2(z) = 2e^{-z}$ dominates the growth of the coefficients $A_0(z) = e^z - 2$ and $A_1(z) = -e^z$ in the sector $\frac{2\pi}{3} \le \arg z \le \frac{4\pi}{3}$, where $A_2(z)$ satisfies

$$\begin{aligned}|A_2(z)| &= 2e^{-r\cos\theta} \ge 2\exp\left(\left(1 - \frac{(\theta-\pi)^2}{2}\right)r\right) \ge \exp\left(\left(1 - \frac{\pi^2}{18}\right)r\right) \\ &= \exp\left(\frac{4}{3}\alpha r\right) \ge \exp\big((1+\delta)\alpha r\big), \quad z = re^{i\theta}.\end{aligned}$$

Therefore, the conditions of Theorem 9.2.22 are fulfilled in $\frac{2\pi}{3} \le \arg z \le \frac{4\pi}{3}$ with $s = 2$ and $j = 0$.

Example 9.2.24 ([287]). The function $f(z) = \exp(z^2/2 + z) + z + 1$ is a solution of

$$f''' + \big(\exp(-z^2/2 - z) - z - 1\big)f'' - f' - (z+1)f = 0.$$

The coefficient $A_2(z) = \exp(-z^2/2 - z) - z - 1$ dominates the growth of all the other coefficients in the sectors $\frac{\pi}{3} \le \arg z \le \frac{2\pi}{3}$ and $\frac{4\pi}{3} \le \arg z \le \frac{5\pi}{3}$. Here we have $s = 2$ and $j = 1$ in Theorem 9.2.22.

As observed in [87], it follows from Theorem 9.2.25 that if we switch the two coefficients in Example 9.2.21, then every nontrivial solution of the equation

$$f'' + (1 - e^z)f' + (e^z - be^{-z} + b - 2)f = 0 \tag{9.2.13}$$

has infinite order. Another way to see this is by means of Theorem 9.2.9. Indeed, denoting $A_0(z) = e^z - be^{-z} + b - 2$, we see that $j = 0$ in Theorem 9.2.9, and, consequently, all nontrivial solutions f of (9.2.13) satisfy

$$(2/\pi + o(1))r = T(r, A_0) \lesssim \log T(r, f).$$

Theorem 9.2.25 ([87]). *Suppose that for real constants α, β, θ_1, θ_2 such that $\alpha > 0$, $\beta > 0$, and $\theta_1 < \theta_2$, we have*

$$|B(z)| \ge \exp((1 + o(1))\alpha|z|^\beta) \quad \textit{and} \quad |A(z)| \le \exp(o(1)|z|^\beta)$$

as $z \to \infty$ in $\theta_1 \le \theta \le \theta_2$. Then every nontrivial solution of (9.2.5) *has infinite order.*

It is easy to construct examples of Theorem 9.2.25. For instance [87], every nontrivial solution of $f'' + e^{P(z)}f' + e^{-P(z)}f = 0$ has infinite order when $P(z)$ is a nonconstant polynomial. For generalizations and extensions of Theorem 9.2.25 to equation (9.2.1), see [21] and [168].

The following result shows that in many cases of Theorem 9.2.25, it suffices that $B(z)$ dominates the growth of $A(z)$ on one ray instead of in a sector.

Theorem 9.2.26 ([120, Theorem 2.1]). *Let $A(z)$ and $B(z)$ be entire such that $\rho(A) = \rho(B) \in (0, \infty)$, and assume that both are of finite type. If* (9.2.5) *possesses a nontrivial finite-order solution, then*

$$h_B(\theta) \leq \max\{0, h_A(\theta)\}, \quad \theta \in [-\pi, \pi). \tag{9.2.14}$$

In particular, if there exists $\theta_0 \in [-\pi, \pi)$ such that $\max\{0, h_A(\theta_0)\} < h_B(\theta_0)$, then any nontrivial solution of (9.2.5) *has infinite order.*

Proof. Write (9.2.5) as (9.2.6) and then use a logarithmic derivative estimate [86, Corollary 1] to see that (9.2.14) holds for almost every θ. Since $A(z)$ and $B(z)$ are of finite type, the indicator functions are continuous by Theorem 4.2.6, and so (9.2.14) holds for every θ. The remaining statement is a trivial consequence of (9.2.14). □

Theorem 9.2.27 ([304]). *Let $A(z)$ be an entire function of c. r. g. such that the set $E = \{\theta \in [0, 2\pi) : h_A(\theta) = 0\}$ has linear measure zero, and let $B(z)$ be an entire function such that $\rho(A) \neq \rho(B)$ and $T(r, B) \sim \log M(r, B)$ as $r \to \infty$ outside a set of finite logarithmic measure. Then any nontrivial solution of* (9.2.5) *has infinite order.*

Laine and Wu proved the following result.

Theorem 9.2.28 ([167]). *If $A(z)$ and $B(z) \not\equiv 0$ are entire functions satisfying $\rho(B) < \rho(A) < \infty$ and if*

$$T(r, A) \sim \log M(r, A) \tag{9.2.15}$$

as $r \to \infty$ outside a set G of finite logarithmic measure, then any nontrivial solution of (9.2.5) *has infinite order.*

Murai [215] showed that condition (9.2.15) holds for an entire function with Fejér gaps, which gives the following result.

Corollary 9.2.29 ([167]). *If $A(z)$ and $B(z) \not\equiv 0$ are entire functions satisfying $\rho(B) < \rho(A) < \infty$ and if $A(z)$ has Fejér gaps, then any nontrivial solution of* (9.2.5) *has infinite order.*

If a set G has finite logarithmic measure, then clearly $\overline{\log\text{dens}}(G) = 0$. Hence Theorem 9.2.30 by Kwon and Kim [164] is an improvement of Theorem 9.2.28, because it allows for a larger exceptional set. We offer a more detailed proof than the original reasoning in [164]; it directly uses a result that allows us to control the logarithmic densities of sets.

Theorem 9.2.30 ([164]). *If $A(z)$ and $B(z) \not\equiv 0$ are entire functions satisfying $\rho(B) < \rho(A) < \infty$ and if* (9.2.15) *holds as $r \to \infty$ outside a set G satisfying*

$$\overline{\log\text{dens}}(G) < \frac{\rho(A) - \rho(B)}{\rho(A)},$$

then any nontrivial solution f of (9.2.5) *has infinite order.*

Proof. For $d \in (0, 1/4)$, let

$$I_d(r) = \{\theta \in [0, 2\pi) : \log|A(re^{i\theta})| < (1-d)\log M(r, A)\}.$$

From (1.2.3) we find that the complement $G^c = [1, \infty) \setminus G$ of G satisfies

$$\underline{\log\mathrm{dens}}(G^c) = 1 - \overline{\log\mathrm{dens}}(G) > 1 - \frac{\rho(A) - \rho(B)}{\rho(A)} = \frac{\rho(B)}{\rho(A)}.$$

Hence for every $\varepsilon > 0$, there exist a small $\gamma > 0$ and a set $F \subset [1, \infty)$ such that

$$\underline{\log\mathrm{dens}}(F) \geq \frac{\rho(B)}{\rho(A)} + \gamma, \tag{9.2.16}$$

and for $r \in F$,

$$\begin{aligned}\left(1 - \frac{\varepsilon}{2\pi}\right)\log M(r, A) &< T(r, A) = m(r, A)\\ &= \frac{1}{2\pi}\left(\int_{I_d(r)} + \int_{I_d^c(r)}\right)\log^+|A(re^{i\theta})|\,d\theta\\ &< \frac{1}{2\pi}((1-d)\,\mathrm{meas}(I_d(r)) + \mathrm{meas}(I_d^c(r)))\log M(r, A)\\ &= \left(1 - \frac{d}{2\pi}\,\mathrm{meas}(I_d(r))\right)\log M(r, A),\end{aligned}$$

that is, $\mathrm{meas}(I_d(r)) < \varepsilon/d$ for $r \in F$.

Next, we apply Fuchs' small arcs lemma (Lemma 7.2.6) to $A(z)$, denoting the order of $A(z)$ by ρ. Set $\zeta = \gamma/3$ and then choose $\delta > 0$ so small that

$$K(\rho, \zeta)\delta\log(1/\delta) < d,$$

where $K(\rho, \zeta) > 0$ is the constant in Lemma 7.2.6; see (7.2.7) for a similar reasoning. Then for every interval $J \subset \mathbb{R}$ of length δ, we have

$$r\int_J \left|\frac{A'(re^{i\theta})}{A(re^{i\theta})}\right| d\theta < dT(r, A), \quad r \in E,$$

where $E \subset (1, \infty)$ satisfies

$$\underline{\log\mathrm{dens}}(E) > 1 - \gamma/3. \tag{9.2.17}$$

We will prove in (9.2.20) that $F \cap E$ has positive upper logarithmic density. Hence if $\phi \in [0, 2\pi)$, then for all sufficiently large $r \in F \cap E$, there exists $\psi \notin I_d(r)$ such that $\phi - \delta < \psi < \phi$ and

$$\log|A(re^{i\phi})| = \log|A(re^{i\psi})| + \int_\psi^\phi \frac{d}{d\theta}\log|A(re^{i\theta})|\,d\theta$$

$$\geq (1-d)\log M(r,A) - r\int_{\psi}^{\phi}\left|\frac{A'(re^{i\theta})}{A(re^{i\theta})}\right| d\theta$$

$$\geq (1-2d)\log M(r,A). \tag{9.2.18}$$

Choose $a = \rho(B) + \gamma\rho(A)/3$ and $b = \rho(B) + \gamma\rho(A)/4$. Then $\rho(B) < b < a < \rho(A)$ independently of whether $\rho(B) = 0$ or $\rho(B) \neq 0$. Define the set

$$H = \{r \geq r_0 : \log M(r,A) \geq r^a\},$$

where $r_0 > 1$ is chosen so that $M(r_0,A) > 1$. Then arguing similarly as in the proof of Theorem 2.5.2 (see Problem 2.5.8), we obtain

$$\overline{\log\operatorname{dens}}(H) \geq 1 - \frac{a}{\rho(A)}.$$

Moreover, from (9.2.16), (9.2.17), and (1.2.3), we obtain

$$\overline{\log\operatorname{dens}}(F^c) \leq 1 - \frac{\rho(B)}{\rho(A)} - \gamma \quad \text{and} \quad \overline{\log\operatorname{dens}}(E^c) \leq \gamma/3.$$

Setting $J = F \cap E \cap H$, we claim that

$$H = J \cup (H \cap (F^c \cup E^c)). \tag{9.2.19}$$

To prove (9.2.19), let $x \in H$. We have two possibilities.

(1) $x \in F$ and $x \in E$. In this case, $x \in F \cap E \cap H = J$.

(2) Either $x \notin F$ or $x \notin E$ (or both). In this case, $x \in F^c \cup E^c$, and since $x \in H$ by assumption, it follows that $x \in H \cap (F^c \cup E^c)$.

This proves $H \subset J \cup (H \cap (F^c \cup E^c))$. The reverse inclusion $J \cup (H \cap (F^c \cup E^c)) \subset H$ is trivial because both J and $H \cap (F^c \cup E^c)$ are subsets of H. This proves (9.2.19).

Next, using identity (9.2.19), we obtain

$$\int_{H\cap(0,r)} \frac{dt}{t} \leq \int_{J\cap(0,r)} \frac{dt}{t} + \int_{H\cap(F^c\cup E^c)\cap(0,r)} \frac{dt}{t},$$

which implies

$$\int_{J\cap(0,r)} \frac{dt}{t} \geq \int_{H\cap(0,r)} \frac{dt}{t} - \int_{H\cap(F^c\cup E^c)\cap(0,r)} \frac{dt}{t}$$

$$\geq \int_{H\cap(0,r)} \frac{dt}{t} - \int_{F^c\cap(0,r)} \frac{dt}{t} - \int_{E^c\cap(0,r)} \frac{dt}{t}.$$

Consequently,

$$\begin{aligned}\overline{\log\operatorname{dens}}(F\cap E) &\geq \overline{\log\operatorname{dens}}(J)\\ &\geq \overline{\log\operatorname{dens}}(H) - \overline{\log\operatorname{dens}}(F^c) - \overline{\log\operatorname{dens}}(E^c)\\ &\geq -\frac{a}{\rho(A)} + \frac{\rho(B)}{\rho(A)} + \frac{2\gamma}{3} = \frac{\gamma}{3} > 0.\end{aligned} \tag{9.2.20}$$

We conclude from (9.2.18) that for every z satisfying $|z| = r \in J$, we have

$$\log|A(z)| \geq (1-2d)r^a \geq \frac{r^a}{2}.$$

Let $f \not\equiv 0$ be a solution of (9.2.5). Since $\rho(B) < \rho(A)$, we see from (9.2.5) that $\rho(f) > 0$, which obviously means that $f' \not\equiv 0$. Then dividing (9.2.5) by f' gives

$$\left|\frac{f''(z)}{f'(z)}\right| \geq |A(z)| - |B(z)|\left|\frac{f(z)}{f'(z)}\right|.$$

By [163, Lemma 3], there exists $R > 0$ such that for all $r \geq R$, there exists a point z_r with $|z_r| = r$ and

$$\left|\frac{f(z_r)}{f'(z_r)}\right| \leq r.$$

From formula (1.11) in [86, Theorem 4], we obtain that there exists a set $E_0 \subset [0,\infty)$ of finite linear measure such that

$$\left|\frac{f''(z)}{f'(z)}\right| \leq rT(2r,f)^2, \quad r \notin E_0.$$

Finally, putting the above estimates together, we obtain that for every $r \in J\setminus E_0$ such that $r \geq R$, there exists a point z_r with $|z_r| = r$ and

$$\exp\left(\frac{r^a}{2}\right) - r\exp(r^b) \leq rT(2r,f)^2,$$

where $a > b > 0$. This shows that $\rho(f) = \infty$. □

In addition to Theorem 9.2.19, Long and coauthors proved several results on Question 9.2.18. An example is the following result.

Theorem 9.2.31 ([297]). *Let $A(z)$ be a nontrivial solution of $w''+P(z)w = 0$, where $P(z) \not\equiv 0$ is a polynomial of degree n, and let $B(z)$ be a transcendental entire function satisfying*

$$\underline{\rho}(B) < \frac{1}{2} + \frac{1}{2(n+1)}$$

and such that $\rho(A) \neq \rho(B)$. Then every nontrivial solution of (9.2.5) *has infinite order.*

More conditions on the lower order of the coefficients $A(z)$ and $B(z)$ guaranteeing that all nontrivial solutions of (9.2.5) have infinite order can be found in [196]. The same conclusion can be obtained by having extremal functions for Yang's inequality, extremal functions for Denjoy's conjecture, special functions, and other types of functions as coefficients; see [194, 197, 297] and the references therein.

For more results and examples concerning Questions 9.2.13 and 9.2.14, see [21, 87, 101, 130, 143, 161, 168, 190, 195, 198, 199, 200, 257, 297] and the references therein.

9.3 The equation $f'' + A(z)f = 0$

Consider a linear differential equation of the form

$$f'' + A(z)f = 0, \tag{9.3.1}$$

where $A(z) \not\equiv 0$ is an entire function. With their 1982 landmark paper [12], Bank and Laine ushered in a new era in complex differential equations, which led to numerous significant investigations; see [166] for details and references. An important feature in [12] was the observation that the product of two linearly independent solutions of (9.3.1) satisfies an interesting and useful nonlinear differential equation. Before this observation, it was well known that the quotient of two linearly independent solutions of (9.3.1) satisfies an equation involving the *Schwarzian.*

Definition 9.3.1. The *Schwarzian derivative* of a nonconstant meromorphic function $w(z)$ is defined as

$$S_w(z) = \left(\frac{w''}{w'}\right)' - \frac{1}{2}\left(\frac{w''}{w'}\right)^2 = \frac{w'''}{w'} - \frac{3}{2}\left(\frac{w''}{w'}\right)^2.$$

Theorem 9.3.2. *Let f_1 and f_2 be linearly independent solutions of* (9.3.1), *and set $w = f_1/f_2$. Then w satisfies the equation*

$$S_w(z) = 2A(z). \tag{9.3.2}$$

Proof. From (9.3.1) and Abel's identity we know that the Wronskian $W(f_1, f_2) = c$, a nonzero constant. Then

$$w' = -\frac{c}{f_2^2} \quad \text{and} \quad \frac{w''}{w'} = -2\frac{f_2'}{f_2}.$$

Taking another derivative and using (9.3.1), we obtain

$$\left(\frac{w''}{w'}\right)' = -2\frac{f_2''}{f_2} + 2\left(\frac{f_2'}{f_2}\right)^2 = 2A(z) + \frac{1}{2}\left(\frac{w''}{w'}\right)^2,$$

which gives (9.3.2) □

The next result gives equation (9.3.3) for the product of two linearly independent solutions of (9.3.1), and (9.3.3) is appropriately called the *Bank–Laine equation.*

Theorem 9.3.3 ([12]). *Let f_1 and f_2 be linearly independent solutions of* (9.3.1), *and set $E = f_1 f_2$. Then E satisfies the differential equation*

$$4A(z) = \left(\frac{E'}{E}\right)^2 - 2\frac{E''}{E} - \left(\frac{c}{E}\right)^2 \tag{9.3.3}$$

with the nonzero constant $c = W(f_1, f_2)$.

Proof. Since $c = f_1 f_2' - f_1' f_2$ and $E = f_1 f_2$, we obtain

$$\frac{c}{E} = \frac{f_2'}{f_2} - \frac{f_1'}{f_1} \quad \text{and} \quad \frac{E'}{E} = \frac{f_1'}{f_1} + \frac{f_2'}{f_2}.$$

Adding these two equations gives

$$2\frac{f_2'}{f_2} = \frac{c}{E} + \frac{E'}{E}. \tag{9.3.4}$$

Differentiating (9.3.4) yields

$$2\frac{f_2''}{f_2} - 2\left(\frac{f_2'}{f_2}\right)^2 = -c\frac{E'}{E^2} + \frac{E''}{E} - \left(\frac{E'}{E}\right)^2. \tag{9.3.5}$$

Substituting $A = -f_2''/f_2$ and (9.3.4) into (9.3.5), we obtain

$$-2A - 2\left[\frac{1}{2}\left(\frac{c}{E} + \frac{E'}{E}\right)\right]^2 = -c\frac{E'}{E^2} + \frac{E''}{E} - \left(\frac{E'}{E}\right)^2,$$

which simplifies to (9.3.3). □

Bank and Laine [12] observed that the product function E in Theorem 9.3.3 satisfies the third-order linear differential equation

$$E''' + 4A(z)E' + 2A'(z)E = 0, \tag{9.3.6}$$

which can be seen by rewriting equation (9.3.3) as

$$4AE^2 = (E')^2 - 2E''E - c^2 \tag{9.3.7}$$

and then differentiating.

Ishizaki [145] showed that when f_1, f_2 are linearly independent solutions of (9.3.1), the functions $f_1^2, f_1 f_2, f_2^2$ form a fundamental solution base for equation (9.3.6). More generally, for integer $n \geq 3$, the functions $f_1^{n-1}, f_1^{n-2} f_2, \ldots, f_1 f_2^{n-2}, f_2^{n-1}$ form a fundamental solution base for some linear differential equation of order n with entire coefficients [51]. When $n = 4$, such an equation is

$$f^{(4)} + 10A(z)f'' + 10A'(z)f' + (3A''(z) + 9A(z)^2)f = 0.$$

Bank and Laine proved the next result, where the product function $E = f_1f_2$ was used in the proof of (a) and (c).

Theorem 9.3.4 ([12, 13]). *Let $A(z)$ be a transcendental entire function.*

(a) *If f_1 and f_2 are linearly independent solutions of* (9.3.1), *where $\rho(A)$ is finite and not a positive integer, then*

$$\max\{\lambda(f_1), \lambda(f_2)\} \geq \rho(A).$$

(b) *If f is a nontrivial solution of* (9.3.1) *where $\lambda(A) < \rho(A)$, then*

$$\lambda(f) \geq \rho(A).$$

(c) *If f_1 and f_2 are linearly independent solutions of* (9.3.1) *such that*

$$\max\{\lambda(f_1), \lambda(f_2)\} < \infty,$$

then any nontrivial solution of (9.3.1) *that is not a constant multiple of f_1 or f_2 satisfies $\lambda(f) = \infty$.*

A method for constructing equations of the form (9.3.1) that possess two linearly independent zero-free solutions where $A(z)$ is a transcendental entire function can be found in [12, p. 356]. In general, this method always gives that $\rho(A)$ is a positive integer or ∞. As for a concrete example [12, p. 356], we see that

$$f_1(z) = \exp\left(\frac{e^z - z}{2}\right) \quad \text{and} \quad f_2(z) = \exp\left(-\frac{e^z + z}{2}\right)$$

are linearly independent solutions of (9.3.1), where $A(z) = -(e^{2z} + 1)/4$.

Theorem 9.3.4(a) led to the *Bank–Laine conjecture* [12], which stated that

$$\max\{\lambda(f_1), \lambda(f_2)\} = \infty$$

whenever $\rho(A)$ is finite and not a positive integer. This conjecture attracted a lot of interest and served as a catalyst for many new research results. Bergweiler and Eremenko [26, 27] showed that the Bank–Laine conjecture is false. Langley [178] showed that the Bank–Laine conjecture holds when the set of finite critical and asymptotic values of the coefficient function $A(z)$ is bounded.

For the version of Theorem 9.3.4 when $A(z)$ is meromorphic, see [13], where the fundamental differences between the meromorphic and entire cases are explained. For a polynomial coefficient $A(z)$, we have the following result.

Theorem 9.3.5 ([12]). *Let f_1 and f_2 be linearly independent solutions of* (9.3.1), *where $A(z)$ is a nonconstant polynomial. If $E = f_1 f_2$, then*

$$\rho(E) = \lambda(E) = \frac{\deg A + 2}{2}.$$

Proof. Since $A(z)$ is a nonconstant polynomial, we know that $A(z) \to \infty$ as $z \to \infty$. Hence from equation (9.3.3) we see that $E(z)$ cannot be a polynomial. Then by applying Wiman–Valiron theory to (9.3.3), we can deduce that $\rho(E) = (k+2)/2$, where $k = \deg A$.

We now assume that $\lambda(E) < \rho(E)$. Rewrite (9.3.3) in the form

$$\frac{c^2}{E^2} = \left(\frac{E'}{E}\right)^2 - 2\frac{E''}{E} - 4A(z). \tag{9.3.8}$$

Since $A(z)$ is a polynomial, we have $m(r, A) = O(\log r)$ as $r \to \infty$. Then since $m(r, 1/E^2) = 2m(r, 1/E)$, we obtain from (9.3.8) and Lemma 4.6.10 that

$$2m(r, 1/E) = S(r, E) + O(\log r). \tag{9.3.9}$$

By our assumption, we can choose a constant a such that $\lambda(E) < a < \rho(E)$. This implies

$$N(r, 1/E) = o(r^a), \quad r \to \infty. \tag{9.3.10}$$

Since $T(r, E) = T(r, 1/E) + O(1)$ by the first fundamental theorem, from (9.3.9) and (9.3.10) it follows that

$$T(r, E) = o(r^a), \quad r \to \infty,\ r \notin F,$$

where $F \subset [0, \infty)$ is a set of finite linear measure. Then by Lemma 1.2.9,

$$T(r, E) = o(r^a), \quad r \to \infty.$$

It follows that $\rho(E) \le a < \rho(E)$, a contradiction. Therefore, our assumption is false, and we must have $\lambda(E) = \rho(E) = (k+2)/2$. □

We can see that Theorem 9.3.5 also follows from the asymptotic integration results in Section 8.3. For generalizations of Theorem 9.3.5 to higher-order linear differential equations, see [284].

Bank and Laine [12] showed that if f_1 and f_2 are linearly independent solutions of (9.3.1) where $A(z)$ is transcendental and $\rho(A) < 1/2$, then $\lambda(E) = \infty$, where $E = f_1 f_2$. Rossi and Shen proved the following result.

Theorem 9.3.6 ([233, 255]). *Let f_1 and f_2 be linearly independent solutions of* (9.3.1) *where* $1/2 \le \rho(A) < 1$, *and set $E = f_1 f_2$. If $\rho(A) = 1/2$ then $\lambda(E) = \infty$, whereas*

$$\frac{1}{\rho(A)} + \frac{1}{\lambda(E)} \le 2 \quad \text{if } \frac{1}{2} < \rho(A) < 1. \tag{9.3.11}$$

The first inequality in (9.3.11) is sharp; see [26, 27]. For an extension of Theorem 9.3.6, see [234, Theorem 2].

Theorem 9.3.7 ([129, Theorem 2.2]). *Let $\alpha \in (0,1]$, and let A be a transcendental entire function satisfying*

$$T(r,A) \sim \alpha \log M(r,A) \tag{9.3.12}$$

as $r \to \infty$ outside a set G with $\underline{\text{logdens}}(G) = \gamma < 1$. Suppose further that one of the following holds:

$$(1)\ \rho(A) \notin \mathbb{N}, \quad (2)\ \underline{\rho}(A) < \rho(A), \quad (3)\ \rho(A) < \frac{1-\gamma}{2(1-\alpha)},$$

where it is understood that the last fraction equals ∞ when $\alpha = 1$. If E is a product of two linearly independent solutions of (9.3.1), *then*

$$\lambda(E) \ge \frac{1-\gamma}{2(1-\alpha)}.$$

In particular, if $\alpha = 1$, then $\lambda(E) = \infty$.

As observed in [129], if A is Mittag-Leffler's function of order ρ, where

$$\frac{1}{2} < \rho < \frac{2+\pi}{2\pi},$$

then A satisfies (9.3.12) for $\alpha = \frac{1}{\rho\pi}$; see [108, p. 19]. Such functions A are examples of entire functions such that $\rho(A) < \frac{1}{2(1-\alpha)}$, which corresponds to assumption (3).

As for further examples of entire functions A satisfying (9.3.12), let $\rho \in (0, 1/2]$ and $\alpha > 0$, and let A be the entire function f of order ρ in Theorem 7.4.6. Then

$$\log M(r,A) \sim \frac{\alpha\pi r^\rho}{\sin(\rho\pi)},$$

and by (7.4.11),

$$\log|A(re^{i\theta})| \sim \frac{\alpha\pi r^\rho}{\sin(\rho\pi)} \cos(\rho\theta), \quad -\pi < \theta < \pi.$$

Therefore,

$$T(r,A) = m(r,A) \sim \frac{ar^{\rho}}{2\sin(\rho\pi)} \int_{-\pi}^{\pi} \cos(\rho\theta)\, d\theta = \frac{ar^{\rho}}{\rho}.$$

This shows that A satisfies (9.3.12) for $a = \frac{\sin(\rho\pi)}{\rho\pi}$.

Each of the conditions (1)–(3) is necessary since the equation

$$f'' + (e^{z} - 1/16)f = 0$$

has two linearly independent solutions f_1 and f_2 such that $\lambda(f_1 f_2) = 0$; see [166, p. 107]. Here the coefficient $A(z) = e^{z} - 1/16$ satisfies (9.3.12) for $a = 1/\pi$ without an exceptional set, and $\underline{\rho}(A) = \rho(A) = 1$.

The lower bound of $\lambda(E)$ in Theorem 9.3.7 does not depend on $\rho(A)$, and we can see that $\lambda(E)$ can be arbitrarily large, provided only that a is close enough to 1 independently of the magnitude of $\rho(A)$.

In Chapter 7, we compared the growth of $T(r,g)$ and $\log M(r,g)$ for an entire function g. We now recall that *Petrenko's deviation of g with respect to ∞* [129] is the quantity

$$\beta^{+}(\infty,g) = \limsup_{r\to\infty} \frac{\log M(r,g)}{T(r,g)}.$$

It is clear that $1 \le \beta^{+}(\infty,g) \le \infty$, where the extreme values can be reached, since $\beta^{+}(\infty, \exp(e^{z})) = \infty$ by Example 4.6.2 and $\beta^{+}(\infty,g) = 1$ for functions g in Sections 7.1 and 7.2. We then obtain the following immediate consequence of Theorem 9.3.7.

Corollary 9.3.8. *Let A be a transcendental entire function, and set $a = 1/\beta^{+}(\infty,A)$. Suppose further that one of* (1)–(3) *with $\gamma = 0$ in Theorem* 9.3.7 *holds. If E is a product of two linearly independent solutions of* (9.3.1), *then*

$$\lambda(E) \ge \frac{\beta^{+}(\infty,A)}{2(\beta^{+}(\infty,A) - 1)}.$$

In particular, if $\beta^{+}(\infty,A) = 1$, then $\lambda(E) = \infty$.

9.4 An example of Bergweiler about c. r. g

Note that Theorem 8.6.2 is related to a question posed by Gol'dberg and Ostrovskii in 1994 [103, p. 300], [89, Question 5.5].

Question 9.4.1. If the coefficients $A_0(z), \ldots, A_{n-1}(z)$ are of c. r. g., then is it true that any transcendental finite-order solution of (9.0.1) will also be of c. r. g.?

In a recent paper [25], Bergweiler answered this question in the negative.

Theorem 9.4.2 ([25]). *For $0 < \sigma < \rho < 1$, there exist entire functions $A(z)$ and $B(z)$ of order ρ and of c. r. g., and an entire function f of order σ and lower order 0, such that*

$$f'' + A(z)f' + B(z)f = 0. \tag{9.4.1}$$

Since f is not of r. g., it is not of p. r. g., and by Lemma 7.5.2, it is not of c. r. g. either. In a private e-mail correspondence, Bergweiler has suggested the following more general form of Theorem 9.4.2, which turns out to be a consequence of Theorem 9.4.2.

Corollary 9.4.3. *For $0 < \sigma < \rho < \infty$, there exist entire functions $A(z)$ and $B(z)$ of order ρ and of c. r. g., and an entire function f of order σ and of lower order 0, such that* (9.4.1) *holds.*

Proof. By Theorem 9.4.2, we may suppose that $\rho \geq 1$. Let n be the smallest integer such that $\rho < n$, and define $\sigma_0 = \sigma/n$. Let f_0 be the function f in the proof of Theorem 9.4.2 with $\rho(f_0) = \sigma_0$. Then the function $f(z) = f_0(z^n)$ is entire, and $\rho(f) = n\rho(f_0) = \sigma$. Then construct the coefficients $A(z)$ and $B(z)$ of order ρ and of c. r. g. as in the proof of Theorem 9.4.2, with the aid of the "new" function f. □

The discussion above leads us to state a question, see [117] and [190].

Question 9.4.4. If the coefficients $A_0(z), \ldots, A_{n-1}(z)$ are exponential polynomials, then is it true that all transcendental finite-order solutions of (9.0.1) are of c. r. g.?

Recalling from Section 7.5 that exponential polynomials are of c. r. g., we see that the assumption in Question 9.4.4 is stronger than that in Question 9.4.1.

We now proceed to give the proof of Theorem 9.4.2. The mathematical ideas come from the original work [25], but more details have been added for the convenience of the reader. To begin with, we state an auxiliary result on meromorphic functions with prescribed principal parts.

Lemma 9.4.5. *Let $\{u_n\}$ be an infinite bounded sequence in $\mathbb{C}\setminus\{0\}$, and let $\{z_n\}$ be an infinite sequence in $\mathbb{C} \setminus \{0\}$ with exponent of convergence $\sigma \in (0,1)$. Then*

$$g(z) = \sum_{n=1}^{\infty} \frac{u_n}{z - z_n} \tag{9.4.2}$$

is a meromorphic function having its poles precisely at the points z_n. Moreover, the proximity function of g satisfies $m(r,g) = o(1)$.

By the assumptions in Lemma 9.4.5, it follows that

$$\sum_{n=1}^{\infty} \left| \frac{u_n}{z_n} \right| < \infty. \tag{9.4.3}$$

Therefore, Lemma 9.4.5 is a special case of Theorem 6.1 in [82, p. 252].

Proof of Lemma 9.4.5. First, we show that g in (9.4.2) is a meromorphic function. Let $R > 0$ be a fixed positive constant, and set

$$g(z) = \sum_{|z_n|<2R} \frac{u_n}{z - z_n} + \sum_{|z_n|\ge 2R} \frac{u_n}{z - z_n} =: S_1(z) + S_2(z).$$

The expression $S_1(z)$ is a finite sum and represents a rational meromorphic function. We now show that $S_2(z)$ converges uniformly on $|z| \le R$. By using (9.4.3), we obtain that for $|z| \le R$,

$$\begin{aligned}|S_2(z)| &\le \sum_{|z_n|\ge 2R} \frac{|u_n|}{|z - z_n|} = \sum_{|z_n|\ge 2R} \frac{|u_n|}{|z_n||1 - \frac{z}{z_n}|} \\ &\le 2 \sum_{|z_n|\ge 2R} \left|\frac{u_n}{z_n}\right| < \infty,\end{aligned}$$

which implies that $S_2(z)$ converges uniformly on $|z| \le R$ by the M-test. Hence $S_2(z)$ is analytic on $|z| \le R$. Since R can be arbitrarily large, it follows that g is a meromorphic function that has poles precisely at the points z_n.

It remains to prove that $m(r, g) = o(1)$. To this end, let $p \in (0, 1)$. Since $\log^+ x \le x^p/p$ for $x \ge 0$, it suffices to prove that

$$\int_0^{2\pi} |g(re^{i\theta})|^p \, d\theta = o(1), \quad r \to \infty. \tag{9.4.4}$$

By Kolokolnikov's inequality [48, Lemma 3.1.2],

$$\int_0^{2\pi} |g(re^{i\theta})|^p \, d\theta = O\left(\frac{n(R, g)}{r}\right), \quad 0 < r < R,$$

where $n(R, g)$ denotes the number of poles z_n of g in $|z| \le R$. Since the exponent of convergence of $\{z_n\}$ is $\sigma \in (0, 1)$, we obtain (9.4.4) by choosing $R = 2r$. □

The above proof of (9.4.4) uses Kolokolnikov's inequality, which is a deep result. We next give an alternate proof, which does not appeal to Kolokolnikov's inequality.

Alternate proof of (9.4.4). We break the function g into parts as follows:

$$g(z) = \left(\sum_{|z_n|<r} + \sum_{|z_n|=r} + \sum_{r<|z_n|}\right) \frac{u_n}{z - z_n} =: g_1(z) + g_2(z) + g_3(z), \quad |z| = r.$$

Next, we suppose that $p \in (\sigma, 1)$ instead of $p \in (0, 1)$. It suffices to prove that

$$\int_0^{2\pi} |g_j(re^{i\theta})|^p \, d\theta = o(1), \quad r \to \infty, \; j = 1, 2, 3, \tag{9.4.5}$$

because (9.4.4) then follows from the fact that

$$|g(z)|^p \le |g_1(z)|^p + |g_2(z)|^p + |g_3(z)|^p,$$

which is a consequence of the following technical lemma. □

Lemma 9.4.6 ([61, p. 57]). *If $a > 0$, $b > 0$, and $0 < p < 1$, then*

$$(a+b)^p \le a^p + b^p.$$

Proof. Define

$$h(x) = \frac{(1+x)^p}{1+x^p}, \quad x \ge 1.$$

Since

$$h'(x) = p(1+x)^{p-1}(1+x^p)^{-2}(1-x^{p-1}) \ge 0,$$

we find that h is a nondecreasing function. In addition, since $h(x) \to 1$ as $x \to \infty$, it follows by continuity that $h(x) \le 1$ for $x \ge 1$. Without loss of generality, we may suppose that $b \ge a$. Then by the properties of h,

$$(a+b)^p = a^p(1+(b/a))^p \le a^p(1+(b/a)^p) = a^p + b^p,$$

and the statement is now proved. □

Proof of (9.4.5). First consider g_2. We focus on one point z_n at a time and use the estimate $\cos\varphi \le 1 - \varphi^2/2 + \varphi^4/24$. Since now $|z| = |z_n|$, we have

$$\begin{aligned}\int_0^{2\pi} \frac{d\theta}{|z-z_n|^p} &= \frac{1}{|z_n|^p}\int_{-\pi}^{\pi} \frac{d\varphi}{|1-e^{i\varphi}|^p} = \frac{1}{|z_n|^p}\int_{-\pi}^{\pi} \frac{d\varphi}{(2-2\cos\varphi)^{p/2}} \\ &= \frac{2}{|z_n|^p}\int_0^{\pi} \frac{d\varphi}{(2-2\cos\varphi)^{p/2}} \le \frac{2}{|z_n|^p}\int_0^{\pi} \frac{d\varphi}{\varphi^p(1-\varphi^2/12)^{p/2}} \\ &\le \frac{2}{|z_n|^p}\int_0^{\pi} \frac{d\varphi}{\varphi^p(1-\pi^2/12)^{p/2}} \le \frac{2\cdot 3^p}{|z_n|^p}\int_0^{\pi} \frac{d\varphi}{\varphi^p} = \frac{K}{|z_n|^p},\end{aligned}$$

where $K = \frac{2\cdot 3^p\cdot \pi^{1-p}}{1-p}$. Therefore, since $\{u_n\}$ is a bounded sequence, there exists a constant $C > 0$ such that

$$\int_0^{2\pi} |g_2(re^{i\theta})|^p\, d\theta \le C \sum_{|z_n|=r} \int_0^{2\pi} \frac{d\theta}{|z-z_n|^p} \le CK \sum_{|z_n|=r} \frac{1}{|z_n|^p} = o(1), \quad r \to \infty.$$

This proves (9.4.5) for $j = 2$.

Next consider g_3, in which case $r < |z_n| =: r_n$. Now

$$\begin{aligned}\int_0^{2\pi}\frac{d\theta}{|z_n - z|^p} &= \frac{1}{|z_n|^p}\int_{-\pi}^{\pi}\frac{d\varphi}{|1-\frac{r}{r_n}e^{i\varphi}|^p}\\ &= \frac{1}{|z_n|^p}\left(\int_{|\varphi|\le 1-r/r_n} + \int_{1-r/r_n<|\varphi|\le\pi}\right)\frac{d\varphi}{|1-\frac{r}{r_n}e^{i\varphi}|^p}\\ &=: \frac{1}{|z_n|^p}(I_1+I_2).\end{aligned}$$

It is clear that

$$I_1 \le \frac{1}{(1-\frac{r}{r_n})^p}\int_{|\varphi|\le 1-r/r_n} d\varphi = 2\left(1-\frac{r}{r_n}\right)^{1-p} \le 2. \tag{9.4.6}$$

If $1-\frac{r}{r_n} < |\varphi| \le \pi$, then the estimate $\cos\varphi \le 1-\varphi^2/2+\varphi^4/24$ gives

$$\begin{aligned}\left|1-\frac{r}{r_n}e^{i\varphi}\right|^2 &= 1-2\frac{r}{r_n}\cos\varphi + \left(\frac{r}{r_n}\right)^2 = \left(1-\frac{r}{r_n}\right)^2 + 2\frac{r}{r_n}(1-\cos\varphi)\\ &\ge \left(1-\frac{r}{r_n}\right)^2 + \frac{r}{r_n}\varphi^2\left(1-\frac{\varphi^2}{12}\right) \ge \left(1-\frac{r}{r_n}\right)^2 + \frac{r}{9r_n}\varphi^2\\ &= \left(1-\frac{r}{r_n}\right)^2\left(1+\frac{r}{9r_n}\left(\frac{\varphi}{1-\frac{r}{r_n}}\right)^2\right).\end{aligned}$$

By a change of variable $x = \frac{\varphi}{1-\frac{r}{r_n}}$, $dx = \frac{d\varphi}{1-\frac{r}{r_n}}$, it follows that

$$\begin{aligned}I_2 &= 2\int_{1-r/r_n}^{\pi}\frac{d\varphi}{|1-\frac{r}{r_n}e^{i\varphi}|^p}\\ &\le \frac{2}{(1-\frac{r}{r_n})^p}\int_{1-r/r_n}^{\pi}\frac{d\varphi}{\left(1+\frac{r}{9r_n}\left(\frac{\varphi}{1-\frac{r}{r_n}}\right)^2\right)^{p/2}}\\ &= 2\left(1-\frac{r}{r_n}\right)^{1-p}\int_1^{\frac{\pi}{1-\frac{r}{r_n}}}\frac{dx}{(1+\frac{r}{9r_n}x^2)^{p/2}}.\end{aligned}$$

If $0 < \frac{r}{r_n} < \frac{1}{2}$, then

$$I_2 \le 2\left(1-\frac{r}{r_n}\right)^{1-p}\int_1^{2\pi} dx \le 4\pi,$$

whereas if $\frac{1}{2} \leq \frac{r}{r_n} < 1$, then

$$I_2 \leq 2 \cdot (18)^{p/2} \left(1 - \frac{r}{r_n}\right)^{1-p} \int_1^{\frac{\pi}{1-\frac{r}{r_n}}} \frac{dx}{x^p} \leq \frac{2 \cdot \sqrt{18} \cdot \pi^{1-p}}{1-p} \leq \frac{10\pi}{1-p}.$$

In both cases, we have

$$I_2 \leq \frac{10\pi}{1-p}. \tag{9.4.7}$$

From (9.4.6) and (9.4.7), it follows that

$$\begin{aligned}\int_0^{2\pi} |g_3(re^{i\theta})|^p \, d\theta &\leq C \sum_{r<|z_n|} \int_0^{2\pi} \frac{d\theta}{|z_n - z|^p} \\ &\leq C\left(2 + \frac{10\pi}{1-p}\right) \sum_{r<|z_n|} \frac{1}{|z_n|^p} = o(1), \quad r \to \infty.\end{aligned}$$

This proves (9.4.5) for $j = 3$.

It remains to prove (9.4.5) for $j = 1$. We now have $r > |z_n| = r_n$, so that

$$\begin{aligned}\int_0^{2\pi} \frac{d\theta}{|z_n - z|^p} &= \frac{1}{|z|^p} \int_{-\pi}^{\pi} \frac{d\varphi}{|1 - \frac{r_n}{r} e^{i\varphi}|^p} \\ &= \frac{1}{|z|^p} \left(\int_{|\varphi| \leq 1 - r_n/r} + \int_{1 - r_n/r < |\varphi| \leq \pi} \right) \frac{d\varphi}{|1 - \frac{r_n}{r} e^{i\varphi}|^p} \\ &=: \frac{1}{r^p}(J_1 + J_2).\end{aligned}$$

Similarly as in (9.4.6) and (9.4.7), we obtain

$$J_1 \leq 2 \quad \text{and} \quad J_2 \leq \frac{10\pi}{1-p}.$$

This gives us

$$\begin{aligned}\int_0^{2\pi} |g_1(re^{i\theta})|^p \, d\theta &\leq \frac{C}{r^p} \sum_{|z_n|<r} \int_{-\pi}^{\pi} \frac{d\varphi}{|1 - \frac{r_n}{r} e^{i\varphi}|^p} \\ &\leq \frac{C}{r^p}\left(2 + \frac{10\pi}{1-p}\right) n(r, g) = o(1), \quad r \to \infty,\end{aligned}$$

where $n(t, g)$ denotes the number of points z_n in $|z| \leq t$. □

We are now ready to prove the main result of this section.

Proof of Theorem 9.4.2. Let $\{z_n\}$ be any sequence of distinct points in $\mathbb{C} \setminus \{0\}$ tending to ∞ such that the exponent of convergence of $\{z_n\}$ is $\sigma \in (0, 1)$. Then by Proposition 3.5.2 and Theorem 3.5.3,

$$f(z) = \prod_{n=1}^{\infty} \left(1 - \frac{z}{z_n}\right) \tag{9.4.8}$$

is an entire function of order σ and genus $p = 0$, and f has zeros precisely at the points z_n. Later on in the proof, we make a specific choice for the points z_n such that all of the above properties hold and there exists a constant $C > 0$ such that

$$\left|\frac{f''(z_n)}{f'(z_n)^2}\right| \le C, \quad n \in \mathbb{N}. \tag{9.4.9}$$

For now, to maintain the generality of this reasoning, we consider (9.4.9) as an assumption. Set $u_n = -f''(z_n)/f'(z_n)^2$. Since the exponent of convergence of $\{z_n\}$ is $\sigma < 1$, from (9.4.9) we find that

$$\sum_{n=1}^{\infty} \left|\frac{u_n}{z_n}\right| \le \sum_{n=1}^{\infty} \frac{C}{|z_n|} = O(1) + \sum_{|z_n| \ge 1} \frac{C}{|z_n|} \le O(1) + \sum_{|z_n| \ge 1} \frac{C}{|z_n|^{(1+\sigma)/2}} < \infty,$$

which gives (9.4.3). Let g be given by (9.4.2), and set $A_0 = fg$. Near each zero point z_n, the function f has a Taylor expansion

$$f(z) = (z - z_n) f'(z_n) + O((z - z_n)^2).$$

Therefore,

$$A_0(z_n) = u_n f'(z_n) = -\frac{f''(z_n)}{f'(z_n)}, \quad n \in \mathbb{N}.$$

This can be rewritten as

$$f''(z_n) + A_0(z_n) f'(z_n) = 0, \quad n \in \mathbb{N},$$

and hence the function

$$B_0 = -\frac{f'' + A_0 f'}{f}$$

is entire. The three entire functions f, A_0, B_0 defined above are now related to each other by means of

$$f'' + A_0(z) f' + B_0(z) f = 0. \tag{9.4.10}$$

By Lemma 9.4.5 and the first fundamental theorem,

$$T(r,g) = N(r,g) + o(1) = N\left(r, \frac{1}{f}\right) + o(1) \leq T(r,f) + O(1).$$

Hence $\rho(g) \leq \rho(f) = \sigma$ by (7.1.2). Then for all $\varepsilon > 0$,

$$T(r,A_0) \leq T(r,f) + T(r,g) = O(r^{\sigma+\varepsilon}),$$

giving us $\rho(A_0) \leq \sigma$. Moreover, since $\rho(f) = \rho(f') = \rho(f'')$ by Theorem 2.3.1, it follows from (7.1.2) and the first fundamental theorem that $\rho(B_0) \leq \sigma$.

Let H be any entire function of order $\rho \in (\sigma, 1)$ and of c. r. g. such that $h_H(\theta) > 0$ for all $\theta \in \mathbb{R}$; see Theorem 7.5.9. Since $\rho > \sigma$, it follows from [183, pp. 159–160] that the Phragmén–Lindelöf indicator of Hf or Hf' equals $h_H(\theta)$. Since H is of c. r. g., it follows that the functions Hf and Hf' are both of c. r. g.

Define $A = A_0 + Hf$ and $B = B_0 - Hf'$. Since $\rho > \sigma$, it follows that

$$\rho(A) = \rho(B) = \rho(H) = \rho.$$

Moreover, since the indicator of H is positive, we deduce that $h_A(\theta) = h_B(\theta) = h_H(\theta)$ for every $\theta \in \mathbb{R}$. In particular, both A and B are of c. r. g. From (9.4.10) we see that the entire functions f, A, B are related to each other by means of

$$f'' + A(z)f' + B(z)f = 0. \tag{9.4.11}$$

Remark. As pointed out in [25], if the indicator of H is not positive everywhere, then the functions A and B might not be of c. r. g. However, a suitable constant $c \in \mathbb{C} \setminus \{0\}$ exists for which $A = A_0 + cHf$ and $B = B_0 - cHf'$ are of c. r. g. [80]. Moreover, these new functions A and B still satisfy (9.4.11).

Up to this point, (9.4.9) has been an assumption. We now proceed to choose the zeros $\{z_n\}$ of the function f in (9.4.8) in a specific way so that $\rho(f) = \sigma$, $\underline{\rho}(f) = 0$ and that (9.4.9) is valid. Then the reasoning above can be followed up to (9.4.11), which completes the proof. □

Let $\{r_k\}$ be a strictly increasing sequence of positive numbers tending to ∞ such that there exists a strictly increasing sequence $\{n_k\}$ of natural numbers satisfying $n_k \sim r_k^\sigma$ as $k \to \infty$. We assume that the growth of the sequence $\{r_k\}$ is fast enough, so that the asymptotic calculations below will work. For example, $r_k^\sigma = 2^{2^{2^k}} = n_k$ will work.

Consider the formal product

$$f(z) = \prod_{k=1}^{\infty}\left(1 - \left(\frac{z}{r_k}\right)^{n_k}\right). \tag{9.4.12}$$

We proceed to verify that f in (9.4.12) is entire and that it has all the required properties. Then the zeros of f are of the form $\xi = \omega_k r_k$, where $\omega_k^{n_k} = 1$. In other words, on each circle $|z| = r_k$, the product f has n_k equally spaced zeros and no other zeros. Each zero of f is simple, as the zeros of solutions of a second-order linear differential equation (9.4.11) should be.

For $\varepsilon > 0$, set

$$\Sigma_1 = \sum_{k=1}^{\infty} \frac{n_k}{r_k^{\sigma+\varepsilon}} \quad \text{and} \quad \Sigma_2 = \sum_{k=1}^{\infty} \frac{n_k}{r_k^{\sigma-\varepsilon}}.$$

Since

$$\frac{n_k}{r_k^{\sigma+\varepsilon}} \sim \frac{1}{r_k^{\varepsilon}}, \qquad k \to \infty,$$

and since $\sum_{k=1}^{\infty} \frac{1}{r_k^{\varepsilon}} < \infty$ as $\{r_k\}$ tends to ∞ sufficiently fast, it follows that the series Σ_1 converges by the limit comparison test. Similarly, the series Σ_2 diverges. This shows that the zero sequence of f has exponent of convergence $\lambda(f) = \sigma \in (0,1)$ and genus $p = 0$. By Proposition 3.5.2 and Theorem 3.5.3, f is entire, and $\rho(f) = \lambda(f) = \sigma$.

Let $n(r)$ denote the number of zeros ξ of f in the disc $|z| \le r$. Then

$$n_k \le n(r_k) = \sum_{\ell=1}^{k} n_\ell \le k n_k.$$

Using $n_k \sim r_k^{\sigma}$ and the fact that $\{r_k\}$ tends to ∞ sufficiently fast together with Theorem 3.1.6, it follows that

$$\sigma = \lim_{k\to\infty} \frac{\log n(r_k)}{\log r_k} \le \limsup_{r\to\infty} \frac{\log n(r)}{\log r} = \lambda(f) = \sigma.$$

Similarly,

$$n_k \le n(r_{k+1} - 1) = \sum_{\ell=1}^{k} n_\ell \le k n_k,$$

and

$$0 \le \liminf_{r\to\infty} \frac{\log n(r)}{\log r} \le \liminf_{k\to\infty} \frac{\log n(r_{k+1} - 1)}{\log(r_{k+1} - 1)} \le \liminf_{k\to\infty} \frac{\log(k n_k)}{\log(r_{k+1}/2)} = 0.$$

This proves $\underline{\rho}(f) = 0$ by liminf analogues of Theorems 3.1.6 and 3.5.3. Thus it remains to prove the following lemma.

Lemma 9.4.7. *The entire function f in* (9.4.12) *satisfies* (9.4.9).

Proof. Using an inductive argument as in the proof of the estimate (3.3.3), we see that

$$\left|\prod_{j=1}^{k-1}\left(1-\left(\frac{r_j}{z}\right)^{n_j}\right)-1\right| \le \prod_{j=1}^{k-1}\left(1+\left|\frac{r_j}{z}\right|^{n_j}\right)-1$$

for all $k \ge 2$ and $z \ne 0$. This implies

$$\begin{aligned}\left|\prod_{j=1}^{k-1}\left(1-\left(\frac{r_j}{z}\right)^{n_j}\right)-1\right| &\le \exp\left(\sum_{j=1}^{k-1}\log\left(1+\left|\frac{r_j}{z}\right|^{n_j}\right)\right)-1\\ &\le \exp\left(\sum_{j=1}^{k-1}\left|\frac{r_j}{z}\right|^{n_j}\right)-1\end{aligned}$$

for all $k \ge 2$ and $z \ne 0$. Let $\varepsilon > 0$. If $z \in \mathbb{C}$ is such that $|z| \sim r_k$ and $k \to \infty$, then $|z| > r_k/2$ for all k large enough, and it follows that

$$\begin{aligned}\left|\prod_{j=1}^{k-1}\left(1-\left(\frac{r_j}{z}\right)^{n_j}\right)-1\right| &\le \exp\left(\sum_{j=1}^{k-1}\left(\frac{2r_j}{r_k}\right)^{n_j}\right)-1\\ &\overset{n_j\ge 1}{\le} \exp\left(\frac{2(k-1)r_{k-1}}{r_k}\right)-1 < \varepsilon\end{aligned}$$

whenever k is large enough, depending on ε. This is certainly true for the suggested choice $r_k^\sigma = 2^{2^{2^k}} = n_k$, but this is also true for many other choices of the sequence $\{r_k\}$. Hence the above reasoning gives

$$\prod_{j=1}^{k-1}\left(1-\left(\frac{r_j}{z}\right)^{n_j}\right) \sim 1,$$

and, consequently,

$$\begin{aligned}\frac{\prod_{j=1}^{k-1}\left(1-\left(\frac{z}{r_j}\right)^{n_j}\right)}{(-1)^{k-1}\prod_{j=1}^{k-1}\left(\frac{z}{r_j}\right)^{n_j}} &= (-1)^{k-1}\prod_{j=1}^{k-1}\left(\left(\frac{r_j}{z}\right)^{n_j}-1\right)\\ &= \prod_{j=1}^{k-1}\left(1-\left(\frac{r_j}{z}\right)^{n_j}\right) \sim 1.\end{aligned} \tag{9.4.13}$$

By (9.4.13),

$$\begin{aligned}f(z) &\sim \prod_{j=1}^{k}\left(1-\left(\frac{z}{r_j}\right)^{n_j}\right) = \left(1-\left(\frac{z}{r_k}\right)^{n_k}\right)\prod_{j=1}^{k-1}\left(1-\left(\frac{z}{r_j}\right)^{n_j}\right)\\ &\sim \left(1-\left(\frac{z}{r_k}\right)^{n_k}\right)(-1)^{k-1}\prod_{j=1}^{k-1}\left(\frac{z}{r_j}\right)^{n_j}.\end{aligned} \tag{9.4.14}$$

A simple differentiation gives, for every z,

$$zf'(z) = -\sum_{j=1}^{\infty} n_j \left(\frac{z}{r_j}\right)^{n_j} \prod_{k\neq j}\left(1-\left(\frac{z}{r_k}\right)^{n_k}\right),$$

and, consequently, for every $z \neq z_\ell$,

$$\begin{aligned}\frac{zf'(z)}{f(z)} &= \sum_{j=1}^{\infty} n_j \frac{(\frac{z}{r_j})^{n_j}}{(\frac{z}{r_j})^{n_j}-1}\\ &= \sum_{j=1}^{k-1} n_j + \sum_{j=1}^{k-1} \frac{n_j}{(\frac{z}{r_j})^{n_j}-1} + n_k \frac{(\frac{z}{r_k})^{n_k}}{(\frac{z}{r_k})^{n_k}-1} + \sum_{j=k+1}^{\infty} n_j \frac{(\frac{z}{r_j})^{n_j}}{(\frac{z}{r_j})^{n_j}-1}.\end{aligned}$$

Since it is assumed that $|z| \sim r_k$ and $k \to \infty$, we have

$$\frac{zf'(z)}{f(z)} = \sum_{j=1}^{k-1} n_j + n_k \frac{(\frac{z}{r_k})^{n_k}}{(\frac{z}{r_k})^{n_k}-1} + o(1). \tag{9.4.15}$$

Recall that the zeros of f are of the form $\xi = \omega_k r_k$, where $\omega_k^{n_k} = 1$. For

$$z = \omega_k r_k (1 + \zeta/n_k)$$

with $|\zeta| \leq 1$, we have

$$(z/r_k)^{n_k} = (1 + \zeta/n_k)^{n_k} \to e^{\zeta}, \quad k \to \infty. \tag{9.4.16}$$

Moreover, if $|\zeta| \leq 1$, then $z = \xi(1+\zeta/n_k)$ satisfies $|z-\xi| \leq r_k/n_k$. Since we may assume that $\sum_{j=1}^{k-1} n_j = o(n_k)$, it follows by substituting $z = \xi(1+\zeta/n_k)$ into (9.4.15) that

$$\frac{zf'(z)}{f(z)} = \left(\frac{e^{\zeta}}{e^{\zeta}-1} + o(1)\right) n_k, \quad k \to \infty. \tag{9.4.17}$$

Thus for a large k (or for a large $|\xi|$), the disc $D_\xi := \{z : |z-\xi| \leq r_k/n_k\}$ does not contain any zeros of f'.

By (9.4.14) and (9.4.16), if $z \in \partial D_\xi$ and $k \to \infty$, then $|z| \sim r_k$, and

$$|f(z)| \sim \left|1 - \left(\frac{z}{r_k}\right)^{n_k}\right| \cdot \prod_{j=1}^{k-1} \left|\frac{z}{r_j}\right|^{n_j} \sim |1-e^{\zeta}| \cdot \prod_{j=1}^{k-1} \left(\frac{r_k}{r_j}\right)^{n_j},$$

and hence by (9.4.17),

$$|f'(z)| = \frac{1}{|z|} \cdot |f(z)| \cdot \left|\frac{zf'(z)}{f(z)}\right| \sim \frac{1}{r_k} \cdot |1-e^{\zeta}| \cdot \prod_{j=1}^{k-1} \left(\frac{r_k}{r_j}\right)^{n_j} \cdot \frac{|e^{\zeta}| n_k}{|e^{\zeta}-1|}$$

$$= \frac{1}{r_k} \cdot \prod_{j=1}^{k-1} \left(\frac{r_k}{r_j}\right)^{n_j} \cdot |e^{\zeta}| n_k.$$

Hence by Cauchy's differentiation formula,

$$\left|\frac{f''(\xi)}{f'(\xi)^2}\right| = \left|\frac{1}{2\pi i} \int_{\partial D_\xi} \frac{dz}{f'(z)(z-\xi)^2}\right| \leq \frac{n_k}{r_k} \max_{z \in \partial D_\xi} \frac{1}{|f'(z)|}$$

$$\sim \frac{1}{|e^{\zeta}|} \prod_{j=1}^{k-1} \left(\frac{r_j}{r_k}\right)^{n_j} \overset{|\zeta| \leq 1}{\leq} e \prod_{j=1}^{k-1} \left(\frac{r_j}{r_k}\right)^{n_j} = o(1), \quad k \to \infty.$$

This completes the proof of (9.4.9). □

10 Nonlinear differential equations

This chapter is focused on entire solutions of nonlinear differential equations with entire function coefficients. Entire solutions often have independent interest from meromorphic solutions, and the form of some equations prevents the possibility of a solution having a pole. We discuss a few classes of equations, while other classes of equations can be found in the referenced papers.

10.1 Differential equations and trigonometric identities

Some differential equations arise from the consideration of trigonometric identities. For instance, the well-known identity

$$2\sin z\cos z = \sin(2z)$$

suggests asking whether $f(z) = \pm\sin z$ are the only entire solutions of the nonlinear differential equation $2f(z)f'(z) = \sin(2z)$? Yi and Zhang answered this question by proving the following result.

Theorem 10.1.1 ([306]). *The only entire solutions of the differential equation*

$$2f(z)f'(z) = \sin(2z) \tag{10.1.1}$$

are the four solutions $f(z) = \pm\sin z, \pm i\cos z$.

Proof. Let f be an entire solution of equation (10.1.1). Integration of (10.1.1) gives

$$f(z)^2 = -\frac{1}{2}\cos(2z) + C,$$

where C is some constant. Then from the identity $\cos(2z) = 2\cos^2 z - 1$ we obtain

$$f(z)^2 + \cos^2 z = (f(z) - i\cos z)(f(z) + i\cos z) = K, \tag{10.1.2}$$

where K is some constant. If $K = 0$, then from (10.1.2) we get $f(z) = \pm i\cos z$.

Now suppose that $K \neq 0$. Then by (10.1.2) both $f(z) - i\cos z$ and $f(z) + i\cos z$ have no zeros, and since from (10.1.2) it is clear that $\rho(f) = 1$, it follows from (10.1.2) that

$$f(z) - i\cos z = \alpha e^{\mu z} \quad \text{and} \quad f(z) + i\cos z = \beta e^{-\mu z}, \tag{10.1.3}$$

where α, β, μ are constants. Eliminating $f(z)$ in the equations in (10.1.3) gives

$$2i\cos z = ie^{iz} + ie^{-iz} = \beta e^{-\mu z} - \alpha e^{\mu z}.$$

https://doi.org/10.1515/9783112249130-010

It follows that we have one of the following two cases: (a) $\mu = -i, \beta = i, \alpha = -i$, or (b) $\mu = i$, $\alpha = -i, \beta = i$. Then from (10.1.3) we see that case (a) gives $f(z) = -\sin z$ and case (b) gives $f(z) = \sin z$. This proves the theorem. □

The next result gives all the entire solutions of a differential equation related to the identity $\cos^2 z - \sin^2 z = \cos(2z)$.

Theorem 10.1.2 ([94]). *The only entire solutions of the differential equation*

$$f(z)^2 - f'(z)^2 = \cos(2z) \tag{10.1.4}$$

are the four solutions $f(z) = \pm\cos z, \pm i \sin z$.

Proof. Let f be an entire solution of (10.1.4). Differentiation of (10.1.4) gives

$$ff' - f'f'' = -\sin(2z), \tag{10.1.5}$$

and a second differentiation gives

$$(f')^2 + ff'' - (f'')^2 - f'f''' = -2\cos(2z). \tag{10.1.6}$$

We differentiate (10.1.6) to get

$$3f'f'' + ff''' - 3f''f''' - f'f^{(4)} = 4\sin(2z). \tag{10.1.7}$$

Eliminating $\sin(2z)$ from (10.1.5) and (10.1.7), and then simplifying, we obtain

$$f'''(3f'' - f) = f'(4f - f^{(4)} - f''). \tag{10.1.8}$$

We now prove that f'''/f' is an entire function. Suppose that $f'(z_0) = 0$ for some $z_0 \in \mathbb{C}$. From (10.1.8), either $f'''(z_0) = 0$ or $3f''(z_0) = f(z_0)$. Suppose that $f'''(z_0) \neq 0$. Then $3f''(z_0) = f(z_0)$, and from (10.1.6) we obtain $f(z_0)^2 = -9\cos(2z_0)$. But from (10.1.4) we have $f(z_0)^2 = \cos(2z_0)$, which means that $f(z_0) = 0$. Hence $f^2 - (f')^2$ has a multiple zero at z_0, which contradicts (10.1.4) because $\cos(2z)$ has only simple zeros. This contradiction means that $f'''(z_0) = 0$. Since $\sin(2z)$ has only simple zeros, and $f'(z_0) = 0$, it follows from (10.1.5) that $f''(z_0) \neq 0$. Therefore, any zero of f' must be both a simple zero of f' and a zero of f''', which proves that f'''/f' is an entire function.

Next, from (10.1.4) and (10.1.6) we have

$$(f'')^2 + f'f''' = 2f^2 - (f')^2 + ff''. \tag{10.1.9}$$

By dividing equation (10.1.9) by f^2 and then applying Wiman–Valiron theory, we can deduce that $\rho(f) = 1$; see Section 4.4, especially Example 4.4.9.

Now set $\alpha = f'''/f'$. Since α is an entire function and $\rho(f) = 1 = \rho(f')$, we obtain from Lemma 4.6.10 that

$$T(r,a) = m(r,a) = m\left(r, \frac{f'''}{f'}\right) = O(\log r)$$

as $r \to \infty$. Hence a must be a polynomial by Theorem 4.6.3.

Now rewrite $a = f'''/f'$ in the form

$$f''' - af' = 0, \tag{10.1.10}$$

and observe from Proposition 8.3.1 that $\rho(f') = (n+2)/2$, where $n = \deg a$. Since $\rho(f') = 1$, we get $n = 0$, and thus a must be a constant. From $a = f'''/f'$ and (10.1.8) we have

$$f^{(4)} + (3a+1)f'' - (a+4)f = 0. \tag{10.1.11}$$

Since $f^{(4)} = af''$ by (10.1.10), equation (10.1.11) reduces to

$$(4a+1)f'' - (a+4)f = 0. \tag{10.1.12}$$

Differentiating (10.1.12) and using (10.1.10) yield

$$a(4a+1)f' - (a+4)f' = 0,$$

which gives $a(4a+1) = a+4$. Hence $a = \pm 1$.

If $a = 1$, then (10.1.12) gives $f'' - f = 0$, whose general solution is $f(z) = c_1 e^z + c_2 e^{-z}$, where c_1, c_2 are constants. Substituting this expression into (10.1.4), we obtain $4c_1c_2 = \cos(2z)$, which is impossible. Therefore $a = -1$.

In the case where $a = -1$, (10.1.12) gives $f'' + f = 0$, whose general solution is $f(z) = c_1 e^{iz} + c_2 e^{-iz}$, where c_1, c_2 are constants. Substituting this expression into (10.1.4), we obtain $4c_1^2 = 1$, $4c_2^2 = 1$, which shows that (10.1.4) has exactly four entire solutions, namely, $f(z) = \pm\cos z, \pm i \sin z$. □

Regarding Theorem 10.1.2, we mention that L. Gao and J. Gao [75] showed that the equation $f(z)^4 - f'(z)^4 = \cos(2z)$ has exactly eight entire solutions, which are $f(z) = \alpha \cos z$ and $f(z) = \beta \sin z$, where α and β are constants satisfying $\alpha^4 = 1$ and $\beta^4 = -1$. It is natural to consider more general terms on the left and right sides of equation (10.1.4). L. Gao and J. Gao proved the following generalization of Theorem 10.1.2.

Theorem 10.1.3 ([75]). *Let a, b, c, d be nonzero constants with $9ad^2 + 4 \neq 0$. Then the differential equation*

$$f(z)^2 + af'(z)^2 = be^{dz} + ce^{-dz} \tag{10.1.13}$$

possesses entire solutions if and only if either $ad^2 + 1 = 0$ or $ad^2 - 4 = 0$. Moreover, we have the following two situations:

(a) *If* $ad^2 + 1 = 0$, *then the entire solutions of* (10.1.13) *are the functions*

$$f(z) = r_k e^{dz} + s_k e^{-dz} + t_k, \quad k = 1, 2, 3, 4,$$

where the t_k *are the four roots of* $t^4 + bc = 0$, $r_k = b/(2t_k)$, *and* $s_k = c/(2t_k)$.

(b) *If* $ad^2 - 4 = 0$, *then the entire solutions of* (10.1.13) *are the functions*

$$f(z) = u_j e^{dz/2} + v_k e^{-dz/2},$$

where the u_j $(j = 1, 2)$ *are the square roots of* $b/2$, *and where the* v_k $(k = 1, 2)$ *are the square roots of* $c/2$.

Remark. If the condition $9ad^2 + 4 \neq 0$ is removed from the hypothesis in Theorem 10.1.3, then the theorem will not hold, that is, we will not necessarily get conclusion (a) or (b) in Theorem 10.1.3. As an example [75], if $a = -1/9$ and $d = 2$, then $9ad^2 + 4 = 0$, and $f(z) = e^z + e^{-3z}$ is a solution of the equation

$$f(z)^2 - \frac{1}{9}f'(z)^2 = \frac{8}{9}e^{2z} + \frac{8}{3}e^{-2z},$$

which is not the situation in (a) or (b).

In 2004, Yang and Li considered the identity $\sin(3z) = 3\sin z - 4\sin^3 z$, which is the origin of the following result.

Theorem 10.1.4 ([298]). *The only entire solutions of the differential equation*

$$4f(z)^3 + 3f''(z) = -\sin(3z)$$

are

$$f_1(z) = \sin z, \quad f_2(z) = \frac{\sqrt{3}}{2}\cos z - \frac{1}{2}\sin z, \quad f_3(z) = -\frac{\sqrt{3}}{2}\cos z - \frac{1}{2}\sin z.$$

Later in 2023, Z. Gao, Liu, and L. Gao considered the identity $\cos(3z) = \cos^3 z - 3\sin^2 z\cos z$ and proved the next result.

Theorem 10.1.5 ([77]). *The only entire solutions of the equation*

$$f(z)^3 - 3f(z)f'(z)^2 = \cos(3z)$$

have the form $f(z) = c_1e^{iz} + c_2e^{-iz}$, *where* c_1 *and* c_2 *are constants satisfying* $c_1^3 = 1/8$ *and* $c_2^3 = 1/8$.

For an extension of Theorem 10.1.5 in a similar manner to Theorem 10.1.7, see [77]. For results of the same nature and related results, see [187, 188].

We next consider the question of existence of entire solutions of the above differential equations when the coefficients in the equation are perturbed in some manner. For instance, as part of a more general theorem, Zhang and Yi perturbed equation (10.1.1) and proved the following result, which contrasts with Theorem 10.1.1.

Theorem 10.1.6 ([306]). *The differential equation*

$$2f(z)f'(z) = \sin(2z) + p(z), \tag{10.1.14}$$

where $p(z) \not\equiv 0$ is a polynomial, does not possess an entire solution.

The next result is an extension of Theorem 10.1.4.

Theorem 10.1.7 ([94]). *Let $p(z)$ and $q(z)$ be polynomials with $p(z)q(z) \not\equiv 0$, and let $\alpha(z)$ be a nonconstant entire function. If the equation*

$$p(z)f(z)^3 + q(z)f''(z) = -\sin(\alpha(z)) \tag{10.1.15}$$

admits an entire solution f, then $\alpha(z) = 3az+b$ for some $a, b \in \mathbb{C}$, and p and q are constants satisfying $27p = 4q^3a^6$. Moreover, f has the form

$$f(z) = c_1e^{iaz} + c_2e^{-iaz},$$

where c_1 and c_2 are constants satisfying

$$c_1^3 = -e^{2bi}c_2^3, \quad 2ipc_1^3 = -e^{bi}, \quad \textit{and} \quad 3pc_1c_2 = qa^2.$$

Remark. There exist examples of Theorem 10.1.7 in addition to those in Theorem 10.1.4:

(1) $f(z) = 2i \sin z$ is a solution of the equation

$$\frac{1}{2}if(z)^3 - \frac{3}{2}if''(z) = -\sin(3z).$$

(2) $f(z) = e^{iz} + ((1+\sqrt{3}i)/2)e^{-iz}$ is a solution of the equation

$$\frac{1}{2}if(z)^3 + \frac{3i(1+\sqrt{3}i)}{4}f''(z) = -\sin(3z).$$

(3) $f(z) = -\sin z$ satisfies

$$4f(z)^3 + 3f''(z) = -\sin(3z+\pi).$$

For more results of a similar nature to the results in this section, plus related results, see the above references, [44, 76, 186, 191, 202], and the references in all these papers.

10.2 Titchmarsh's equation

Titchmarsh considered the following differential equation in his classic book [273]:

$$f(z)f''(z) = -\sin^2 z. \tag{10.2.1}$$

By using Laguerre's theorem, Titchmarsh [273, pp. 266–268] showed that equation (10.2.1) has no real finite-order entire solutions other than $f(z) = \pm \sin z$. In 2019, Li, Lü, and Yang extended and generalized this result by showing that any entire solution of (10.2.1) must be real and of finite order, and by obtaining the following result.

Theorem 10.2.1 ([189]). *Suppose that f is an entire function satisfying*

$$f(z)f''(z) = p(z)\sin^2 z, \tag{10.2.2}$$

where $p(z) \not\equiv 0$ is a polynomial with real coefficients and real zeros. Then p must be a nonzero constant, and $f(z) = a \sin z$, where a is a constant satisfying $a^2 = -p$.

Proof. Let f be an entire solution of (10.2.2). Since $f \not\equiv 0$, we rewrite (10.2.2) as

$$\frac{1}{f^2} = \frac{f''}{f} \cdot \frac{1}{p(z)\sin^2 z}. \tag{10.2.3}$$

We know that $m(r, f''/f) = S(r,f)$ by Lemma 4.6.10. Also, by (10.2.2), the zeros of f occur at the zeros of $p(z)\sin^2 z$, which means that $N(r, f''/f) \le N(r, 1/f) \le N(r, 1/\sin^2 z) + O(\log r) = O(r)$. Therefore, $T(r, f''/f) = O(r) + S(r,f)$. Then by (10.2.3), we have

$$T(r, 1/f^2) \le O(r) + T(r, 1/\sin^2 z) + S(r,f).$$

Using the first fundamental theorem gives $T(r, 1/f^2) = T(r, f^2) + O(1) = 2T(r,f) + O(1)$ and $T(r, 1/\sin^2 z) = 2T(r, \sin z) + O(1) = O(r)$. This yields $T(r,f) = O(r) + S(r,f)$. Then from Lemma 1.2.9, $T(r,f) = O(r)$ as $r \to \infty$. Hence $\rho(f) \le 1$. Since $\rho(f) = \rho(f'')$, the reverse inequality $\rho(f) \ge 1$ is clear by (10.2.2). Thus $\rho(f) = 1$.

We will now show that the zeros of f must be simple zeros, except perhaps for at most finitely many zeros. To this end, assume that this is not true. Then f has an infinite number of multiple zeros. Since all the zeros of $p(z)$ and $\sin z$ are real, we see from (10.2.2) that all the zeros of f must be real and that all the zeros of f are integer multiples of π except for at most finitely many. Since all the zeros of $\sin z$ are simple and $p(z)$ is a polynomial, it follows from (10.2.2) that all the multiple zeros of f must be double zeros, except for at most a finite number. Thus f has an infinite number of real double zeros. We can assume without loss of generality that f has an infinite number of positive double zeros. Choose a large enough integer $m > 0$ such that $m\pi$ is a double zero of f and $p(z + m\pi)$ is a polynomial whose zeros all lie on the negative real axis. Then $f(z + m\pi)$ has a double zero at $z = 0$. We can also assume that $p(x + m\pi) > 0$ for all $x \in [0, \infty)$. By substituting $g(z) = f(z + m\pi)$ into equation (10.2.2) we see that we can assume that f has a double zero at $z = 0$ and $p(x) > 0$ for all $x \in [0, \infty)$.

We will now show that f is real on the real axis, i. e., is a real entire function. Differentiate (10.2.2) twice and then substitute $z = 0$ into the resulting equation to get

$$f''(0)^2 = 2p(0),$$

where we used $f(0) = f'(0) = 0$. Since $p(0) > 0$, we see that $f''(0)$ must be a nonzero real number. Next, differentiate (10.2.2) three times and substitute $z = 0$ into the resulting equation. This yields

$$4f''(0)f'''(0) = 6p'(0).$$

Since $f''(0)$ and $p'(0)$ are real numbers with $f''(0) \neq 0$, we obtain that $f'''(0)$ is a real number. Then differentiating (10.2.2) four times, substituting $z = 0$ into the resulting equation, and then isolating the term that contains $f''(0)f^{(4)}(0)$, we obtain that $f^{(4)}(0)$ is a real number. By continuing this inductive process, we can deduce that $f^{(k)}(0) \in \mathbb{R}$ for all $k = 0, 1, 2, \ldots$. Hence from the MacLaurin series for f we obtain that f is a real entire function.

Since f has all real zeros from (10.2.2), and since $\rho(f) = 1$, the hypothesis of Laguerre's theorem is satisfied. Therefore, f' has all real zeros, and also, f' is a real entire function with $\rho(f') = 1$. Thus the hypothesis of Laguerre's theorem is satisfied for f' as well.

From the above, f has a double zero at $z = 0$. At $z = \pi$, there are two possibilities from (10.2.2): either $f(\pi) = 0$ or $f(\pi) \neq 0$. We separate these two cases.

Case 1. Suppose first that $f(0) = f'(0) = 0$ and $f(\pi) = 0$. Then by Laguerre's theorem there exists a point x_1 satisfying $0 < x_1 < \pi$ such that $f'(x_1) = 0$. Since $f'(0) = f'(x_1) = 0$, it again follows from Laguerre's theorem that there exists a point x_2 satisfying $0 < x_2 < x_1$ such that $f''(x_2) = 0$. Then from (10.2.2) we have $\sin x_2 = 0$, which is impossible since $0 < x_2 < \pi$. This contradiction shows that case 1 cannot occur.

Case 2. Suppose that $f(0) = f'(0) = 0$ and $f(\pi) \neq 0$. Then by (10.2.2) there exists an integer $k \geq 1$ such that

$$f(\pi) \neq 0, \ldots, f(k\pi) \neq 0, \quad f''(\pi) = \cdots = f''(k\pi) = 0, \quad f((k+1)\pi) = 0.$$

Since $f'(0) = f'((k+1)\pi) = 0$, by Laguerre's theorem there exists a simple zero x_3 of f'' satisfying $0 < x_3 < (k+1)\pi$. By (10.2.2) we must have $x_3 = m\pi$ for some m satisfying $1 \leq m \leq k$. Since $f(\pi) \neq 0, \ldots, f(k\pi) \neq 0$, we have $f(x_3) \neq 0$. Then from (10.2.2) it follows that x_3 must be a double zero of f'', which is a contradiction. Hence case 2 cannot occur either.

Since cases 1 or 2 cannot occur, our original assumption must be false, which means that f has all simple zeros, except for at most finitely many.

It follows from (10.2.2) that the function $\sin z/f$ has at most finitely many poles. Then $N(r, \sin z/f) = O(\log r)$ as $r \to \infty$. From (10.2.2) we have

$$\left(\frac{\sin z}{f}\right)^2 = \frac{1}{p(z)} \cdot \frac{f''}{f}.$$

Since f has finite order, we obtain by Lemma 4.6.10 that $m(r, f''/f) = O(\log r)$ as $r \to \infty$. Hence it follows that $m(r, \sin z/f) = O(\log r)$. Thus $T(r, \sin z/f) = O(\log r)$, which means that $\sin z/f$ is a rational function by Theorem 4.6.3. Thus $f(z) = a(z)\sin z$, where $a(z)$ is a rational function, which when substituted into (10.2.2) gives

$$(a(a'' - a) - p)\sin z + 2aa'\cos z = 0.$$

Since a and p are rational functions, it follows that $a(a'' - a) - p \equiv 0$ and $aa' \equiv 0$. This implies that a and p are constants satisfying $p = -a^2$. □

Equation (10.2.2) has the form

$$ff^{(k)} = H(z), \tag{10.2.4}$$

where $H(z)$ is an entire function with $H(z) \not\equiv 0$. Theorem 10.2.1 (with $p(z)$ a real nonzero constant), Theorem 10.1.1 and other results led C.-C. Yang to raise the following conjecture in a talk at Steklov Institute of Mathematics, Moscow, in 2016 (http://www.mathnet.ru/eng/conf800), which also appeared as Conjecture 1.1 in [283].

C.-C. Yang's Conjecture. *If f is an entire solution of* (10.2.4), *where $H(z)$ is a periodic entire function, then f is a periodic function.*

The following more general conjecture was posed by Liu, Wei, and Yu [192]: If n and k are positive integers, and f is an entire solution of $f^n f^{(k)} = H(z)$, where $H(z)$ is a periodic entire function, then f is a periodic function. For partial progress on this generalized conjecture and variations of these conjectures, see [192, 193]. For other related results concerning Yang's conjecture, see the references in [94, 193].

Historical notes

Edward Charles Titchmarsh (1899–1963) was an English mathematician, who was the Savilian Chair of Geometry at Oxford for thirty years. He made important contributions to various fields in analysis and wrote well-known books on the theory of functions, Fourier integrals, eigenfunction expansions, and the Riemann zeta-function.

10.3 A growth result

More generally than (10.2.4), consider entire solutions of nonlinear *monomial differential equations* of the form

$$f^{n_0}(f')^{n_1}(f'')^{n_2}\cdots(f^{(k)})^{n_k} = H(z), \tag{10.3.1}$$

where $H(z)$ is entire with $H(z) \not\equiv 0$, $k \geq 1$, $n_0 \geq 1$, and $n_k \geq 1$. The following result gives upper and lower bounds for the growth of entire solutions of equations of the form (10.3.1).

Theorem 10.3.1 ([94]). *If f is an entire solution of* (10.3.1), *then*

$$\frac{1}{q}T(r,H) + S(r,f) \le T(r,f) \le \frac{1}{n_0}T(r,H) + S(r,f), \tag{10.3.2}$$

where $q = n_0 + n_1 + \cdots + n_k$. In particular, $\rho(f) = \rho(H)$.

The left inequality in (10.3.2) is the best possible. For instance, $f(z) = e^z$ satisfies (10.3.1) with $H(z) = e^{qz}$, which reduces the left inequality in (10.3.2) to an equality. The results in Theorems 10.2.1 and 10.1.1 also reduce the left inequality in (10.3.2) to an equality. However, there are examples for which the left inequality in (10.3.2) does not reduce to an equality; see Example 10.3.2, where an example of Toppila is used. Furthermore, using the same example of Toppila, Example 10.3.3 shows that the factor $1/n_0$ in the right inequality in (10.3.2) cannot be replaced by $1/n_j$ for any j satisfying $1 \le j \le k$ and $n_j > 0$. We first prove Theorem 10.3.1 and then discuss these examples.

Proof of Theorem 10.3.1. Let f be an entire solution of equation (10.3.1). From (4.6.17) we know that

$$T(r,u') \le T(r,u) + S(r,u) \tag{10.3.3}$$

for any nonconstant entire function u. From (10.3.3) we see that an $S(r,u')$ term is also an $S(r,u)$ term. Therefore, from (10.3.1) we can deduce that

$$T(r,H) \le qT(r,f) + S(r,f), \tag{10.3.4}$$

where $q = n_0 + n_1 + \cdots + n_k$.

Next, we prove a lower bound for $T(r,H)$. Rewriting (10.3.1) in the form

$$f^q = \frac{f^{q-n_0}}{(f')^{n_1}\cdots(f^{(k)})^{n_k}}H(z),$$

we obtain

$$qT(r,f) \le T\left(r, \frac{f^{q-n_0}}{(f')^{n_1}\cdots(f^{(k)})^{n_k}}\right) + T(r,H).$$

By the first fundamental theorem,

$$qT(r,f) \le T(r,w) + T(r,H) + O(1),$$

where

$$w = \frac{(f')^{n_1}\cdots(f^{(k)})^{n_k}}{f^{q-n_0}}.$$

Since $q - n_0 = n_1 + n_2 + \cdots + n_k$, we have

$$w = \left(\frac{f'}{f}\right)^{n_1} \cdots \left(\frac{f^{(k)}}{f}\right)^{n_k}. \tag{10.3.5}$$

Thus $m(r, w) = S(r,f)$, which gives

$$qT(r,f) \le N(r, w) + T(r, H) + S(r,f). \tag{10.3.6}$$

We now estimate $N(r, w)$ in terms of $N(r, 1/H)$. Let z_0 be a pole of w. Then by (10.3.5), z_0 is a zero of f, and by (10.3.1), z_0 is a zero of H. Let α denote the multiplicity of the pole z_0 of w, let s denote the multiplicity of the zero z_0 of f, and let β denote the multiplicity of the zero z_0 of H.

Case 1. First, suppose that $s > k$. Then by (10.3.5),

$$\alpha = \sum_{j=1}^{k} jn_j = kq - \sum_{j=0}^{k-1}(k - j)n_j,$$

and by (10.3.1),

$$\beta = \sum_{j=0}^{k}(s - j)n_j \ge \sum_{j=0}^{k-1}(s - j)n_j \ge \sum_{j=0}^{k-1}(k - j)n_j.$$

Hence

$$\frac{\alpha}{\beta} \le \frac{kq}{\sum_{j=0}^{k-1}(k - j)n_j} - 1.$$

Since $\sum_{j=0}^{k-1}(k - j)n_j \ge kn_0$, this yields

$$\frac{\alpha}{\beta} \le \frac{q}{n_0} - 1, \quad s > k. \tag{10.3.7}$$

Case 2. Second, suppose that $1 \le s \le k$. Then by (10.3.5),

$$\alpha \le \sum_{j=1}^{s-1} jn_j + s\sum_{j=s}^{k} n_j = sq - \sum_{j=0}^{s-1}(s - j)n_j,$$

and by (10.3.1),

$$\beta \ge \sum_{j=0}^{s-1}(s - j)n_j.$$

Hence

$$\frac{\alpha}{\beta} \le \frac{sq}{\sum_{j=0}^{s-1}(s - j)n_j} - 1.$$

Since $\sum_{j=0}^{s-1}(s-j)n_j \geq sn_0$, this yields

$$\frac{\alpha}{\beta} \leq \frac{q}{n_0} - 1, \quad 1 \leq s \leq k. \tag{10.3.8}$$

Since (10.3.7) and (10.3.8) cover all the possible values of s, we obtain

$$\frac{\alpha}{\beta} \leq \frac{q}{n_0} - 1$$

at every pole z_0 of w. It follows that

$$N(r,w) \leq \left(\frac{q}{n_0} - 1\right) N\left(r, \frac{1}{H}\right). \tag{10.3.9}$$

Now substituting (10.3.9) into (10.3.6), we get

$$qT(r,f) \leq \left(\frac{q}{n_0} - 1\right) N\left(r, \frac{1}{H}\right) + T(r,H) + S(r,f).$$

Since $N(r,1/H) \leq T(r,H) + O(1)$, this reduces to

$$T(r,f) \leq \frac{1}{n_0} T(r,H) + S(r,f). \tag{10.3.10}$$

Combining (10.3.4) and (10.3.10) gives (10.3.2), which proves the theorem. □

We now use an example of Toppila in our discussion of the double inequality (10.3.2). From (10.3.2) we have $T(r,H) \leq qT(r,f) + S(r,f)$ for any entire solution f of (10.3.1). As mentioned above, there are examples of (10.3.1) that possess entire solutions satisfying the equality

$$T(r,H) = qT(r,f) + S(r,f). \tag{10.3.11}$$

However, the following example shows that (10.3.11) does not always hold.

Example 10.3.2. This example was presented in [94]. Toppila [276] created an entire function g of order one that satisfies the following strict inequality: For an absolute constant c satisfying $0 < c < 1$,

$$T(r,g') < cT(r,g) \quad \text{for all } r > 0. \tag{10.3.12}$$

Consider the differential equation

$$ff^{(k)} = gg^{(k)}. \tag{10.3.13}$$

Obviously, $f = g(z)$ is an entire solution of (10.3.13). From (10.3.3) we deduce that

$$T(r,g^{(k)}) \leq T(r,g') + S(r,g').$$

Then from (10.3.12) and (10.3.3) we obtain by (10.3.13) with $H = gg^{(k)}$ that

$$T(r,H) = T(r, gg^{(k)}) \le T(r, g') + T(r,g) + S(r,g') \le (1+c)T(r,g) + S(r,g),$$

which shows that (10.3.11) does not hold for this example, because $1 + c < q = 2$.

Although Toppila's function g satisfies (10.3.12) for all values of $r > 0$, an earlier example of Hayman [107] can also be used to show that (10.3.11) does not hold for every equation (10.3.1). See also the later examples of Langley [172], which are discussed at the end of this section.

We next use Toppila's example to show that the factor $1/n_0$ in the right inequality in (10.3.2) cannot be replaced by $1/n_j$ for any j such that $1 \le j \le k$ and $n_j > 0$.

Example 10.3.3 ([94]). Consider the equation

$$f^{n_0}(f')^{n_1}(f'')^{n_2}\cdots(f^{(k)})^{n_k} = g^{n_0}(g')^{n_1}(g'')^{n_2}\cdots(g^{(k)})^{n_k}, \tag{10.3.14}$$

where g is Toppila's transcendental entire function, which satisfies (10.3.12), $k \ge 1$, $n_0 \ge 1$, and $n_k \ge 1$. Obviously, $f = g(z)$ is a solution of (10.3.14). Now let j be any fixed integer among $\{1, 2, \dots, k\}$, where $n_j > 0$. Set

$$H = g^{n_0}(g')^{n_1}(g'')^{n_2}\cdots(g^{(k)})^{n_k} \quad \text{and} \quad q = n_0 + n_1 + \cdots + n_k.$$

From (10.3.3) we can see that

$$\begin{aligned} T(r,H) &\le (q - n_j)T(r,g) + T(r, (g^{(j)})^{n_j}) + S(r,g) \\ &= (q - n_j)T(r,g) + n_j T(r, (g^{(j)})) + S(r,g) \\ &\le (q - n_j)T(r,g) + n_j T(r, g') + S(r,g') + S(r,g). \end{aligned}$$

Then from (10.3.12) we obtain

$$T(r,H) \le (q - n_j + cn_j)T(r,g) + S(r,g), \tag{10.3.15}$$

where $0 < c < 1$ is an absolute constant. Now choose the integers $n_0, n_1, \dots, n_k$ so that

$$\frac{q}{n_j} < 2 - c. \tag{10.3.16}$$

It is easy to see that the integers $n_0, n_1, \dots, n_k$ can be chosen so that (10.3.16) holds, because $2 - c$ is an absolute constant larger than 1, and q/n_j can be made as close to 1 as desired by letting n_j be arbitrarily large and keeping the other integers fixed.

If we now assume that

$$T(r,g) \le \frac{1}{n_j}T(r,H) + S(r,g), \tag{10.3.17}$$

then substituting (10.3.15) into (10.3.17), we obtain

$$T(r,g) \le \left(\frac{q}{n_j} - 1 + c\right)T(r,g) + S(r,g),$$

or

$$\left(2 - c - \frac{q}{n_j}\right)T(r,g) \leq S(r,g),$$

which is a contradiction because $2 - c - q/n_j > 0$ by (10.3.16). This contradiction shows that (10.3.17) does not hold, which means that we cannot replace $1/n_0$ in the right inequality in (10.3.2) with $1/n_j$ for any j satisfying $1 \leq j \leq k$ and $n_j > 0$.

Remark. (1) For a growth result of a similar nature to Theorem 10.3.1 on entire solutions of monomial difference-differential equations, see [75, Theorem 4].

(2) Recalling from Example 10.3.2 that Toppila's function g satisfies $\rho(g) = 1$ and (10.3.12), we mention the following result of Langley [172]: For each constant $\gamma \in (1/2, \infty)$, there exist an entire function h satisfying $\rho(h) = \gamma$ and a constant $K = K(\gamma) > 1$ such that

$$\liminf_{r\to\infty} \frac{T(r,h)}{T(r,h')} \geq K(\gamma).$$

10.4 Binomial differential equations

Consider entire solutions of nonlinear *binomial differential equations* of the form

$$A(z)f^{m_0}(f')^{m_1}\cdots(f^{(p)})^{m_p} + B(z)f^{n_0}(f')^{n_1}\cdots(f^{(k)})^{n_k} = H(z), \tag{10.4.1}$$

where A, B, and H are entire functions such that $A(z)B(z)H(z) \not\equiv 0$,

$$\max\{m_0, m_p\} \geq 1, \quad \max\{n_0, n_k\} \geq 1, \quad \max\{m_0, n_0\} \geq 1, \quad \max\{m_p, n_k\} \geq 1,$$

and where it is assumed that the left-hand side of (10.4.1) does not reduce to

$$(A(z) + B(z))f^{m_0}(f')^{m_1}\cdots(f^{(p)})^{m_p}.$$

An analogous growth result to Theorem 10.3.1 cannot hold for nonlinear binomial equations of the form (10.4.1), since $f(z) = \sin z$ satisfies $f^2 + (f')^2 = 1$.

C.-C. Yang's conjecture in Section 10.2 is about the monomial equation (10.2.4). In contrast, the next example shows that such a property does not hold for nonlinear binomial equations of the form (10.4.1).

Example 10.4.1. The differential equation

$$(f')^2 - ff'' = e^{2z} \tag{10.4.2}$$

is satisfied by $f(z) = ze^z$. Observe that (10.4.2) is a nonlinear binomial equation where the right-hand side has the periodic coefficient $H(z) = e^{2z}$ but where the solution $f(z) = ze^z$ is not periodic.

Example 10.4.1 is contained in Theorem 10.4.5. We now give background leading to Theorem 10.4.5.

In his seminal paper [104], Hayman proved the following result.

Theorem 10.4.2 ([104]). *If f is a meromorphic function such that f and f'' have only a finite number of zeros and poles, then*

$$f(z) = R(z)e^{P(z)}, \tag{10.4.3}$$

where $R(z)$ is a rational function, and $P(z)$ is a polynomial. Of these the only functions for which f and f'' have no zeros are e^{az+b} and $(az+b)^{-n}$, where n is a positive integer, and a and b are constants.

Langley [173] showed that the hypothesis that f has finitely many poles can be removed from Theorem 10.4.2, which extended a result of Mues [212]. For related results, see [24, 173, 174].

Regarding Theorem 10.4.2, Hayman observed ([104, p. 34], [106, p. 77]) that if f has no zeros, then $f = 1/g$, where g is an entire function, and

$$f'' = \frac{2(g')^2 - gg''}{g^3},$$

which leads to the question of when can a differential polynomial $2(g')^2 - gg''$ in an entire function g have no zeros? Mues proved the next result.

Theorem 10.4.3 ([213]). *Let g be an entire function, and let $a \neq 1$ be a constant. If the differential polynomial*

$$gg'' - a(g')^2$$

has no zeros, then we have one of the following three cases:

(a) $g(z) = e^{\alpha z+\beta}$, *where $\alpha \neq 0$ and β are constants.*

(b) $g(z) = Az + B$, *where $A \neq 0$ and B are constants, and $a \neq 0$.*

(c) $g(z) = \alpha z^2 + \beta z + \gamma$, *where α, β, γ are constants satisfying $\beta^2 - 4\alpha\gamma \neq 0$ and $a = 1/2$.*

Mues [213, p. 327] showed that the case $a = 1$ in Theorem 10.4.3 is exceptional by observing that $g(z) = \sin z$ satisfies $gg'' - (g')^2 = -1$ and by giving the next example.

Example 10.4.4 ([213]). Let $h(z)$ be an entire function, and let $w(z)$ be an entire function that satisfies

$$w''(z) = h''(z) + e^{2h(z)}.$$

Then the function g defined by $g = e^{w-h}$ satisfies the equation

$$gg'' - (g')^2 = e^{2w(z)}.$$

These results and examples lead naturally to questions about differential equations of the form $ff'' - a(f')^2 = H(z)$. The next result gives an explicit form of all the entire solutions of equations of the form (10.4.4), where a is allowed to be a polynomial $a(z)$, and where the case $a(z) \equiv 1$ is addressed.

Theorem 10.4.5 ([98]). *Let f be an entire solution of a differential equation of the form*

$$ff'' - a(z)(f')^2 = b(z)e^{2c(z)}, \tag{10.4.4}$$

where $a(z), b(z), c(z)$ are polynomials such that $b(z) \not\equiv 0$ and $c(z)$ is nonconstant. Then we have one of the following situations:

(a) *$b(z)$ is nonconstant, and $f(z) = p(z)e^{c(z)}$, where $p(z)$ is a polynomial.*

(b) *$a(z) \equiv \tau$ for some constant $\tau \neq 1$, $b(z) \equiv \mu$ for some constant $\mu \neq 0$, $c(z) = \lambda z$ for some constant $\lambda \neq 0$, and*

$$f(z) = \alpha e^{\lambda z}, \tag{10.4.5}$$

where α is a constant satisfying $\alpha^2\lambda^2(1-\tau) = \mu$.

(c) *$a(z) \equiv 1$, $b(z) \equiv \mu$ for some constant $\mu \neq 0$, $c(z) = \lambda z^2 + \gamma z$, and*

$$f(z) = \alpha e^{\lambda z^2 + \gamma z}, \tag{10.4.6}$$

where α, λ, γ are constants such that $2\lambda\alpha^2 = \mu$.

(d) *$a(z) \equiv 1$, $b(z) \equiv \mu$ for some constant $\mu \neq 0$, $c(z) = \lambda z$ for some constant $\lambda \neq 0$, and one of the following three cases occur:*

$$f(z) = (\alpha z + \beta)e^{\lambda z}, \tag{10.4.7}$$

where α and β are constants such that $\alpha^2 = -\mu$;

$$f(z) = \alpha e^{2\lambda z} + \beta, \tag{10.4.8}$$

where α and β are constants satisfying $4\alpha\beta\lambda^2 = \mu$; and

$$f(z) = \alpha e^{\gamma z} + \beta e^{(2\lambda - \gamma)z}, \tag{10.4.9}$$

where α, β, γ are constants satisfying

$$\gamma(2\lambda - \gamma) \neq 0 \quad \text{and} \quad 4\alpha\beta(\lambda - \gamma)^2 = \mu.$$

Remark. We make three observations regarding Theorem 10.4.5.

(1) A calculation can be made to verify that case (c) in Theorem 10.4.5 is equivalent to Example 10.4.4 when $h(z)$ is a constant.

(2) Case (b) in Theorem 10.4.5 is contained in Theorem 10.4.3. Case (a) in Theorem 10.4.5 when $a(z) \equiv 0$ is contained in Langley's theorem [173].

(3) Ozawa [224] considered the zeros of differential polynomials of the form $ff'' - a(z)(f')^2$ for particular choices of transcendental entire functions $a(z)$.

The differential polynomials

$$ff'' - a(z)(f')^2, \quad f'f'' - a(z)f^2, \quad ff' - a(z)(f'')^2, \tag{10.4.10}$$

where $a(z)$ is a polynomial, have the common characteristic that they are second-order homogeneous binomial differential polynomials of degree two, where each of f, f', f'' appears in exactly one term. Regarding the first differential polynomial in (10.4.10), Theorem 10.4.5 gives an explicit form of all entire solutions of equation (10.4.4). Regarding the second two differential polynomials in (10.4.10), consider the following two analogous differential equations:

$$f'f'' - a(z)f^2 = b(z)e^{2c(z)}, \tag{10.4.11}$$

$$ff' - a(z)(f'')^2 = b(z)e^{2c(z)}, \tag{10.4.12}$$

where $a(z), b(z), c(z)$ are polynomials such that $b(z) \not\equiv 0$ and $c(z)$ is nonconstant. Discussion and posed questions concerning equations (10.4.11) and (10.4.12) in [98] include the following two questions.

Question 10.4.6. Can explicit forms be found for all the entire solutions of equation (10.4.11)?

Question 10.4.7. Can explicit forms be found for all the entire solutions of equation (10.4.12)?

In other words, is it possible to obtain an analogous result to Theorem 10.4.5 for equation (10.4.11) or (10.4.12)? For partial progress on these questions, see [78, 201, 294].

Functions of the form $f(z) = p(z)e^{c(z)}$, where $p(z) \not\equiv 0$ is a polynomial, are obviously solutions of equations of the form (10.4.11) and (10.4.12). The following examples give other forms of entire solutions of these two equations.

Example 10.4.8 ([98]). Let $\alpha, \beta, \lambda, \mu$ be nonzero constants satisfying $\lambda^3 = \mu^3$ and $\lambda \neq \mu$. Then the function

$$f(z) = \alpha e^{\lambda z} + \beta e^{\mu z}$$

satisfies the equations

$$f'f'' - \lambda^3 f^2 = \alpha\beta\lambda(\mu - \lambda)(\mu + 2\lambda)e^{(\lambda+\mu)z}$$

and

$$ff' - \frac{1}{\lambda^3}(f'')^2 = \frac{\alpha\beta}{\lambda}(\lambda - \mu)(\lambda + 2\mu)e^{(\lambda+\mu)z}.$$

Example 10.4.9 ([98]). If α, β, λ are nonzero constants, then $f(z) = \alpha e^{\lambda z} + \beta$ satisfies the equation

$$ff' - \frac{1}{\lambda^3}(f'')^2 = \alpha\beta\lambda e^{\lambda z}.$$

Example 10.4.10 ([294]). For any constant $K \neq 0$, the function $f(z) = e^{-2z} + Ke^{z}$ satisfies the equation

$$f'f'' - f^2 = -9e^{-4z}.$$

Example 10.4.11 ([78]). The function $f(z) = e^{z} + e^{-2z}$ satisfies the equation

$$ff' + \frac{1}{8}(f'')^2 = \frac{9}{8}e^{2z}.$$

We mention that parallel studies on finding the form of entire solutions of equations of the form

$$\begin{aligned} a(z)ff'' + b(z)(f')^2 &= c(z)e^{2q(z)}, \\ a(z)f'f'' + b(z)f^2 &= c(z)e^{2q(z)}, \end{aligned}$$

where $a(z)$, $b(z)$, $c(z)$, $q(z)$ are polynomials such that $a(z)b(z)c(z) \not\equiv 0$ and $q(z)$ is nonconstant, can be found in [201, 301].

For results and examples related to those in this section, see the above referenced papers, [76, 294, 295], and the references therein. Although it was observed earlier that there cannot exist an analogous result to Theorem 10.3.1 for nonlinear binomial differential equations, we mention that [295, Theorem 1.2] has a similar nature to Theorem 10.3.1 for a particular subclass of equations of the form (10.4.1).

11 Further measures of growth

The most common ways to measure the growth of an entire function f is with the order of growth $\rho(f)$ and with the type $\tau(f)$ when $\rho(f) \in (0,\infty)$, which were introduced in Section 2.1. In this chapter, we discuss some other ways to measure the growth of f.

In Section 11.1, entire functions of zero order are presented in terms of logarithmic order. The concept of logarithmic order is useful for difference operators and q-difference equations. Entire functions of finite φ-order are discussed in Section 11.2. This concept forms a bridge between the logarithmic order of growth and the usual order of growth via the restriction $\log r \le \varphi(r) \le r$.

The next two sections focus on growth parameters that are useful in expressing the growth of entire solutions of linear differential equations in terms of the growth of the coefficients. Section 11.3 is devoted to entire functions of finite hyper-order and of finite (p,q)-order. The various orders of growth that have been discussed so far are all defined in terms of a limit superior. In Section 11.4, we illustrate that the convergence of certain weighted integrals involving either the characteristic function $T(r,f)$ or the logarithmic maximum modulus $\log M(r,f)$ can also be used to express the growth of an entire function f in a natural way.

The last two sections of this chapter discuss the proximate and relative orders, of which the former is a classical topic invented by Valiron in the 1920s, whereas the latter was invented more recently by Bernal in the 1980s.

11.1 Logarithmic order

One possible way to express the growth of entire functions of zero order is by means of logarithmic order and logarithmic type, which we now define.

Definition 11.1.1 ([23, 47]). The *logarithmic order* of growth (log-order) of an entire function f is given by

$$\rho_{\log}(f) = \limsup_{r\to\infty} \frac{\log\log M(r,f)}{\log\log r}$$

when f is nonconstant, and $\rho_{\log}(f) = 0$ when f is a constant. If $\rho_{\log}(f) \in (0,\infty)$, then the *logarithmic type* (log-type) of f is

$$\tau_{\log}(f) = \limsup_{r\to\infty} \frac{\log M(r,f)}{(\log r)^{\rho_{\log}(f)}}.$$

More generally, we say that the log-order of a nondecreasing unbounded function ψ : $(0,\infty) \to (0,\infty)$ is defined by

$$\rho_{\log}(\psi(r)) = \limsup_{r\to\infty} \frac{\log \psi(r)}{\log\log r}.$$

If ψ is nondecreasing and bounded, then we set $\rho_{\log}(\psi) = 0$.

https://doi.org/10.1515/9783112249130-011

Constant functions have log-order zero, whereas nonconstant polynomials have log-order one. There are no entire functions with log-order belonging to the open interval $(0,1)$, but there exist transcendental entire functions of log-order one [47, Theorem 7.3]. Note that finite log-order implies zero usual order, whereas usual order > 0 implies infinite log-order.

Example 11.1.2. Let $\rho > 1$ be an arbitrary constant, and let $\phi(r) = (\log r)^\rho$ for $r > 1$. Then there exists an entire function f satisfying $\log M(r,f) \sim \phi(r)$ by Theorem 7.3.1. This gives $\rho_{\log}(f) = \rho$. In other words, for any $\rho > 1$, there exists an entire function f satisfying $\rho_{\log}(f) = \rho$. Of course, such a function is transcendental, not only because of Theorem 7.3.1, but also because of Lemma 2.1.5.

For an entire function f, the expression $\log M(r,f)$ can be replaced with $T(r,f)$ in the definition of log-order by Lemma 7.1.1 and the proof of Lemma 7.1.2. This convention allows us to consider the log-order of meromorphic functions as well.

The concept of log-order is useful for Askey–Wilson difference operators acting on entire functions [50] and for q-difference equations [286]. We will briefly discuss the latter topic.

Theorem 11.1.3 ([286]). *For $n \in \mathbb{N}$, let $a_0(z), \dots, a_n(z)$ be entire functions of finite log-order, and let $q \in \mathbb{C} \setminus \{0\}$ be such that $|q| \neq 1$. If f is a meromorphic solution of the homogeneous linear q-difference equation*

$$\sum_{k=0}^{n} a_k(z) f(q^k z) = 0 \tag{11.1.1}$$

and if there exists an integer $j \in \{0, \dots, n\}$ such that

$$\rho_{\log}(a_j) > \max_{k \neq j} \rho_{\log}(a_k),$$

then $\rho_{\log}(f) = \rho_{\log}(a_j) + 1$.

Earlier, Bergweiler, Ishizaki, and Yanagihara [32] had proved that a meromorphic solution f of a nonhomogeneous linear q-difference equation corresponding to (11.1.1) with rational coefficients has log-order $\rho_{\log}(f) = 2$.

The proof of Theorem 11.1.3 in [286] is based on a growth estimate for the logarithmic q-difference $m(r, f(qz)/f(z))$ and on an estimate for the minimum modulus of an entire function of finite log-order in the spirit of Theorem 3.8.1.

Example 11.1.4 ([286]). Given a positive integer m, we construct an entire function of log-order m that satisfies a certain q-difference equation.

For $q \in \mathbb{C}$ such that $|q| > 1$, define

$$W_1(z) = 1 - z \quad \text{and} \quad W_2(z) = \prod_{n=1}^{\infty} W_1(z/q^n) = \prod_{n=1}^{\infty}\left(1 - \frac{z}{q^n}\right).$$

Since $\sum_{n=1}^{\infty}|q|^{-n} < \infty$, the product $W_2(z)$ converges uniformly on compact subsets of $\mathbb{C}$ and represents an entire function by Theorem 3.3.10. Moreover, it is clear that $W_2(z)$ satisfies

$$W_2(qz) = \prod_{n=0}^{\infty}\left(1 - \frac{z}{q^n}\right) = (1-z)\prod_{n=1}^{\infty}\left(1 - \frac{z}{q^n}\right) = W_1(z)W_2(z).$$

Since $\rho_{\log}(W_1) = 1$, if follows by Theorem 11.1.3 that $\rho_{\log}(W_2) = 2$. Similarly, we find that the function

$$W_3(z) = \prod_{n=1}^{\infty} W_2(z/q^n) = \prod_{n=1}^{\infty}\prod_{k=1}^{\infty}\left(1 - \frac{z}{q^{n+k}}\right)$$

is entire and satisfies

$$W_3(qz) = \prod_{n=1}^{\infty}\prod_{k=0}^{\infty}\left(1 - \frac{z}{q^{n+k}}\right) = \prod_{n=1}^{\infty}\left(1 - \frac{z}{q^n}\right)\prod_{k=1}^{\infty}\left(1 - \frac{z}{q^{n+k}}\right) = W_2(z)W_3(z),$$

and hence $\rho_{\log}(W_3) = 3$. Proceeding inductively, for any integer $m \geq 2$, the function

$$W_m(z) = \prod_{n=1}^{\infty} W_{m-1}(z/q^n)$$

is entire and satisfies the q-difference equation

$$W_m(qz) = W_{m-1}(z)W_m(z)$$

and has logarithmic order $\rho_{\log}(W_m) = m$.

Next, we will illustrate that the log-order and log-type of an entire function can be expressed in terms of its Taylor coefficients. The situation is somewhat similar to the case of the usual order and type. The proof of Theorem 11.1.5 is a laborious computation and will be skipped since we have already seen several treatments of Taylor coefficients in this book. The details can be found in the original reference [23].

Theorem 11.1.5 ([23]). *If $f(z) = \sum_{n=0}^{\infty} a_n z^n$ is entire, then*

$$\rho_{\log}(f) = 1 + \limsup_{r\to\infty} \frac{\log n}{\log\log(\frac{1}{\sqrt[n]{|a_n|}})}.$$

Moreover, if $\rho = \rho_{\log}(f) \in (1,\infty)$, then

$$\tau_{\log}(f) = \frac{(\rho-1)^{\rho-1}}{\rho^\rho} \limsup_{r\to\infty} \frac{n}{(\log\frac{1}{\sqrt[n]{|a_n|}})^{\rho-1}}.$$

Example 11.1.6 ([23, Example 5.2]). Let $q \in (0,1)$ and $\alpha > 1$, and let

$$f(z) = \sum_{n=0}^{\infty} a_n z^n, \quad a_n \in \mathbb{C}.$$

Then the following examples in (a), (b), (c) are entire functions with the indicated growth properties:

(a) If $a_n = q^{n^{\frac{\alpha}{\alpha-1}}}$, then f has log-order α and log-type

$$\tau_{\log}(f) = \frac{(\alpha-1)^{\alpha-1}}{\alpha^{\alpha}(\log(1/q))^{\alpha-1}}.$$

(b) If $a_n = q^{e^n}$, then f has log-order one. The log-type of f does not emerge from Theorem 11.1.5, but it can be calculated directly. Indeed, if r satisfies $\log r = e^k$ for some large $k \in \mathbb{N}$, then

$$\frac{\log M(r,f)}{\log r} = \frac{\log \sum_{n=0}^{\infty} q^{e^n} r^n}{\log r} \geq \frac{\log\left(q^{e^k} r^k\right)}{\log r} = k - \log\frac{1}{q}.$$

By letting $k \to \infty$, we find that f has log-type infinity (because $\rho_{\log}(f) = 1$).

(c) If $a_n = q^{n\log^2 n}$, then f has infinite log-order. By using Theorem 2.2.2, we see that the usual order of f is $\rho(f) = 0$. Thus there are entire functions of infinite log-order and zero usual order.

The remaining part of this section is devoted to comparing the minimum modulus $L(r,f)$ to the maximum modulus $M(r,f)$ of entire functions f of finite log-order.

Theorem 11.1.7 ([37, Theorem 3.6.1]). *Let f be an entire function satisfying*

$$\log M(r,f) = O(\log^2 r). \tag{11.1.2}$$

Then $\log L(r,f) \sim \log M(r,f)$ as $r \to \infty$ in a set of upper linear density one.

The maximal growth rate for f in (11.1.2) is log-order two and finite log-type. The following theorem is a new generalization of Theorem 11.1.7 to functions of finite log-order. For relations between densities of sets, we refer to (1.2.4).

Theorem 11.1.8. *Let f be an entire function of log-order $q \in (1,\infty)$. Then there exists a set $J \subset [1,\infty)$ with $\overline{\log\log\operatorname{dens}}(J) \geq 1/q$ and $\overline{\log\operatorname{dens}}(J) = 1$ such that $\log L(r,f) \sim \log M(r,f)$ as $r \to \infty$ in J.*

Four auxiliary results are needed. See [23] for a parallel discussion and recall Definition 3.5.8 for the genus of an entire function.

Lemma 11.1.9 ([37, Lemma 3.5.1]). *Let f be an entire function of genus zero satisfying $f(0) = 1$. Then*

$$N(r) \leq \log M(r,f) \leq N(r) + Q(r), \tag{11.1.3}$$

where

$$Q(r) = r\int_r^{\infty} \frac{n(t)}{t^2}\,dt, \tag{11.1.4}$$

and where $N(r) = N(r,1/f)$ and $n(t) = n(t,1/f)$.

Proof. The first inequality in (11.1.3) follows directly from Jensen's theorem, see (3.2.9). Let $\{z_n\}$ denote the zeros of f, listed according to multiplicities and organized with respect to increasing moduli, and set $r_n = |z_n|$. Then by the Hadamard factorization,

$$f(z) = \prod_{n=1}^{\infty}(1 - z/z_n). \tag{11.1.5}$$

For $0 < r < R$, the Riemann–Stieltjes integration and integration-by-parts formula (1.5.4) give us

$$\frac{n(R)}{R} - \frac{n(r)}{r} + \int_r^R \frac{n(t)}{t^2}\,dt = \int_r^R \frac{dn(t)}{t} = \sum_{r<|z_n|<R} \frac{1}{|z_n|}.$$

Since f is of genus zero, the sum on the right remains bounded as $R \to \infty$, and then the sum tends to zero as $r \to \infty$. Since $\frac{n(R)}{R}$ and $\int_r^R \frac{n(t)}{t^2}\,dt$ on the left are both nonnegative and since $\int_r^R \frac{n(t)}{t^2}\,dt$ is nondecreasing in R, it follows that both quantities remain bounded as $R \to \infty$. Thus we can see that

$$n(r) = O(r) \quad \text{and} \quad \int_r^\infty \frac{n(t)}{t^2}\,dt = o(1), \quad r \to \infty.$$

We can say more than $n(r) = O(r)$. Since we can assume without loss of generality that $n(r) \geq 1$, we have

$$\frac{n(r)}{r} = n(r)\int_r^\infty \frac{dt}{t^2} \leq \int_r^\infty \frac{n(t)}{t^2}\,dt = o(1), \tag{11.1.6}$$

which gives $n(r) = o(r)$. Now for $r = |z|$, we again use Riemann–Stieltjes integration and integrations by parts to obtain

$$\begin{aligned}\log M(r,f) &\leq \log\prod_{n=1}^{\infty}(1 + r/r_n) = \sum_{n=1}^{\infty}\log(1 + r/r_n)\\ &= \int_0^\infty \log(1 + r/t)\,dn(t) = r\int_0^\infty \frac{n(t)}{t(t+r)}\,dt\\ &= r\left(\int_0^r + \int_r^\infty\right)\frac{n(t)}{t(t+r)}\,dt \leq N(r) + Q(r).\end{aligned}$$

This proves the second inequality in (11.1.3). □

For the rest of this section, $Q(r)$ will denote the function in (11.1.4), where $f(z)$ is an entire function of genus zero satisfying $f(0) = 1$. The next auxiliary result is a generalization of [37, Lemma 3.5.8].

Lemma 11.1.10. *Let f be an entire function of genus zero satisfying $f(0) = 1$, and let $q > 1$. Then $\rho_{\log}(f) \le q$ if and only if $\rho_{\log}(Q(r)) \le q-1$.*

Proof. Suppose first that $\rho_{\log}(f) \le q$. Then $N(r)$ has log-order $\le q$ by (11.1.3). Let $\varepsilon > 0$. Assuming that $r > 1$ and $n(r) \ge 1$, and using

$$n(r)\log r = n(r)\int_r^{r^2}\frac{dt}{t} \le \int_r^{r^2}\frac{n(t)}{t}\,dt \le N(r^2),$$

we obtain

$$n(r) \le \frac{N(r^2)}{\log r} \le O(\log^{q-1+\varepsilon} r). \tag{11.1.7}$$

Since $h(t) = (\log t)^{q-1+\varepsilon}t^{-1/2}$ is strictly decreasing for all sufficiently large positive t, we can assume that r is large enough so that $h(t) \le h(r)$ for all $t \ge r$. Using this, (11.1.4), and (11.1.7), we obtain

$$\begin{aligned} Q(r) &= r\int_r^{\infty}\frac{n(t)}{t^2}\,dt = O\left(r\int_r^{\infty}\frac{\log^{q-1+\varepsilon} t}{t^2}\,dt\right) \\ &= O\left(r^{1/2}\log^{q-1+\varepsilon} r\int_r^{\infty}\frac{dt}{t^{3/2}}\right) = O(\log^{q-1+\varepsilon} r). \end{aligned}$$

Since $\varepsilon > 0$ is arbitrary, it follows that $\rho_{\log}(Q(r)) \le q-1$.

Conversely, suppose that $\rho_{\log}(Q(r)) \le q-1$. Then

$$n(r) = rn(r)\int_r^{\infty}\frac{dt}{t^2} \le Q(r) \le O(\log^{q-1+\varepsilon} r),$$

and so

$$N(r) = \int_1^{r}\frac{n(t)}{t}\,dt + O(\log r) = O(\log^{q+\varepsilon} r).$$

The statement $\rho_{\log}(f) \le q$ now follows from the second inequality in (11.1.3). □

The next lemma is a modification of [121, Corollary 3.3].

Lemma 11.1.11. *Let f and g be entire functions with $\rho_{\log}(g) < \rho_{\log}(f) < \infty$, and let $\varphi : (0,\infty) \to (0,\infty)$ be a nondecreasing unbounded function such that $\log\varphi(r) = o(\log\log r)$. Then the set*

$$F = \{r \geq 1 : \log M(r,g)\varphi(r) \leq \log M(r,f)\}$$

satisfies $\overline{\log\log\operatorname{dens}}(F) \geq 1 - \rho_{\log}(g)/\rho_{\log}(f)$ *and* $\overline{\log\operatorname{dens}}(F) = 1$.

Proof. If $\overline{\log\log\operatorname{dens}}(F) < 1 - \rho_{\log}(g)/\rho_{\log}(f)$, then there exists $\varepsilon > 0$ such that

$$\overline{\log\log\operatorname{dens}}(F) < \frac{\rho_{\log}(f) - \rho_{\log}(g) - 2\varepsilon}{\rho_{\log}(f) - \varepsilon} \quad \text{and} \quad \frac{\rho_{\log}(f) - \varepsilon}{\rho_{\log}(g) + \varepsilon} > 1.$$

Set $\alpha = \frac{\rho_{\log}(f)-\varepsilon}{\rho_{\log}(g)+\varepsilon}$. Since $\alpha > 1/(1 - \overline{\log\log\operatorname{dens}}(F))$ and

$$\log M(r,f) < \log M(r,g)\varphi(r) \leq \varphi(r)(\log r)^{\rho_{\log}(g)+\varepsilon}, \quad r \in [R,\infty) \setminus F,$$

for R large enough, we obtain by Lemma 1.2.7 with $\psi(r) = \log\log r$ that there exists R_0 large enough such that

$$\log M(r,f) \leq \varphi(e^{\log^{\alpha} r})(\log r)^{\alpha(\rho_{\log}(g)+\varepsilon)}, \quad r \geq R_0 \geq R.$$

But then $\rho_{\log}(f) \leq \rho_{\log}(f) - \varepsilon$, a contradiction.

If $\overline{\log\operatorname{dens}}(F) < 1$, then we choose $\alpha > 1/(1 - \overline{\log\operatorname{dens}}(F))$ and $\varepsilon > 0$ such that $\rho_{\log}(g) + 2\varepsilon < \rho_{\log}(f)$. Similarly as above, we use Lemma 1.2.7 with $\psi(r) = \log r$ to obtain

$$\log M(r,f) \leq \varphi(r^{\alpha})(\alpha\log r)^{\rho_{\log}(g)+\varepsilon}, \quad r \geq R_0 \geq R.$$

This again leads to $\rho_{\log}(f) \leq \rho_{\log}(f) - \varepsilon$, a contradiction. □

Lemma 11.1.12 ([37, Lemma 3.5.10]). *Let f be an entire function of genus 0 satisfying $f(0) = 1$, and let $\alpha > 1$. Then there exists a nondecreasing function $\Delta(r)$, which tends to ∞ arbitrarily slowly, such that for large values of R,*

$$\log|f(z)| \geq \log M(\alpha R,f) - Q(\alpha R)\Delta(R), \quad |z| = r < R,$$

whenever z lies outside a certain collection of discs whose circular projection $E \subset [0,\infty)$ satisfies $\overline{\operatorname{dens}}(E) = 0$.

Proof. The function f is of the form (11.1.5). Then

$$|f(z)| = \prod_{n=1}^{\infty}|1 - z/z_n| = \prod_{r_n \leq \alpha R}|1 - z/z_n| \cdot \prod_{r_n > \alpha R}|1 - z/z_n| = P_1P_2,$$

where $r_n = |z_n|$ and $r = |z| < R$. Note that $|1 - z/z_n| \geq |1 - r/r_n|$ for all n and recall that $n(r) = o(r)$. Using $-\log(1-x) \geq x$ for $0 < x < 1$ and $n(r) = o(r)$, from (11.1.6) we obtain

$$\begin{aligned}\log P_2 &\geq \log \prod_{r_n > aR} (1 - r/r_n) = \int_{aR}^{\infty} \log(1 - r/t)\, dn(t) \\ &= -n(aR) \log\left(1 - \frac{r}{aR}\right) - r \int_{aR}^{\infty} \frac{n(t)}{t(t-r)}\, dt \\ &\geq \frac{rn(aR)}{aR} - \frac{aR}{a-1} \int_{aR}^{\infty} \frac{n(t)}{t^2}\, dt = \frac{rn(aR)}{aR} - \frac{Q(aR)}{a-1}.\end{aligned}$$

For P_1, we use the Boutroux–Cartan lemma (Lemma 3.9.2): If z lies outside a certain collection of discs the sum of whose radii is $2H$, then

$$\begin{aligned}\log P_1 &\geq n(aR) \log \frac{H}{e} + \log \prod_{r_n \leq aR} r_n^{-1} \\ &= n(aR) \log \frac{H}{e} - \int_0^{aR} (\log t)\, dn(t) \\ &= n(aR) \log \frac{H}{eaR} + N(aR).\end{aligned}$$

Putting these two estimates together and using (11.1.3), we obtain, outside the exceptional discs,

$$\begin{aligned}\log|f(z)| &\geq n(aR)\left(\frac{r}{aR} + \log \frac{H}{eaR}\right) + N(aR) - \frac{Q(aR)}{a-1} \\ &\geq n(aR) \cdot \log \frac{H}{eaR} + \log M(aR, f) - \frac{aQ(aR)}{a-1}.\end{aligned}$$

We choose the exceptional discs in any way such that $H = H(R) = o(R)$. In particular, we choose the discs in such a way that $H = H(R) = o(1)R$, where $o(1)$ is a nonincreasing function of R that approaches zero arbitrarily slowly as $R \to \infty$. Since $H(R) = o(R)$, it follows that if $E \subset [0, \infty)$ denotes the circular projection of these discs, then $\overline{\text{dens}}(E) = 0$. Then for large values of R, the coefficient of $n(aR)$ will be negative. From (11.1.6) we find that $n(r) \leq Q(r)$, and hence

$$\log|f(z)| \geq \log M(aR, f) - Q(aR)\left(\frac{a}{a-1} - \log \frac{H(R)}{eaR}\right).$$

Set

$$\Delta(R) = \frac{a}{a-1} - \log \frac{H(R)}{eaR}. \tag{11.1.8}$$

Since $H(R) = o(1)R$, where the function $o(1)$ is nonincreasing and approaches zero arbitrarily slowly, it follows that $\Delta(R)$ is a nondecreasing function that tends to ∞ arbitrarily slowly as $R \to \infty$. Hence the statement follows. □

Proof of Theorem 11.1.8. We begin by observing that the assumption $\rho_{\log}(f) \in (1,\infty)$ implies that f is transcendental and of zero (usual) order. As such, f has infinitely many zeros z_n by Corollary 3.5.10, which we may assume to be listed according to their multiplicities and organized with respect to increasing modulus.

(1) Suppose that $f(0) = 1$. By Jensen's formula, we easily obtain

$$N(r) \le \log M(r,f).$$

Moreover,

$$n(r) = \frac{n(r)}{\log 2}\int_r^{2r}\frac{dt}{t} \le \frac{1}{\log 2}\int_r^{2r}\frac{n(t)}{t}\,dt \le 2N(2r).$$

From these two estimates together with the assumption $\rho_{\log}(f) < \infty$, it follows that $\int_0^\infty \frac{n(t)}{t^2}\,dt$ converges. Then by Lemma 3.1.4, the sum $\sum_{n=1}^\infty |z_n|^{-1}$ converges, which means that f has genus zero. Hence the previously obtained lemmas involving genus zero can be applied to f.

Let $G \subset [e,\infty)$ denote the circular projection of the discs in Lemma 11.1.12. The portion of the projection in $[0,e]$ is irrelevant. Then $\overline{\operatorname{dens}}(G) = 0$, and by Lemma 1.2.6,

$$0 \le \overline{\log\log\operatorname{dens}}(G) \le \overline{\log\operatorname{dens}}(G) \le \overline{\operatorname{dens}}(G) = 0$$

and $\underline{\log\log\operatorname{dens}}(G^c) = 1$, where G^c denotes the complement of G in $[e,\infty)$. Since

$$\frac{\int_{G^c\cap[e,r]}\frac{dt}{t\log t}}{\log\log r} \le \frac{\int_e^r \frac{dt}{t\log t}}{\log\log r} = 1, \quad r > e,$$

it follows that

$$\lim_{r\to\infty}\frac{\int_{G^c\cap[e,r]}\frac{dt}{t\log t}}{\log\log r} = 1. \tag{11.1.9}$$

Let $Q(r)$ be the function in (11.1.4), let $\varphi : (0,\infty) \to (0,\infty)$ be a nondecreasing unbounded function such that $\log\varphi(r) = o(\log\log r)$, and define

$$F = \{r \ge 1 : Q(r)\varphi(r) \le \log M(r,f)\}.$$

Note that $\log M(r,f) > 0$ for $r \ge 1$ due to the assumption $f(0) = 1$. It is assumed that $\rho_{\log}(f) = q \in (1,\infty)$, and hence $\rho_{\log}(Q(r)) \le q-1$ by Lemma 11.1.10. Then by Lemma 11.1.11, with $Q(r)$ in place of $\log M(r,g)$, we obtain

$$\overline{\log\log\operatorname{dens}}(F) \geq 1 - (q-1)/q = 1/q \quad \text{and} \quad \overline{\log\operatorname{dens}}(F) = 1. \tag{11.1.10}$$

For $\gamma > 1$, define

$$F_\gamma = \{r \geq 1/\gamma : Q(\gamma r)\varphi(\gamma r) \leq \log M(\gamma r, f)\}.$$

Observe that $F_\gamma = \{s/\gamma : s \in F\}$ and $F = \{\gamma s : s \in F_\gamma\}$. We have

$$\frac{\int_{F_\gamma \cap [1/\gamma, r]} \frac{dt}{t}}{\log r} = \frac{\log(\gamma r)}{\log r} \cdot \frac{\int_{F \cap [1,\gamma r]} \frac{dt}{t}}{\log(\gamma r)}$$

and

$$\frac{\int_{F_\gamma \cap [e/\gamma, r]} \frac{dt}{t\log t}}{\log\log r} = \frac{\log\log(\gamma r)}{\log\log r} \cdot \frac{\int_{F \cap [e,\gamma r]} \frac{dt}{t\log t}}{\log\log(\gamma r)}.$$

Thus, using (11.1.10), we obtain

$$\overline{\log\log\operatorname{dens}}(F_\gamma) = \overline{\log\log\operatorname{dens}}(F) \geq 1/q \quad \text{and} \quad \overline{\log\operatorname{dens}}(F_\gamma) = \overline{\log\operatorname{dens}}(F) = 1.$$

Now define $J = G^c \cap F_\gamma \subset [e, \infty)$. Using the set-theoretic identity $A \cap B = A \setminus (A \setminus B)$, we get

$$\begin{aligned} J \cap [e,r] &= (G^c \cap [e,r]) \cap (F_\gamma \cap [e,r]) \\ &= (G^c \cap [e,r]) \setminus ((G^c \cap [e,r]) \setminus (F_\gamma \cap [e,r])) \supset (G^c \cap [e,r]) \setminus ([e,r] \setminus (F_\gamma \cap [e,r])) \\ &= (G^c \cap [e,r]) \setminus (F_\gamma^c \cap [e,r]), \end{aligned}$$

and, consequently,

$$\overline{\log\log\operatorname{dens}}(J) \geq \limsup_{r\to\infty} \left(\frac{\int_{G^c \cap [e,r]} \frac{dt}{t\log t}}{\log\log r} - \frac{\int_{F_\gamma^c \cap [e,r]} \frac{dt}{t\log t}}{\log\log r} \right).$$

Thus by (11.1.9), Problem 1.1.12, and Lemma 1.2.6,

$$\begin{aligned} \overline{\log\log\operatorname{dens}}(J) &\geq \lim_{r\to\infty} \frac{\int_{G^c \cap [e,r]} \frac{dt}{t\log t}}{\log\log r} + \limsup_{r\to\infty} \left(-\frac{\int_{F_\gamma^c \cap [e,r]} \frac{dt}{t\log t}}{\log\log r} \right) \\ &= 1 - \liminf_{r\to\infty} \frac{\int_{F_\gamma^c \cap [e,r]} \frac{dt}{t\log t}}{\log\log r} \geq 1 - (1 - 1/q) = 1/q. \end{aligned}$$

Similarly, we can prove that $\overline{\log\operatorname{dens}}(J) = 1$. Let $\alpha > 1$ and $R = \alpha r$, where $r = |z|$. Then the conclusion of Lemma 11.1.12 reads as

$$\log|f(z)| \geq \log M(\alpha^2 r, f) - Q(\alpha^2 r)\Delta(\alpha r), \quad r \notin G \cup [0, e],$$

where $\Delta(r)$ increases arbitrarily slowly to ∞ as $r \to \infty$. Therefore, we may apply the function $\varphi(r) = \Delta(r)^2$ in the definition of the set F_γ with $\gamma = \alpha^2$. Then

$$
\begin{aligned}
Q(\alpha^2 r)\Delta(\alpha r) \le Q(\alpha^2 r)\Delta(\alpha^2 r) &= \frac{Q(\alpha^2 r)\varphi(\alpha^2 r)}{\sqrt{\varphi(\alpha^2 r)}} \overset{r \in F_\gamma}{\le} \frac{\log M(\alpha^2 r, f)}{\sqrt{\varphi(\alpha^2 r)}} \\
&= o(\log M(\alpha^2 r, f)), \quad r \in J.
\end{aligned}
$$

This gives us

$$
\begin{aligned}
\log M(r,f) \ge \log|f(z)| &\ge (1 - o(1)) \log M(\alpha^2 r, f) \\
&\ge (1 - o(1)) \log M(r,f), \quad |z| = r \in J,
\end{aligned}
$$

and the proof is now complete in the case where $f(0) = 1$.

(2) Now suppose that $f(0) \neq 1$. Then there exist constants $C \in \mathbb{C} \setminus \{0\}$ and $k \in \mathbb{Z}$ such that $g(z) = Cz^k f(z)$ is entire and satisfies $g(0) = 1$. Since $\rho_{\log}(f) = q > 1$ and since polynomials have log-order one, we have $\rho_{\log}(g) = q$. Hence we may apply the reasoning in case (1) to the function g and obtain

$$
\log M(r,g) \ge \log|g(z)| \ge (1 - o(1)) \log M(r,g), \quad |z| = r \in J.
$$

Moreover, $\log|g(z)| = \log|f(z)| + O(\log r)$, $r \in J$, and since f and g are both transcendental, we have $\log r = o(\log M(r,f))$ and $\log r = o(\log M(r,g))$ as $r \to \infty$ by Lemma 2.1.5. Thus the statement for f follows. □

More interesting properties of the log-order will be discussed in Section 11.2 as special cases of the concept of general φ-order. For further properties on the log-order, see [23, 47].

11.2 φ-order of growth

The φ-order was introduced in 2009 for meromorphic functions in the unit disc and was used as a growth indicator for solutions of linear differential equations [54]. In 2021 the corresponding definition of φ-order for meromorphic functions in the complex plane was used in connection with q-difference equations [125].

Definition 11.2.1 ([125]). Let $\varphi : (R_0, \infty) \to (0, \infty)$ be a nondecreasing unbounded function, where $R_0 \ge 0$ is a fixed number. The *φ-order* of a nondecreasing unbounded function $T : (R_0, \infty) \to (0, \infty)$ is the number $\rho_\varphi(T)$ defined by

$$
\rho_\varphi(T) = \limsup_{r \to \infty} \frac{\log T(r)}{\log \varphi(r)}.
$$

If T is nondecreasing and bounded, then we set $\rho_\varphi(T) = 0$.

Definition 11.2.2. For a nonconstant meromorphic function f, the φ-order of f is defined by $\rho_\varphi(f) = \rho_\varphi(T(r,f))$, where $T(r,f)$ is the Nevanlinna characteristic function of f. If f is a constant, then we set $\rho_\varphi(f) = 0$.

Clearly, the classical order $\rho(f)$ and the logarithmic order $\rho_{\log}(f)$ follow as special cases of φ-order by choosing $\varphi(r) = r$ and $\varphi(r) = \log r$, respectively. Thus the φ-order can be considered as a general growth scale between $\rho_{\log}(f)$ and $\rho(f)$ when imposing the global growth restriction

$$\log r \leq \varphi(r) \leq r, \quad r > R_0. \tag{11.2.1}$$

The φ-order provides a continuum of order measures between the classical order and the logarithmic order. In this regard, see also Example 11.1.6(c).

For an entire function f, Lemma 7.1.1 allows us to use $T(r,f)$ and $\log M(r,f)$ interchangeably in the definitions of the usual order and the log-order. The same is true for the φ-order $\rho_\varphi(f)$ of f:

$$\rho_\varphi(f) = \limsup_{r\to\infty} \frac{\log\log M(r,f)}{\log\varphi(r)}, \tag{11.2.2}$$

provided that $\varphi(r)$ is *subadditive*. This means that

$$\varphi(a+b) \leq \varphi(a) + \varphi(b) \quad \text{for all } a, b > R_0.$$

Note that $\varphi(r) = r$ is trivially subadditive for $r > 0$, and since $a + b \leq ab$ for $a, b \geq 2$, it is easy to see that $\varphi(r) = \log r$ is subadditive for $r \geq 2$. Many results involving the φ-order require the subadditivity of $\varphi(r)$ [125]. The next result is well known.

Proposition 11.2.3. *If $g : [0,\infty) \to [0,\infty)$ is a concave function, then g is subadditive.*

Proof. We can see that there is no loss of generality in assuming that $g(0) \geq 0$. From the concave inequality (1.3.2) and the condition $g(0) \geq 0$, we get for $x \geq 0$ and $t \in [0,1]$,

$$g(tx) = g(tx + (1-t)\cdot 0) \geq tg(x) + (1-t)g(0) \geq tg(x).$$

Thus by choosing $t = \frac{a}{a+b}$ or $t = \frac{b}{a+b}$, where $a, b \geq 0$, we obtain

$$\begin{aligned} g(a+b) &= \frac{a}{a+b}g(a+b) + \frac{b}{a+b}g(a+b) \\ &\leq g\left(\frac{a}{a+b}(a+b)\right) + g\left(\frac{b}{a+b}(a+b)\right) \\ &= g(a) + g(b), \end{aligned} \tag{11.2.3}$$

which proves the statement. □

Following Definition 11.2.1, for the rest of this section, we will assume that $\varphi : (R_0, \infty) \to (0, \infty)$ is a nondecreasing unbounded function, where $R_0 \geq 0$ is a fixed number.

The next result contains the usual order case [37, p. 14] and the logarithmic order case [47, Theorem 1.1] as special cases.

Lemma 11.2.4 ([125]). *Suppose that $\varphi(r)$ is differentiable. Let $T : (R_0, \infty) \to (0, \infty)$ be an integrable, nondecreasing, and unbounded function. Then T is of φ-order $\rho \in [0, \infty)$ if and only if the integral*

$$\int_{R_0}^{\infty} \frac{\varphi'(t)T(t)}{\varphi(t)^{\mu+1}}\,dt \tag{11.2.4}$$

is convergent for $\mu > \rho$ and divergent for $\mu < \rho$. Moreover, if $\rho_\varphi(T) = \infty$, then the integral in (11.2.4) is divergent for every $\mu \in \mathbb{R}$.

Proof. We prove the result in three steps.

(1) Suppose that $\rho_\varphi(T) = \rho < \infty$. Let $\varepsilon > 0$ be fixed. For $\mu = \rho + 2\varepsilon > \rho$, there exists $r_\varepsilon > R_0$ such that $T(r) < \varphi(r)^{\rho+\varepsilon}$ for $r > r_\varepsilon$. It follows that

$$\int_{R_0}^{\infty} \frac{\varphi'(t)T(t)}{\varphi(t)^{\mu+1}}\,dt = \int_{R_0}^{r_\varepsilon} \frac{\varphi'(t)T(t)}{\varphi(t)^{\mu+1}}\,dt + \int_{r_\varepsilon}^{\infty} \frac{\varphi'(t)T(t)}{\varphi(t)^{\mu+1}}\,dt \leq O(1) + \int_{r_\varepsilon}^{\infty} \frac{\varphi'(t)}{\varphi(t)^{1+\varepsilon}}\,dt < \infty,$$

that is, the integral (11.2.4) is convergent for $\mu > \rho$. Next, we prove that the integral (11.2.4) is divergent for $\mu < \rho$. If $\mu \leq 0$, then clearly

$$\int_{R_0}^{\infty} \frac{\varphi'(t)T(t)}{\varphi(t)^{\mu+1}}\,dt \geq \int_{R_0}^{\infty} \frac{\varphi'(t)T(t)}{\varphi(t)}\,dt \geq \int_{R_0}^{\infty} \frac{\varphi'(t)}{\varphi(t)}\,dt = \infty.$$

Here we have made the technical assumptions that $\varphi(r) \geq 1$ and $T(r) \geq 1$ for all $r > R_0$, which are allowed because $\varphi(r)$ and $T(r)$ are unbounded and nondecreasing. Suppose then on the contrary to our claim that $\rho > 0$ and the integral (11.2.4) is convergent for some $\mu \in (0, \rho)$. Now

$$\frac{1}{\varphi(r)^\mu} = -\int_r^{\infty} d(\varphi(t)^{-\mu}) = \int_r^{\infty} \frac{\mu\varphi'(t)}{\varphi(t)^{\mu+1}}\,dt.$$

By the assumptions on T, we obtain

$$\frac{T(r)}{\mu\varphi(r)^\mu} = T(r)\int_r^{\infty} \frac{\varphi'(t)}{\varphi(t)^{\mu+1}}\,dt \leq \int_r^{\infty} \frac{\varphi'(t)T(t)}{\varphi(t)^{\mu+1}}\,dt < \infty. \tag{11.2.5}$$

Hence $\rho = \rho_\varphi(T) \le \mu < \rho$, a contradiction. This completes step (1).

(2) Suppose that the integral (11.2.4) is convergent for $\mu > \rho$ and divergent for $\mu < \rho$. We deduce from (11.2.5) that $\rho_\varphi(T) \le \mu = \rho + \varepsilon$ for any positive number ε, and hence $\rho_\varphi(T) \le \rho$. If $\rho_\varphi(T) < \rho$, then we can write $\rho_\varphi(T) = \rho - 2\varepsilon$ for some $\varepsilon > 0$. From step (1) the integral (11.2.4) is convergent for $\mu = \rho_\varphi(T) + \varepsilon = \rho - \varepsilon < \rho$, a contradiction. Thus $\rho_\varphi(T) = \rho$.

(3) Finally, suppose that $\rho_\varphi(T) = \infty$. A simple modification of the reasoning in step (1) shows that (11.2.4) diverges for every $\mu \in \mathbb{R}$. □

An entire function f satisfies $\rho(f') = \rho(f)$ by Theorem 2.3.1. This identity is well known for meromorphic functions as well [299, p. 36]. The identity $\rho_{\log}(f') = \rho_{\log}(f)$ for a meromorphic f is known by [47, Theorem 6.1]. It is therefore desirable to preserve the φ-order in differentiation. To achieve this, we need to impose some restrictions on φ that are automatically true when $\varphi(r) = r$ or $\varphi(r) = \log r$.

Before stating and proving our result, we state an auxiliary result that is a modification of Chuang's inequality, which follows easily from the proof of [307, Theorem 2.6.1] (see inequality (2.6.5) in [307], where K is defined in (2.6.3)).

Lemma 11.2.5. *Let f be a meromorphic function with $f(0) \ne \infty$. Then for $0 < r < R < \infty$, we have*

$$T(r,f) \le \frac{R}{R-r}\left(1 + \log\frac{R}{R-r}\right)T(R,f') + \log^+ R. \tag{11.2.6}$$

Lemma 11.2.6 ([125]). *The equality $\rho_\varphi(f') = \rho_\varphi(f)$ holds for a transcendental meromorphic function f, provided that one of the following two conditions holds:*

(a) *$\varphi(r)$ is subadditive.*

(b) *There exists a nondecreasing function $s : (R_0, \infty) \to (0, \infty)$, where $R_0 \ge 1$, which satisfies the following properties: $r < s(r) \le r^2$ for $r > R_0$,*

$$\rho_\varphi\left(\frac{s(r)}{s(r)-r}\right) = 0, \quad \textit{and} \quad \liminf_{r\to\infty} \frac{\log\varphi(r)}{\log\varphi(s(r))} = 1. \tag{11.2.7}$$

If $\varphi(r)$ is either of the subadditive functions r or $\log r$, then typical choices for $s(r)$ are $s(r) = 2r$ and $s(r) = r^2$, respectively.

Proof of Lemma 11.2.6. If $f(0) = \infty$, then there exists $k \in \mathbb{N}$ such that the function $g(z) = z^k f(z)$ satisfies $g(0) \ne \infty$, and $T(r,g) = (1 + o(1))T(r,f)$ by Lemmas 2.1.5 and 7.1.1. Hence we may suppose that $f(0) \ne \infty$.

(a) Suppose that $\varphi(r)$ is subadditive. Choosing $R = 2r$ in (11.2.6), we have

$$T(r,f) \le 4T(2r,f') + \log^+(2r) = (4 + o(1))T(2r,f').$$

On the other hand, estimate (4.6.16) implies $T(r,f') \leq 3T(r,f)$ for all r outside a possible exceptional set $E \subset (0,\infty)$ of finite linear measure. The set E can be avoided by means of Lemma 1.2.8, giving us

$$T(r,f') \leq 3T(2r,f), \quad r \geq R_0.$$

Since the subadditivity of φ gives $\varphi(2r) \leq 2\varphi(r)$, the equality $\rho_\varphi(f') = \rho_\varphi(f)$ follows from the two obtained inequalities.

(b) Suppose that a function $s(r)$ with the given assumptions exists. Choosing $R = s(r)$ in (11.2.6), we have

$$T(r,f) \lesssim \frac{s(r)}{s(r)-r}\left(1 + \log\frac{s(r)}{s(r)-r}\right)T(s(r),f') + \log^+ r.$$

Then assumptions (11.2.7) give us

$$\rho_\varphi(f) \leq 2\rho_\varphi\left(\frac{s(r)}{s(r)-r}\right) + \limsup_{r\to\infty}\frac{\log T(s(r),f')}{\log\varphi(s(r))}\cdot\frac{\log\varphi(s(r))}{\log\varphi(r)} \leq \rho_\varphi(f').$$

An estimate by Gol'dberg and Grinshtein [48, Corollary 3.2.3] yields

$$\begin{aligned} T(r,f') &\lesssim T(r,f) + \log^+ T(s(r),f) + \log^+\frac{s(r)}{r(s(r)-r)} \\ &\lesssim T(s(r),f) + \log^+\frac{s(r)}{s(r)-r}. \end{aligned}$$

Therefore, $\rho_\varphi(f') \leq \rho_\varphi(f)$, and the equality $\rho_\varphi(f') = \rho_\varphi(f)$ follows. □

Although the restriction $s(r) \leq r^2$ was used in the above proof, the condition $s(r) \leq r^c$ for any $c > 1$ would have worked just as well. It was observed in [125] that auxiliary functions $s(r)$ that grow faster than r^2 are not particularly interesting.

The φ-exponent of convergence for sequences can be defined analogously to the usual exponent of convergence. It turns out that the φ-order of $n(r,a,f)$ equals the φ-exponent of convergence of the a-points of f. See [125] for more details.

Finally, note that an expression for the φ-order of an entire function in terms of its Taylor coefficients does not seem to be known.

11.3 Hyper-order, iterated order, and (p,q)-order

In this section we will recall some growth parameters that are suitable for entire functions of infinite usual order of growth.

Definition 11.3.1 ([222]). The *hyper-order* and *hyper-lower order* of an entire function f are given, respectively, by

$$\rho_2(f) = \limsup_{r\to\infty} \frac{\log\log\log M(r,f)}{\log r} \quad \text{and} \quad \underline{\rho}_2(f) = \liminf_{r\to\infty} \frac{\log\log\log M(r,f)}{\log r}$$

when f is nonconstant, and $\rho_2(f) = 0 = \underline{\rho}_2(f)$ when f is a constant.

It is easy to see that $\rho_2(f) = 0$ whenever f is entire of finite order, and that $\rho_2(\exp(e^z)) = \underline{\rho}_2(\exp(e^z)) = 1$, since $M(r, \exp(e^z)) = \exp(e^r)$.

For meromorphic functions, the hyper-orders in Definition 11.3.1 can be defined by using the characteristic $T(r,f)$ in place of $\log M(r,f)$. For $c \in \mathbb{C} \setminus \{0\}$, the so-called logarithmic difference $m(r, f(z + c)/f(z))$ can be estimated for meromorphic functions f of hyper-order $\rho_2(f) < 1$, which has a crucial impact in the theory of complex difference equations [45, 100].

For entire functions, the quantities $T(r,f)$ and $\log M(r,f)$ can be used interchangeably in the definitions of $\rho_2(f)$ and $\underline{\rho}_2(f)$. The hyper-type and hyper-lower type can also be defined in an obvious way, but these two concepts seem to appear quite rarely.

In general, it is known that the solutions of equations of the form

$$f^{(n)} + A_{n-1}(z)f^{(n-1)} + \cdots + A_1(z)f' + A_0(z)f = 0$$

with entire coefficients of finite order of growth are of finite hyper-order of growth [33, 46, 156]. But what if the coefficients are of finite hyper-order of growth, or what if the coefficients are allowed to grow even faster? One possible answer lies in using the iterated logarithm in the definition of order of growth. This leads to the definition of the *iterated k-order* [33, 243] of a nonconstant entire function f by means of

$$\rho_k(f) = \limsup_{r\to\infty} \frac{\log_{k+1} M(r,f)}{\log r},$$

where $\log_1 x = \log x$, $\log_2 x = \log\log x$, and $\log_{k+1} x = \log(\log_k x)$. It turns out that if the coefficients are entire functions of finite iterated k-order, then the solutions are of finite iterated $(k + 1)$-order [33, 156]. The lower-order counterpart of the iterated k-order can be defined in terms of the limit inferior.

For a simple example, we mention that if $g(z) = \exp(\exp(e^z))$, then $\rho_3(g) = 1$ and $\rho_2(g) = \infty$.

Example 11.3.2. There are entire functions f for which $\rho_k(f) = \infty$ for all $k \in \mathbb{N}$. The idea of the construction lies in defining the *iterated k-exponent of convergence*

$$\lambda_k(f) = \limsup_{r\to\infty} \frac{\log_k N(r, 1/f)}{\log r}$$

for the zeros of a nonconstant function f. Then let f be a Weierstrass product whose number of zeros increases as r increases in such a way that $\lambda_k(f) = \infty$ for all $k \in \mathbb{N}$. By Jensen's theorem, we have $N(r, 1/f) \leq \log M(r,f) + O(\log r)$, completing the construction.

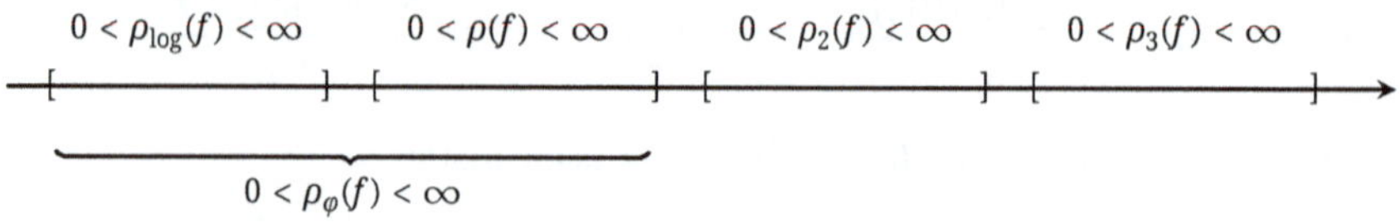

Figure 11.1: The log-order, the usual order, and the iterated k-orders divide entire functions f into "growth boxes" such that there is a gap between two consecutive "boxes". The φ-order closes the gap between the log-order and the usual order. The growth increases to the right, and the figure continues to the right indefinitely. There are nothing but constant functions to the left of the first "box".

A generalization of the type for entire functions of finite iterated k-order of growth (including a variation of the Weierstrass product for such functions) can be found in the paper [149] by Jank and Volkmann.

In addition to being in the numerator, the iterated logarithm can also be applied in the denominator. This leads to the definition of (p,q)-orders as follows.

Definition 11.3.3 ([154]). Let p, q be integers such that $p \geq q \geq 0$, and set $\log_0 x = x$. Then the (p,q)-*order* and (p,q)-*lower order* of an entire function f are given, respectively, by

$$\rho(p,q)(f) = \limsup_{r\to\infty} \frac{\log_p M(r,f)}{\log_q r} \quad \text{and} \quad \underline{\rho}(p,q)(f) = \liminf_{r\to\infty} \frac{\log_p M(r,f)}{\log_q r}$$

if f is nonconstant, and $\rho(p,q)(f) = 0 = \underline{\rho}(p,q)(f)$ if f is a constant.

With Figure 11.1 in mind, it is natural to ask whether the (p,q)-order could be used to close the gaps between two consecutive "growth boxes". Let us take a look at the gap between the usual order and the hyper-order:

$$\begin{aligned} \rho = \rho(f) < \infty \quad &\Rightarrow \quad \log\log M(r,f) \leq (\rho+\varepsilon)\log r, \\ \rho = \rho_2(f) < \infty \quad &\Rightarrow \quad \log\log M(r,f) \leq r^{\rho+\varepsilon}, \\ \rho = \rho(3,2)(f) < \infty \quad &\Rightarrow \quad \log\log M(r,f) \leq (\log r)^{\rho+\varepsilon}, \end{aligned}$$

where $\varepsilon > 0$ is small and $r > 0$ is sufficiently large. From this diagram, we see that the $(3,2)$-order covers part of the gap between the usual order and the hyper-order, but it does not do this continuously in the way the φ-order does between the log-order and the usual order. In the next section, we give a general growth order that enables us to express the growth of entire functions in a continuous manner without dividing functions into "growth boxes".

11.4 (Ψ, φ)-order and integrated (Ψ, ω)-order

The usual order, log-order, hyper-order, iterated k-orders, (p,q)-order, and φ-order are special cases of the following definition. This definition does not divide entire functions

f into "growth boxes" as in the case with some of the aforementioned orders, but offers a continuous way to measure the growth of f instead.

Definition 11.4.1 ([52]). Let $\Psi : [0, \infty) \to [0, \infty)$ and $\varphi : (0, \infty) \to (0, \infty)$ be continuous, nondecreasing, and unbounded functions. Then the (Ψ, φ)-*order* and (Ψ, φ)-*lower order* of an entire function f are given, respectively, by

$$\rho_{\Psi,\varphi}(f) = \limsup_{r\to\infty} \frac{\Psi(\log M(r,f))}{\log \varphi(r)} \quad \text{and} \quad \underline{\rho_{\Psi,\varphi}}(f) = \liminf_{r\to\infty} \frac{\Psi(\log M(r,f))}{\log \varphi(r)}$$

when f is nonconstant, and $\rho_{\Psi,\varphi}(f) = 0 = \underline{\rho_{\Psi,\varphi}}(f)$ when f is a constant.

For a large class of functions $\phi(r) \neq O(\log r)$ of moderate or arbitrarily rapid growth, Theorem 7.3.1 by Clunie says that there exists an entire function f for which $\log M(r,f) \sim \phi(r)$ as $r \to \infty$. By continuity,

$$\frac{\Psi(\log M(r,f))}{\log \varphi(r)} \sim \frac{\Psi(\phi(r))}{\log \varphi(r)}, \qquad r \to \infty.$$

Depending on ϕ, specific functions Ψ and φ can now be chosen so that the above quotient has the desired growth. In layman's terms, the function Ψ presses the gas pedal, whereas φ applies the brakes.

The growth of entire functions can also be expressed in terms of integrals. To begin with, let us consider Lemma 11.2.4 for the usual order and the log-order. An entire function f is of

- finite order ρ if and only if the integral $\int_{R_0}^{\infty} \frac{\log M(r,f)}{r^{\mu+1}}\, dr$ is convergent for $\mu > \rho$ and divergent for $\mu < \rho$;
- finite logarithmic order ρ if and only if the integral $\int_{R_0}^{\infty} \frac{\log M(r,f)}{r \log^{\mu+1} r}\, dr$ is convergent for $\mu > \rho$ and divergent for $\mu < \rho$.

Common to both orders, which are defined in terms of the limit superior, is that a certain weighted integral of $\log M(r,f)$ either converges or diverges. This brings us to the conclusion that other weighted integrals of $\log M(r,f)$ can also be used to measure the growth of f.

Definition 11.4.2 ([52]). Let $\Psi : [0, \infty) \to [0, \infty)$ be a continuous, nondecreasing, and unbounded function, and let $\omega \in L^1(0, \infty)$. Then the *integrated* (Ψ, ω)-*order* of an entire function f is the quantity

$$\int_0^{\infty} \Psi(\log^+ M(r,f))\omega(r)\, dr.$$

The integrated (Ψ, ω)-order has no natural lower order counterpart. The assumption $\omega \in L^1(0, \infty)$ in Definition 11.4.2 guarantees that constant functions are always of finite integrated (Ψ, ω)-order. This differs from the situation with limsup orders, where the order of a constant function is assumed to be zero.

The usual order $\rho(f)$ of an entire function f has several fundamental properties, the equality $\rho(f') = \rho(f)$ being one of them. If any of these properties is desired for the (Ψ, φ)-order or for the integrated (Ψ, ω)-order, then certain restrictions on the functions Ψ, ω, φ need to be assumed [52].

11.5 Proximate order

Let f be an entire function of finite order $\rho > 0$. In the definition of the type, the growth of $\log M(r,f)$ is compared to the growth of r^ρ. This can be further refined. For example, Lindelöf [183] compared the growth of $\log M(r,f)$ to

$$r^\rho (\log r)^{\alpha_1} (\log_2 r)^{\alpha_2} \cdots (\log_p r)^{\alpha_p},$$

where $\log_j r$ denotes the iterated logarithm, and $\alpha_1, \dots, \alpha_p$ are real constants. More generally, it is convenient to define a certain class of slowly varying functions $L(r)$ (to be defined below) and then compare the growth of $\log M(r,f)$ to functions of the form $r^\rho L(r)$. This approach was taken by Valiron, who introduced the concept of proximate order of growth.

Definition 11.5.1 ([183, p. 32], [279, p. 64]). A differentiable function $\rho : (0, \infty) \to (0, \infty)$ satisfying the conditions

$$\lim_{r\to\infty} \rho(r) = \rho \in [0, \infty) \quad \text{and} \quad \lim_{r\to\infty} r\rho'(r) \log r = 0$$

is called a *proximate order*. For an entire function f, if the quantity

$$\sigma_f = \limsup_{r\to\infty} \frac{\log M(r,f)}{r^{\rho(r)}}$$

is different from zero and infinity, then $\rho(r)$ is called a *proximate order of* f, and σ_f is called the type of the function f with respect to the proximate order $\rho(r)$.

The proximate order and the corresponding type of a given entire function f are not uniquely determined. For example, if $\rho(r)$ is a proximate order of f, then $\rho_0(r) = \rho(r) + \frac{\log c}{\log r}$ for $c > 0$ is also a proximate order of f. Moreover,

$$r^{\rho_0(r)} = r^{\rho(r)} \cdot r^{\frac{\log c}{\log r}} = r^{\rho(r)} \exp\left(\frac{\log c}{\log r} \cdot \log r\right) = cr^{\rho(r)},$$

so that the original type is now divided by c.

With respect to an arbitrary proximate order, an entire function f may be of minimal, mean, or maximal type. With respect to its own proximate order, f is always of mean type by definition.

Valiron has proved that every entire function of finite positive order has a proximate order [279, p. 64]. This proof can also be found in Cartwright's book [43, p. 54]. Thus, in this expanded framework, every entire function of finite positive order can be regarded as a function of mean type.

Definition 11.5.2 ([248, p. 1]). A function $L : (0, \infty) \to (0, \infty)$ is called *slowly varying* if

$$\lim_{r\to\infty} \frac{L(\lambda r)}{L(r)} = 1$$

for every $\lambda > 0$.

We state a useful lemma on proximate orders without proof.

Lemma 11.5.3 ([183, pp. 32–33]). *If $\rho(r)$ is a proximate order, then the function $r^{\rho(r)-\rho}$ is slowly varying. In addition, if $\rho > 0$, then the function $r^{\rho(r)}$ is strictly increasing for all sufficiently large values of $r > 0$.*

Denote the slowly varying function $r^{\rho(r)-\rho}$ in Lemma 11.5.3 by $L(r)$. With this notation, we see that the growth of $\log M(r,f)$ in the definition of the proximate order is comparable to that of $r^{\rho}L(r)$, as mentioned above.

From $r^{\rho(r)} = r^{\rho}L(r)$, we get

$$\rho(r) = \rho + \frac{\log L(r)}{\log r}.$$

This gives a better view of what proximate orders look like, and this identity can be used to construct concrete examples of proximate orders.

Proximate orders can be used to generalize the theory of entire functions f of finite positive order. We state two examples in this direction.

(1) A *generalized Phragmén–Lindelöf indicator function* of f is given by

$$h_f(\theta) = \limsup_{r\to\infty} \frac{\log|f(re^{i\theta})|}{r^{\rho(r)}},$$

where $\rho(r)$ is some proximate order of f [183, p. 70]. Using the generalized indicator leads to better estimates for the growth of f in sectors. As an example, we mention Bernstein's estimate for the minimum modulus in [183, p. 73]. Moreover, analogously to the case of the usual indicator function in Corollary 4.2.7, the quantity $\max_{\alpha\le\theta\le\beta} h_f(\theta)$ equals the type σ_f of f inside the angle $\alpha \le \arg z \le \beta$ [183, p. 72].

(2) The concept of angular density in Definition 7.5.5 can be generalized as follows. Let $Z = \{w_n\}$ be an infinite zero sequence of f. We say that Z has the *angular density of index $\rho(r)$* if for all but a countable set of values θ_1, θ_2, the limit

$$\lim_{r\to\infty} \frac{n_Z(r,\theta_1,\theta_2)}{r^{\rho(r)}}$$

exists as a finite number [183, p. 89]. The generalized angular density can be used to obtain better information on the zero distribution of f.

For further properties of proximate orders, we refer the reader to the book [183] of Levin.

11.6 Relative order

Following the notation in [34], we let $F(r)$, $G(r)$, and $H(r)$ denote the respective maximum moduli of given entire functions f, g, and h.

Recall from Lemma 2.1.8 that the usual order $\rho(f)$ of an entire function f can be expressed alternatively as

$$\rho(f) = \inf\left\{\beta > 0 : F(r) \leq e^{r^\beta} \text{ for all } r \geq r(\beta)\right\}.$$

If such a $\beta > 0$ does not exist, then $\rho(f) = \infty$. In this definition, the growth of $F(r)$ is compared to the growth of the exponential function. The closer to one the order of growth is, the more the growth of $F(r)$ resembles the growth of the exponential e^r. The key idea in the concept of *relative order*, introduced by Bernal in 1988, is to replace the exponential function with another function.

Definition 11.6.1 ([34]). If f and g are entire functions, then the *order of f relative to g* is the following number:

$$\rho_g(f) = \inf\left\{\beta > 0 : F(r) \leq G(r^\beta) \text{ for all } r \geq r(\beta)\right\}.$$

If such a $\beta > 0$ does not exist, then $\rho_g(f) = \infty$.

Note that for a nonconstant entire function f, the function $F(r)$ is strictly increasing, continuous, and unbounded. Therefore, it has an inverse function $F^{-1} : (|f(0)|, \infty) \to (0, \infty)$, which is also strictly increasing, continuous, and unbounded.

Theorem 11.6.2 ([34, Teorema 1]). *If f, g, and h are nonconstant entire functions, then the following statements hold:*

(a) $\rho_g(f) = \limsup_{r\to\infty} \frac{\log G^{-1}(F(r))}{\log r} = \limsup_{r\to\infty} \frac{\log G^{-1}(r)}{\log F^{-1}(r)}$.

(b) *If g is a polynomial and f is transcendental, then $\rho_g(f) = \infty$.*

(c) *If f and g are both polynomials, then $\rho_g(f) = \frac{\deg f}{\deg g}$.*

(d) *If $F(r) \leq G(r)$, then $\rho_h(f) \leq \rho_h(g)$.*

(e) *If $G(r) \leq H(r)$, then $\rho_g(f) \geq \rho_h(f)$.*

Proof. Statements (b) and (c) are trivial, whereas (d) and (e) follow from the first identity in (a). Thus it suffices to prove (a).

To prove (a), suppose first that $\rho = \rho_g(f) < \infty$. Let $\varepsilon > 0$. By the definition of relative order, there exists $r(\varepsilon) > 1$ such that

$$G^{-1}(F(r)) \le r^{\rho+\varepsilon}, \quad r \ge r(\varepsilon).$$

This gives us

$$\limsup_{r\to\infty} \frac{\log G^{-1}(F(r))}{\log r} \le \rho + \varepsilon,$$

where we may let $\varepsilon \to 0^+$. Again by the definition of relative order, there exists a strictly increasing sequence $\{r_n\}$ of positive real numbers tending to infinity such that

$$G^{-1}(F(r_n)) \ge r_n^{\rho-\varepsilon}, \quad n \in \mathbb{N}.$$

This yields

$$\limsup_{r\to\infty} \frac{\log G^{-1}(F(r))}{\log r} \ge \limsup_{n\to\infty} \frac{\log G^{-1}(F(r_n))}{\log r_n} \ge \rho - \varepsilon,$$

where we may let $\varepsilon \to 0^+$. This proves the first identity in (a) under the assumption that $\rho < \infty$.

Suppose then that $\rho = \rho_g(f) = \infty$ and assume on the contrary that

$$\lambda := \limsup_{r\to\infty} \frac{\log G^{-1}(F(r))}{\log r} < \infty.$$

Then for any $\varepsilon > 0$, there exists $r(\varepsilon) > 1$ such that

$$F(r) \le G(r^{\lambda+\varepsilon}), \quad r \ge r(\varepsilon).$$

But this implies that $\rho \le \lambda + \varepsilon < \infty$, a contradiction. Thus $\lambda = \infty$, and, consequently, $\rho = \lambda$. This completes the proof of the first identity in (a).

The second identity in (a) follows from the first one by noticing that as $r \to \infty$, we have $F(r) \to \infty$ if and only if $F^{-1}(r) \to \infty$. □

If $g(z) = e^z$, then $G(r) = e^r$, and $G^{-1}(r) = \log r$. Thus the first identity in Theorem 11.6.2(a) reduces to the definition of the usual order of f.

Definition 11.6.3. Let g be a nonconstant entire function. We say that g satisfies property (A) if for any constant $\sigma > 1$, $G(r)^2 \le G(r^\sigma)$ for all $r \ge r(\sigma)$.

The function $g(z) = e^z$ trivially satisfies property (A). If g is a nonconstant polynomial, then it does not satisfy property (A).

The relative order satisfies similar properties as the usual order. Some of these properties are stated in the following two theorems.

Theorem 11.6.4 ([34, Teorema 3]). *Let f, g, f_1, and f_2 be nonconstant entire functions, and let $P(z) \not\equiv 0$ be a polynomial. Then the following statements hold:*

(a) $\rho_g(f_1 + f_2) \leq \max\{\rho_g(f_1), \rho_g(f_2)\}$, *where the equality holds if* $\rho_g(f_1) \neq \rho_g(f_2)$.

(b) *If f is transcendental, then* $\rho_g(Pf) = \rho_g(f)$.

(c) *If g satisfies property (A), then*

$$\rho_g(f_1 f_2) \leq \max\{\rho_g(f_1), \rho_g(f_2)\},$$

where the equality holds if $\rho_g(f_1) \neq \rho_g(f_2)$.

Theorem 11.6.5 ([34, Teorema 4]). *Suppose that f and g are nonconstant entire functions such that at least one of them is transcendental. Then*

$$\rho_g(f') = \rho_{g'}(f) = \rho_{g'}(f') = \rho_g(f).$$

Recall from Theorem 2.2.2 that the order of an entire function $f(z) = \sum_{n=0}^{\infty} a_n z^n$ can be expressed in terms of its Taylor coefficients by means of

$$\rho(f) = \limsup_{n\to\infty} \frac{n \log n}{-\log |a_n|}.$$

An analogue of this for the relative order can be found in [34, Teorema 7]. For further properties of relative orders, we refer the reader to Bernal [34].

Based on Theorem 11.6.2, the concept of relative order seems to be particularly useful for expressing the growth of composite entire functions. Indeed, the existing literature on this topic is quite extensive, the amount of research papers being too numerous to be cited here.

12 More topics on entire functions

In this last chapter, we discuss the following unrelated topics in the theory of entire functions:

(1) asymptotic values, including Denjoy's conjecture, which was proved by Ahlfors,
(2) Bank–Laine functions, which is a class of entire functions that originated from the equation $f'' + A(z)f = 0$, where $A(z)$ is entire,
(3) entire functions whose zeros and one-points lie on a finite number of rays, which includes recent work of Bergweiler, Eremenko, and Hinkkanen,
(4) entire solutions of Fermat-type equations, which have parallels to number theory, and
(5) properties of multiple a-points of entire functions.

12.1 Asymptotic values

If f is an entire function such that $f(z) \to a \in \widehat{\mathbb{C}}$ as $z \to \infty$ along some continuous path γ tending to infinity, then a is called an *asymptotic value* of f. For instance, 0 is an asymptotic value of $f(z) = e^z$, since $e^z \to 0$ as $z \to \infty$ along any horizontal half-line in the left half-plane.

We have already encountered asymptotic values in the Phragmén–Lindelöf theory. Indeed, Theorem 4.1.4 shows that if an entire function f has an asymptotic value a on two rays emanating from the origin and if f is bounded between the two rays, then a is an asymptotic value of f in the corresponding closed sector. Theorem 4.1.5 in turn is about the uniqueness of asymptotic values.

Example 12.1.1. For an integer $q > 0$, define then entire function f by $f(z) = \int_0^z \exp(-t^q)\,dt$. The function f has q asymptotic values

$$\exp(2\pi ki/q)\int_0^\infty \exp\left(-t^q\right)dt, \quad k = 0,\ldots,q-1,$$

as $z \to \infty$ along the rays $\arg z = 2\pi k/q$. This can be seen by rotating the line of integration to $t = re^{2\pi ki/q}$, $dt = e^{2\pi ki/q}\,dr$, in which case

$$f(z) \to e^{2\pi ki/q}\int_0^\infty \exp\left(-r^q e^{2\pi ki}\right)dr = e^{2\pi ki/q}\int_0^\infty \exp\left(-r^q\right)dr$$

as $z \to \infty$ along $\arg z = 2\pi k/q$.

Example 12.1.1 deals with finite asymptotic values. However, by the following theorem, every nonconstant entire function has ∞ as an asymptotic value. It is rather surprising that a rigorous proof for this well-known result is not easy to find. The proof below is a modification of Yu's [302] solution to [240, Exercise 12.12].

https://doi.org/10.1515/9783112249130-012

Theorem 12.1.2 ([240, Exercise 12.12], [302]). *If f is a nonconstant entire function, then ∞ is an asymptotic value of f.*

Proof. If f is a nonconstant polynomial, then f tends to ∞ along any ray emanating from the origin. Hence we may suppose that f is transcendental.

Define $E_1 = \{z \in \mathbb{C} : |f(z)| > 1\}$. If $E_1 = \emptyset$, then f must be a constant function by Liouville's theorem, a contradiction. If $E_1 = \mathbb{C}$, then f omits all the points in the closed unit disc, which contradicts Picard's theorem. Hence $E_1 \neq \mathbb{C}$, and E_1 has finite boundary points. The set E_1 is open by continuity and consists of one or more connected components, which are disjoint from each other, and whose union is E_1. Let F_1 be one of the components of E_1. By continuity, F_1 is an open set, and $|f(z)| \geq 1$ on ∂F_1. Suppose that $|f(z_0)| > 1$ for some $z_0 \in \partial F_1$, in which case $z_0 \in E_1$. By continuity there exists a small radius $\delta > 0$ such that $|f(z)| > 1$ for all $z \in D(z_0, \delta)$, in which case $D(z_0, \delta) \subset E_1$. Since all the components of E_1 are disjoint from each other, we have that $D(z_0, \delta)$ intersects with F_1 only, which is not possible because z_0 is a boundary point of F_1. Hence we have $|f(z)| = 1$ on ∂F_1.

We proceed to prove that F_1 is unbounded. Assume that this is not the case. Since F_1 is open and connected and since f is entire, we may apply the maximum modulus principle to conclude that the maximum modulus of f in $\overline{F}_1$ is attained on ∂F_1. Hence $|f(z)| \leq 1$ for all $z \in \overline{F}_1$, which contradicts with the fact that $F_1 \subseteq E_1$. Consequently, every component of E_1 is unbounded.

For every $n \in \mathbb{N}$, we now define

$$E_n = \{z \in \mathbb{C} : |f(z)| > n\}.$$

Since f is unbounded in E_1 above, we have $E_n \neq \emptyset$ for every $n \in \mathbb{N}$. Clearly, $E_{n+1} \subseteq E_n$ for every $n \in \mathbb{N}$. Arguing similarly as above, we see that every component of every set E_n is unbounded.

So far, the component F_1 of E_1 is fixed. We proceed to choose a particular component F_n of each E_n as follows. Let G be any component of E_2. Arguing similarly as above, we see that G is open and $|f(z)| = 2$ on ∂G. Clearly, $G \subseteq E_2 \subseteq E_1$, and hence G must lie entirely in a component of E_1. We choose for F_2 a component G of E_2 that lies entirely in F_1. By continuing in this way we obtain an infinite sequence $\{F_n\}$ of nonempty nested open sets such that F_n is a component of E_n and

$$F_1 \supseteq F_2 \supseteq F_3 \supseteq \cdots.$$

For each $k \in \mathbb{N}$, pick a point $z_k \in F_k$. By the nesting property of the sets F_k we obtain that $z_n \in F_k$ for all $n \geq k$. Since each F_k is open and connected, F_k is a domain. In particular, this means that z_k can be connected to z_{k+1} by a continuous path γ_k that lies entirely in F_k. Note that the definition of E_k states that $|f(z)| > k$ for all $z \in F_k \subset E_k$, and hence $|f(z)| > k$ on γ_k. Let γ be the continuous union of the continuous paths γ_k. It is clear that f tends to ∞ along γ. □

Theorem 12.1.3 ([142, p. 161]). *If $a \in \mathbb{C}$ is a Picard value of a nonconstant entire function f, then a is an asymptotic value of f.*

Proof. We apply Theorem 12.1.2 to the nonconstant entire function $F(z) = 1/(f(z) - a)$ and conclude that ∞ is an asymptotic value of F. This is equivalent to saying that a is an asymptotic value of f. □

We now state the famous *Denjoy conjecture*, which was proved by Ahlfors.

Theorem 12.1.4 ([2], [305, § 4.2]). *If f is a transcendental entire function of finite order ρ, then f has at most 2ρ distinct finite asymptotic values.*

The original conjecture appeared in 1907. In 1921, Carleman proved that the number of distinct finite asymptotic values is at most $5\rho/2$. The proof for the sharp upper bound 2ρ was given by Ahlfors in 1929, and since then, the conjecture has been known as the Denjoy–Carleman–Ahlfors theorem.

A weaker version of Denjoy's conjecture for rays is easy to prove, and since it gives an intuitive picture of the conjecture, we will give the proof below.

Theorem 12.1.5 ([142, p. 161]). *Let f be an entire function of finite order ρ. Then there are at most 2ρ rays emanating from the origin along which f has distinct finite asymptotic values.*

Proof. If $\rho < 1/2$, then Corollary 4.1.3 shows that f is unbounded along every ray emanating from the origin. Therefore, f cannot have a finite asymptotic value along any such ray, and the statement follows. Hence we may assume that $\rho \geq 1/2$.

Suppose now that f has two distinct finite asymptotic values, respectively, on two consecutive rays L_1 and L_2 that emanate from the origin. Let us assume that the angle between L_1 and L_2 is $< \frac{\pi}{\rho}$. Since f has finite asymptotic values along L_1 and L_2, it is bounded on these rays. Then by Theorem 4.1.1, the function f is bounded in the corresponding closed sector determined by L_1 and L_2. But then by Theorem 4.1.5, the asymptotic values along L_1 and L_2 must be equal, which is a contradiction because the asymptotic values are distinct. Hence the angle between L_1 and L_2 must be $\geq \frac{\pi}{\rho}$. Since $2\rho \cdot \frac{\pi}{\rho} = 2\pi$, it follows that there exists at most 2ρ rays along which f has distinct finite asymptotic values. □

Example 12.1.6 ([305, p. 210]). For an integer $q > 0$, define

$$f(z) = \int_0^z \frac{\sin(t^q)}{t^q}\, dt,$$

where the integrand is an abbreviated notation for an obvious power series. Since we observe that f' is entire of order q, it follows that f is also entire of order q. Moreover, f has the $2q$ finite asymptotic values

$$e^{k\pi i/q} \int_0^\infty \frac{\sin(r^q)}{r^q}\, dr, \quad k = 1, \ldots, 2q,$$

along the rays

$$\arg z = \frac{k\pi i}{q}, \quad k = 1, \ldots, 2q.$$

This illustrates the sharpness of Theorems 12.1.5 and 12.1.4.

For the next theorem of Sons, recall Fejér gaps from Definition 7.2.2.

Theorem 12.1.7 ([262]). *Suppose that $f(z) = \sum_{n=0}^{\infty} a_n z^{\lambda_n}$ is entire and has Fejér gaps. If f has finite lower order, then f has no finite asymptotic values.*

In Section 9.2, we discussed solutions of linear differential equations that have an asymptotic value in a sector. Langley [180] recently proved some results concerning the asymptotic values of Bank–Laine functions, which have connections to solutions of equations of the form $f'' + A(z)f = 0$, where $A(z)$ is entire. We discuss this class of functions in the next section.

Historical notes

Lars Ahlfors (1907–1996, Finland) was a student of Rolf Nevanlinna. While working on and eventually proving Denjoy's conjecture, Ahlfors isolated himself for a long period of time and was nowhere to be found, which made everyone very worried [182, p. 89]. In 1936, Ahlfors was awarded the Fields medal (jointly with Jesse Douglas), which was awarded for the first time that year.

12.2 Bank–Laine functions

We now discuss a class of entire functions that arose from the studies of Bank and Laine about the linear differential equation

$$f'' + A(z)f = 0, \tag{12.2.1}$$

where $A(z)$ is an entire function. These studies of Bank and Laine were discussed in Section 9.3. Recall that when f_1 and f_2 are linearly independent solutions of (12.2.1), the product $E = f_1 f_2$ satisfies the Bank–Laine equation (9.3.3), which can be rewritten as

$$4AE^2 = (E')^2 - 2E''E - c^2, \tag{12.2.2}$$

where c is the nonzero constant $c = W(f_1, f_2)$. Observe that from equation (12.2.2) we have $E'(z_0) = \pm c$ whenever $E(z_0) = 0$. Since $c = W(f_1, f_2)$, there is no loss in generality to assume that $c = 1$.

Definition 12.2.1. A *Bank–Laine function* (or *BL-function*) is an entire function $E(z)$ such that $E'(z_0) = \pm 1$ whenever $E(z_0) = 0$.

Example 12.2.2. A simple BL-function is $E(z) = \sin z$. Observe that if $A(z)$ is defined by (12.2.2) with $c = 1$ and $E(z) = \sin z$, then $A(z) \equiv 1/4$, and the equation

$$f'' + \frac{1}{4} f = 0$$

is satisfied by

$$f_1(z) = \sqrt{2}\sin(z/2) \quad \text{and} \quad f_2(z) = \sqrt{2}\cos(z/2),$$

where $E = f_1 f_2 = \sin z$. This is not a coincidence, see Theorem 12.2.4.

Example 12.2.3 ([85]). For any positive integer k,

$$E(z) = (\sin z)\exp\left(\frac{2iz^k}{\pi^{k-1}}\right)$$

is a nonreal BL-function of order k, whereas

$$E(z) = (\sin z)e^{i \sin z}$$

is a nonreal BL-function of infinite order. Also,

$$E(z) = (\sin z)e^{\sin z}$$

is a real BL-function of infinite order. For a discussion of these three functions in the context of the next result, see [85].

Theorem 12.2.4 ([13]). (a) *If f_1 and f_2 are linearly independent solutions of* (12.2.1) *satisfying $W(f_1, f_2) = 1$, then $E = f_1 f_2$ is a BL-function, and*

$$4A(z) = \left(\frac{E'}{E}\right)^2 - 2\frac{E''}{E} - \left(\frac{1}{E}\right)^2. \tag{12.2.3}$$

(b) *Conversely, if E is a BL-function, then the function A defined by* (12.2.3) *is an entire function, and E is the product of two linearly independent solutions f_1 and f_2 of* (12.2.1), *where $A(z)$ is the function in* (12.2.3), *such that $W(f_1, f_2) = 1$.*

Proof. Part (a) follows from Theorem 9.3.3, since (9.3.3) gives (12.2.3), and by (12.2.3), $E'(z_0) = \pm 1$ whenever $E(z_0) = 0$.

To prove part (b), suppose that E is a BL-function. From (12.2.3) we have

$$A = \frac{h}{4E^2}, \quad \text{where } h = (E')^2 - 2E''E - 1.$$

It is easy to see that h has a zero of multiplicity at least two whenever E has a zero. Since E is a BL-function, it has only simple zeros, which means that $A(z)$ is an entire function.

Now define w_1 and w_2 as follows:

$$w_1 = \frac{1}{2}\left(\frac{E'}{E} - \frac{1}{E}\right) \quad \text{and} \quad w_2 = \frac{1}{2}\left(\frac{E'}{E} + \frac{1}{E}\right). \tag{12.2.4}$$

The only poles of w_1 and w_2 occur at zeros of E. Let z_0 be a zero of E. Since $E'(z_0) = \pm 1$, we see from inspection of (12.2.4) that either (i) w_1 is analytic at z_0, and w_2 has a simple pole with residue $= 1$ at z_0, or (ii) w_1 has a simple pole with residue $= 1$ at z_0, and w_2 is analytic at z_0. Thus w_1 and w_2 are meromorphic functions whose poles are all simple with residue $= 1$. It follows that w_1 and w_2 are logarithmic derivatives of entire functions g_1 and g_2, i. e.,

$$\frac{g_1'}{g_1} = \frac{1}{2}\left(\frac{E'}{E} - \frac{1}{E}\right) \quad \text{and} \quad \frac{g_2'}{g_2} = \frac{1}{2}\left(\frac{E'}{E} + \frac{1}{E}\right). \tag{12.2.5}$$

Differentiating the first equation in (12.2.5) gives

$$\frac{g_1''}{g_1} - \left(\frac{g_1'}{g_1}\right)^2 = \frac{1}{2}\left(\frac{E''}{E} - \left(\frac{E'}{E}\right)^2 + \frac{E'}{E^2}\right). \tag{12.2.6}$$

Substituting the first equation in (12.2.5) into equation (12.2.6) and simplifying, we obtain

$$\frac{g_1''}{g_1} = -\frac{1}{4}\left(\frac{E'}{E}\right)^2 + \frac{1}{2}\frac{E''}{E} + \frac{1}{4}\left(\frac{1}{E}\right)^2.$$

Then from (12.2.3) we have $g_1'' + Ag_1 = 0$. Similarly, $g_2'' + Ag_2 = 0$.

Adding the two equations in (12.2.5) and then integrating, we obtain $E = bg_1g_2$ for some constant $b \neq 0$. Then subtracting the two equations in (12.2.5), we obtain $W(g_1, g_2) = 1/b$. Thus if we set $f_1 = bg_1$ and $f_2 = g_2$, then $E = f_1f_2$, $W(f_1, f_2) = 1$, and f_1, f_2 satisfy equation (12.2.1) with A defined by (12.2.3), where A is an entire function. This proves part (b). □

If $A(z)$ is a polynomial, then by Proposition 8.3.1, all the solutions of (12.2.1) are entire functions of finite order, which would make the product $E(z)$ of any two linearly independent solutions of (12.2.1) be of finite order as well. Such solutions have well-known properties of growth and distribution of zeros, see Section 8.3. On the other hand, it is possible to have a finite order E and a transcendental A.

Example 12.2.5 ([12, 178]). The equation

$$f'' - \left(e^{2z} + \frac{1}{4}\right)f = 0 \tag{12.2.7}$$

has the solutions

$$f_1(z) = e^{-z/2}\exp\left(-e^z\right) \quad \text{and} \quad f_2(z) = e^{-z/2}\exp\left(e^z\right),$$

where $f_1 f_2 = e^{-z}$ and $\rho(f_1 f_2) = 1$. On the other hand, equation (12.2.7) also possesses the solutions

$$g_1(z) = e^{-z/2}\sinh(e^z) \quad\text{and}\quad g_2(z) = e^{z/2}\cosh(e^z),$$

where $g_1 g_2 = \frac{1}{2}\sinh(2e^z)$ and $\lambda(g_1 g_2) = +\infty$.

Example 12.2.6 ([12, 124]). There exist equations of the form (12.2.1) that possess a zero-free solution base. We now discuss a general method for finding such a zero-free example. Let

$$f_j(z) = \exp\big(g(z) + (-1)^j h(z)\big), \quad j = 1, 2,$$

where g and h are given entire functions, and h is nonconstant. Substituting f_1, f_2 into (12.2.1) and simplifying, we see that g and h depend on one another by means of

$$h''/h' = -2g'.$$

Hence $h'(z) = Ce^{-2g(z)}$ for some $C \in \mathbb{C}\setminus\{0\}$. In fact, since $W(f_1, f_2) = 2h'e^{2g}$, we must have $W(f_1, f_2) = 2C$. The product function $E(z) = f_1(z)f_2(z) = e^{2g(z)}$ is zero-free, and defining $A(z)$ by means of

$$4A(z) = \left(\frac{E'}{E}\right)^2 - 2\frac{E''}{E} - \left(\frac{2C}{E}\right)^2,$$

we obtain

$$A = -g'' - (g')^2 - (h')^2.$$

Thus $\{f_1, f_2\}$ is a zero-free solution base for $f'' + A(z)f = 0$. We can also observe that when g is a polynomial and h is transcendental, then $A(z)$ is transcendental and $E(z)$ is of finite order.

We next state an existence result that was first proved by Šeda in 1959, and then in 1985, Shen gave a different proof by using BL-functions. For some background concerning this result, see [119].

Theorem 12.2.7 ([247, 254]). *Let $\{a_n\}$ and $\{b_n\}$ be two sequences of distinct complex numbers with no finite limit points such that $\{a_n\}$ and $\{b_n\}$ have no points in common. Then there exists an entire function $A(z)$ such that equation* (12.2.1) *possesses two linearly independent solutions f_1 and f_2 whose exact zeros are $\{a_n\}$ and $\{b_n\}$, respectively.*

By multiplying the two solutions f_1 and f_2 in Theorem 12.2.7 with suitable nonzero constants, we may suppose that their Wronskian satisfies $W(f_1, f_2) = f_1 f_2' - f_1' f_2 = 1$. Then by Theorem 12.2.4(a), the product $E = f_1 f_2$ is a BL-function. Moreover, we have

$$\begin{aligned} E'(a_n) &= 2f_1(a_n)f_2'(a_n) - (f_1(a_n)f_2'(a_n) - f_1'(a_n)f_2(a_n)) = -W(f_1, f_2)(a_n) = -1,\\ E'(b_n) &= 2f_1'(b_n)f_2(b_n) + (f_1(b_n)f_2'(b_n) - f_1'(b_n)f_2(b_n)) = W(f_1, f_2)(b_n) = 1. \end{aligned}$$

Langley has several papers on BL-functions [175, 176, 178–180]. By inspecting Shen's proof of Theorem 12.2.7, we can see that even if the sequences have finite exponent of convergence, the BL-function E that is created may still have infinite order. The reader may come to the same conclusion by studying Šeda's proof of Theorem 12.2.7 in [119, Section 6].

We now discuss BL-functions whose zeros are all real. The next result is a combination of results from [60] and [178].

Theorem 12.2.8 ([60, 178]). *Let E be a BL-function of finite order with infinitely many real zeros and no other zeros. Then the counting function satisfies $n(r, 0, E) \neq o(r)$ as $r \to \infty$. If in addition, all the zeros of E are positive, then $\lambda(E) \geq 3/2$.*

Theorem 12.2.8 is sharp. Specifically, we cannot have $n(r, 0, E) = o(r)$ from the example $E(z) = \sin z$, and the second statement $\lambda(E) \geq 3/2$ is sharp from an example constructed in [178] by using quasiconformal methods.

The next result shows that if a couple of conditions are added to the hypothesis of Theorem 12.2.8, then the conclusion can be significantly strengthened.

Theorem 12.2.9 ([179]). *Let E be a real BL-function of finite order with only real zeros, where the zeros are neither bounded above nor bounded below, and let A be defined by* (12.2.3). *Then one of the following statements holds:*

(a) *There exist $\eta, \omega_1, \omega_2 \in \mathbb{R}$ such that $\eta \sin(\omega_1 - \omega_2) \neq 0$ and*

$$A = \eta^2, \quad E(z) = \pm \frac{\sin(\eta z - \omega_1)\sin(\eta z - \omega_2)}{\eta \sin(\omega_1 - \omega_2)}. \tag{12.2.8}$$

(b) *A is transcendental and $\lambda(E) \geq 3$. Moreover, if $\lambda(E) = 3$, then $\rho(E) = \rho(A) = 3$.*

A simple example [179] of (a) is given by

$$A(z) = 1/4, \quad E(z) = \sin z = 2\sin(z/2)\cos(z/2) = -2\sin(z/2)\sin(z/2 - \pi/2).$$

The sharpness of (b) is shown in [179] by using quasiconformal techniques to construct an example that satisfies $\lambda(E) = \rho(E) = \rho(A) = 3$. In addition, the function

$$E(z) = \frac{1}{\pi} e^{2\pi i z^2} \sin(\pi z)$$

shows that the condition that E is real in the hypothesis cannot be removed. Note that if $A(z)$ is a nonconstant polynomial, then (12.2.1) cannot possess two linearly independent solutions where each solution has only real zeros, see [85, 133, 266].

Following [31], we say that an infinite real sequence with no finite accumulation points is *one-sided* if it is bounded from above or below and *two-sided* otherwise.

Theorem 12.2.10 ([31]). *Let E be a BL-function of finite order with only real zeros, and let A be given by* (12.2.3).

(a) *If the zeros of E form an infinite one-sided sequence, then there exists $n \in \mathbb{N} \setminus \{1\}$ such that $\lambda(E) = \rho(E) = \rho(A) = n - 1/2$.*

(b) *If the zeros of E form an infinite two-sided sequence and A is nonconstant, then there exists $n \in \mathbb{N} \setminus \{1\}$ such that $\lambda(E) = \rho(E) = \rho(A) = 2n - 1$.*

Keeping in mind that we always have $\rho(A) \leq \rho(E)$ from (12.2.3), we proceed to state a result for the case where the zeros of E lie on finitely many rays.

Theorem 12.2.11 ([31]). *Let A and E be entire functions satisfying* (12.2.3), *and suppose that the zeros of E lie on finitely many rays emanating from the origin. Then $\rho(E) < \infty$ if and only if $\rho(A) < \infty$.*

The polynomials $E(z) = \pm z$ and $E(z) = \pm(z^2 + z)$ are well-known BL-functions. The following result seems to be new and gives the form of all polynomial BL-functions.

Theorem 12.2.12. *Suppose that E is a polynomial BL-function. Then $\deg E \leq 2$, and E has one of the following forms:*

(a) *$E(z)$ is a nonzero constant.*

(b) *$E(z) = \pm z + c$, where c is a constant.*

(c) *$E(z) = \pm\frac{1}{a-b}(z - a)(z - b)$, where a and b are distinct constants.*

Proof. Obviously, if E is a nonzero constant, then it satisfies the definition of a BL-function, which gives part (a).

Now suppose that E is a nonconstant polynomial BL-function. Then by Theorem 12.2.4(b), $E = f_1 f_2$, where f_1, f_2 are linearly independent entire solutions of

$$f'' + A(z)f = 0, \tag{12.2.9}$$

where $A(z)$ is an entire function. Since E is a polynomial, the functions f_1 and f_2 must have the form

$$f_1(z) = P_1(z)e^{g(z)} \quad \text{and} \quad f_2(z) = P_2(z)e^{-g(z)},$$

where P_1, P_2 are polynomials satisfying $P_1P_2 \not\equiv 0$, and g is an entire function. Since f_1, f_2 are linearly independent solutions of (12.2.9), their Wronskian is a nonzero constant, that is,

$$W(f_1, f_2) = f_1f_2' - f_1'f_2 = c \in \mathbb{C} \setminus \{0\},$$

which gives

$$P_1P_2' - P_1'P_2 - 2P_1P_2g' = c. \tag{12.2.10}$$

Since P_1, P_2 are polynomials and g' is an entire function, it follows from (12.2.10) that g must be a polynomial. Moreover, since the degree of P_1P_2 is strictly larger than the degrees of both P_1P_2' and $P_1'P_2$, it follows from (12.2.10) that we must have $g' \equiv 0$. Therefore, g is a constant, which means that f_1 and f_2 are polynomials. Hence by (12.2.9), $A(z) \equiv 0$ and $f_1'' \equiv f_2'' \equiv 0$. Then $\deg f_1 \leq 1$, $\deg f_2 \leq 1$, and $\deg E \leq 2$.

Suppose that $E(z) = az + b$, where $a \neq 0$, and b is a constant. Then $E(-b/a) = 0$, and $E'(-b/a) = a = \pm 1$. Therefore, $E(z) = \pm z + b$, which gives part (b).

Last, suppose that $E(z) = C(z-a)(z-b)$, where $C \neq 0$ and a, b are constants. Then $E'(z) = C(2z - a - b)$. We must have

$$E'(a) = C(a-b) = \pm 1 \quad \text{and} \quad E'(b) = C(b-a) = \pm 1.$$

Thus $a \neq b$. It follows that either $a - b = -(b-a)$, which is true, or $a - b = b - a$, which gives $a = b$, a contradiction. Hence $C = 1/(a-b)$ or $C = 1/(b-a)$, and

$$E(z) = \pm \frac{1}{a-b}(z-a)(z-b).$$

This gives part (c). □

Definition 12.2.13 ([66]). A sequence $\{z_n\}$ of distinct points in $\mathbb{C}$ is called a *Bank–Laine sequence* (or a *BL-sequence*) if it has a finite exponent of convergence and if there is an entire function $A(z)$ of finite order such that the differential equation (12.2.1) has linearly independent solutions f_1, f_2, where the zeros of $E = f_1 f_2$ are precisely the z_n, i. e., every z_n is a zero of E, and every zero of E is a z_n.

Example 12.2.14 ([66]).
(a) Any finite sequence $z_1, \ldots, z_n$ of distinct points is a BL-sequence. To see this, let $P(z)$ be the polynomial $P(z) = (z - z_1) \cdots (z - z_n)$, and let $E = Pe^Q$, where Q is the Lagrange interpolation polynomial satisfying $Q(z_j) = -\log P'(z_j)$ for $j \in \{1, \ldots, n\}$. Note that $P'(z_j) \neq 0$ because the zeros z_j are simple. Now $E'(z_j) = P'(z_j)e^{Q(z_j)} = 1$ for $j \in \{1, \ldots, n\}$.
(b) The integers $0, \pm 1, \pm 2, \ldots$ form a BL-sequence. To see this, define $E(z) = \frac{1}{\pi}\sin(\pi z)$. Then E has zeros precisely at the integers, and if ζ is an integer, then $E'(\zeta) = \cos(\pi\zeta) = \pm 1$.

We use the next lemma in the proof of Theorem 12.2.16.

Lemma 12.2.15. *In Definition* 12.2.13, *we have* $\rho(E) < \infty$.

Proof. We use observations on page 354 of [12]. The Bank–Laine equation (12.2.2) can be rewritten as

$$E^2 = c^2\left(\left(\frac{E'}{E}\right)^2 - 2\frac{E''}{E} - 4A\right)^{-1}.$$

Using the first fundamental theorem and the lemma on the logarithmic derivative, we can see that

$$T(r,E) = O\left(\overline{N}\left(r, \frac{1}{E}\right) + T(r,A)\right) + S(r,E).$$

Since in Definition 12.2.13, we have $\lambda(E) < \infty$ and $\rho(A) < \infty$, it follows that $\rho(E) < \infty$. □

In view of Example 12.2.14(b), it is somewhat surprising that the nonzero integers do not form a BL-sequence. Therefore, a subsequence of a BL-sequence need not be a BL-sequence.

Theorem 12.2.16 ([66]). *The nonzero integers do not form a BL-sequence.*

Proof. Suppose on the contrary that the nonzero integers form a BL-sequence. Then from Definition 12.2.13 and Lemma 12.2.15, we deduce that there exists an entire function E of finite order such that E has a zero at every nonzero integer and no other zeros, and such that $E'(z) = \pm 1$ at every zero z of E. The entire function $\frac{\sin(\pi z)}{z}$ has the same zeros as E, so that

$$E(z) = \frac{\sin(\pi z)}{z} e^{Q(z)}$$

for some polynomial $Q(z)$. If z is a zero of E, then $z \neq 0$, and

$$\sin(\pi z) = 0 \quad \text{and} \quad E'(z) = \pm\frac{\pi}{z} e^{Q(z)}.$$

Therefore, the polynomial $Q(z)$ satisfies $e^{Q(n)} = \pm\frac{n}{\pi}$ for every $n \in \mathbb{N}$. This means, in particular, that

$$\operatorname{Re} Q(n) = \log\left(\frac{n}{\pi}\right), \quad n \in \mathbb{N}. \tag{12.2.11}$$

However, we also have $Q(z) = a_m z^m + \cdots + a_1 z + a_0$ for some $a_0, \ldots, a_m \in \mathbb{C}$, so that

$$\operatorname{Re} Q(n) = \sum_{j=0}^{m} n^j \operatorname{Re} a_j, \quad n \in \mathbb{N}.$$

Thus $\operatorname{Re} Q(n)$ is a polynomial in n, and $\operatorname{Re} Q(n) \sim a_m n^m$ as $n \to \infty$. This growth rate does not match that in (12.2.11). This contradiction proves the statement. □

By Theorem 12.2.7, any sequence of distinct points with no finite limit points is a zero sequence of a BL-function. Hence the reason why the nonzero integers do not form a BL-sequence is that the associated coefficient function $A(z)$ is of infinite order.

12.3 Zeros and ones on finitely many rays

In this section, we consider entire functions f whose a-points and b-points all lie on a finite number of rays emanating from the origin, where a and b are distinct complex numbers. By considering $g(z) = (z - a)/(b - a)$, we see that there is no loss in generality to assume that $a = 0$ and $b = 1$.

Some results in this section were proved under similar but less restrictive hypotheses, such as allowing the zeros and ones of an entire function (a) to be “close” to a finite number of rays, (b) to “accumulate” in a finite number of directions, or (c) to lie on a finite number of rays except possibly for finitely many. For simplicity, we will keep the discussion to the zeros and ones all lying on a finite number of rays, and those who are interested in more general results of a similar nature with less restrictive hypotheses can find them in the referenced papers.

We will discuss recent as well as older results. In this section, a *ray* always means a ray emanating from the origin. We begin with a classical theorem of Edrei that is contained in a general result.

Theorem 12.3.1 ([62]). *Let f be an entire function whose zeros and one-points all lie on a finite number of rays. If $\mu > 0$ is the smallest angle between the rays, then the order $\rho(f)$ is finite and satisfies $\rho(f) \le \pi/\mu$.*

As Edrei noted [62], the functions

$$f(z) = \cos(z^{n/2}), \quad n = 1, 2, 3, \dots,$$

show that in Theorem 12.3.1, the inequality $\rho(f) \le \pi/\mu$ can be an equality.

Although the zeros of an entire function can be arbitrarily assigned, the zeros and one-points of an entire function cannot be arbitrarily assigned, see Nevanlinna [217]. In fact, to each infinite sequence $\{a_n\}$ of complex numbers, there exists a discrete infinite sequence $\{b_n\}$ of complex numbers, where $\{a_n\}$ and $\{b_n\}$ have no points in common, such that there does not exist an entire function whose zeros occur exactly at the $\{a_n\}$ and whose one-points occur exactly at the $\{b_n\}$, see [238]. Thus it is natural to ask the following:

Question 12.3.2. What is possible and what is not possible regarding entire functions whose zeros and ones lie on a finite number of rays?

We discuss several results that address this general question. In the 1920s, Biernacki [35] and Milloux [209] studied entire functions of finite order whose zeros all lie on a line and whose one-points all lie on some other line. Under different hypotheses, they showed that the lines must be parallel. The following recent result of Bergweiler, Eremenko, and Hinkkanen completely describes the situation where the lines intersect.

Theorem 12.3.3 ([29]). *Suppose that the zeros of an entire function f all lie on a line L_1 and the one-points of f all lie on a different line L_2 that intersects L_1. Then f has one of the following three forms:* (1) $f(z) = e^{az+b}$, (2) $f(z) = 1 - e^{az+b}$, *or* (3) *f is a polynomial of degree at most two ($a, b \in \mathbb{C}$).*

Regarding parallel lines, Baker [8] and Kobayashi [160] showed that if f is a transcendental entire function whose zeros all lie on a line L_1 and whose ones all lie on a different line L_2 parallel to L_1, then f has the form $f(z) = Q(e^{az})$, where $Q(z)$ is a polynomial and a is a nonzero real constant.

As discussed in [29], the proof of Theorem 12.3.3 is essentially contained in Milloux's paper [209], where some details had been omitted. Theorem 12.3.3 is a corollary of more general results in [29].

The next result can be seen by combining Theorem 12.3.1 with a corollary of a result of Biernacki [36, p. 539].

Theorem 12.3.4. *There does not exist a transcendental entire function f whose zeros all lie on one ray and whose one-points all lie on another ray.*

On the other hand, Bergweiler, Eremenko, and Hinkkanen recently constructed examples of entire functions whose zeros all lie on one ray and whose ones all lie on two other rays, which were the first nonelementary examples of entire functions whose zeros and ones all lie on a finite number of rays.

Theorem 12.3.5 ([29]). *Given any integer $m \geq 3$, there exists an entire function of order $1/2 + 1/m$ whose zeros are all positive and whose one-points all lie on the two rays*

$$\arg z = \pm \frac{2\pi}{m+2}.$$

It is worth noting the contrast between Theorems 12.3.5 and 12.3.4. Two observations in [29] are as follows: (i) Theorem 12.3.3 shows that Theorem 12.3.5 does not hold when $m = 2$ except in the trivial case where f has no zeros and f is of the form $f(z) = e^{az}$. (ii) By taking $f(z^n)$, where n is a positive integer, with f as in Theorem 12.3.5, we obtain an entire function whose zeros all lie on n rays and whose one-points all lie on $2n$ rays, which are distinct from the rays where the zeros lie.

The next result gives a property of the examples in Theorem 12.3.5.

Theorem 12.3.6 ([29]). *Let S_1, S_2, S_3 be three distinct rays. Let f be a transcendental entire function whose zeros all lie on S_1 and whose one-points all lie on S_2 and S_3, where there are an infinite number of zeros and an infinite number of ones. Then the angle between S_1 and S_2 equals the angle between S_1 and S_3, and these equal angles are less than $\pi/2$.*

Remark. Theorem 12.3.4 does not hold for meromorphic functions by the following example of Bergweiler and Eremenko [28]. Let $\mathrm{Ai}(z)$ be the Airy integral, and set

$$h(z) = e^{\pi i/3} \frac{\mathrm{Ai}(e^{2\pi i/3} z)}{\mathrm{Ai}(e^{-2\pi i/3} z)}.$$

Then h is a meromorphic function, and by using special properties of the Airy integral, it is shown in [28] that all the zeros of h lie on $\arg z = \pi/3$, all the ones of h lie on the negative real axis, and all the poles of h lie on $\arg z = -\pi/3$.

Edrei considered the special class of entire functions whose zeros and ones are all real. We obtain the next result from Theorem 12.3.1.

Corollary 12.3.7 ([62]). *If f is an entire function with all real (resp., positive) zeros and all real (resp., positive) ones, then $\rho(f) \le 1$ (resp., $\rho(f) \le 1/2$).*

Simple examples are $\sin z$, $\cos z$, and $\cos\sqrt{z}$. Edrei [62] proved several properties of this class of functions.

Theorem 12.3.8 ([62]). *Let f be an entire function that has all real zeros and all real ones.*

(a) *If $f(z)$ is not real for some real value of z, then f has one of the following two forms:*

$$f(z) = \frac{\sin(\xi z + \nu)e^{i(\xi z + \gamma)}}{\sin(\nu - \gamma)}, \quad \sin(\nu - \gamma) \ne 0,$$

where ξ, ν, γ are real constants;

$$f(z) = \frac{\sin(p(\xi z + \nu))e^{i(p-1)(\xi z + \nu)}}{\sin(\xi z + \nu)},$$

where p is an integer ($p \ne 0, 1$), and ξ, ν are real constants.

(b) *Suppose that f is a real entire function. If μ is a real value satisfying $0 \le \mu \le 1$, then all the roots of the equation $f(z) = \mu$ are real.*

(c) *If there exists an unbounded sequence $\{\omega_n\}$ of complex numbers such that all the roots of the equations*

$$f(z) = \omega_n, \quad n = 1, 2, 3, \ldots,$$

are real, then $f(z)$ is a polynomial of degree at most two.

We close this section with a few related remarks. Following [7], a complex number ξ is called a *linearly distributed value* of an entire function f if there is a line L in the complex plane on which all the solutions of $f(z) = \xi$ lie. For example, 0 is a linearly distributed value for $f(z) = \sin z$. Baker proved the following characterization of the exponential function.

Theorem 12.3.9 ([7]). *If f is an entire function for which every complex value is linearly distributed, then f either is a polynomial of degree at most two or has the form $f(z) = a + be^{cz}$, where a, b, c are constants.*

Various characterizations of the exponential function and related results can be found in [7, 8, 158–160].

Last, regarding a less restrictive property, we mention that for information about entire functions whose zeros all lie on a finite number of rays, see [208, 264] and the references therein.

12.4 Fermat-type equations

Consider the functional equation

$$f^n + g^n = 1, \tag{12.4.1}$$

where f and g are entire functions and n is a positive integer. When $n = 2$, all the nonconstant entire solutions of (12.4.1) are sine and cosine functions.

Theorem 12.4.1 ([148]). *If f and g are nonconstant entire functions that satisfy*

$$f^2 + g^2 = 1, \tag{12.4.2}$$

then for some nonconstant entire function $w(z)$, we have

$$f(z) = \cos w(z) \quad \textit{and} \quad g(z) = \sin w(z).$$

Proof. If f and g are nonconstant entire solutions of (12.4.2), then

$$f^2 + g^2 = (f - ig)(f + ig) = 1.$$

Thus $f - ig$ and $f + ig$ are entire functions that have no zeros. It follows from Theorem 3.4.1 that $f - ig = e^u$ and $f + ig = e^v$, where u and v are entire functions. Solving these two equations for f and g gives

$$f = \frac{e^u + e^v}{2} \quad \text{and} \quad g = \frac{e^v - e^u}{2i}.$$

Substituting these expressions into (12.4.2) yields $e^{u+v} = 1$. By using $e^u = e^{-v}$, we obtain $f(z) = \cos(-iv(z))$ and $g(z) = \sin(-iv(z))$. This proves the theorem. □

Theorem 12.4.2 ([148], [211, pp. 135-136]). *Equation* (12.4.1) *does not possess nonconstant entire solutions f and g when $n \geq 3$.*

Proof. Suppose on the contrary that f and g are nonconstant entire solutions of (12.4.1) for some $n \geq 3$. Then

$$g^n\left(\left(\frac{f}{g}\right)^n + 1\right) = g^n\left(\frac{f}{g} + a_1\right)\left(\frac{f}{g} + a_2\right)\cdots\left(\frac{f}{g} + a_n\right) = 1, \tag{12.4.3}$$

where $a_1, \ldots, a_n$ are n distinct nonzero constants. Observe first that if z_0 is a zero of g, then $f(z_0) \neq 0$ by (12.4.1), which means that the meromorphic function f/g has a pole at z_0. Therefore, it follows from (12.4.3) that f/g cannot take any of the n distinct values $-a_1, \ldots, -a_n$. Since $n \geq 3$, this means that f/g has at least three Picard values, which contradicts Theorem 4.6.13 (Picard's great theorem), unless f/g is a rational function. Since a nonconstant rational function does not have any Picard values, we obtain that f/g is a constant. But if $f/g = c$ for some $c \in \mathbb{C}$, then from (12.4.1) we obtain $(c^n + 1)g^n = 1$, which is impossible because g is nonconstant. This proves the theorem. □

Remark. Theorem 12.4.2 brings to mind the famous "Fermat last theorem", which was proved by Wiles [291] and Taylor and Wiles [271], and which can be stated as follows: The equation $x^n + y^n = 1$ cannot be satisfied by nonzero rational numbers x and y when $n \geq 3$.

Next, it is natural to consider the functional equation

$$f^n + g^n + h^n = 1, \tag{12.4.4}$$

where f, g, h are entire functions and n is a positive integer. In 1971, Toda proved a general theorem, which contains the following result.

Theorem 12.4.3 ([275]). *Equation* (12.4.4) *cannot possess nonconstant entire solutions* f, g, h *when* $n \geq 7$.

After Toda proved Theorem 12.4.3, several authors gave alternate proofs, see [73, 84, 111, 146, 169]. On the other hand, there exist nonconstant entire solutions of (12.4.4) for any $1 \leq n \leq 5$; see the references regarding [89, Question 3.3].

Example 12.4.4 ([97]). The equation $f^5 + g^5 + h^5 = 1$ possesses the entire solutions

$$\begin{aligned} f(z) &= \frac{1}{3}\big((2-\sqrt{6})e^z + 1 + (2+\sqrt{6})e^{-z}\big),\\ g(z) &= \frac{1}{6}\big(\{\sqrt{6}-2+(3\sqrt{2}-2\sqrt{3})i\}e^z + 2 - \{\sqrt{6}+2-(3\sqrt{2}+2\sqrt{3})i\}e^{-z}\big),\\ h(z) &= \frac{1}{6}\big(\{\sqrt{6}-2+(2\sqrt{3}-3\sqrt{2})i\}e^z + 2 - \{\sqrt{6}+2+(3\sqrt{2}+2\sqrt{3})i\}e^{-z}\big). \end{aligned}$$

One open question remains.

Question 12.4.5. Does there exist nonconstant entire solutions f, g, h of (12.4.4) when $n = 6$?

Ishizaki [146] made partial progress on Question 12.4.5 by showing that if f, g, h are nonconstant entire functions that satisfy (12.4.4) when $n = 6$, then there exists a small meromorphic function $a(z)$ with respect to f, g, and h such that

$$W(f^6, g^6, h^6) = a(z)(f(z)g(z)h(z))^4.$$

To say that $a(z)$ is a *small meromorphic function* with respect to f, g, h means that $T(r, a) = S(r, f)$, $T(r, a) = S(r, g)$, $T(r, a) = S(r, h)$.

For the case of polynomials, we have the following result.

Theorem 12.4.6 ([90, 111, 219]). *Equation* (12.4.4) *cannot possess nonconstant polynomial solutions* f, g, h *when* $n \geq 6$.

There are known examples of nonconstant polynomial solutions of (12.4.4) when $1 \leq n \leq 3$; see the references regarding [89, Question 3.4]. For instance, the identity [210]

$$\frac{1}{2}(1+z^{3})^{3}+\frac{1}{2}(1-z^{3})^{3}-3(z^{2})^{3}=1$$

provides such an example when $n = 3$. Two open questions remain.

Question 12.4.7. Does there exist nonconstant polynomial solutions of (12.4.4) when $n = 4$? What about $n = 5$?

It is worth noting that known results on entire solutions of Fermat-type equations of the form (12.4.4) were used in [76] to prove results about the entire solutions of a class of binomial differential equations, see (10.4.1).

Equations (12.4.1) and (12.4.4) are particular cases of the general *Fermat-type equation*

$$f_1^n + f_2^n + \cdots + f_k^n = 1, \tag{12.4.5}$$

where $f_1, f_2, \ldots, f_k$ are meromorphic functions and n and k are positive integers. Following Hayman [111], the results, examples, and open questions concerning equation (12.4.5) can be naturally separated into the following four classes of functions: (a) polynomials, (b) entire functions, (c) rational functions, and (d) meromorphic functions, including all the above material. We make some comments.

(1) For results, examples, and open questions regarding (12.4.5), see [90, 111].
(2) Hayman [90, 111] showed how Cartan's form of the second fundamental theorem can be used to prove all the known nonexistence results for (12.4.5) in all four classes of functions. For this type of problem, Cartan's form of Nevanlinna theory works better than typical Nevanlinna theory because of the presence of zeros of high multiplicity due to terms of the form f^n.
(3) For $k = 3, 4, 5, \ldots$, equations (12.4.5) provide a collection of an infinite number of open questions on the existence of nonconstant solutions in each of the four classes of functions.
(4) Information about the possible degrees of nonconstant polynomial solutions of (12.4.5) can be found in [93, 147].

Last, we mention that Waring's problem has some similarities in nature to these Fermat-type equation problems, see [90, 111].

12.5 Multiplicities of a-points

In this section, we discuss properties of the multiplicities of a-points of entire functions. First, we show that for any given entire function f, there can exist at most one finite value a for which the multiplicities of the a-points of f can tend to infinity.

Theorem 12.5.1. *Let f be a nonconstant entire function. For any $a \in \mathbb{C}$, let $z_n = z_n(a)$ denote the distinct a-points of f, ordered by increasing modulus, and let $m_n = m_n(a) \geq 1$*

denote the multiplicity of z_n. Then there exists at most one finite value a where the sequence $\{z_n\}$ is infinite such that $m_n \to \infty$ as $n \to \infty$.

Proof. We use Nevanlinna theory. Suppose that for some $a \in \mathbb{C}$, the sequence $\{z_n\}$ is infinite and $\{m_n\}$ tends to infinity. Obviously, $\overline{n}(r,a,f)$ is an unbounded function of r. It follows that for any $\varepsilon > 0$, there exists $R(\varepsilon) > 0$ such that

$$n(r,a,f) \geq \frac{1}{\varepsilon}\overline{n}(r,a,f), \quad r \geq R(\varepsilon).$$

Thus from the first fundamental theorem, we see that

$$T(r,f) \geq N(r,a,f) + O(1) \geq \frac{1}{\varepsilon}\overline{N}(r,a,f) + O(1), \quad r \to \infty.$$

Then using the measure $\Theta(a,f)$ defined in (4.6.20), we obtain

$$\Theta(a,f) = 1 - \limsup_{r\to\infty} \frac{\overline{N}(r,a,f)}{T(r,f)} \geq 1 - \varepsilon.$$

Since $\varepsilon > 0$ can be arbitrarily small, we conclude that $\Theta(a,f) = 1$. Since f is entire, we also have $\Theta(\infty,f) = 1$. Hence by Theorem 4.6.21, we obtain that $\Theta(b,f) = 0$ for all values of b satisfying $b \neq a, \infty$. Therefore, from what was just proven it follows that there cannot exist another value $b \in \mathbb{C} \setminus \{a\}$ such that the multiplicities of the b-points of f tend to infinity. □

In the other direction from Theorem 12.5.1, the following example shows that there exist entire functions f that possess a finite value a such that the multiplicities of the a-points of f are unbounded, and yet $\overline{n}(r_n,a,f) \sim n(r_n,a,f)$ on a sequence $\{r_n\}$ tending to infinity.

Example 12.5.2. We construct an entire function g that possesses zeros of unbounded multiplicities, but which also satisfies

$$2^n = \overline{n}(2^n,0,g) = \big(1+o(1)\big)n(2^n,0,g), \quad n \to \infty. \tag{12.5.1}$$

Let g be an entire function that has its sequence of zeros $\{z_n\}$ exactly at the positive integers such that for each $n \geq 1$, g has a zero of multiplicity $n+1$ at 2^n, and where the zeros of g at all the other positive integers are simple. Then $\overline{n}(2^n,0,g) = 2^n$, and

$$\begin{aligned} n(2^n,0,g) &= 1+2+1+3+1+1+1+4+1+1+\cdots+1+5+\cdots \\ &= 1+2+\cdots+n+(2-1)+(2^2-1)+\cdots+(2^{n-1}-1)+\cdots \\ &= \frac{n(n+1)}{2}+(2+2^2+2^3+\cdots+2^{n-1})-(n-1) \\ &= 2(2^{n-1}-1)+O(n^2) = \big(1+o(1)\big)2^n, \quad n\to\infty, \end{aligned}$$

which yields (12.5.1). Observe that in this case the sequence $\{m_n\}$ of multiplicities of the zeros $\{z_n\}$ does not have a limit.

Next, we consider nonconstant entire functions f that possess a value $a \in \mathbb{C}$, where every a-point of f has multiplicity at least two. Such a value a is called a *completely ramified value* or *complete multiple value* of f. There can exist at most two such values.

Theorem 12.5.3. *Any nonconstant entire function can have at most two finite completely ramified values.*

Proof. Let $a \in \mathbb{C}$ be a completely ramified value of some nonconstant entire function f. Then from the definition of completely ramified value, we see that $n(r,a,f) \geq 2\overline{n}(r,a,f)$. Therefore, using the first fundamental theorem, we obtain

$$\overline{N}(r,a,f) \leq \frac{1}{2}N(r,a,f) \leq \frac{1}{2}T(r,f) + O(1).$$

It follows that $\Theta(a,f) \geq 1/2$. Since f is entire, we also have $\Theta(\infty,f) = 1$. Then from Theorem 4.6.21, it follows that there can exist at most two values $a \in \mathbb{C}$ that satisfy $\Theta(a,f) \geq 1/2$. This means that f can have at most two finite completely ramified values. □

Theorem 12.5.3 is sharp because there exist nonconstant entire functions that possess two finite completely ramified values. For instance, $\sin z$ has the two completely ramified values 1 and -1, because the zeros of $\sin z - 1$ and $\sin z + 1$ all have multiplicity two. This is easy to see by using the representation

$$\sin z = \frac{e^{iz} - e^{-iz}}{2i}.$$

Observe that from Theorem 4.6.21, it immediately follows that $\Theta(a, \sin z) = 0$ for any value a satisfying $a \neq 1, -1, \infty$.

Bibliography

[1] Ablowitz, M. J., Fokas, A. S.: *Complex Variables: Introduction and Applications*. Cambridge University Press, Cambridge (1997)

[2] Ahlfors, L.: *Über die asymptotischen Werte der ganzen Funktionen endlicher Ordnung*. Ann. Acad. Sci. Fenn. **32**(6), 1–15 (1929) (German)

[3] Anderson, G. D., Vamanamurthy, M. K., Vuorinen, M.: *Generalized convexity and inequalities*. J. Math. Anal. Appl. **335**(2), 1294–1308 (2007)

[4] Apostol, T. M.: *Mathematical Analysis*, 2nd edn. Addison-Wesley Publishing Co., Reading, Mass.-London–Don Mills, Ont. (1974)

[5] Armitage, D. H.: *Entire functions that tend to zero on every line*. Am. Math. Mon. **114**(3), 251–256 (2007)

[6] Baernstein, A. II: *A generalization of the* $\cos \pi \rho$ *theorem*. Trans. Am. Math. Soc. **193**, 181–197 (1974)

[7] Baker, I. N.: *Entire functions with linearly distributed values*. Math. Z. **86**, 263–267 (1964)

[8] Baker, I. N.: *Entire functions with two linearly distributed values*. Ann. Acad. Sci. Fenn., Ser. A 1 Math. **5**(2), 381–386 (1980)

[9] Bank, S.: *On the value distribution theory for entire solutions of second-order linear differential equations*. Proc. Lond. Math. Soc. (3) **50**, 505–534 (1985)

[10] Bank, S.: *A note on the zeros of solutions of* $w'' + P(z)w = 0$ *where P is a polynomial*. Appl. Anal. **25**(1–2), 29–41 (1987)

[11] Bank, S.: *Addendum to "A note on the zeros of solutions of* $w'' + P(z)w = 0$ *where P is a polynomial"*. Appl. Anal. **30**, 1–3, 245–248 (1988)

[12] Bank, S., Laine, I.: *On the oscillation theory of* $f'' + Af = 0$ *where A is entire*. Trans. Am. Math. Soc. **273**(1), 351–363 (1982)

[13] Bank, S., Laine, I.: *On the zeros of meromorphic solutions of second-order linear differential equations*. Comment. Math. Helv. **58**, 656–677 (1983)

[14] Bank, S., Laine, I., Langley, J.: *On the frequency of zeros of solutions of second order linear differential equations*. Result. Math. **10**(1–2), 8–24 (1986)

[15] Barry, P. D.: *The minimum modulus of small integral and subharmonic functions*. Proc. Lond. Math. Soc. (3) **12**, 445–495 (1962)

[16] Barry, P. D.: *Some theorems related to the* $\cos \pi \rho$ *theorem*. Proc. Lond. Math. Soc. (3) **21**, 334–360 (1970)

[17] Bartle, R. G., Sherbert, D. R.: *Introduction to Real Analysis*, 4th edn. John Wiley & Sons, Inc., New York (2011)

[18] Bateman, H. and the staff of the Bateman Manuscript Project: *Higher Transcendental Functions*. Volume II. McGraw Hill Book Publishing Co. (1953). Reprinted editions 1981, 1985

[19] Bateman, H. and the staff of the Bateman Manuscript Project: *Higher Transcendental Functions*. Volume III. McGraw Hill Book Publishing Co. (1955). Robert E. Krieger Publishing Co., Reprinted edition 1981

[20] Beardon, A., Minda, D.: *On the pointwise limit of complex analytic functions*. Am. Math. Mon. **110**(4), 289–297 (2003)

[21] Belaïdi, B.: *Estimation of the hyper-order of entire solutions of complex linear ordinary differential equations whose coefficients are entire functions*. Electron. J. Qual. Theory Differ. Equ. **2002**, 5 (2002), 8 pp.

[22] Belaïdi, B., Hamouda, S.: *Orders of solutions of an n-th order linear differential equation with entire coefficients*. Electron. J. Differ. Equ. **2001**, 61 (2001), 5 pp.

[23] Berg, C., Pedersen, H. L.: Logarithmic order and type of indeterminate moment problems. In: *Difference Equations, Special Functions and Orthogonal Polynomials*, pp. 51–79. World Sci. Publ., Hackensack, NJ (2007) (English summary). With an appendix by Walter Hayman

[24] Bergweiler, W.: *On the zeros of certain homogeneous differential polynomials*. Arch. Math. **64**, 199–202 (1995)

https://doi.org/10.1515/9783112249130-013

[25] Bergweiler, W.: *A question of Goldberg and Ostrovskii concerning linear differential equations with coefficients of completely regular growth*. Proc. Am. Math. Soc. **151**(5), 2097–2101 (2023)

[26] Bergweiler, W., Eremenko, A.: *On the Bank–Laine conjecture*. J. Eur. Math. Soc. **19**(6), 1899–1909 (2017)

[27] Bergweiler, W., Eremenko, A.: *Quasiconformal surgery and linear differential equations*. J. Anal. Math. **137**(2), 751–812 (2019)

[28] Bergweiler, W., Eremenko, A.: *Meromorphic functions with three radially distributed values*. Comput. Methods Funct. Theory **21**, 671–696 (2021)

[29] Bergweiler, W., Eremenko, A., Hinkkanen, A.: *Entire functions with two radially distributed values*. Math. Proc. Camb. Philos. Soc. **165**, 93–108 (2018)

[30] Bergweiler, W., Eremenko, A., Langley, J. K.: *Real entire functions of infinite order and a conjecture of Wiman*. Geom. Funct. Anal. **13**(5), 975–991 (2003)

[31] Bergweiler, W., Eremenko, A., Rempe, L.: *Second order linear differential equations with a basis of solutions having only real zeros*. J. Anal. Math. **152**(1), 53–108 (2024)

[32] Bergweiler, W., Ishizaki, K., Yanagihara, N.: *Meromorphic solutions of some functional equations*. Methods Appl. Anal. **5**(3), 248–258 (1998)

[33] Bernal, L. G.: *On growth k-order of solutions of a complex homogeneous linear differential equation*. Proc. Am. Math. Soc. **101**(2), 317–322 (1987)

[34] Bernal, L. G.: *Orden relativo de crecimiento des funciones enteras*. Collect. Math. **39**, 209–229 (1988) (Spanish)

[35] Biernacki, M.: *Sur le déplacement des zéros des fonctions entières par leur derivation*. C. R. Acad. Sci. Paris **1975**, 18–20 (1922) (French)

[36] Biernacki, M.: *Sur le theorie des fonctions entieres*. Bull. Acad. Pol. Sci. Lett. Lett. Cl. Sci. Math. Natur. Ser. A (1929), 529–590 (French)

[37] Boas, R. Jr.: *Entire Functions*. Academic Press Inc., New York (1954)

[38] Boas, R. Jr.: *Fundamental sets of entire functions*. Ann. Math. (2) **47**, 21–32 (1946)

[39] Buck, C.: *On the distribution of the zeros of an entire function*. J. Indian Math. Soc. **16**, 147–149 (1952)

[40] Calvis, D. T., Edwards, C. H., Penney, D. E.: *Differential Equations & Linear Algebra*, 4th edn. Pearson, Boston (2018)

[41] Carroll, T., Drasin, D., Hinkkanen, A., Langley, J. K.: *Walter K. Hayman FRS (1926–2020), A biographical sketch*. Comput. Methods Funct. Theory **21**(4), 535–542 (2021)

[42] Cartan, H.: *Elementary Theory of Analytic Functions of One or Several Complex Variables*. Addison-Wesley Co., Reading, MA (1963)

[43] Cartwright, M. L.: *Integral Functions*. Cambridge Tracts in Mathematics and Mathematical Physics, vol. 44. Cambridge at the University Press (1956)

[44] Chen, M.-F., Huang, X.-M.: *Entire solutions of nonlinear differential equations derived from trigonometric identitities*. J. Contemp. Math. Anal. **60**(5), 450–458 (2025)

[45] Chen, Z.: *Complex Differences and Difference Equations*. Mathematics Monograph Series, vol. 29. Science Press, Beijing (2014)

[46] Chen, Z.-X., Yang, C.-C.: *Some further results on the zeros and growths of entire solutions of second order linear differential equations*. Kodai Math. J. **22**(2), 273–285 (1999)

[47] Chern, P. T.-Y.: *On meromorphic functions with finite logarithmic order*. Trans. Am. Math. Soc. **358**(2), 473–489 (2006)

[48] Cherry, W., Ye, Z.: *Nevanlinna's Theory of Value Distribution. The Second Main Theorem and its Error Terms*. Springer Monographs in Mathematics. Springer-Verlag, Berlin (2001)

[49] Chiang, Y.-M., Feng, S.: *On the Nevanlinna characteristic of* $f(z+\eta)$ *and difference equations in the complex plane*. Ramanujan J. **16**(1), 105–129 (2008)

[50] Chiang, Y.-M., Feng, S.: *Nevanlinna theory of the Askey–Wilson divided difference operator*. Adv. Math. **329**, 217–272 (2018)

[51] Chyzhykov, I., Gröhn, J., Heittokangas, J., Rättyä, J.: *Oscillation of solutions of LDE's in domains conformally equivalent to unit disc*. J. Geom. Anal. **32**, 3 (2022)

[52] Chyzhykov, I., Gröhn, J., Heittokangas, J., Rättyä, J.: *Description of growth and oscillation of solutions of LDE's*. J. Aust. Math. Soc. **115**(2), 145–171 (2023)

[53] Chyzhykov, I., Gundersen, G. G., Heittokangas, J.: *Linear differential equations and logarithmic derivative estimates*. Proc. Lond. Math. Soc. (3) **86**(3), 735–754 (2003)

[54] Chyzhykov, I., Heittokangas, J., Rättyä, J.: *Finiteness of φ-order of solutions of linear differential equations in the unit disc*. J. Anal. Math. **109**, 163–198 (2009)

[55] Clunie, J.: *On integral functions having prescribed asymptotic growth*. Can. J. Math. **17**, 396–404 (1965)

[56] Clunie, J., Kövari, T.: *On integral functions having prescribed asymptotic growth II*. Can. J. Math. **20**, 7–20 (1968)

[57] Conway, J. B.: *Functions of One Complex Variable*, 2nd edn. Graduate Texts in Mathematics, vol. 11. Springer-Verlag, New York–Berlin (1978)

[58] Dickson, D. G.: *Asymptotic distribution of zeros of exponential sums*. Publ. Math. (Debr.) **11**, 297–300 (1964)

[59] Ding, J., Heittokangas, J., Wen, Z.-T.: *A continuous transition from $\mathcal{E}$-sets to R-sets and beyond*. J. Geom. Anal. **33**(9), 278 (2023), 31 pp.

[60] Drasin, D., Langley, J. K.: Bank–Laine functions via quasiconformal surgery. In: *Transcendental Dynamics and Complex Analysis*. London Math. Soc. Lecture Notes, vol. 348, pp. 165–178. Cambridge University Press (2008)

[61] Duren, P. L.: *Theory of H^p Spaces*. Pure and Applied Mathematics, vol. 38. Academic Press, New York–London (1970)

[62] Edrei, A.: *Meromorphic functions with three radially distibuted values*. Trans. Am. Math. Soc. **78**, 276–293 (1955)

[63] Edrei, A., Fuchs, W. H. J.: *Entire and meromorphic functions with asymptotically prescribed characteristic*. Can. J. Math. **17**, 383–395 (1965)

[64] Edwards, S., Hellerstein, S.: *Non-real zeros of derivatives of real entire functions and the Pólya–Wiman conjectures*. Complex Var. Theory Appl. **47**(1), 25–57 (2002)

[65] Egorychev, E.: *Integral Representations and Combinations of Combinatorial Sums*. Translations Math. Mono., vol. 59. Amer. Math. Soc., Providence, RI (1984)

[66] Elzaidi, S. M.: *On Bank–Laine sequences*. Complex Var. Theory Appl. **38**(3), 201–220 (1999)

[67] Eremenko, A., Merenkov, S.: *Nevanlinna functions and real zeros*. Ill. J. Math. **49**(4), 1093–1110 (2005)

[68] Fenton, P.: A glance at Wiman–Valiron theory. In: *Complex Analysis and Dynamical Systems II*. Contemp. Math. Israel Math. Conf. Proc., vol. 382, pp. 131–139. Amer. Math. Soc., Providence, RI (2005) (English summary)

[69] Frei, M.: *Sur l'ordre des solutions entiëres d'une équations diffërentielle linëaire*. C. R. Acad. Sci. Paris **236**, 38–40 (1953) (French)

[70] Frei, M.: *Über die Lösungen linearer Differentialgleichungen mit ganzen Funktionen als Koeffizienten*. Comment. Math. Helv. **35**, 201–222 (1961) (German)

[71] Frei, M.: *Über die subnormalen Lösungen der Differentialgleichung* $w'' + e^{-z}w' + konst \cdot w = 0$. Comment. Math. Helv. **36**, 1–8 (1961) (German)

[72] Fuchs, W. H. J.: *Proof of a conjecture of G. Pólya concerning gap series*. Ill. J. Math. **7**, 661–667 (1963)

[73] Fujimoto, H.: *On meromorphic maps into the complex projective space*. J. Math. Soc. Jpn. **26**, 272–288 (1974)

[74] Gamelin, T.: *Complex Analysis*. Undergraduate Texts in Mathematics. Springer-Verlag, New York (2001)

[75] Gao, L., Gao, J.: *Trigonometric identities and entire solutions of non-linear binomial differential equations*. Mediterr. J. Math. **20**, 167 (2023), 15 pp.

[76] Gao, L., Jiang, X., Liu, K., Song, C.: *Entire solutions of certain type binomial differential equations*. J. Math. Anal. Appl. **556**(2), 130245 (2026)

[77] Gao, Z., Liu, M., Gao, L.: *On entire solutions of certain types of nonlinear differential equations*. ScienceAsia **49**, 685–692 (2023)

[78] Gao, L., Song, C.: *Exponential polynomials as solutions of certain type binomial differential equations*. Anal. Math. Phys. **14**(4), 91 (2024), 19 pp.

[79] Gil', M.: *Localization and Perturbation of Zeros of Entire Functions*. Lecture Notes in Pure and Applied Mathematics, vol. 258. CRC Press, Boca Raton, FL (2010)

[80] Gol'dberg, A. A., Eremenko, A. E., Ostrovskii, I. V.: *The sum of entire functions of completely regular growth*. Sov. J. Contemp. Math. Anal. **18**(1), 1–12 (1983)

[81] Gol'dberg, A. A., Levin, B. Ya., Ostrovskii, I. V.: *Entire and Meromorphic Functions*. Sovrem. Probl. Mat. Fund. Naprav. [Current Problems in Mathematics. Fundamental Directions], vol. 85. Akad. Nauk SSSR, Vsesoyuz. Inst. Nauchn. i Tekhn. Inform., Moscow (1991)

[82] Gol'dberg, A. A., Ostrovskii, I. V.: *Value Distribution of Meromorphic Functions*. Translations of Mathematical Monographs, vol. 236. American Mathematical Society, Providence, RI (2008)

[83] Gorenflo, R., Kilbas, A. A., Mainardi, F., Rogosin, S. V.: *Mittag-Leffler Functions, Related Topics and Applications*. Springer Monographs in Mathematics. Springer, Heidelberg (2014)

[84] Green, M. L.: *Some Picard theorems for holomorphic maps to algebraic varieties*. Am. J. Math. **97**, 43–75 (1975)

[85] Gundersen, G. G.: *On the real zeros of solutions of* $f'' + A(z)f = 0$ *where* $A(z)$ *is entire*. Ann. Acad. Sci. Fenn., Ser. A 1 Math. **11**(2), 275–294 (1986)

[86] Gundersen, G. G.: *Estimates for the logarithmic derivative of a meromorphic function, plus similar estimates*. J. Lond. Math. Soc. (2) **37**(1), 88–104 (1988)

[87] Gundersen, G. G.: *Finite order solutions of second order linear differential equations*. Trans. Am. Math. Soc. **305**(1), 415–429 (1988)

[88] Gundersen, G. G.: *Solutions of* $f'' + P(z) = 0$ *that have almost all real zeros*. Ann. Acad. Sci. Fenn., Math. **26**, 483–488 (2001)

[89] Gundersen, G. G.: *Research questions on meromorphic functions and complex differential equations*. Comput. Methods Funct. Theory **17**(2), 195–209 (2017)

[90] Gundersen, G. G., Hayman, W. K.: *The strength of Cartan's version of Nevanlinna theory*. Bull. Lond. Math. Soc. **36**, 433–454 (2004)

[91] Gundersen, G. G., Heittokangas, J. M., Wen, Z.-T.: *Contour integral solutions of linear differential equations which include a generalization of the Airy integral*. Houst. J. Math. **47**(2), 321–351 (2021)

[92] Gundersen, G. G., Heittokangas, J., Zemirni, A.: *Asymptotic integration theory for* $f'' + P(z)f = 0$. Expo. Math. **40**(1), 94–126 (2022)

[93] Gundersen, G. G., Ishizaki, K., Kimura, N.: *Restrictions on meromorphic solutions of Fermat-type equations*. Proc. Edinb. Math. Soc. **63**, 654–665 (2020)

[94] Gundersen, G. G., Lü, W.-R., Ng, T.-W., Yang, C.-C.: *Entire solutions of differential equations that are related to trigonometric identities*. J. Math. Anal. Appl. **507**, 1 (2022)

[95] Gundersen, G. G., Steinbart, E. M.: *A generalization of the Airy integral for* $f'' - z^n f = 0$. Trans. Am. Math. Soc. **337**(2), 737–755 (1993)

[96] Gundersen, G. G., Steinbart, E., Wang, S.: *The possible orders of solutions of linear differential equations with polynomial coefficients*. Trans. Am. Math. Soc. **350**(3), 1225–1247 (1998)

[97] Gundersen, G. G., Tohge, K.: Entire and meromorphic solutions of $f^5 + g^5 + h^5 = 1$. In: *Symposium on Complex Differential and Functional Equations*. Univ. Joensuu Dept. Math. Rep. Ser., vol. 6, pp. 57–67 (2004)

[98] Gundersen, G. G., Yang, C.-C.: *Entire solutions of binomial differential equations*. Comput. Methods Funct. Theory **21**, 605–617 (2017)

[99] Hahn, L.-S., Epstein, B.: *Classical Complex Analysis*. Jones and Bartlett Publishers International, London, UK (1996)

[100] Halburd, R., Korhonen, R., Tohge, K.: *Holomorphic curves with shift-invariant hyperplane preimages*. Trans. Am. Math. Soc. **366**(8), 4267–4298 (2014)

[101] Hamouda, S., Belaïdi, B.: *On the growth of solutions of* $w^{(n)} + e^{-z}w' + Q(z)w = 0$ *and some related extensions*. Hokkaido Math. J. **35**(3), 573–586 (2006)

[102] Hardy, G. H.: *A Course of Pure Mathematics*, 10th edn. Cambridge University Press, Cambridge (1967)

[103] Havin, V. P., Nikolski, N. K. (eds.): *Linear and Complex Analysis, Problem book 3. Part II*. Lecture Notes in Math., vol. 1574. Springer-Verlag, Berlin (1994)

[104] Hayman, W. K.: *Picard values of meromorphic functions and their derivatives*. Ann. Math. **70**, 9–42 (1959)

[105] Hayman, W. K.: *Slowly growing integral and subharmonic functions*. Comment. Math. Helv. **34**, 74–84 (1960)

[106] Hayman, W. K.: *Meromorphic Functions*. Oxford Mathematical Monographs. Clarendon Press, Oxford (1964)

[107] Hayman, W. K.: *On the characteristic of functions meromorphic in the plane and of their integrals*. Proc. Lond. Math. Soc. (3) **14a**, 93–128 (1965)

[108] Hayman, W. K.: *Research Problems in Function Theory*. The Athlone Press University of London, London (1967)

[109] Hayman, W. K.: *The local growth of power series: a survey of the Wiman–Valiron method*. Can. Math. Bull. **17**(3), 317–358 (1974)

[110] Hayman, W. K.: *The minimum modulus of integral functions of order one*. J. Anal. Math. **28**, 171–212 (1975)

[111] Hayman, W. K.: *Waring's Problem für analytische Funktionen*. Bayer. Akad. Wiss., Math.-Nat.wiss. Kl. Sitz.ber. **1984**, 1–13 (1985) (German)

[112] Hayman, W. K., Lingham, E. F.: *Research Problems in Function Theory*, 50th anniversary edn. Problem Books in Mathematics. Springer, Cham (2019)

[113] Hayman, W. K., Stewart, F. M.: *Real inequalities with applications to function theory*. Proc. Camb. Philol. Soc. **50**, 250–260 (1954)

[114] Heins, M.: *Entire functions with bounded minimum modulus; subharmonic function analogues*. Ann. Math. (2) **49**, 200–213 (1948)

[115] Heittokangas, J., Ishizaki, K., Tohge, K., Wen, Z.-T.: *Zero distribution and division results for exponential polynomials*. Isr. J. Math. **227**(1), 397–421 (2018)

[116] Heittokangas, J., Ishizaki, K., Tohge, K., Wen, Z.-T.: *Dual exponential polynomials and a problem of Ozawa*. Proc. R. Soc. Edinb. **152**, 701–719 (2022)

[117] Heittokangas, J., Ishizaki, K., Tohge, K., Wen, Z.-T.: *Value distribution of exponential polynomials and their role in the theories of complex differential equations and oscillation theory*. Bull. Lond. Math. Soc. **55**, 1–77 (2023)

[118] Heittokangas, J., Korhonen, R., Rättyä, J.: *Growth estimates for solutions of linear complex differential equations*. Ann. Acad. Sci. Fenn. **29**(1), 233–246 (2004)

[119] Heittokangas, J., Laine, I.: *Solutions of $f'' + A(z)f = 0$ with prescribed sequences of zeros*. Acta Math. Univ. Comen. **LXXIV**(2), 287–307 (2005)

[120] Heittokangas, J., Laine, I., Tohge, K., Wen, Z.-T.: *Completely regular growth solutions of second order complex linear differential equations*. Ann. Acad. Sci. Fenn., Math. **40**(2), 985–1003 (2015)

[121] Heittokangas, J., Latreuch, Z., Wang, J., Zemirni, A.: *On the limits of real-valued functions in sets involving ψ-density, and applications*. J. Math. Anal. Appl. **507**(1), 125787 (2022), 15 pp.

[122] Heittokangas, J., Pulkkinen, S., Yu, H., Zemirni, M. A.: *Standard solutions of complex linear differential equations*. Ann. Fenn. Math. **49**(2), 663–683 (2024)

[123] Heittokangas, J., Rättyä, J.: *Zero distribution of solutions of complex linear differential equations determines growth of coefficients*. Math. Nachr. **284**(4), 412–420 (2011)

[124] Heittokangas, J., Tohge, K.: *A unit disc analogue of the Bank–Laine conjecture does not hold*. Ann. Acad. Sci. Fenn. **36**(1), 341–351 (2011)

[125] Heittokangas, J., Wang, J., Wen, Z.-T., Yu, H.: *Meromorphic functions of finite φ-order and linear q-difference equations*. J. Differ. Equ. Appl. **27**(9), 1280–1309 (2021)

[126] Heittokangas, J., Wen, Z.-T.: *Generalization of Pólya's zero distribution theory for exponential polynomials, and sharp results for asymptotic growth*. Comput. Methods Funct. Theory **21**(2), 245–270 (2021)

[127] Heittokangas, J., Wen, Z.-T.: *The asymptotic number of zeros of exponential sums in critical strips*. Monatshefte Math. **194**(2), 261–273 (2021)

[128] Heittokangas, J., Yu, H., Zemirni, M. A.: *On the number of linearly independent admissible solutions to linear differential and linear difference equations*. Can. J. Math. **73**(6), 1556–1591 (2021)

[129] Heittokangas, J., Zemirni, M. A.: *On Petrenko's deviations and second order differential equations*. Kodai Math. J. **44**, 181–193 (2021)

[130] Hellerstein, S., Miles, J., Rossi, J.: *On the growth of solutions of certain linear differential equations*. Ann. Acad. Sci. Fenn., Ser. A 1 Math. **17**(2), 343–365 (1992)

[131] Hellerstein, S., Rossi, J.: *Zeros of meromorphic solutions of second order linear differential equations*. Math. Z. **192**, 603–612 (1986)

[132] Hellerstein, S., Rossi, J.: *On the distribution of zeros of solutions of second-order differential equations*. Complex Var. Theory Appl. **13**(1–2), 99–109 (1989)

[133] Hellerstein, S., Shen, L.-C., Williamson, J.: *Real zeros of derivatives of meromorphic functions and solutions of second order differential equations*. Trans. Am. Math. Soc. **285**, 759–776 (1984)

[134] Hellerstein, S., Shen, L. C., Williamson, J.: *Reality of the zeros of an entire function and its derivatives*. Trans. Am. Math. Soc. **275**(1), 319–331 (1983)

[135] Hellerstein, S., Williamson, J.: *Derivatives of entire functions and a question of Pólya*. Trans. Am. Math. Soc. **227**, 227–249 (1977)

[136] Hellerstein, S., Williamson, J.: *Derivatives of entire functions and a question of Pólya II*. Trans. Am. Math. Soc. **234**(2), 497–503 (1977)

[137] Helmrath, W., Nikolaus, J.: *Ein elementarer Beweis bei der Anwendung der Zentralindexmethode auf Differentialgleichungen*. Complex Var. Theory Appl. **3**, 387–396 (1984) (German)

[138] Hille, E.: *Lectures on Ordinary Differential Equations*. Addison-Wesley Publ. Co., Reading, Mass.–London–Don Mills, Ont. (1969)

[139] Hille, E.: *Analytic Function Theory, Volume II*, 2nd edn. AMS Chelsea Publishing, Providence, RI (1987)

[140] Hille, E.: *Ordinary Differential Equations in the Complex Domain*. Dover Publications, Inc., Mineola, NY (1997). Reprint of the 1976 original

[141] Hirschman, I. I., Widder, D. V.: *The Convolution Transform*. Princeton University Press, Princeton, NJ (1955)

[142] Holland, A. S. B.: *Introduction to the Theory of Entire Functions*. Pure and Applied Mathematics, vol. 56. Academic Press (Harcourt Brace Jovanovich, Publishers), New York–London (1973)

[143] Huang, Z.-B., Laine, I., Lin, J.-L.: *Growth of solutions of some second order differential equations with entire coefficients*. Bull. Korean Math. Soc. **62**(1), 1–14 (2025)

[144] Ince, E. L.: *Ordinary Differential Equations*. Dover Publications, New York (1944)

[145] Ishizaki, K.: *On the complex oscillation of linear differential equations of third order*. Complex Var. Theory Appl. **24**(3–4), 289–300 (1994)

[146] Ishizaki, K.: *A note on the functional equation $f^n + g^n + h^n = 1$ and some complex differential equations*. Comput. Methods Funct. Theory **2**(1), 67–85 (2002)

[147] Ishizaki, K., Kimura, N.: *Entire and meromorphic solutions of the functional equation $f^n + g^n + h^n = 1$ and differential equations*. Comput. Methods Funct. Theory **19**(1), 157–172 (2019)

[148] Iyer, G.: *On certain functional equations*. J. Indian Math. Soc. **3**, 312–315 (1939)

[149] Jank, G., Volkmann, L.: *On the type of entire functions of finite iterated order*. Complex Var. Theory Appl. **1**(2–3), 195–208 (1982/83)

[150] Jank, G., Volkmann, L.: *Einföhrung in die Theorie der ganzen und meromorphen Funktionen mit Anwendungen auf Differeentialgleichungen*. Birkhäuser, Basel–Boston (1985) (German)

[151] Janković, S., Ostrogorski, T.: *Convex additively slowly varying functions*. J. Math. Anal. Appl. **274**(1), 228–238 (2002)

[152] Jeffreys, H., Jeffreys, B.: *Methods of Mathematical Physics*. Cambridge Univ. Press, Cambridge (1999). Reprint of the third (1956) edition

[153] Juneja, O. P.: *On the coefficients of an entire series of finite order*. Arch. Math. (Basel) **21**, 374–378 (1970)

[154] Juneja, O. P., Kapoor, G. P., Bajpai, S. K.: *On the (p, q)-order and lower (p, q)-order of an entire function*. J. Reine Angew. Math. **282**, 53–67 (1976)

[155] Juneja, O. P., Singh, P.: *On the growth of an entire series with gaps*. J. Math. Anal. Appl. **30**, 330–334 (1970)

[156] Kinnunen, L.: *Linear differential equations with solutions of finite iterated order*. Southeast Asian Bull. Math. **22**(4), 385–405 (1998)

[157] Kjellberg, B.: *A theorem on the minimum modulus of entire functions*. Math. Scand. **12**, 5–11 (1963)

[158] Kobayashi, T.: *On a characteristic property of the exponential function*. Kodai Math. Semin. Rep. **29**(1–2), 130–156 (1977)

[159] Kobayashi, T.: *Entire functions with three linearly distributed values*. Kodai Math. J. **1**(1), 133–158 (1978)

[160] Kobayashi, T.: *An entire function with linearly distibuted values*. Kodai Math. J. **2**(1), 54–81 (1979)

[161] Kumar, S., Saini, M.: *Order and hyper-order of solutions of second order linear differential equations*. Bull. Malays. Math. Sci. Soc. **44**(2), 1035–1056 (2021)

[162] Kwon, K.-H.: *Nonexistence of finite order solutions of certain second order linear differential equations*. Kodai Math. J. **19**(3), 378–387 (1996)

[163] Kwon, K.-H.: *On the growth of entire functions satisfying second order linear differential equations*. Bull. Korean Math. Soc. **33**(3), 487–496 (1996)

[164] Kwon, K.-H., Kim, J. H.: *Maximum modulus, characteristic, deficiency and growth of solutions of second order linear differential equations*. Kodai Math. J. **24**, 344–351 (2001)

[165] Kövari, T.: *The gap-theorem for entire functions of finite order*. Mich. Math. J. **12**(2), 133–140 (1965)

[166] Laine, I.: *Nevanlinna Theory and Complex Differential Equations*. De Gruyter Studies in Mathematics, vol. 15. Walter de Gruyter & Co., Berlin (1993)

[167] Laine, I., Wu, P.: *Growth of solutions of second order linear differential equations*. Proc. Am. Math. Soc. **128**(9), 2693–2703 (2000)

[168] Laine, I., Yang, R.: *Finite order solutions of complex differential equations*. Electron. J. Differ. Equ. **2004**, 65 (2004), 8 pp.

[169] Lang, S.: *Complex Analysis*. Grad. Texts in Math., vol. 103. Springer-Verlag, New York (1985)

[170] Langer, R. E.: *On the zeros of exponential sums and integrals*. Bull. Am. Math. Soc. **37**(4), 213–239 (1931)

[171] Langley, J. K.: *On complex oscillation and a problem of Ozawa*. Kodai Math. J. **9**, 430–439 (1986)

[172] Langley, J. K.: *On the deficiencies of composite entire functions*. Proc. Edinb. Math. Soc. **36**, 151–164 (1992)

[173] Langley, J. K.: *Proof of a conjecture of Hayman concerning f and f''*. J. Lond. Math. Soc. **48**(2), 500–514 (1993)

[174] Langley, J. K.: *A lower bound for the number of zeros of a meromorphic function and its second derivative*. Proc. Edinb. Math. Soc. **39**, 171–185 (1996)

[175] Langley, J. K.: *Bank–Laine functions with sparse zeros*. Proc. Am. Math. Soc. **129**, 1969–1978 (2001)

[176] Langley, J. K.: *Composite Bank–Laine functions and a question of Rubel*. Trans. Am. Math. Soc. **354**, 1177–1191 (2001)

[177] Langley, J. K.: *Postgraduate Notes on Complex Analysis*. https://www.maths.nottingham.ac.uk/plp/pmzjkl/pg1.pdf (2014), 199 pp.

[178] Langley, J. K.: *Bank–Laine functions, the Liouville transformation and the Eremenko-Lyubich class*. J. Anal. Math. **141**(1), 225–246 (2020)

[179] Langley, J. K.: *Bank–Laine functions with real zeros*. Comput. Methods Funct. Theory **20**(3–4), 653–665 (2020)

[180] Langley, J. K.: *Asymptotic values, deficiencies and Bank–Laine functions*. Comput. Methods Funct. Theory. https://doi.org/10.1007/s40315-025-00593-4

[181] Lebedev, N. N.: *Special Functions and Their Applications*. Prentice-Hall, Inc., Englewood Cliffs, NJ (1965). Revised English edition Translated and edited by Richard A. Silverman

[182] Lehto, O.: *Korkeat maailmat – Rolf Nevanlinnan elämä*. Otavan Kirjapaino Oy, Keuruu (2001) (Finnish)

[183] Ja, L. B.: *Distribution of Zeros of Entire Functions*. Transl. Math. Monogr., vol. 5. American Mathematical Society, Providence, RI (1980). Translated from the Russian by R. P. Boas, J. M. Danskin, F. M. Goodspeed, J. Korevaar, A. L. Shields and H. P. Thielman. Revised edition.

[184] Ja, L. B., Ostrovskii, I. V.: *The dependence of the growth of an entire function on the distribution of zeros of its derivatives*. Sib. Mat. Zh. **1**, 427–455 (1960). English translation: American Mathematical Society Translations. Series 2 **32** (1963), 323–357

[185] Levinson, N., Redheffer, R. M.: *Complex Variables*. Holden-Day, Inc., San Francisco, Calif.–Cambridge–Amsterdam (1970)

[186] Li, N., Wu, L., Xu, H.: *Entire solutions of non-linear binomial differential equations involving trigonometric identities*. Rev. R. Acad. Cienc. Exactas Fís. Nat., Ser. A Mat. RACSAM **119**, 4 (2025)

[187] Li, P.: *Entire solutions of certain type of differential equations*. J. Math. Anal. Appl. **344**(1), 253–259 (2008)

[188] Li, P.: *Entire solutions of certain type of differential equations II*. J. Math. Anal. Appl. **375**(1), 310–319 (2011)

[189] Li, P., Lü, W.-R., Yang, C.-C.: *Entire solutions of certain types of nonlinear differential equations*. Houst. J. Math. **45**, 431–437 (2019)

[190] Li, X.-Y., Wang, J., Wen, Z.-T.: *Representation of finite order solutions to linear differential equations with exponential sum coefficients*. J. Differ. Equ. **423**, 118–132 (2025)

[191] Liu, H., Mao, Z.: *The exact meromorphic solutions of some nonlinear differential equations*. Acta Math. Sci. Ser. B **44**(1), 103–114 (2024)

[192] Liu, K., Wei, Y. M., Yu, P. Y.: *Generalized Yang's Conjecture on the periodicity of entire functions*. Bull. Korean Math. Soc. **57**(5), 1259–1267 (2020)

[193] Liu, X., Korhonen, R., Liu, K.: *Variations of a conjecture of C.-C. Yang concerning periodicity*. Comput. Methods Funct. Theory **22**, 27–33 (2022)

[194] Long, J. R.: *Applications of value distribution theory in the theory of complex differential equations*. Dissertation, Publications of the Univ. Eastern Finland, Dissertations in Forestry and Natural Sciences, no. 176 (2015).

[195] Long, J. R.: *Growth of solutions of second order complex linear differential equations with entire coefficients*. Filomat **32**(1), 275–284 (2018)

[196] Long, J. R., Heittokangas, J., Ye, Z.: *On the relationship between the lower order of coefficients and the growth of solutions of differential equations*. J. Math. Anal. Appl. **444**, 153–166 (2016)

[197] Long, J. R., Shi, L., Wu, X., Zhang, S.: *On a question of Gundersen concerning the growth of solutions of linear differential equations*. Ann. Acad. Sci. Fenn., Math. **43**, 337–348 (2018)

[198] Long, J. R., Wu, P.: *On the hyper-order of solutions of second order linear differential equations*. Adv. Math. **42**(3), 320–326 (2013)

[199] Long, J. R., Wu, T., Wu, X.: *Growth of solutions of complex differential equations with solutions of another equation as coefficients*. Comput. Methods Funct. Theory **19**(1), 3–16 (2019)

[200] Long, J. R., Wu, P. C., Zhang, Z.: *On the growth of solutions of second order linear differential equations with extremal coefficients*. Acta Math. Sin. Engl. Ser. **29**(2), 365–372 (2013)

[201] Long, J., Xia, M., Xiang, X.: *On a question of Gundersen–Yang concerning entire solutions of binomial differential equations*. arXiv:2501.09250v1 [mathCV] (16 Jan 2025)

[202] Mao, Z., Liu, H.: *Entire solutions of some nonlinear homogeneous differential equations*. Bull. Malays. Math. Sci. Soc. **47**, 6 (2024)

[203] Marden, M.: *On the zeros of the derivative of an entire function*. Am. Math. Mon. **75**, 829–839 (1968)

[204] Markushevich, A. I.: *Theory of Functions of a Complex Variable. Vol. I*. Prentice-Hall, Inc., Englewood Cliffs, NJ (1965). Translated and edited by Richard A. Silverman

[205] Markushevich, A. I.: *Theory of Functions of a Complex Variable. Vol. II*. Prentice-Hall, Inc., Englewood Cliffs, NJ (1965). Revised English edition translated and edited by Richard A. Silverman

[206] Markushevich, A. I.: *Entire Functions*. American Elsevier Publishing Co., Inc., New York (1966). Translated from the Russian by Scripta Technica, Inc. Translation editor: Leon Ehrenpreis

[207] Marsden, J. E.: *Basic Complex Analysis. With the assistance of Michael Buchner, Michael Hoffman and Clifford Risk*. W. H. Freeman and Co., San Francisco, CA (1973)

[208] Miles, J.: *On entire functions of infinite order with radially distributed zeros*. Pac. J. Math. **81**(1), 131–157 (1979)

[209] Milloux, H.: *Sur la distribution des valuers des fonctions entières d'ordre fini, à zéros reels*. Bull. Sci. Math. (2) **51**, 303–319 (1927) (French)

[210] Molluzzo, J.: *Monotonicity of quadrature formulas and polynomial representation*. Doctoral thesis, Yeshiva University (1972)

[211] Montel, P. *Leçons sur les Familles Normales de Fonctions Analytiques et leurs Applications*. Gauthier-Villars, Paris (1927) (French)

[212] Mues, E.: *Über eine Vermutung von Hayman*. Math. Z. **119**, 11–20 (1971) (German)

[213] Mues, E.: *Über die Nullstellen homogener Differentialpolynome*. Manuscr. Math. **23**, 325–341 (1978) (German)

[214] Munroe, M. E.: *Introduction to Measure and Integration*. Addison-Wesley Publishing Co., Inc., Cambridge, MA (1953)

[215] Murai, T.: *The deficiency of entire functions with Fejér gaps*. Ann. Inst. Fourier (Grenoble) **33**(3), 39–58 (1983)

[216] Nevanlinna, R.: *Zur Theorie der meromorphen Funktionen*. Acta Math. **46**, 1–99 (1925) (German)

[217] Nevanlinna, R.: Über die Konstruktion von meromorphen Funktionen mit gegebenen Wertzuordnungen. In: *Festschrift zur Gedächtnisfeier für Karl Weierstrass*, pp. 579–582. Westdeutscher Verlag, Köln–Opladen (1966) (German)

[218] Nevanlinna, R., Paatero, V.: *Introduction to Complex Analysis*. AMS Chelsea Publishing, Providence, RI (2007). Translated from the 1965 German original by T. Kövari and G. S. Goodman. Reprint of the 1982 second edition

[219] Newman, D., Slater, M.: *Waring's problem for the ring of polynomials*. J. Number Theory **11**, 477–487 (1979)

[220] Niculescu, C. P., Persson, L.-E.: *Convex Functions and their Applications. A Contemporary Approach*. CMS Books in Mathematics/Ouvrages de Mathématiques de la SMC, vol. 23. Springer, New York (2006)

[221] Olver, F. W. J.: *Asymptotics and Special Functions*. Computer Science and Applied Mathematics. Academic Press (Harcourt Brace Jovanovich, Publishers), New York–London (1974)

[222] Ozawa, M.: *On the solution of the functional equation* $f \circ g(z) = F(z), III$. Kodai Math. Semin. Rep. **20**, 257–263 (1968)

[223] Ozawa, M.: *On a solution of* $w'' + e^{-z}w' + (az + b)w = 0$. Kodai Math. J. **3**(2), 295–309 (1980)

[224] Ozawa, M.: Zeros of certain differential polynomials. In: Yang, C.-C., Niino, K., Komatu, Y. (eds.) *Analytic Function Theory of One Complex Variable*. Pitman Research Notes in Math. Series, vol. 212, pp. 199–225. Longman Scientific and Technical (1989)

[225] Petrenko, V. P., *Entire Curves*. Kharkov (1984) (Russian)

[226] Pólya, G.: *Sur une question concernant les fonctions entières*. C. R. Acad. Sci., Sér. A–B **158**, 330–333 (1914)

[227] Pólya, G.: *Bermerkung zur Theorie der ganzen Funktionen*. Jahresber. Dtsch. Math.-Ver. **24**, 392–400 (1915)

[228] Pólya, G.: *Geometrisches über die Verteilung der Nullstellen spezieller ganzer Funktionen*. Sitz.-Ber. Bayer. Akad. Wiss. (1920), 285–290.

[229] Ramachandra, K.: *An application of Borel–Carathéodory theorem*. J. Ramanujan Math. Soc. **4**(1), 45–52 (1989)

[230] Rao, M. M., Ren, Z. D.: *Theory of Orlicz Spaces*. Monogr. Textbooks Pure Appl. Math., vol. 146. Marcel Dekker, Inc., New York (1991). xii+449 pp.

[231] Rényi, C.: *On a conjecture of G. Pólya*. Acta Math. Acad. Sci. Hung. **7**, 145–150 (1956)

[232] Ronkin, L. I.: *Functions of Completely Regular Growth*. Mathematics and its Applications (Soviet Series), vol. 81. Kluwer Academic Publishers Group, Dordrecht (1992). Translated from the Russian by A. Ronkin and I. Yedvabnik

[233] Rossi, J.: *Second order differential equations with transcendental coefficients*. Proc. Am. Math. Soc. **97**(1), 61–66 (1986)

[234] Rossi, J.: *The Tsuji characteristic and real zeros of solutions of second order ordinary differential equations*. J. Lond. Math. Soc. (2) **36**(3), 490–500 (1987)

[235] Rossi, J., Wang, S.: *The radial oscillation of solutions to ODE's in the complex domain*. Proc. Edinb. Math. Soc. **39**(2), 491–498 (1996). 3

[236] Royden, H. L., Fitzpatrick, P. M.: *Real Analysis*, 4th edn. Prentice Hall (2010)

[237] Rubel, L. A.: *Entire and Meromorphic Functions. With the assistance of James E. Colliander*. Universitext. Springer-Verlag, New York (1996)

[238] Rubel, L. A., Yang, C.-C.: *Interpolation and unavoidable families of meromorphic functions*. Mich. Math. J. **20**, 289–296 (1973)

[239] Rudin, W.: *Principles of Mathematical Analysis*, 3rd edn. International Series in Pure and Applied Mathematics. McGraw-Hill Book Co., New York–Auckland–Düsseldorf (1976)

[240] Rudin, W.: *Real and Complex Analysis*, 3rd edn. McGraw-Hill Book Co., New York (1987)

[241] Saenko, V. V.: *Integral representation of the Mittag-Leffler function*. Russ. Math. **66**, 43–58 (2022)

[242] Saks, S., Zygmund, A.: *Analytic Functions*. Monografie Matematyczne, Tom XXVIII. Polskie Towarzystwo Matematyczne, Warszawa-Wroclaw (1952). Translated by E. J. Scott

[243] Sato, D.: *On the rate of growth of entire functions of fast growth*. Bull. Am. Math. Soc. **69**, 411–414 (1963)

[244] Schiff, J. L.: *Normal Families*. Universitext. Springer-Verlag, New York (1993)

[245] Schilling, R. L.: *Measures, Integrals and Martingales*, 2nd edn. Cambridge University Press, Cambridge (2017)

[246] Schwengeler, E.: *Geometrisches über die Verteilung der Nullstellen spezieller ganzer Funktionen (Exponentialsummen)*. Diss., Zürich (1925) (German)

[247] Šeda, V.: *On some properties of solutions of the differential equation* $y'' = Q(z)y = 0$ *where* $Q(z) \neq 0$ *is an entire function*. Acta Fac. Rerum Nat. Univ. Comen. Math. **4**, 223–253 (1959)

[248] Seneta, E.: *Regularly Varying Functions*. Lecture Notes in Math., vol. 508. Springer-Verlag, Berlin–New York (1976), v+112 pp.

[249] Shah, S. M.: *On integral functions of perfectly regular growth*. J. Lond. Math. Soc. **14**, 293–302 (1939)

[250] Shah, S. M.: *On the lower order of integral functions*. Bull. Am. Math. Soc. **52**, 1046–1052 (1946)

[251] Shah, S. M.: *On the coefficients of an entire series of finite order*. J. Lond. Math. Soc. **26**, 45–46 (1951)

[252] Sheil-Small, T.: *On the zeros of the derivatives of real entire functions and Wiman's conjecture*. Ann. Math. (2) **129**(1), 179–193 (1989)

[253] Sheil-Small, T.: *Complex Polynomials*. Cambridge Studies in Advanced Mathematics, vol. 75. Cambridge University Press, Cambridge (2002)

[254] Shen, L.-C.: *Construction of a differential equation* $y'' + Ay = 0$ *with solutions having the prescribed zeros*. Proc. Am. Math. Soc. **95**(4), 544–546 (1985)

[255] Shen, L.-C.: *Solution to a problem of S. Bank regarding exponent of convergence of zeros of the solutions of differential equation* $f'' + Af = 0$. Chin. Sci. Bull. **30**(12), 1579–1585 (1985)

[256] Shen, L.-C.: *On the area of the region where an entire function is greater than one*. Proc. Am. Math. Soc. **102**(1), 68–70 (1988)

[257] Shi-An, G., Chen, T.-W., Chen, Z.-X.: *A note concerning the complex oscillation and the growth of solutions for higher order homogeneous linear differential equations*. Complex Var. Theory Appl. **34**(1–2), 109–117 (1997)

[258] Shin, K.-C.: *New polynomials P for which* $f'' + P(z)f = 0$ *has a solution with almost all real zeros*. Ann. Acad. Sci. Fenn., Math. **27**(2), 491–498 (2002)

[259] Sibuya, Y.: *Global Theory of a Second Order Linear Ordinary Differential Equation with a Polynomial Coefficient*. North-Holland Publishing Co./American Elsevier Publishing Co., Inc., Amsterdam–Oxford/New York (1975)

[260] Simmons, G. F.: *Differential Equations with Applications and Historical Notes*. International Series in Pure and Applied Mathematics. McGraw-Hill Book Co., New York (1972)

[261] Singh, S. K., Gopalakrishna, H. S.: *Growth of entire and meromorphic functions*. Comment. Math. Univ. St. Pauli **26**(1), 3–15 (1977/78)
[262] Sons, L. R.: *On the Macintyre conjecture*. Ill. J. Math. **14**, 613–620 (1970)
[263] Steinmetz, N.: *Zur Wertverteilung von Exponentialpolynomen*. Manuscr. Math. **26**(1–2), 155–167 (1978/79) (German)
[264] Steinmetz, N.: *On the order and lower order of entire functions with radially distributed zeros*. Proc. Am. Math. Soc. **87**(3), 449–452 (1983)
[265] Steinmetz, N.: *Exceptional values of solutions of linear differential equations*. Math. Z. **201**(3), 317–326 (1989)
[266] Steinmetz, N.: *Linear differential equations with exceptional fundamental sets II*. Proc. Am. Math. Soc. **117**(2), 355–358 (1993)
[267] Steinmetz, N.: *Laplace contour integrals and linear differential equations*. Comput. Methods Funct. Theory **21**(4), 565–585 (2021)
[268] Swanson, C., Headley, V.: *An extension of Airy's equation*. SIAM J. Appl. Math. **15**, 1400–1412 (1967)
[269] Tao, L., Long, J.-R.: *On growth of solutions of complex differential equation concerning generalized Ozawa's problem*. Houst. J. Math. **48**(3), 451–462 (2022)
[270] Taylor, A. E., Mann, W. R.: *Advanced Calculus*, 3rd edn. John Wiley and Sons, New York (1983)
[271] Taylor, R., Wiles, A.: *Ring-theoretic properties of certain Hecke algebras*. Ann. Math. **141**, 553–572 (1995)
[272] Tijdeman, R.: *On the number of zeros of general exponential polynomials*. Indag. Math. **74**, 1–7 (1971). Nederl. Akad. Wetensch. Proc. Ser. A 33
[273] Titchmarsh, E. C.: *The Theory of Functions*, 2nd edn. Oxford University Press, Oxford (1939)
[274] Titchmarsh, E. C.: *Eigenfunction Expansions Associated with Second-Order Differential Equations. Part I*, 2nd edn. Oxford University Press, London (1962)
[275] Toda, N.: *On the functional equation* $\sum_{i=0}^{p} a_i f_i^{n_i} = 1$. Tôhoku Math. J. **23**, 289–299 (1971)
[276] Toppila, S.: *On Nevanlinna's characteristic function of entire functions and their derivatives*. Ann. Acad. Sci. Fenn., Math. **3**, 131–134 (1977)
[277] Townsend, D.: *On the differentiability of* $T(r,f)$. Int. J. Math. Math. Sci. **5**(2), 351–356 (1982)
[278] Tsuji, M.: *Potential Theory in Modern Function Theory*. Chelsea Publishing Co., New York (1975). Reprinting of the 1959 original
[279] Valiron, G.: *Lectures on the General Theory of Integral Functions*. Chelsea Publishing Company, New York (1949)
[280] Veech, W. A.: *A Second Course in Complex Analysis*. W. A. Benjamin, Inc., New York–Amsterdam (1967)
[281] Voorhoeve, M., van der Poorten, A. J.: *Wronskian determinants and the zeros of certain functions*. Indag. Math. **78**(5), 417–424 (1975). Nederl. Akad. Wetensch. Proc. Ser. A 37
[282] Voorhoeve, M., van der Poorten, A. J., Tijdeman, R.: *On the number of zeros of certain functions*. Indag. Math. **78**(5), 407–416 (1975). Nederl. Akad. Wetensch. Proc. Ser. A 37
[283] Wang, Q., Hu, P. C.: *On zeros and periodicity of entire functions*. Acta Math. Sci. Ser. A Chin. Ed. **38**(2), 209–214 (2018)
[284] Wang, S.: *On the frequency of zeros of a fundamental solution set of complex linear differential equations*. Kodai Math. J. **20**(2), 143–155 (1997)
[285] Wasow, W.: *Asymptotic Expansions for Ordinary Differential Equations*. Dover Publ., New York (1987)
[286] Wen, Z.-T.: *Finite logarithmic order solutions of linear q-difference equations*. Bull. Korean Math. Soc. **51**(1), 83–98 (2014)
[287] Wen, Z.-T., Gundersen, G. G., Heittokangas, J.: *Dual exponential polynomials and linear differential equations*. J. Differ. Equ. **264**(1), 98–114 (2018)
[288] Wen, Z.-T., Ye, Z.: *Wiman–Valiron theorem for q-differences*. Ann. Acad. Sci. Fenn., Math. **41**(1), 305–312 (2016)
[289] Weyl, H.: *Meromorphic Functions and Analytic Curves*. Princeton Univ. Press, Princeton, NJ (1943)
[290] Whittaker, E. T., Watson, G. N.: *A Course of Modern Analysis*, 4th edn. University Press, Cambridge (1927)

[291] Wiles, A.: *Modular elliptic curves and Fermat's Last Theorem*. Ann. Math. **141**, 443–551 (1995)

[292] Wittich, H.: *Über das Anwachsen der Lösungen linearer Differentialgleichungen*. Math. Ann. **124**, 277–288 (1952) (German)

[293] Wittich, H.: *Neure Untersuchungen über eindeutige analytische Funktionen*, Zweite, korrigierte Auflage. Ergebnisse der Mathematik und ihrer Grenzgebiete, vol. 8. Springer-Verlag, Berlin–New York (1968) (German)

[294] Wu, L., Liu, M., Hu, P.: *Entire solutions of certain type binomial differential equations*. Anal. Math. Phys. **13**(2), 25 (2023), 17 pp.

[295] Wu, L., Liu, M., Hu, P.-C.: *On entire solutions of nonlinear binomial differential equations*. Rocky Mt. J. Math. **55**(3), 867–879 (2025)

[296] Wu, S.: *On the location of zeros of solutions of* $f'' + A(z)f = 0$, *where* $A(z)$ *is an entire function*. Math. Scand. **74**, 293–312 (1994)

[297] Wu, X., Long, J., Heittokangas, J., Qiu, K.: *Second-order complex linear differential equations with special functions or extremal functions as coefficients*. Electron. J. Differ. Equ. **2015**, 143 (2015), 15 pp.

[298] Yang, C.-C., Li, P.: *On the transcendental solutions of a certain type of nonlinear differential equations*. Arch. Math. (Basel) **82**(5), 442–448 (2004)

[299] Yang, C.-C., Yi, H. X.: *Uniqueness Theory of Meromorphic Functions*. Mathematics and its Applications, vol. 557. Kluwer Academic Publishers Group, Dordrecht (2003)

[300] Yang, L.: *Value Distribution Theory*. Springer-Verlag/Science Press Beijing, Berlin/Beijing (1993). Translated and revised from the 1982 Chinese original

[301] Yang, S.-S., Liao, L.-W., Lu, X.-Q.: *Entire solutions of certain type binomial differential equations*. Comput. Methods Funct. Theory **25**(4), 785–800 (2025)

[302] Yu, K.-W.: *A Complete Solution Guide to Real and Complex Analysis I* (2019). https://sites.google.com/view/yukitwing/

[303] Zhang, G.: *Infinite growth of solutions of second order complex differential equation*. Open Math. **16**(1), 1233–1242 (2018)

[304] Zhang, G.: *The infinite growth of solutions of second order linear complex differential equations with completely regular growth coefficient*. Bull. Korean Math. Soc. **58**(2), 419–431 (2021)

[305] Zhang, G. H.: *Theory of Entire and Meromorphic Functions. Deficient and Asymptotic Values and Singular Directions*. Translations of Mathematical Monographs, vol. 122. American Mathematical Society, Providence, RI (1993). Translated from the Chinese by Chung-Chun Yang

[306] Zhang, X. B., Yi, H.-X.: *Entire solutions of a certain type of functional-differential equations*. Appl. Math. J. Chin. Univ. **28**(2), 138–146 (2013)

[307] Zheng, J. H.: *Value Distribution of Meromorphic Functions*. Tsinghua University Press/Springer, Beijing/Heidelberg (2010)

Index

https://doi.org/10.1515/9783112249130-014

www.ingramcontent.com/pod-product-compliance
Lightning Source LLC
Chambersburg PA
CBHW061152240926
51846CB00006B/994
* 9 7 8 3 1 1 2 2 4 9 1 2 3 *